100% HONORS GEOMETRY
MATHEMATICS CONTESTS

GEOTOPIA

VOLUME 1

LEARNING GEOMETRY THROUGH
PROOF AND CHALLENGING PROBLEMS

High School Honors Geometry
Math Contests Preparation

AMC 8&10&12 / Math League / MathCON

SINAN KANBIR, Ph.D.
Mathematics Education

MATH TOPIA PRESS

FATIH S. TETIKOGLU, Ph.D.
Mathematics

Paper Book

978-1-7356252-0-1

e–Book

978-1-735-6252-1-8

Graphic Design - Typesetting

Dr Fatih Sabahattin Tetikoglu

fsetik@gmail.com

Preface

What is Geotopia and what makes this book special and unique?

This book grew out of working with geometry honor course students, math contest participants, and geometric puzzle lovers. The book, Geotopia Volume 1, covers triangles and polygon geometry concepts which can be considered the heart of Euclidean geometry. It presents geometric concepts in a student-friendly format and design. Through proofs and carefully crafted problems, it shows how all geometric concepts are well-connected with each other. It convinces students that one geometric idea can be used to conceptualize other ones and there are always multiple ways to approach problems. Book buyers, regardless of their levels, become Geotopian participants. The participants will feel challenged but eventually will feel connected to those geometric concepts. We hope that, with this volume 1 and the volume 2, which will be published soon, you will enjoy the beauty that lies in the study of elementary and advanced Euclidean Geometry.

Possible target audiences of this book

- Geometry and honor geometry course takers.
- Participants in mathematics contests such as AMC 8, AMC 10, AMC 12, MATHCOUNTS, MathCON, and Math Leagues.
- Math Circle students.
- Pre-service and in-service mathematics teachers.
- Math Circle organizers.
- Parents of gifted geometry students.
- General geometry and math enthusiasts.

Structure of the book

This book conceptualizes theorems by going through their proofs. It applies theorems to some cognitively challenging problems via examples and solutions. Further, it expands your understanding through productive struggle through wrestling with related problems as well as previous contest problems (i.e., AMC 8-10-12, MathCON, CWML).

The ends of each of the 12 chapters are separated into three sections: Test 1, some introductory problems and problems similar to those previously presented; test 2, the problems sets are challenging and tougher than Test 1 problems but still consist of solvable and fun puzzles, test 3, consists of previous years' contest problems, mainly American Mathematics Competitions (AMC 8-10-12, MathCON, and CWML).

How to use this book

The book's lectures consist of definitions, theorems, proofs, examples, and solutions. With this structure, we hope that this book will be convenient for independent geometry learners. Students should engage each chapters' concepts by trying themselves first, then carefully unfold given proofs and move to examples as they work through them. Solutions are presented very clearly; however, geometry learners are supposed to explore and discover themselves when they first engage with the problems. There will also be side problems that provide further similar challenges. Students can also access solutions by using barcode readers when they see problems with barcodes. Each example and test problem has multiple entry points and solutions that every Geotopian can begin with. There is no time pressure; students can enjoy their deep geometric thinking over speed.
Notetaking is an essential skill for Geotopians; that's why this book provides them some space both on the sides and below the problems, which makes it a very student-friendly book. We encourage participants to take their own notes and become good mathematical note-takers.

Acknowledgments

We would like to thank Kevin Schoenecker, University of Wisconsin System mathematics professor, for his willingness to proofread this book and always express his enjoyment. Special thanks to Catriona Agg, creator of great geometry puzzles on Twitter, for her permission to use some selected problems on the sides of this book. We also thank the Mathematical Association of America (MAA) for granting us permission to use some previous years' American Mathematics Competitions (AMC) problems as well as MathCON, CWML, and PiMC. We are also thankful to Ava Kumar, Andrew Carratu, and Alexander Svoronos for test solving and reading the drafts of this book and giving very valuable comments as students and geometry learners. Last but not least, we would like to thank University of Wisconsin-Stevens Point Mathematical Sciences department faculty members for giving valuable feedback and always being willing to work problems in the hallways.

TABLE OF CONTENTS

CHAPTER 06

CHAPTER 07

CHAPTER 08

CHAPTER 09

CHAPTER 10

CHAPTER 11

CHAPTER 12

ANSWER KEYS

CHAPTER 1

Note

Euclidean

The term Euclidean refers to everything that can historically or logically be referred to Euclid's monumental treatise The *Thirteen Books of the Elements*, written around the year 300 B.C.

The Euclidean geometry of the plane (Books I-IV) and of the three-dimensional space (Books XI-XIII) is based on five postulates, the first four of which are about the basic objects of plane geometry (point, straight line, circle, and right angle), which can be drawn by straightedge and compass (the so-called Euclidean tools of geometric construction). Euclid's fifth postulate, also known as the parallel postulate, is modified in so-called non-Euclidean geometry.

Euclidean Geometry

A geometry in which Euclid's fifth postulate holds, sometimes also called parabolic geometry. Two-dimensional Euclidean geometry is called plane geometry, and three-dimensional Euclidean geometry is called solid geometry. Hilbert proved the consistency of Euclidean geometry.

www.mathworld.wolfram.com

My Notes

Definition 1.1

Angle, Side, and Vertex

An **angle** is a union of two rays that have a common endpoint. The two rays are called the **sides** of the angle. The common endpoint is called the **vertex** of the angle.

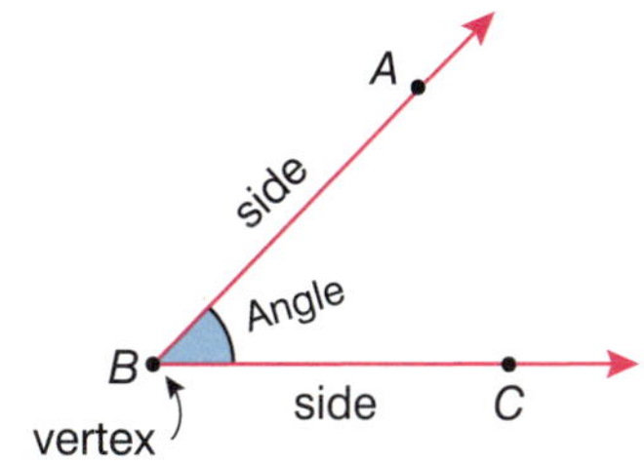

Example 1.1

In the figure,

$m\angle AOB = 20°$

$m\angle BOC = x+10°$, and

$m\angle EOD = 3x-50°$.

If $\angle EOC$ and $\angle AOD$ are congruent, then find x.

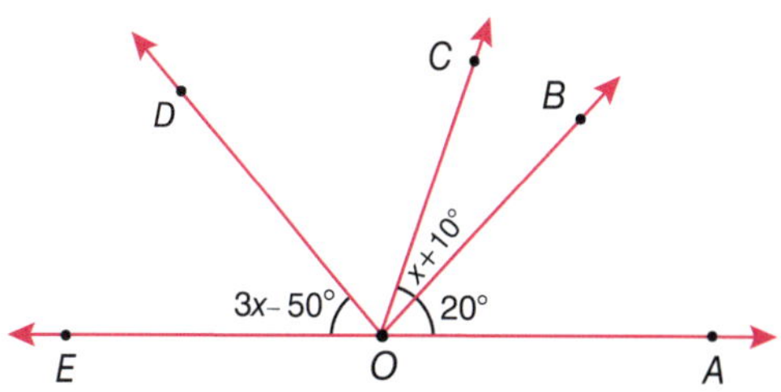

Solution 1.1

If $\angle EOC$ and $\angle AOD$ are congruent, then

$$3x-50°+m\angle DOC = m\angle DOC+ x+10°+20°$$

$$3x-50° = x+10°+20°$$

$$2x = 80°$$

$$x = 40°.$$

Definition 1.2

Angle Bisector

An **angle bisector** is a line ray or line segment that divides an angle into two congruent angles. In the figure, ray OB is a bisector of $\angle AOC$ because

$$m\angle AOB = m\angle BOC \text{ and}$$

$$m\angle AOB = \frac{1}{2}m\angle AOC.$$

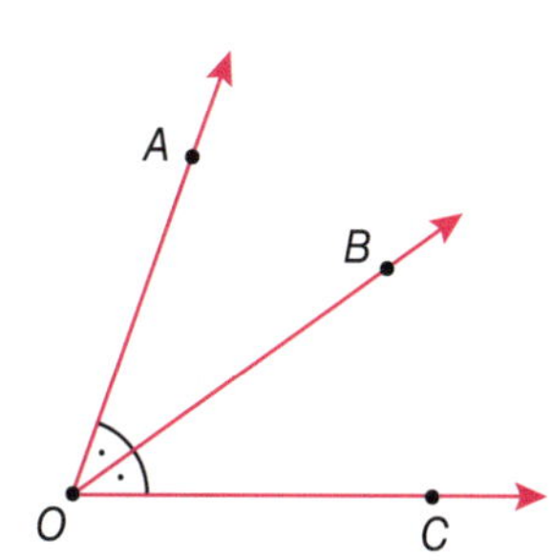

Example 1.2

In the figure,

$m\angle AOC = 16°$ and $m\angle COB = 48°$.

Find the measure of the angle which is formed by the angle bisectors of $\angle AOC$ and $\angle COB$.

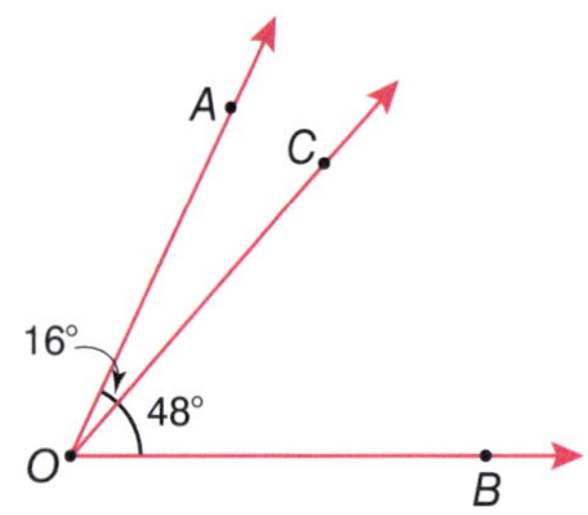

Solution 1.2

If $\overrightarrow{OX}$ is the angle bisector of $\angle AOC$, then $m\angle XOC = 8°$.

If $\overrightarrow{OY}$ is the angle bisector of $\angle COB$, then $m\angle YOC = 24°$.

Therefore, $m\angle XOY = 8°+24° = 32°$.

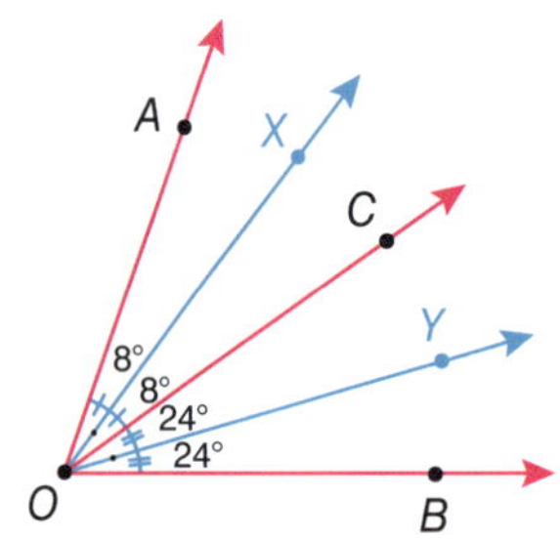

Try This

If $\angle ABC$ is an obtuse angle and

$2\angle ABC = 3\angle XYZ$,

what type angle is XYZ ?

My Notes

Definition 1.3

Types of Angle According to Measure

Angle A is called an **acute** angle if and only if $0<m\angle A<90°$.

Angle A is called a **right** angle if and only if $m\angle A = 90°$.

Angle A is called an **obtuse** angle if and only if $90°<m\angle A<180°$.

Angle A is called a **straight** angle if and only if $m\angle A = 180°$.

Angle A is called a **complete** angle if and only if $m\angle A = 360°$.

Definition 1.4

Complementary Angles

Two angles are called complementary if the sum of their measures is 90°. Each angle is called a complement of the other angle.

In the below figure, $m\angle A+m\angle B = 90°$. By the definition, $\angle A$ and $\angle B$ are complementary angles, $\angle A$ is a complement of $\angle B$, and $\angle B$ is a complement of $\angle A$.

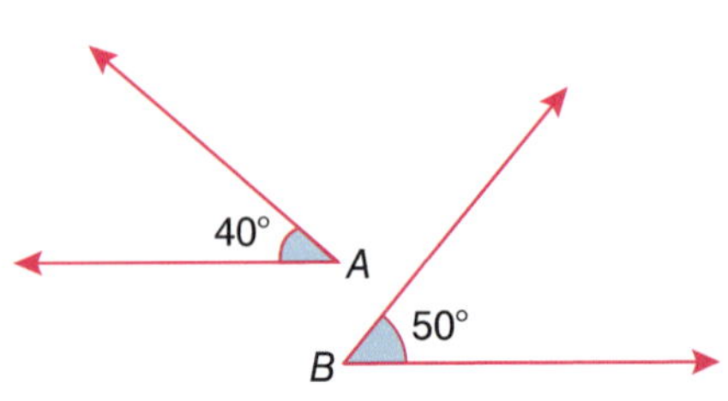

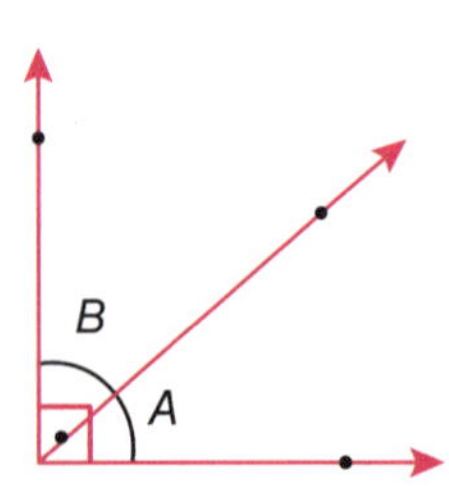

Prove this

Two positive angles formed by a pair of intersecting lines are called vertical angles.

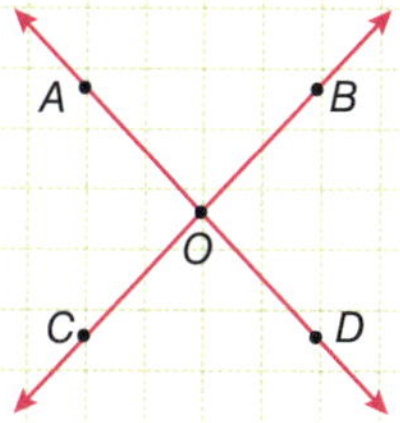

Prove that $m\angle AOB = m\angle COD$.

My Notes

Example 1.3

If the figure, $\angle AOC$ and $\angle COB$ are adjacent complementary angles and rays $\overrightarrow{OK}$ and $\overrightarrow{ON}$ are their respective bisectors. Prove that $m\angle KON = 45°$.

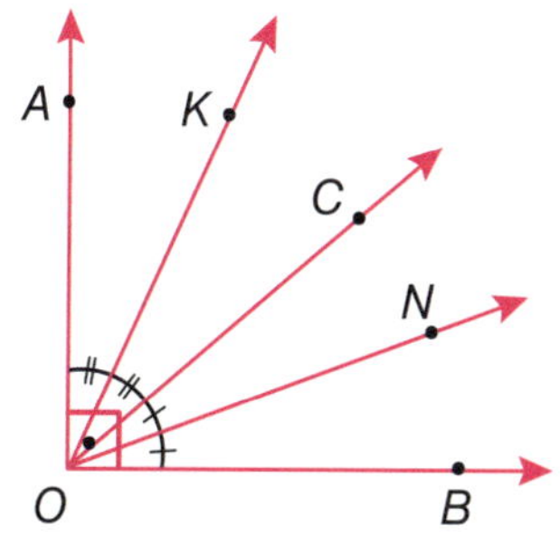

Solution 1.3

$m\angle KON = m\angle KOC + m\angle CON$ (1)

Since rays OK and ON are angle bisectors,

$m\angle KOC = \dfrac{m\angle AOC}{2}$ and $m\angle CON = \dfrac{m\angle COB}{2}$. (2)

Substituting (2) into (1) gives $m\angle KON = \dfrac{m\angle AOC}{2} + \dfrac{m\angle COB}{2} = \dfrac{90°}{2} = 45°$.

Definition 1.5

Supplementary Angles and Supplement of an Angle

Two angles are called supplementary angles if the sum of their measures is 180°. Each angle is called a supplement of the other angle in the figure since $m\angle K + m\angle T = 180°$. By the definition, $\angle K$ and $\angle T$ are supplementary angles, $\angle K$ is a supplement of $\angle T$, and $\angle T$ is a supplement of $\angle K$.

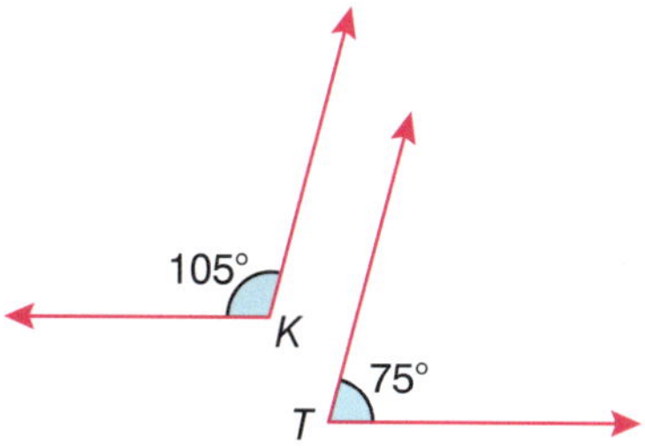

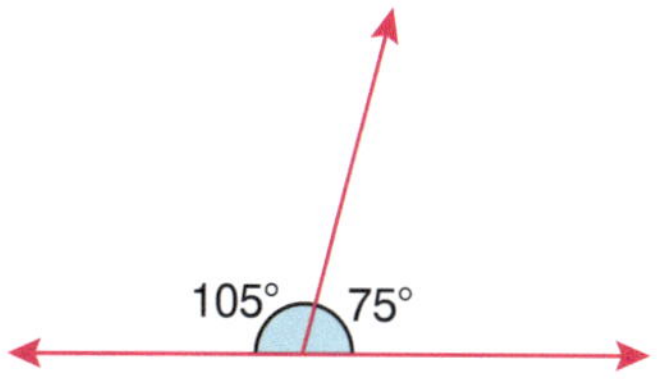

Example 1.4

In the figure, $\angle AOC$ and $\angle COB$ are adjacent supplementary angles, and rays OK and ON are their respective bisectors. Prove that ray OK and ray ON are perpendicular to each other.

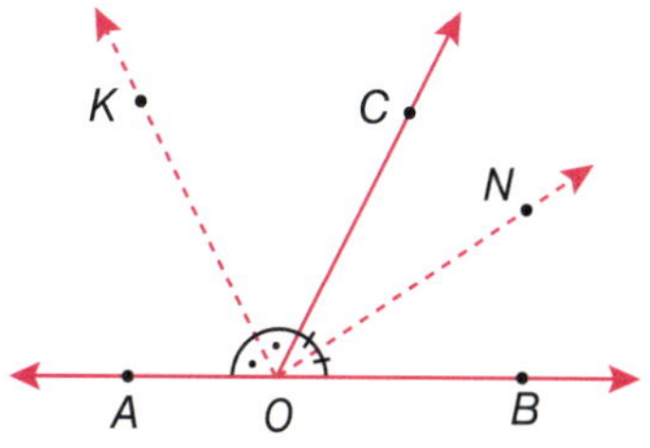

Solution 1.4

$m\angle KON = m\angle KOC + m\angle CON$ (1)

Since rays OK and ON are angle bisectors,

$$m\angle KOC = \frac{m\angle AOC}{2} \text{ and } m\angle CON = \frac{m\angle COB}{2}. \quad (2)$$

Substituting (2) into (1) gives

$$m(\angle KON) = \frac{m\angle AOC}{2} + \frac{m\angle COB}{2}.$$

So, $m(\angle KON) = \dfrac{m\angle AOC + m\angle COB}{2} = \dfrac{180^\circ}{2} = 90^\circ$ Q.E.D.

Example 1.5

In the figure,

$CD \parallel EF \parallel AB$ and

AE is the angle bisector of $\angle CAB$.

If $m\angle ACD = 100^\circ$, then find $m\angle AEF$.

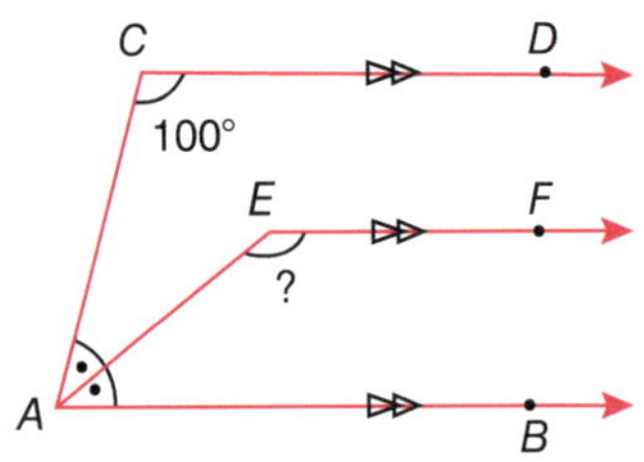

Solution 1.5

If CD and AB are parallel to each other, then $m\angle A + m\angle C = 180^\circ$. So, $m\angle CAB = 80^\circ$.

Since EA is bisector of CAB, $m\angle EAB = 40^\circ$. Since, EF and AB are also parallel to each other, $m\angle EAB + m\angle AEF = 180^\circ$.

Therefore, $40^\circ + m\angle AEF = 180^\circ$

$m\angle AEF = 140^\circ$.

Example 1.6

In the figure,

$a = 3c = 6b$.

What is the value of a?

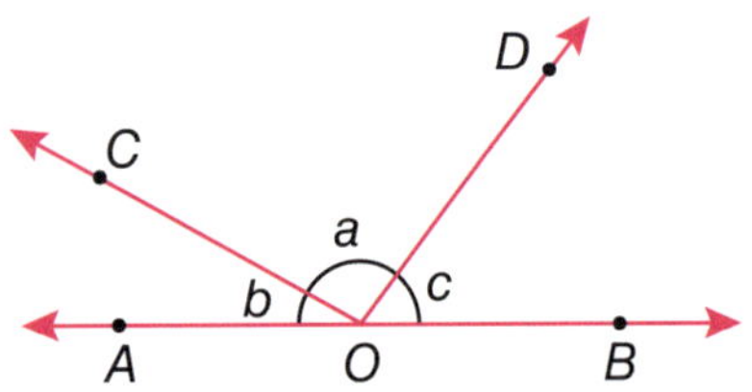

Solution 1.6

By writing a, b, and c in terms of b, we get $a = 6b$, $b = b$, and $c = 2b$.

Since $a+b+c = 180^\circ$, we have $6b+b+2b = 180^\circ \Rightarrow b = 20^\circ$.

Therefore, $a = 6b = 120^\circ$.

Q.E.D

"Q.E.D" (Sometimes written "QED") is an abbreviation for the Latin phrase "quod erat demonstrandum" ("that which was to be demonstrated"), a notation which is often placed at the end of a mathematical proof to indicate its completion. Several symbols are occasionally used as synonyms for Q.E.D. These include a filled square ■

My Notes

Theorem

$L_1 // L_2$

$a+b+c = x+y$

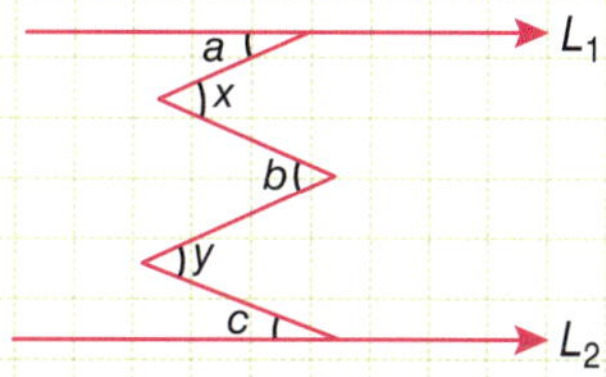

Proof

$L_1//L_2//L_3//L_4//L_5$

$x-a+y-c = b$

$x+y = a+b+c$

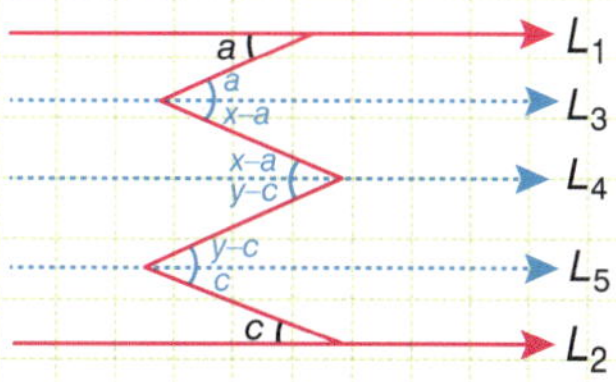

My Notes

Example 1.7

Ray OC lies in the interior of an angle AOB with $m\angle AOB = 124°$. The measure of the angle which is formed by the angle bisectors of $\angle AOB$ and $\angle COB$ is 36°

What is $m\angle COB$?

Solution 1.7

Figure-1

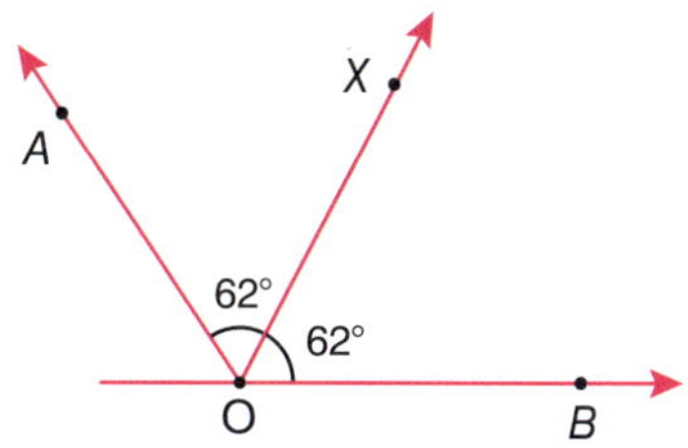

Figure-2

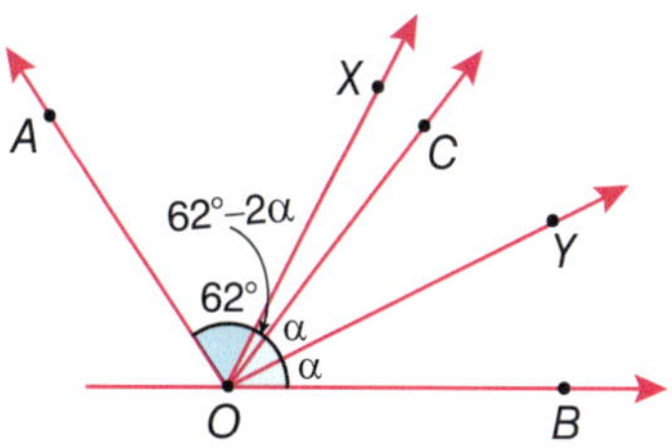

Let $\overrightarrow{OX}$ be the angle bisector of $\angle AOB$. So, ray $\overrightarrow{OC}$ must lie on either $\angle AOX$ or $\angle XOB$.

Let $\overrightarrow{OY}$ be the angle bisector of $\angle COB$ and $m\angle COB = 2\alpha$. So, $m\angle XOY = m\angle YOB = \alpha$.

Since $m\angle COB = 62°$ and $m\angle XOC = m\angle XOB - m\angle COB = 62° - 2\alpha$.

So, $m\angle XOY = m\angle XOC + m\angle COY = 62° - 2\alpha + \alpha = 36° \Rightarrow \alpha = 26°$.

Therefore, $m\angle COB = 52°$.

Definition 1.6

Interior and Exterior Angles of a Triangle

In a triangle *ABC,* the angles *BAC*, *ABC,* and *ACB* are called the interior angles of the triangle. They are written as ∠A, ∠B, and ∠C, respectively. The adjacent supplementary angles of these interior angles are called the exterior angles of the triangle. They are written as ∠A′, ∠B′, and ∠C′, respectively.

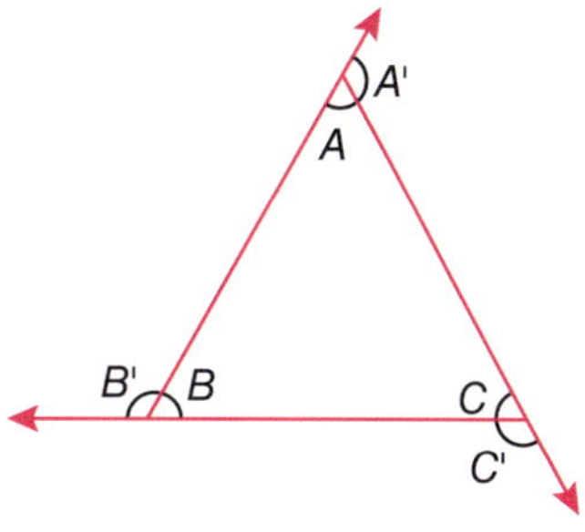

Interior Angles	Exterior Angles
$m\angle A = a$	$m\angle A' = 180° - a$
$m\angle B = b$	$m\angle B' = 180° - b$
$m\angle C = c$	$m\angle C' = 180° - c$

Theorem 1.1

The Triangle Sum Theorem

In a triangle, the three interior angles always add up to 180°.

Proof 1.1

Let $L_1 // L_2$. By the Alternate Interior Angle Theorem, we have

$m\angle 1 = m\angle A$ and $m\angle 2 = m\angle B$.

So, $m\angle 1 + m\angle C + m\angle 2$

$= m\angle A + m\angle B + m\angle C = 180°$.

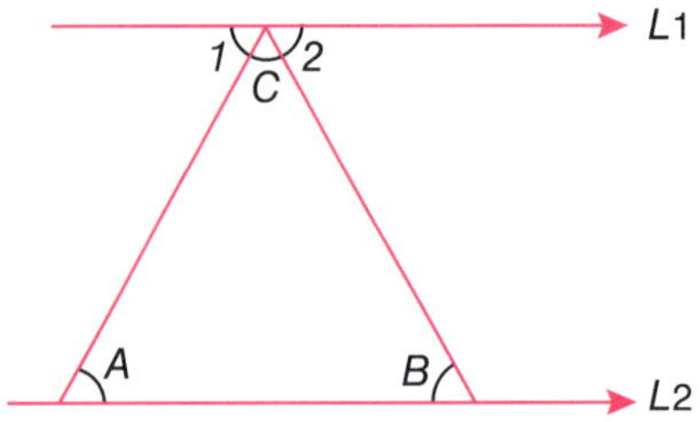

Example 1.8

In a triangle ABC,

$m\angle A = 3x - 10°$, $m\angle B = 2x + 20°$, and $m\angle C = 5x$.

What is x?

Solution 1.8

$m\angle A + m\angle B + m\angle C = 180°$ (Triangle Sum Theorem)

$3x - 10° + 2x + 20° + 5x = 180°$

$10x + 10° = 180°$

$x = 17°$.

Theorem 1.2

Exterior Angle Theorem

$m\angle ACD = m\angle A + m\angle B$

$x = a + b$.

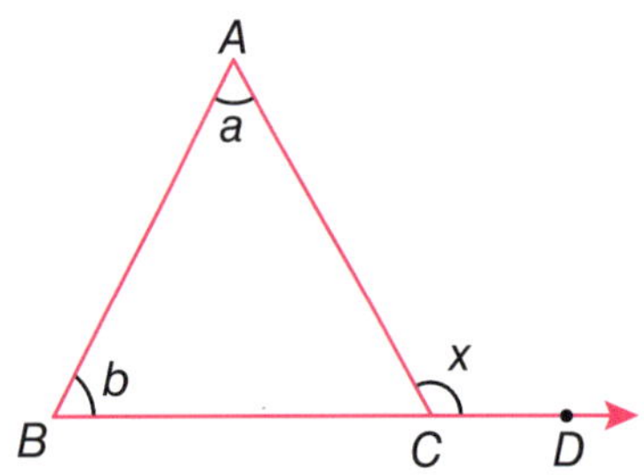

Proof 1.2

$m\angle ACD + m\angle BCA = 180°$ (1)

$m\angle A + m\angle B + m\angle C = 180°$ (2)

From (1) and (2), $m\angle ACD + m\angle C = m\angle A + m\angle B + m\angle C$.

Therefore, $m\angle ACD = m\angle A + m\angle B$.

Try This

In this flat diagram, each identically shaped pencil is formed by a rectangle attached to an isosceles triangle.

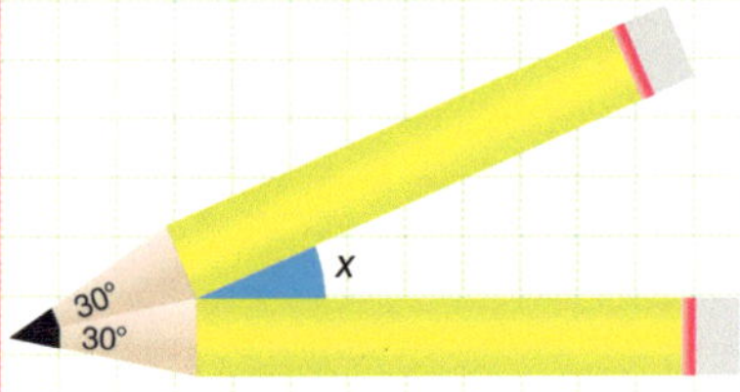

If the angle of each pencil tip in the diagram is 30°, then what is x?

www.brilliant.org

My Notes

Notes

Triangle Exterior Angle Sum

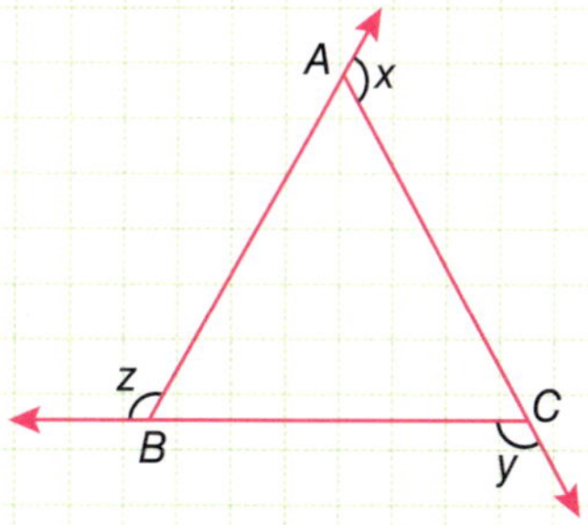

$x+y+z = 360°$

The sum of all three exterior angles of a triangle is 360°.

My Notes

Example 1.9

Prove that the sum of the exterior angles of a triangle is 360°.

Solution 1.9

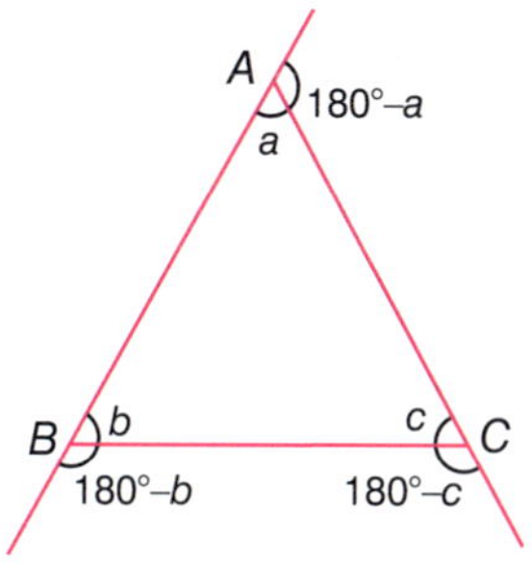

Let a, b, and c be the measures of the interior angles of triangle ABC.

The exterior angles are $180°-a$, $180°-b$, and $180°-c$.

The sum of the exterior angles is

$$\begin{aligned}180°-a+180°-b+180°-c &= 540°-(a+b+c)\\ &= 540°-180°\\ &= 360°.\end{aligned}$$

Example 1.10

In the figure $m\angle ACT = 120°$, $m\angle KAB = 5x$, and $m\angle PBC = 7x$.

a) Find the value of x.

b) Find $m\angle BAC$.

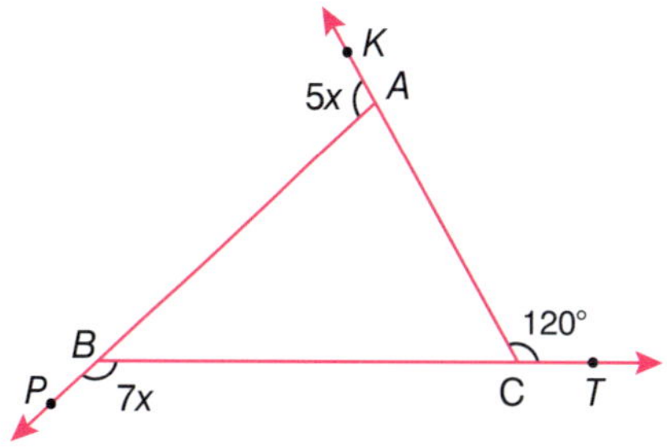

Solution 1.10

a) $m\angle KAB+m\angle PBC+m\angle ACT = 360°$ (Triangle Exterior Angle Sum Theorem)

$$\begin{aligned}5x+7x+120° &= 360°\\ 12x &= 240°\\ x &= 20°.\end{aligned}$$

b) $m\angle KAB+m\angle BAC = 180°$

$$\begin{aligned}(5\cdot 20°)+m\angle BAC &= 180°\\ m\angle BAC &= 80°.\end{aligned}$$

Example 1.11

Two complementary angles, A and B, have measures in the ratio of 3 to 1. What is the ratio of the measure of the complement of angle A to the measure of the complement of angle B?

Solution 1.11

Since the ratio of A to B is 3 to 1, $A = 3B$.

Since $A+B = 90°$, $4B = 90°$. So, $B = 22.5°$ and $A = 3\cdot(22.5°) = 67.5°$.

The complement of angle A is 22.5° and the complement of angle B is 67.5°.

The ratio is $\frac{22.5°}{67.5°} = \frac{1}{3}$.

Alternate solve:

$$\frac{\text{complement of } A}{\text{complement of } B} = \frac{B}{A} = \frac{B}{3B} = \frac{1}{3}.$$

Try This

In triangle ABC, the measure of angle bisectors BAC is 56°. The angle of angles B and C meet at I.

Find the measure of angle BIC.

My Notes

Selected Angle Theorems

Theorem 1.3

The measure of the angle formed by the bisectors of two interior angles of a triangle is 90° more than half of the third angle. That is, $m\angle BIC = 90° + \frac{m\angle A}{2}$.

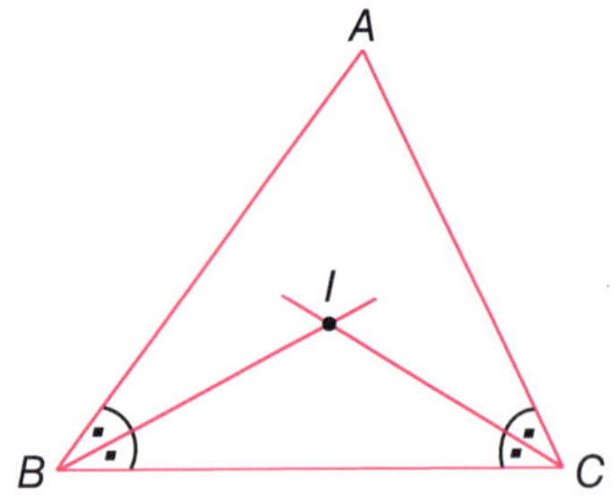

Proof 1.3

From ΔBIC, $\frac{m\angle B}{2} + \frac{m\angle C}{2} + m\angle BIC = 180°$ (1)

From ΔABC, $m\angle A + m\angle B + m\angle C = 180°$ (2)

$\Rightarrow$ Multiply first equation by 2.

We get $m\angle B + m\angle C + 2\, m\angle BIC = 360°$ (3)

(3)–(2) $\Rightarrow 2\, m\angle BIC - m\angle A = 180°$

$$m\angle BIC = 90° + \frac{m\angle A}{2}.$$

Try This

In a triangle with *a, b,* and *c,* the mean of *b* and *c* is *a*.

What is the value of *a*?

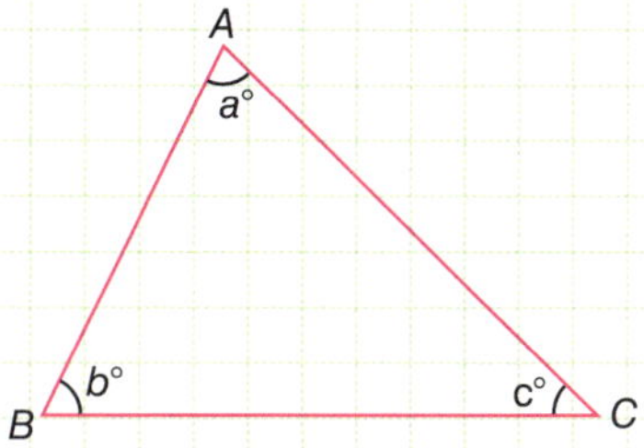

My Notes

Theorem 1.4

$m\angle BDC = 90° - \dfrac{m\angle A}{2}.$

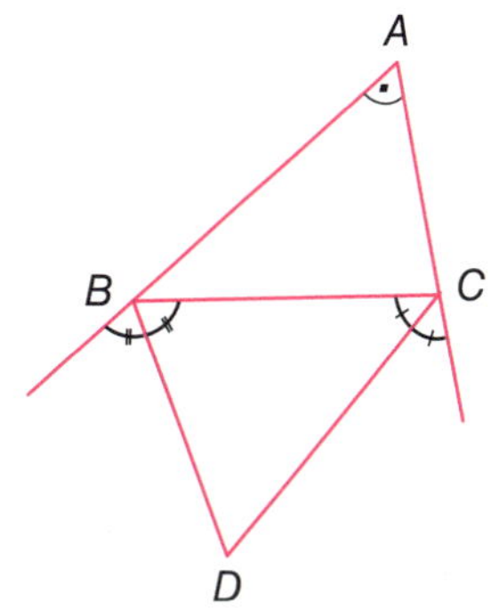

Proof 1.4

Let $m\angle B = 2b$, $m\angle C = 2c$.

From ΔABC, $m\angle A + 180° - 2b + 180° - 2c = 180°$

$m\angle A + 180° = 2b + 2c$. (1)

From ΔBDC, $m\angle D + b + c = 180°$. (2)

Multiply (2) by 2.

$2m\angle D + 2b + 2c = 360°$ (3)

(1) + (3) gives $2m\angle D = 180° - m\angle A$

Therefore, $m\angle D = 90° - \dfrac{m\angle A}{2}.$

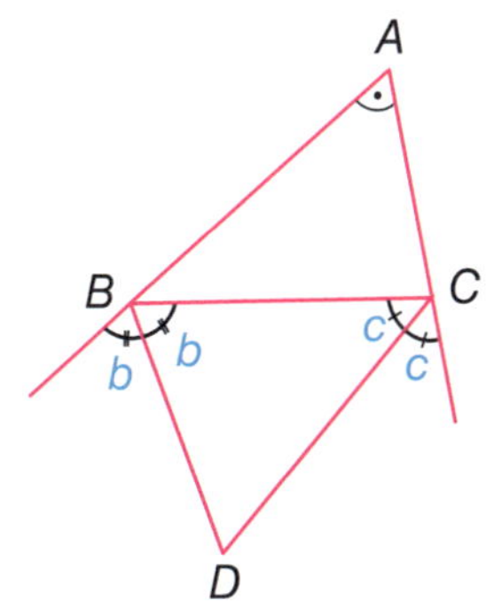

Theorem 1.5

$2m\angle D = m\angle A.$

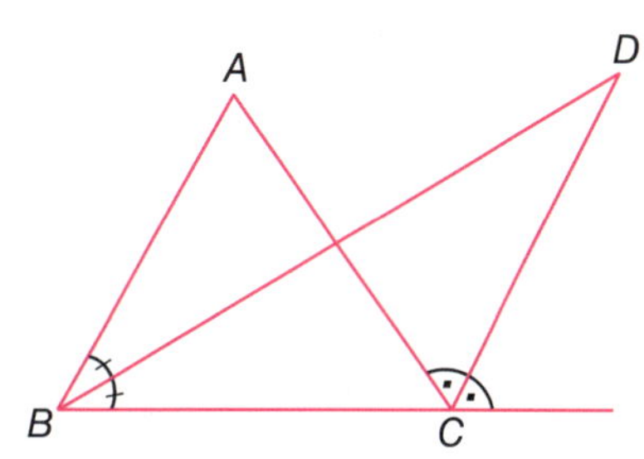

Proof 1.5

Let $m\angle ABD = b$ and $m\angle ACD = c$.

From ΔABC, (1) $2c = 2b + m\angle A$. (Theorem 1.2)

From ΔBCD, (2) $c = b + m\angle D$. (Theorem 1.2)

Multiply (2) by 2 and subtract from (1), we obtain $2m\angle D = m\angle A$.

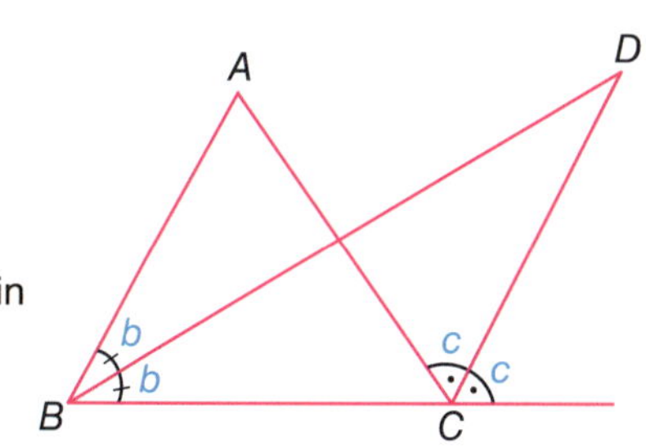

Theorem 1.6

$m\angle BKC = a+b+c$.

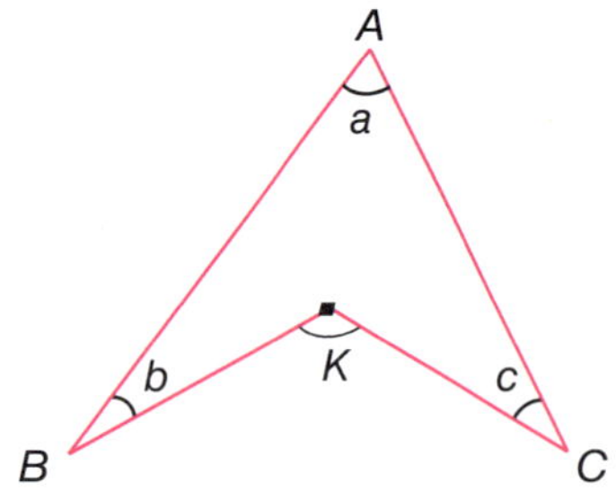

Proof 1.6

Extend BK to AC.

$m\angle KDC = a+b$, (Theorem 1.2)

$m\angle BKC = a+b+c$. (Theorem 1.2)

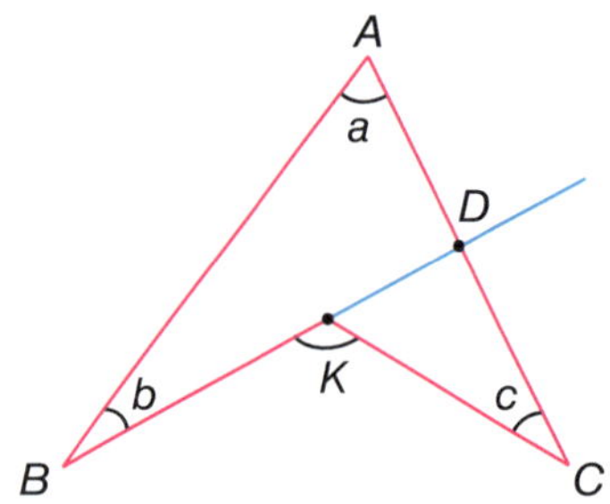

Catriona Agg @Cshearer41
Geometry puzzle

The three shaded triangles are isosceles.

What is the missing angle?

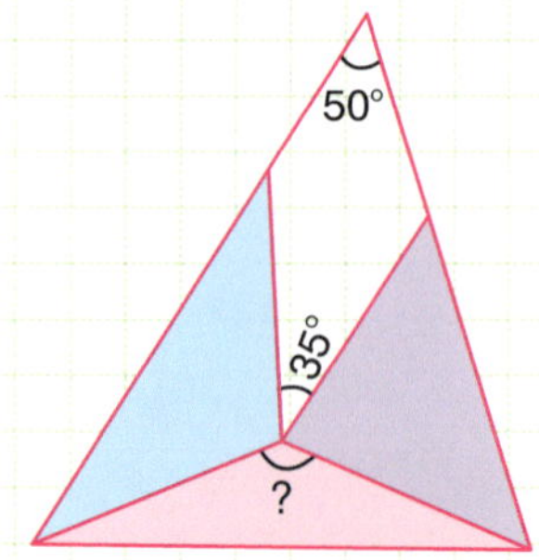

My Notes

Theorem 1.7

AN is the angle bisector of $\angle BAC$ and $AH \perp BC$ prove that $m\angle HAN = \dfrac{|m\angle B - m\angle C|}{2}$.

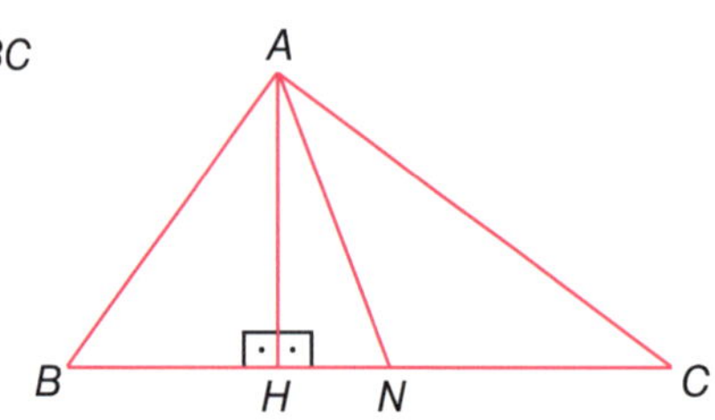

Proof 1.7

Let $x = m\angle HAN$,

$m\angle B = b$ and $m\angle C = c$ (Suppose $b>c$)

$m\angle BAH = 90° - b$,

$m\angle BAN = m\angle NAC = 90° - b + x$, and

$m\angle ANC = 90° - b + x + c$.

So, from ΔAHN

$x + 90° - b + x + c + 90° = 180° \Rightarrow x = \dfrac{b-c}{2}$.

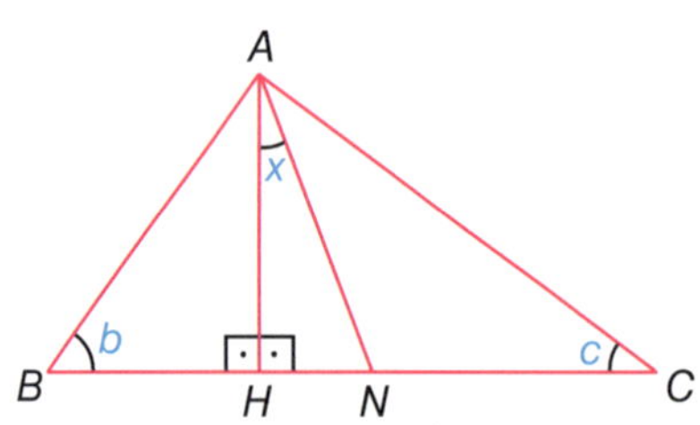

Try This

In a right triangle ABC, the median AM is half of side BC. The angle between AM and the height of the triangle AH is $40°$.

Find $m\angle ACB$.

My Notes

Example 1.12

In the figure, K is the intersection point of the bisectors of the exterior angles at vertex A and vertex B with $m\angle A = 120°$ and $m\angle B = 40°$.

Find $m\angle BKA$.

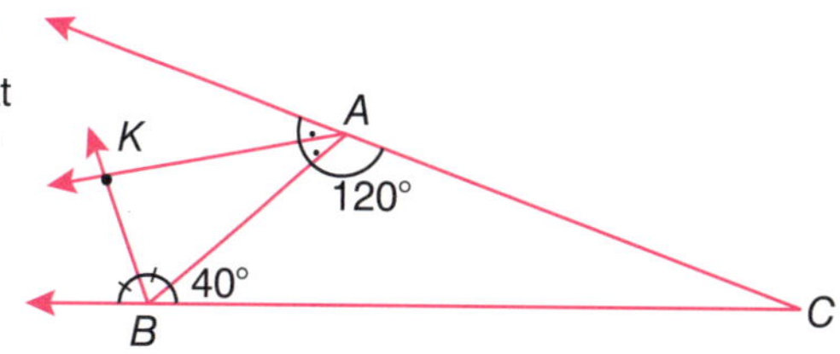

Solution 1.12

$m\angle A + m\angle B + m\angle C = 180°$ (Triangle Angle–Sum Theorem in ΔABC)

$20° + 40° + m\angle C = 180°$

$m\angle C = 20°$ (1)

$m\angle BKA = 90° - \dfrac{m\angle C}{2}$ (Property 3.1b)

$= 90° - \dfrac{20°}{2}$ (Substitute (1))

$= 80°.$

Example 1.13

One of the acute angles in a right triangle measures $20°$.

Find the angle between the altitude and the angle bisector which is drawn from the vertex of the right angle of the triangle.

Solution 1.13

Method 1

From (Theorem 1.7)

$m\angle HAN = x = \dfrac{70° - 20°}{2} = 25°.$

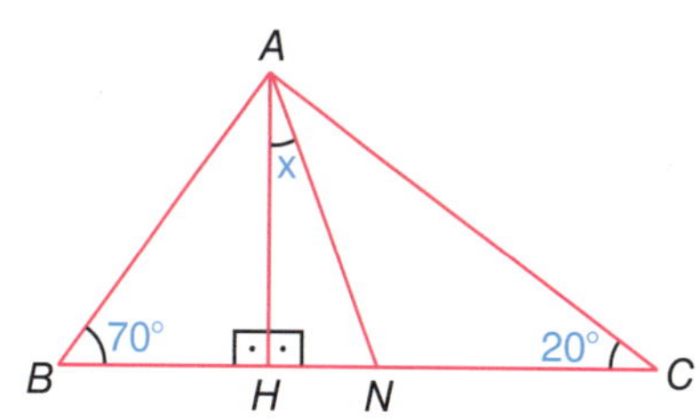

Method 2

AN is the angle bisector of $m\angle A = 90°$.

Since $m\angle B = 70°$,

then $m\angle BAH = 20°$.

Therefore, $m\angle HAN = 45° - 20° = 25°$.

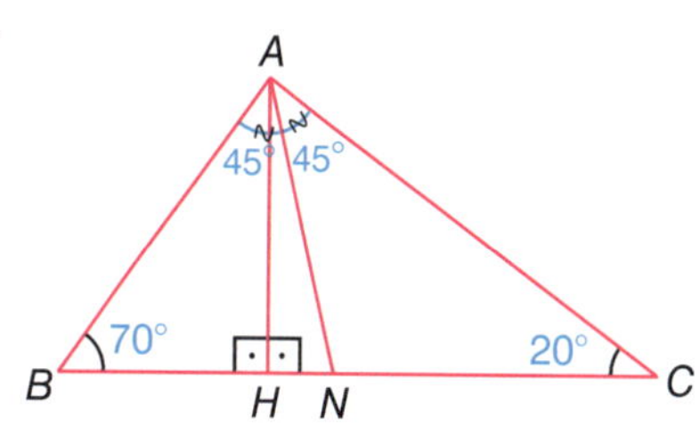

Example 1.14

BD is the angle bisector of $m\angle ABC$

$m\angle ACB = 50°$,

$m\angle ACD = 65°$.

Find $m\angle ADB$.

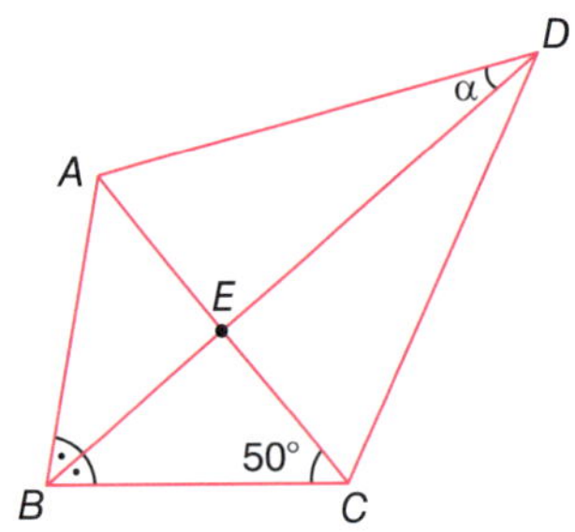

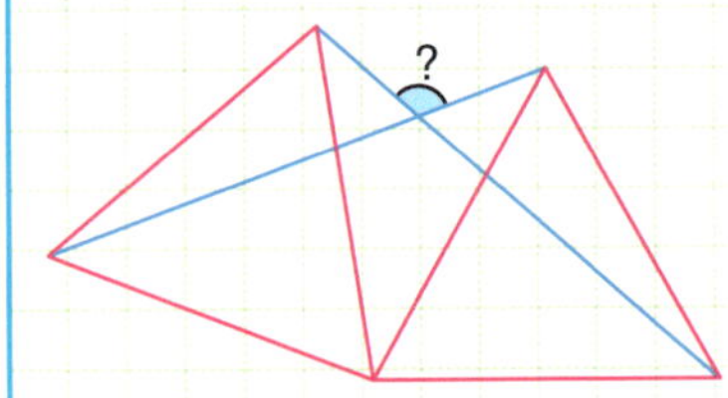

My Notes

Solution 1.14

By extending BC towards point G, we can notice that CD is the angle bisector of $\angle ACG$. AD also has to be the angle bisector of $\angle CAF$ (Since point D is the intersection of two angle bisectors, then the third one must be an angle bisector).

From (Theorem 1.3) $\alpha = \frac{50°}{2} = 25°$.

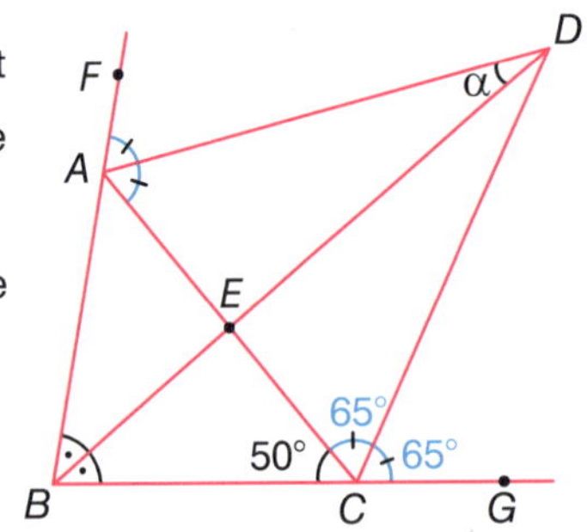

Example 1.15

In the figure, AN is the angle bisector of $\angle BAC$.

$m\angle ANC = 100°$ and $m\angle B = 2m\angle C$.

What is $m\angle B$?

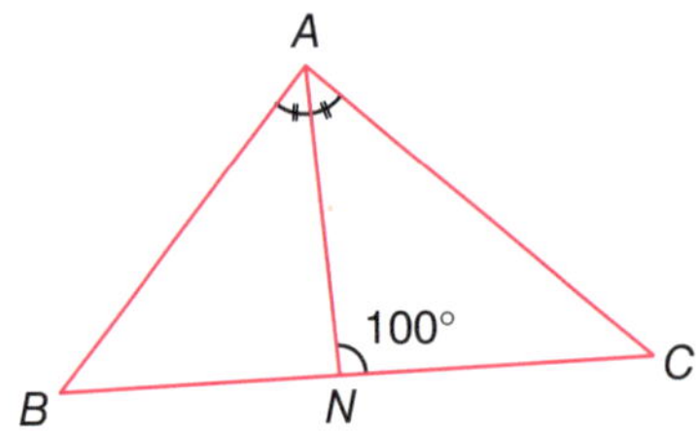

Solution 1.15

Since $m\angle BAC = 180° - 3\alpha$.

Then $m\angle ANC = 90° - \frac{3}{2}\alpha$.

From ΔANB, $100° + 90° - \frac{3}{2}\alpha + \alpha = 180°$.

So, $\frac{\alpha}{2} = 10°$, then $\alpha = 20°$. Therefore, $m\angle B = 40°$.

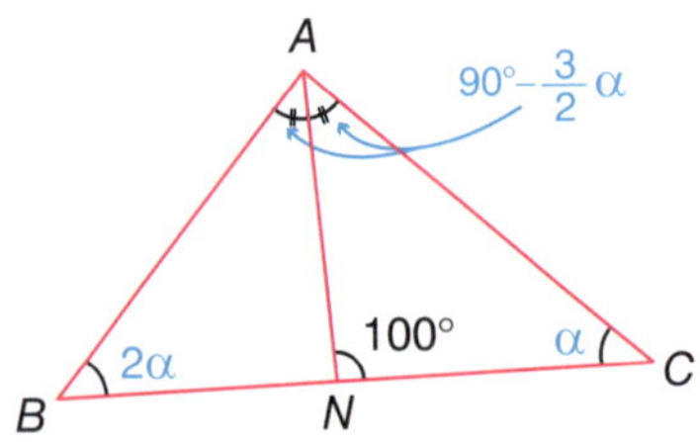

Try This

A sequence of isosceles triangles are constructed starting with

$AB = BC$, then

$BC = CD$, and so on.

If the angle $BAC = 13°$ then how many isosceles triangles can be drawn?

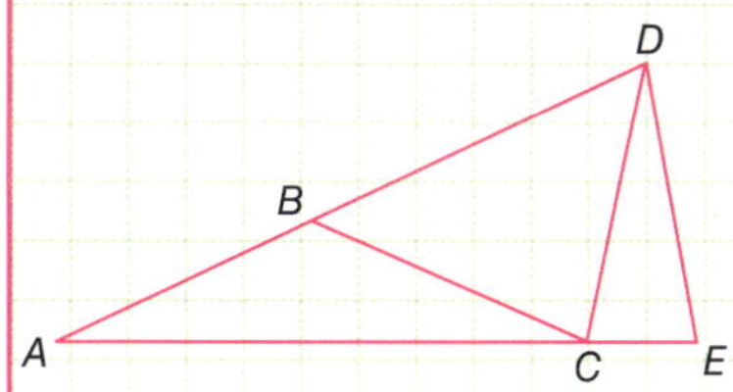

My Notes

Example 1.16

BE and *CE* are angle bisectors. *AD* is the median of *BC*.

Given that $m\angle DAE = 10°$ and

$m\angle BEC = 135°$.

Find $m\angle CAD$.

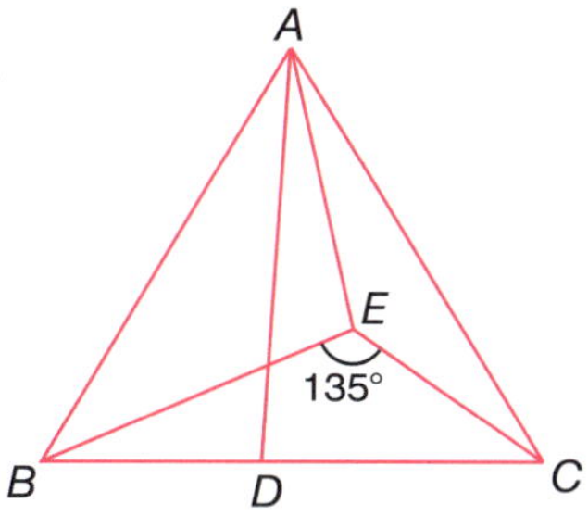

Solution 1.16

From (Theorem 1.13)

$$\frac{m\angle A}{2} + 90° = 135° \Rightarrow m\angle A = 90°$$

AE is also the angle bisector of $\angle A$.

So, $m\angle EAC = 45°$,

$m\angle CAD = 45° + 10° = 55°$.

Triangle Angles Problem by using Isosceles Triangle

Example 1.17

In a triangle *DEF*.

Point *T* is on *DF* such that $DT = DE$.

Given that $m\angle EDT = 40°$

and $m\angle DEF = 85°$.

Find $m\angle TEF$.

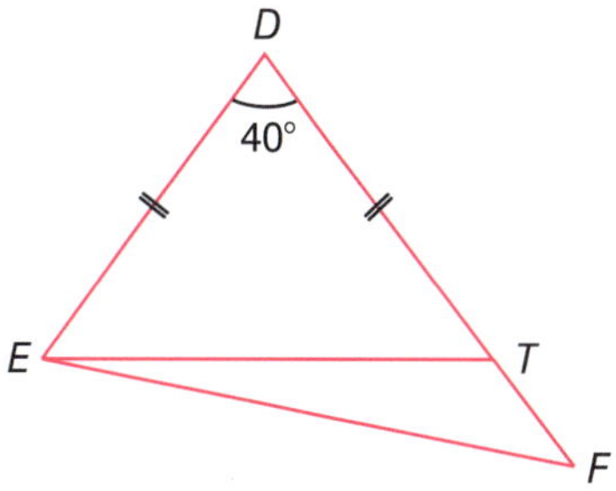

Solution 1.17

By the Triangle Sum Theorem in ΔDET.

$$m\angle EDT + m\angle DET + m\angle DTE = 180°$$

$$40° + 2m\angle DET = 180° \quad (m\angle DET = m\angle DTE)$$

$$m\angle DET = 70°.$$

So, $m\angle TEF = m\angle DEF - m\angle DET = 85° - 70° = 15°$.

Example 1.18

In a triangle *MNP*,

$MS = MP$,

$ST = TP$,

$m\angle M = 94°$, and

$m\angle N = 26°$.

Find $m\angle MST$.

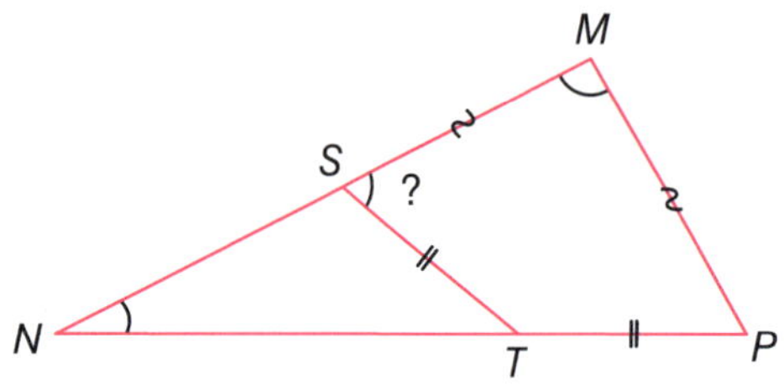

Solution 1.18

When we connect points *M* and *T*,

we get two congruent triangles. So, *MT* is the angle bisector of $\angle NMP$.

$$m\angle MST = 180° - (73° + 47°)$$
$$= 180° - 120° = 60°.$$

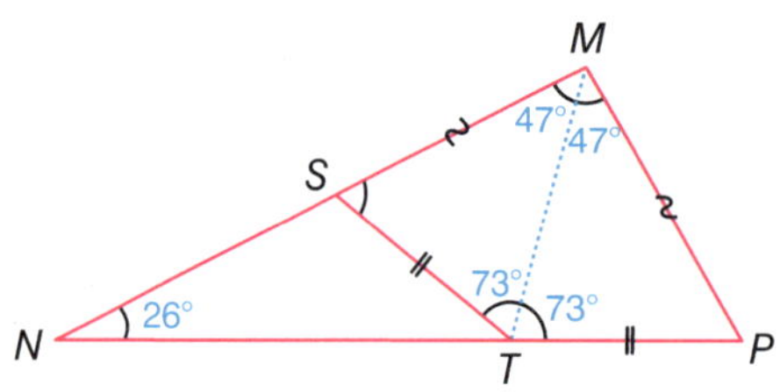

Example 1.19

$AB = AC$,

$BD = CD = CE$,

B, *C*, and *E* are collinear.

Find $m\angle BDE$.

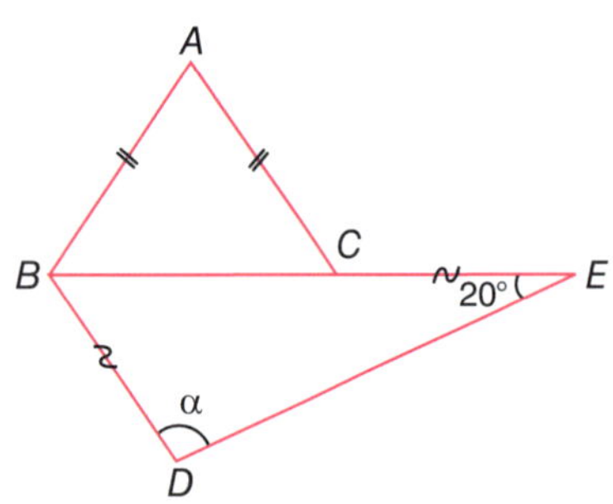

Solution 1.19

When we connect points *D* and point *C*, we get two isosceles triangles namely ΔBDC and ΔCDE.

Since $DC = CE$, then $m\angle BCD = 20° + 20° = 40°$.

Since $BD = CD$, then
$m\angle BDC = 180° - (40° + 40°) = 100°$.

Therefore, $m\angle BDE = 100° + 20° = 120°$.

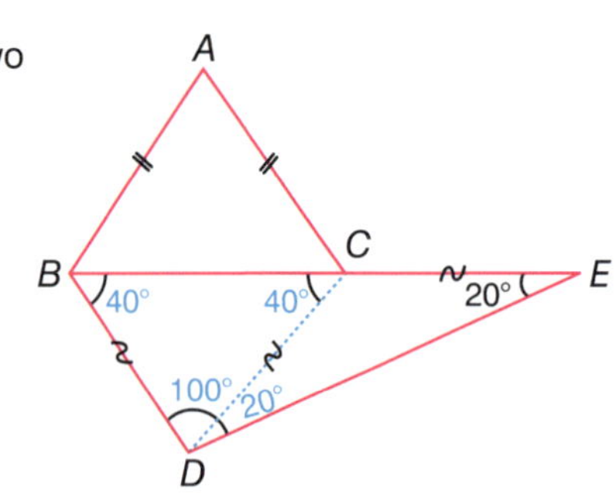

Q.E.D.

In ΔABC, *D*, *E*, *F* are on *AB*, *BC*, *CA*, such that $DE = EF = FD$, $AB = AD$.

Prove that

$$\angle DEB = \frac{1}{2}(\angle ADF + \angle CFE).$$

Try This

In a ΔABC, $BD = AE$ and $AD = EC$,

$m\angle ADE = 25°$, and

$m\angle ABC = 35°$.

Find $m\angle DEC$.

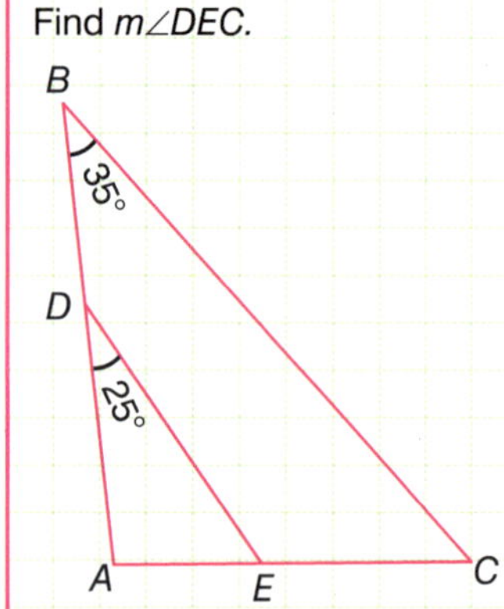

1. Suppose that angles $\angle A$ and $\angle B$ are supplementary angle with

$m\angle A = 4x + 75°$, and

$m\angle B = 5x + 15°$.

Find the measure of $\angle B$.

2. The measures of complement of an angle is $\frac{1}{7}$ of the measures of its supplement.

Find the measure of this angle.

3. In the figure, $AB \perp CD$,

$m\angle DCE = 3x - 10°$,

$m\angle CEA = x$, and

$m\angle EAB = 20°$.

What is the value of x?

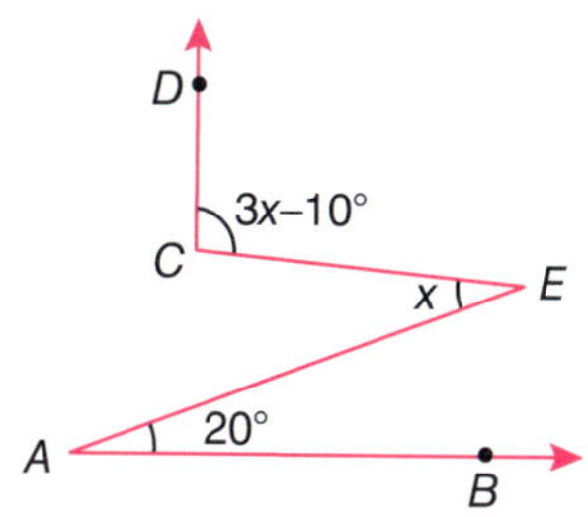

4. In the figure,

$AB = BD$,

$AD = DC$, and

$m\angle DAC = 35°$.

Find $m\angle B$.

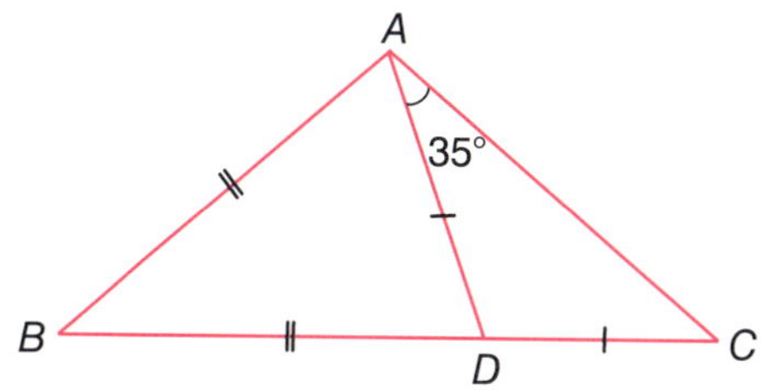

5. AD and BD are bisectors with

$m\angle ABD = 7x + 14$,

$m\angle ACD = 2x + 4$.

Find $m\angle BCD$.

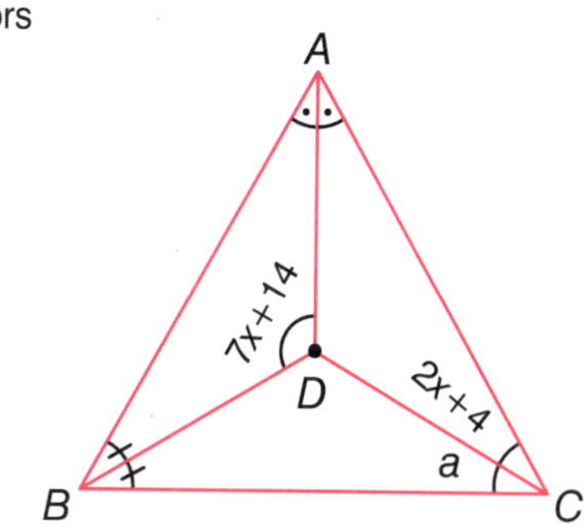

6. In the figure,

$PQ = PS = PR$,

$m\angle SPR = 24°$.

Find $m\angle SQR$.

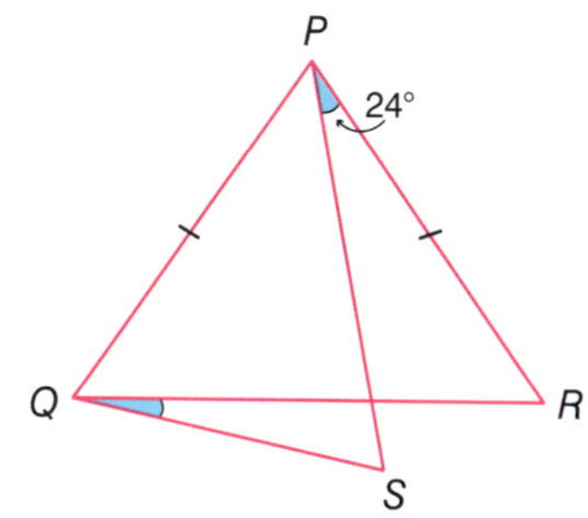

7. Two angles of an isosceles triangle measures 80° and $x°$.

What is the sum of the three possible values of x?

8. In the figure, $DE \perp AC$,

$AB = BD$,

$AE = ED$, and

$m\angle ABC = 50°$.

Find $m\angle C$.

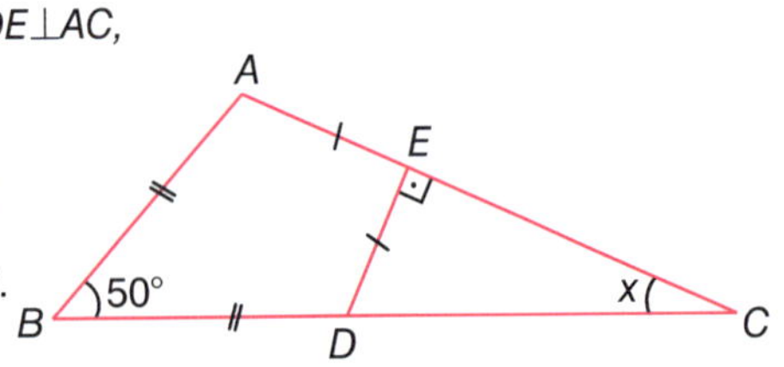

9. ABC is a triangle and point D is a point inside the triangle ABC. ABC and ADC are isosceles triangles. With $AB = BC$ and $AD = DC$. The angle ABC measures 50 degrees, and angle ADC measures 130 degrees.

What is the degree measure of angle BAD?

10. In the figure,

$MK = NK = PK$.

What is $a+b+c$?

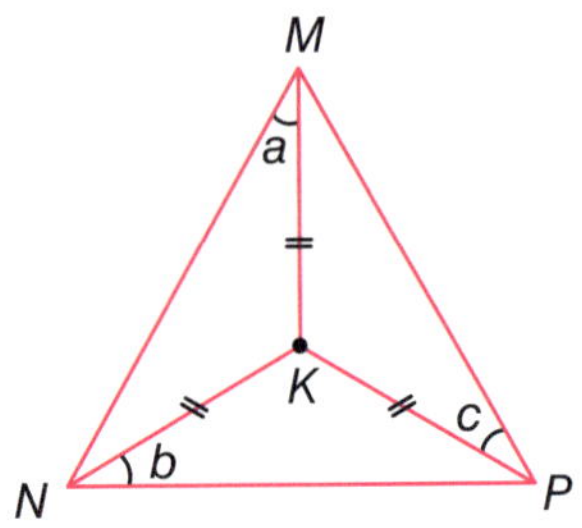

11. $AD = BD$,

$m\angle DAC = x$, and

$m\angle BCE = 2x$.

If $m\angle EAB = 110°$ find the value of x.

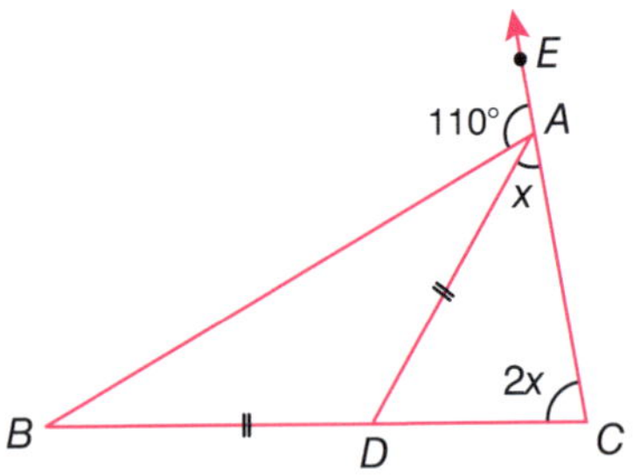

12. One of the acute angle in a right triangle measures 24°.

Find the angle between the altitude and the angle bisector which one drawn from the vertex of the right angle of triangle.

1. Determine the sum of the angles at the star points, $a+b+c+d+e$ in the figure below.

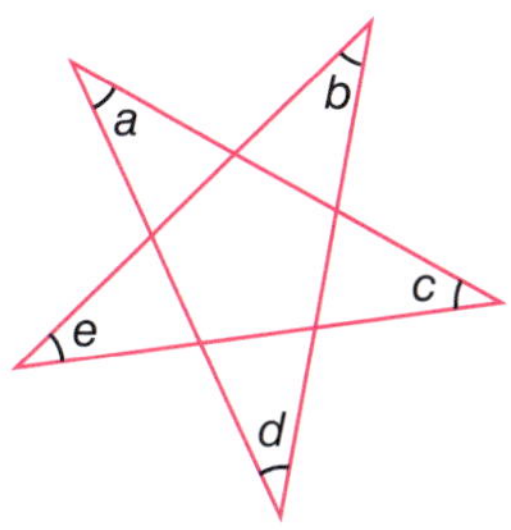

2.

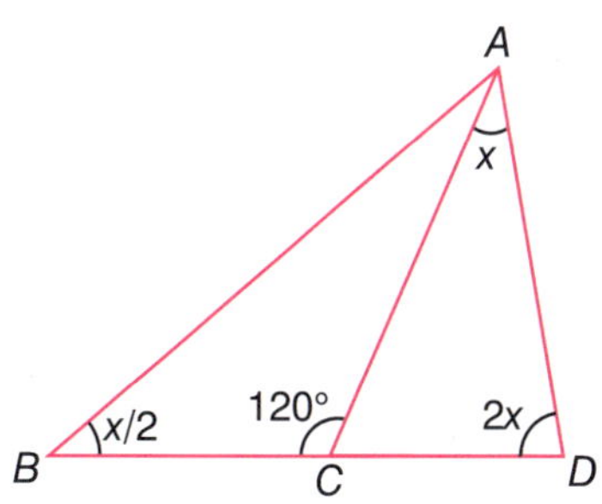

In the figure, measures of $\angle D$ is four times the measures of $\angle ABC$ and the measures of $\angle ADC$ is twice the measures of $\angle CAD$.

If $m\angle ACB = 120°$, what is $m\angle B$?

3. In the figure, ray KC is the bisector $\angle AKD$

$m\angle CKD = 4x-22°$

$m\angle AKD = 4x+6°$ and

$m\angle AKB = 3m\angle BKC$.

Find $m\angle CKB$.

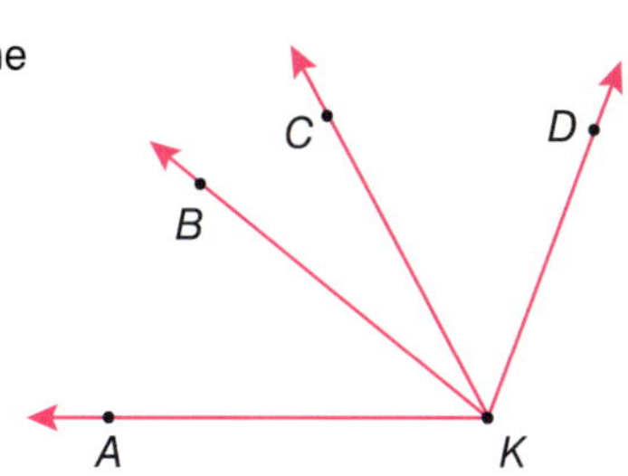

4. AD and CB bisector angles of A and C, respectively

$m\angle AEC = 75°$,

$m\angle B = 30°$.

What is $m\angle ADC$?

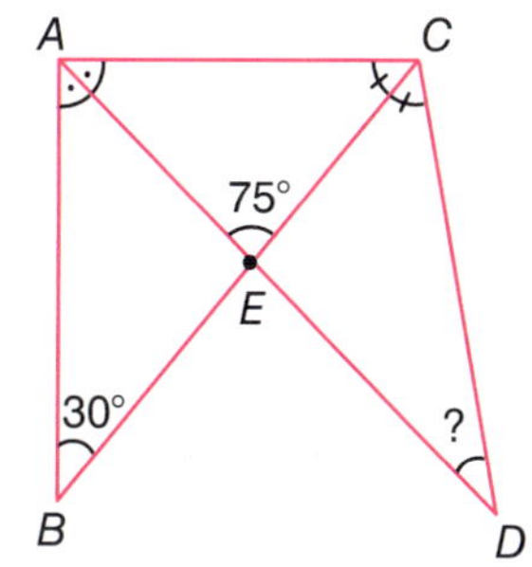

5.

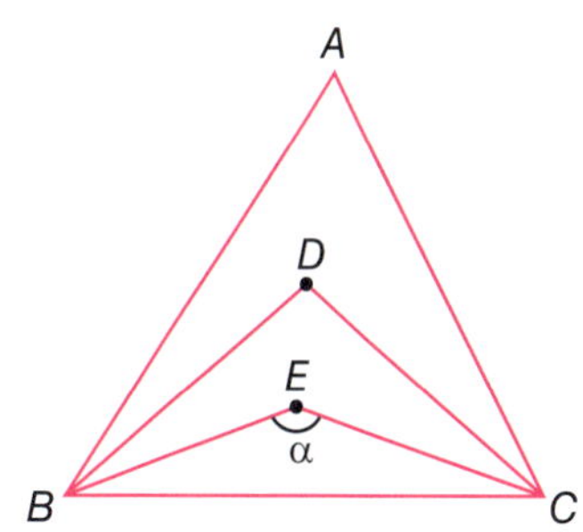

The trisectors of angles B and C, meet at points D and E, as shown. Angle A measures 57° and angle DBE measures 21 degrees .

What is the measure of angle α?

6.

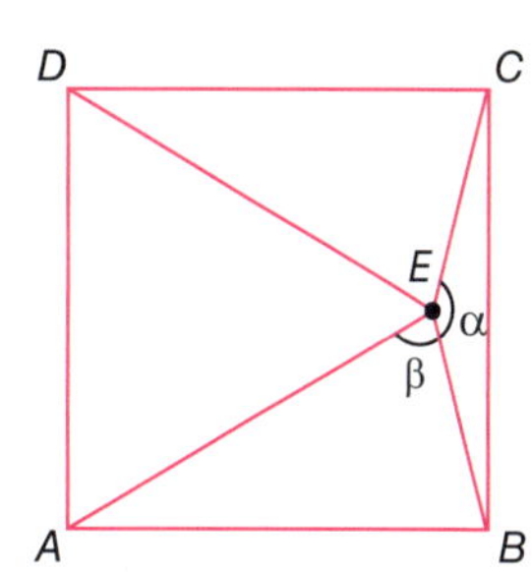

$ABCD$ is a square and ΔADE is an equilateral triangle.

What is the sum of measure of $\angle AEB$ and $\angle CEB$?

7. What is the measures acute angle formed by the hands of the clock 6:40 PM?

8. In triangle ABC, D is on BC so that AD bisects $\angle BAC$. Also $AD = DB$ and $3m\angle ACB = 2m\angle ADB$.

What is the measure of $\angle ACB$?

9. In the figure, AD is an angle bisector,

$CE = CD$,

$m\angle ACD = 60°$.

Find $m\angle ABC$.

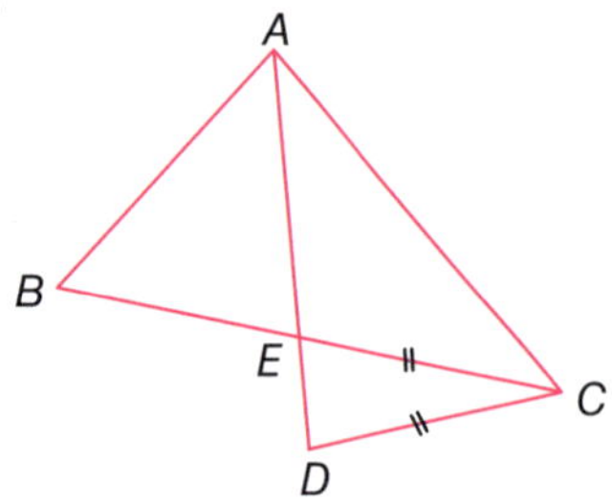

10. Triangles ABC and ADC are isosceles measures with $AD = DC$ and $AB = BC$. Point D is inside triangle ABC, $m\angle ABC = 50°$ and $m\angle ADC = 130°$.

What is the degree measures of $\angle BAD$?

11. In the figure, $ABCD$ is a square and EAB is an equilateral triangle.

Find $m\angle ELB$.

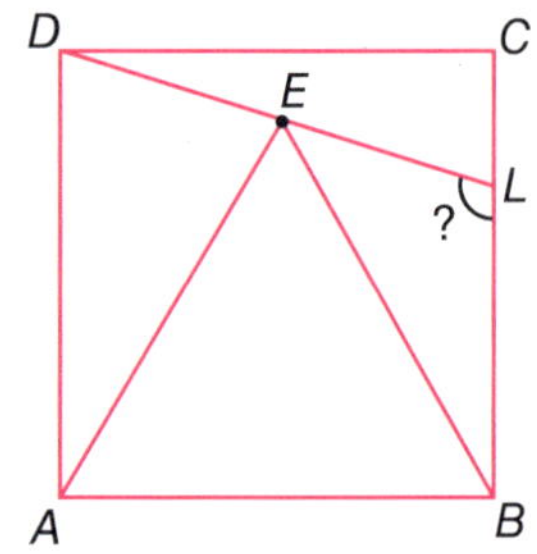

12. In a triangle ABC, there is a point D on side BC. Such that $DC = AD = AB$ if $m\angle ABD = 54°$

Find $m\angle DAC$.

1. (MathCON, 2019-Final#12)

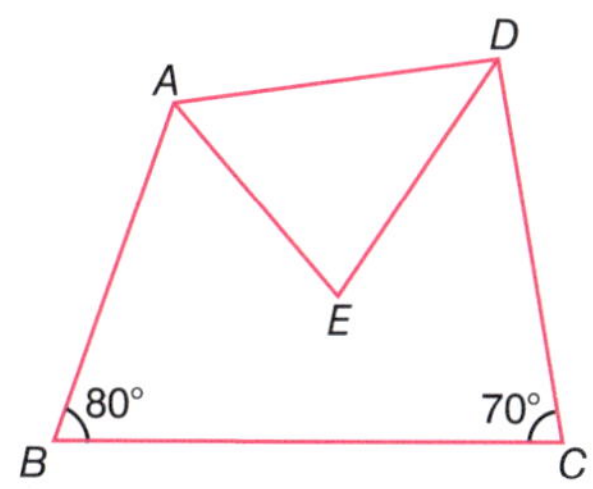

$ABCD$ is a quadrilateral where AE and DE are angle bisectors.

Find the value of $\angle AED$.

A) 60° B) 75° C) 90° D) 100° E) None°

2. (CWML -2016-17-Meet 3#4) Two complementary angles have degree measures $\frac{x+50}{2}$ and $\frac{x+70}{4}$.

The measure of the smaller angle is:

A) 10° B) 20° C) 30° D) 40° E) 50°

3. (2020 AIME I #1)

In ΔABC with $AB = AC$, point D lies strictly between A and C on side AC and point E lies strictly between A and B on side $\overline{AB}$ such that $AE = ED = DB = BC$. The degree measure of $\angle ABC$ is $\frac{m}{n}$, where m and n are relatively prime positive integers.

Find $m+n$.

4. (CWML 2016-17-Meet 3 #6)

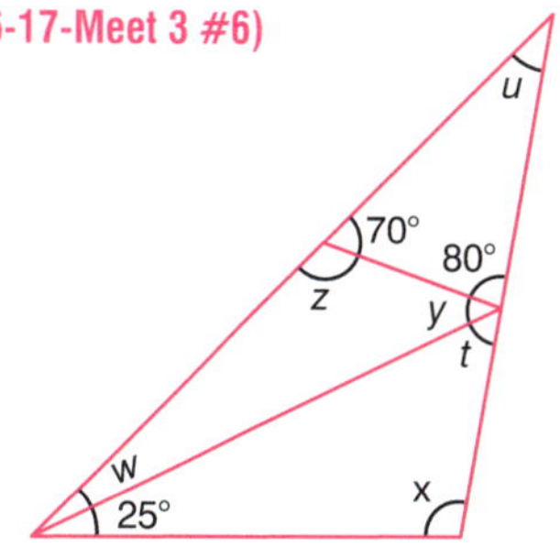

Given the degree measures the in the triangle shown below, determine the value of $x-y$.

5. (2016 AlphaStar PIMC 17 #6)

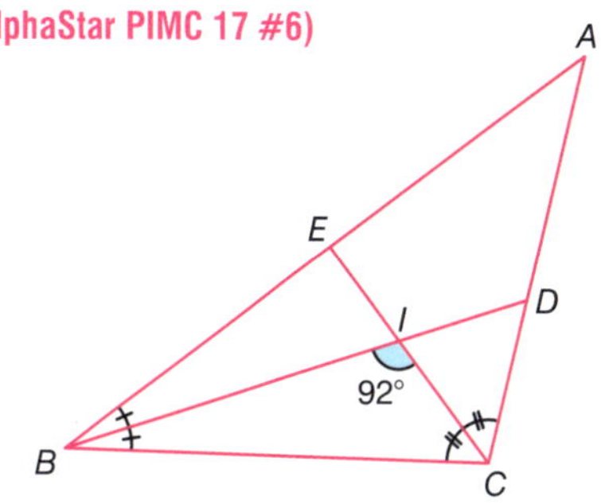

In the figure below, BD and CE are angle bisectors and they intersect at I.

If $\angle BIC = 92°$, what is the measure of $\angle BAC$, in degrees?

6. (2016 AlphaStar PIMC #7) A triangle is isosceles if two of its angles have the same measure. In an isosceles triangle ΔABC, $\angle A = 50°$.

What is the sum of the largest and smallest possible degree measures of $\angle B$?

7. **(2016 AMC10B #7)** The ratio of the measures of two acute angles is 5:4, and the complement of one of these two angles is twice as large as the complement of the other.

What is the sum of the degree measures of the two angles?

A) 75 B) 90 C) 135 D) 150 E) 270

8. **(2011 AMC-12B #10)** Rectangle $ABCD$ has $AB = 6$ and $\angle BC = 3$. Point M is chosen on side AB so that $\angle AMD = \angle CMD$.

What is the degree measure of $\angle AMD$?

A) 15° B) 30° C) 45° D) 60° E) 75°

9. **(2009 AMC 8 #19)** Two angles of an isosceles triangle measure 70°and x.

What is the sum of the three possible values of x?

A) 95 B) 125 C) 140 D) 165 E) 180

10. **(2012 AMC-10A #4)** Let $\angle ABC = 24°$ and $\angle ABD = 20°$.

What is the smallest possible degree measure for angle CBD?

A) 0 B) 2 C) 4 D) 6 E) 12

11. **(2007 AMC-12A #6)** Triangles ABC and ADC are isosceles with $AB = BC$ and $AD = DC$. Point D is inside triangle ABC, angle ABC measures 40 degrees, and angle ADC measures 140 degrees.

What is the degree measure of angle BAD?

A) 20 B) 30 C) 40 D) 50 E) 60

12. **(2010 AMC-10A #14)** Triangle ABC has $AB = 2 \cdot AC$. Let D and E be on $\overline{AB}$ and $\overline{BC}$, respectively, such that $\angle BAE = \angle ACD$. Let F be the intersection of segments AE and CD, and suppose that $\angle CFE$ is equilateral.

What is $\angle ACB$?

A) 60° B) 75° C) 90° D) 105° E) 120°

CHAPTER 2

1 Isosceles Triangles
2 Equilateral Triangles
3 Right Triangles
4 Pythagorean Theorem

Note

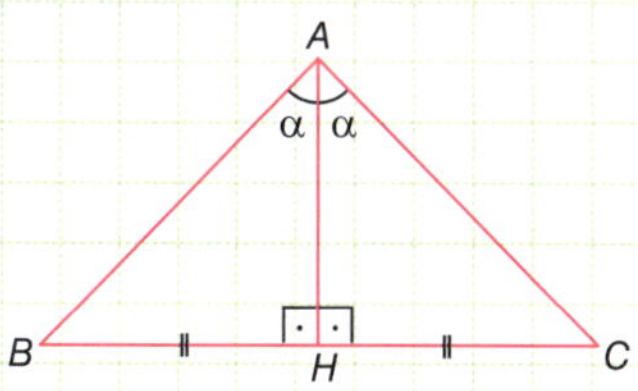

- $AH \perp BC$
- $BH = HC$
- AH is an angle bisector.

If two of these three features come together, then there is always an isosceles triangle.

Therefore, $AB = AC$.

My Notes

Isosceles Triangle

Theorem 2.1

If two sides of a triangle are congruent, then the angles opposite of the sides are also congruent. The bisector of the vertex angle of an isosceles triangle separates the triangle into two congruent triangles.

Proof 2.1

ΔABC is an isosceles triangle with AB = AC,

AD bisects $\angle BAC$.

If $m\angle BAD = m\angle DAC = \alpha$ and $m\angle C = \beta$,

then $\alpha+\beta+90° = 180°$.

Therefore, $m\angle B = \beta$.

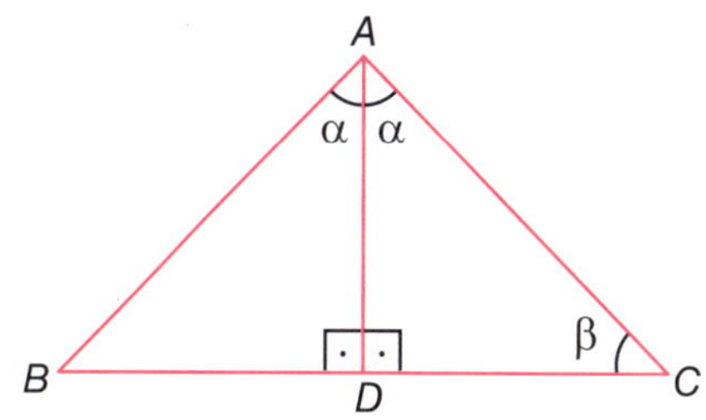

Theorem 2.2

ΔABC is an isosceles triangle with

$AB = AC$.

If $AB // DF$ and $ED // AC$,

then $DF + DE = AB = AC$

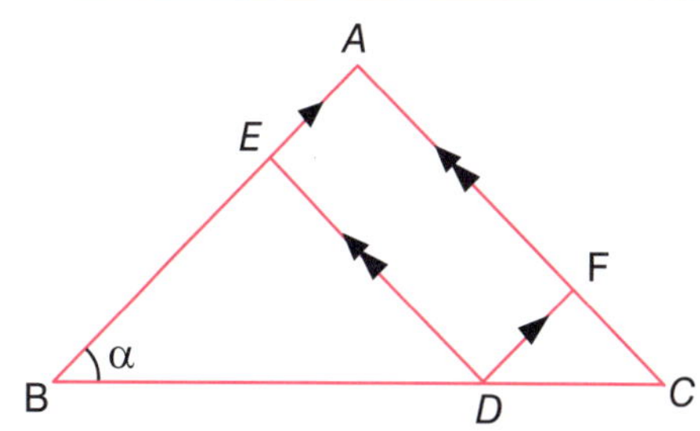

Proof 2.2

Since ΔABC is an isosceles triangle

$m\angle B = m\angle C = \alpha$.

$m\angle EDB = \alpha$ and $ED // AC$ so, $m\angle FDC = \alpha$.

Thus ΔBED and ΔFDC are isosceles triangles

and $BE = x$ and $FC = y$.

Further, $AF = x$, and $EA = y$, so, $EAFD$ is a parallelogram. Therefore, $AB = AC = x+y$.

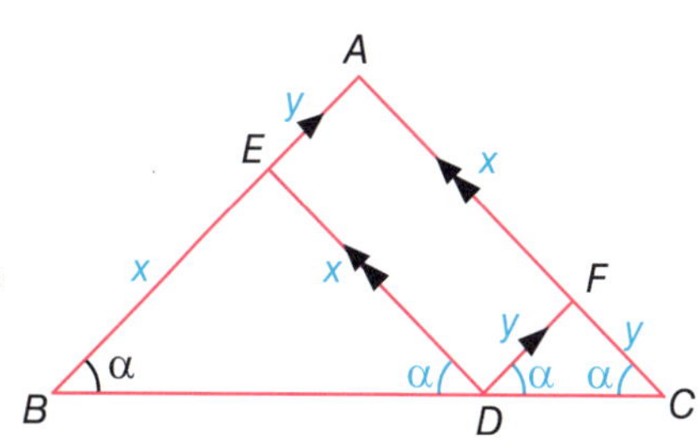

Theorem 2.3

Suppose ΔABC is an isosceles triangle with $AB = AC$ and point E on BC.

H is a perpendicular foot from vertex B.
If $ED \perp AB$ and $EF \perp AC$, then $DE+FE = BH$.

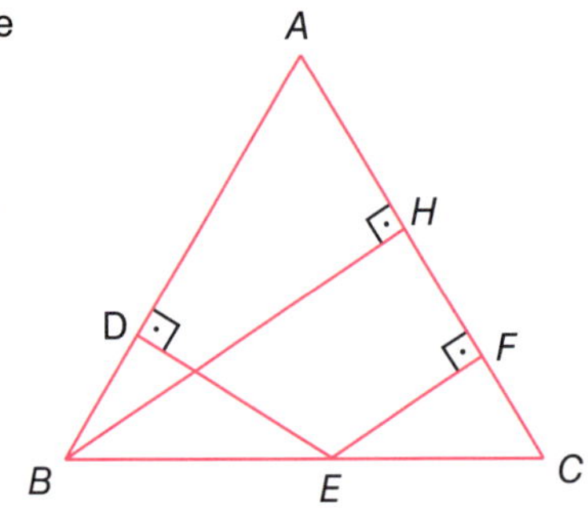

Proof 2.3

Let $DE = x$ and $EF = y$.

Let $BH = h$, and $AB = AC = a$.

AreaΔABC = AreaΔABE + AreaΔAEC

$$\frac{a \cdot h}{2} = \frac{a \cdot x}{2} + \frac{a \cdot y}{2}$$

$$\frac{a \cdot h}{2} = \frac{a \cdot (x+y)}{2}$$

$$\Rightarrow h = x+y.$$

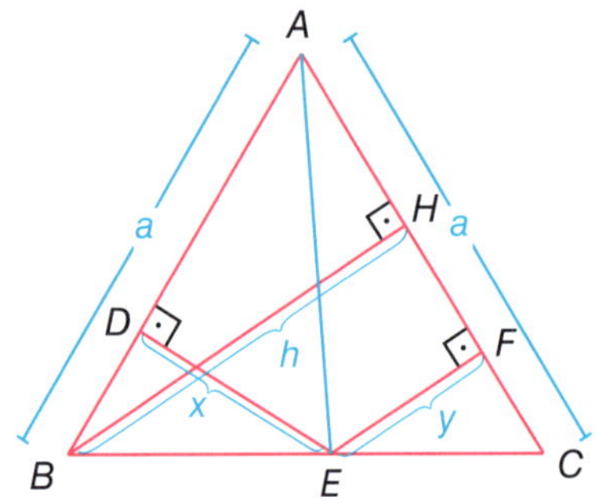

Theorem 2.4

ABC is an equilateral triangle and point P inside the triangle. If points D, E, and F are on the sides of ΔABC and $PD//AC$, $PE//BC$, and $PF//AB$, then $PE+PF+PD = AB = AC = BC$.

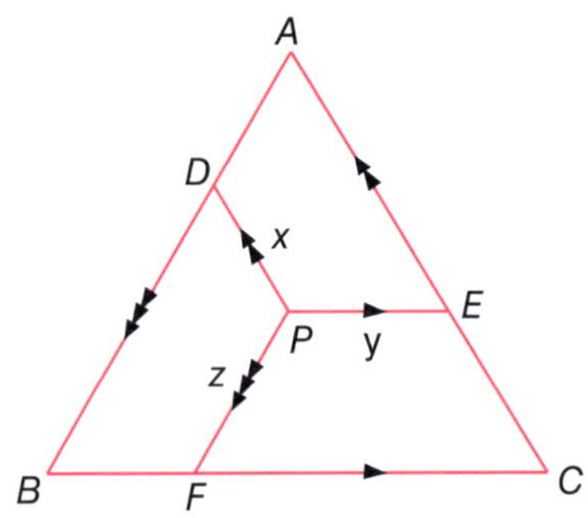

Proof 2.4

Let $DP = x$, $PE = y$, and $FP = z$. Extend DP to BC, and since $DP'//AC$, $m\angle DP'B = 60°$, and since $PF//AB$, $m\angle PFP' = 60°$, $\Delta PFP'$ is an equilateral triangle, $PECP'$ is a parallelogram and $BFPD$ is an isosceles trapezoid.

Therefore, $BC = x+y+z$.

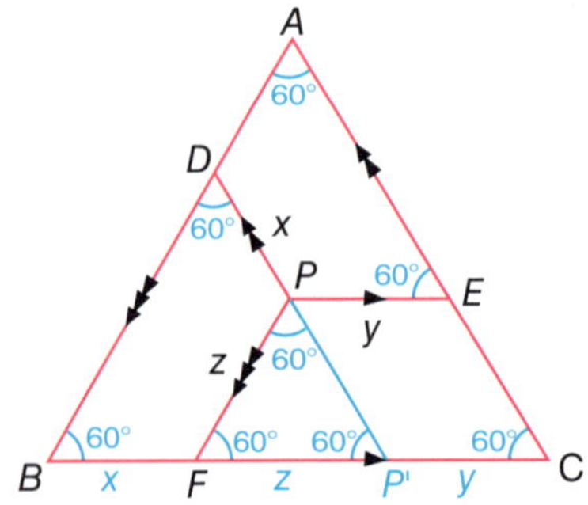

My Notes

My Notes

Theorem 2.5

ABC is an equilateral triangle and *P* is inside the triangle with $DP \perp AB$, $PF \perp AC$, and $PE \perp BC$.

The sum of the lengths $x+y+z$ is equal to an altitude of the triangle.

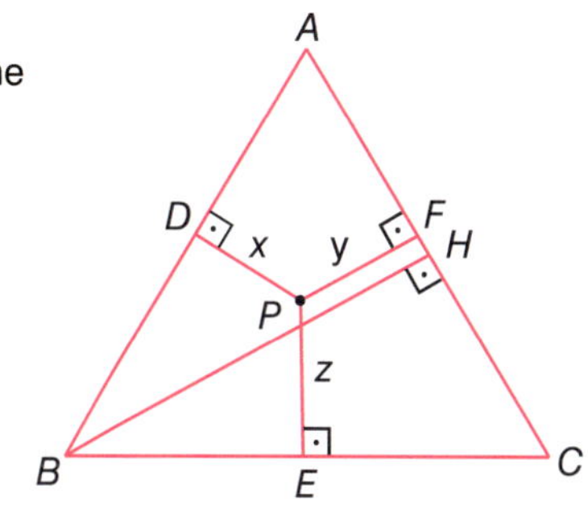

Proof 2.5

The area of ΔABC is the sum of the areas of ΔAPB, ΔAPC, ΔBPC,

$$\frac{a\cdot x}{2}+\frac{a\cdot y}{2}+\frac{a\cdot z}{2}=\frac{a\cdot h}{2}$$

$$\frac{a\cdot(x+y+z)}{2}=\frac{a\cdot h}{2}$$

$$x+y+z=h.$$

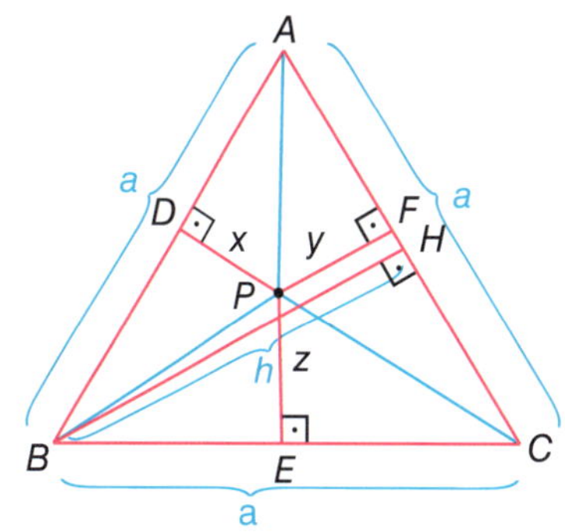

Theorem 2.6

Pythagorean Theorem

If the side lengths of a right triangle are *a, b,* and *c,* where the side of length c is the hypotenuse, then

$$a^2+b^2=c^2.$$

Or, if $a^2+b^2=c^2$ for any triangle with sides *a,b,* and *c,* then the triangle is a right triangle.

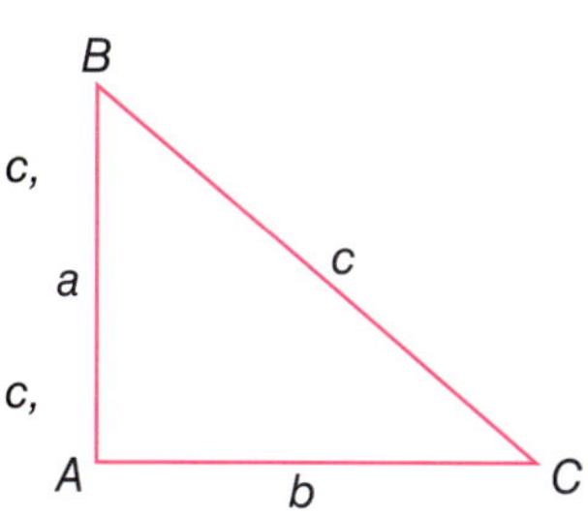

Proof 2.6

Proof 1

ABCD is a square.

The sum of the area the four triangles and the center square is the area of the square.

$$\left(\frac{1}{2}ab\right)+\left(\frac{1}{2}ab\right)+\left(\frac{1}{2}ab\right)+\left(\frac{1}{2}ab\right)+c^2=(a+b)^2$$

$$4\left(\frac{1}{2}ab\right)+c^2=(a+b)^2$$

$$2ab+c^2=a^2+2ab+b^2$$

$$c^2=a^2+b^2.$$

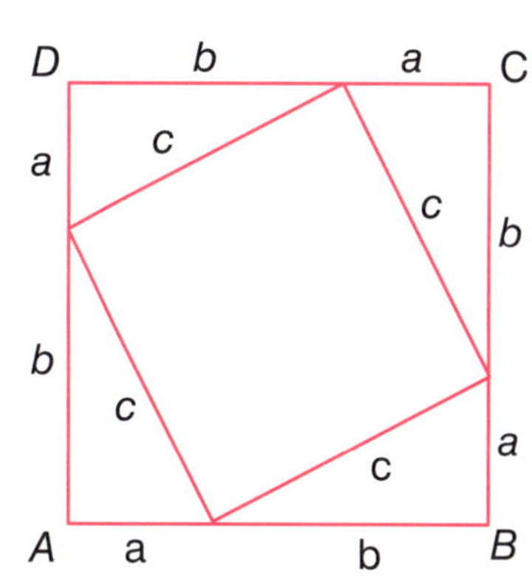

Proof 2 (President Garfield's proof)

The area of trapezoid *ABCD* is equal to the sum of the three triangles.

$$\frac{(a+b)}{2}(a+b) = 2\left(\frac{1}{2}ab\right)+\frac{c^2}{2}$$
$$a^2+2ab+b^2 = 2ab+c^2$$
$$a^2+b^2 = c^2$$

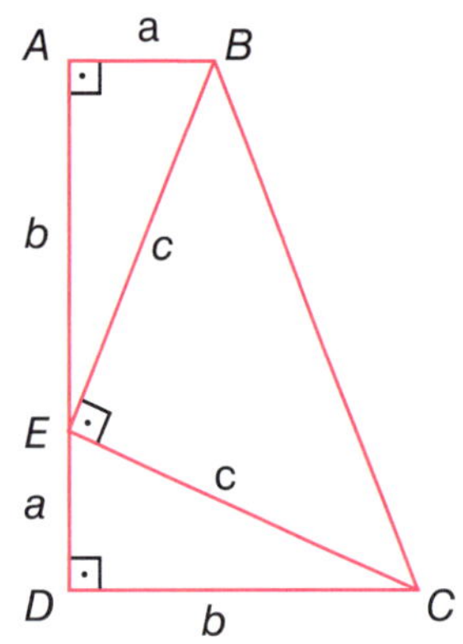

Carnot's Theorem

If a triangle *ABC* with interior point *P* has perpendiculars *PK*, *PN*, and *PT* drawn to sides *BC*, *AC*, and *AB*, respectively then

$$AT^2+BK^2+CN^2 = AN^2+BT^2+CK^2$$

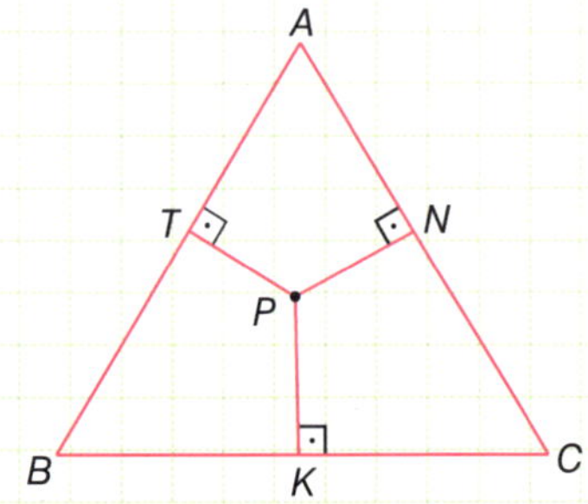

My Notes

Remark 2.1

A triangle is a right triangle if and only if a median is equal to that of the side it bisects.

Example 2.1

Given that the perimeter of a right triangle is $2+\sqrt{6}$ cm and the median to the hypotenuse is 1 cm, find the sum of the two legs.

Solution 2.1

By the remark;

$$a+b+2 = 2+\sqrt{6}.$$

Therefore, $a+b = \sqrt{6}$.

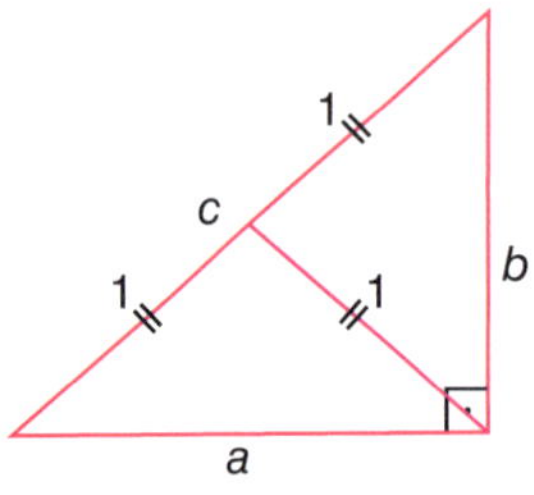

Notes 2.1

Pythagorean Triples

A Pythagorean triple is any triple (a,b,c) of positive integers such that $a^2+b^2 = c^2$.

Here are a few triples:

Try This

- 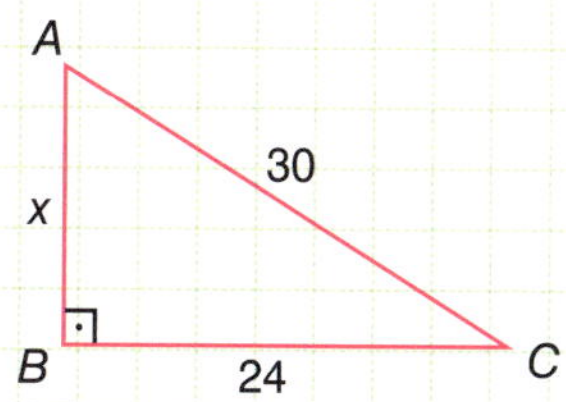

- 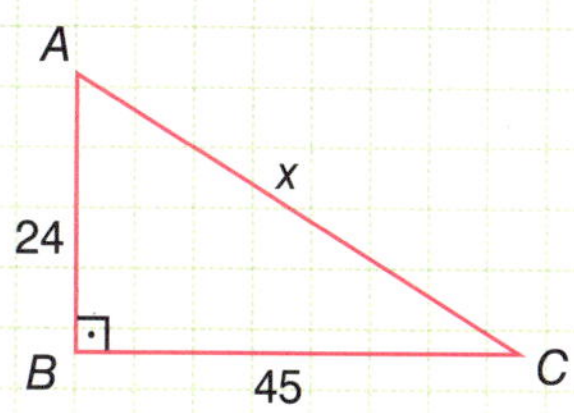

- 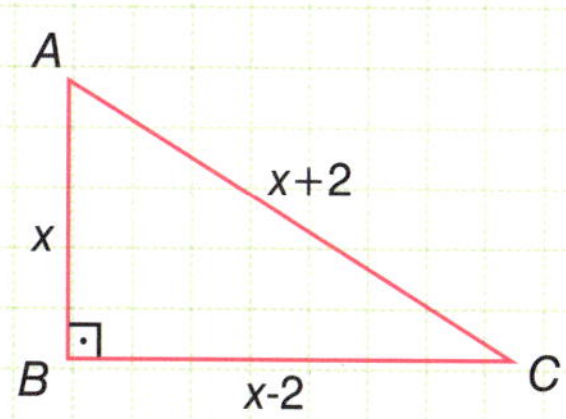

- 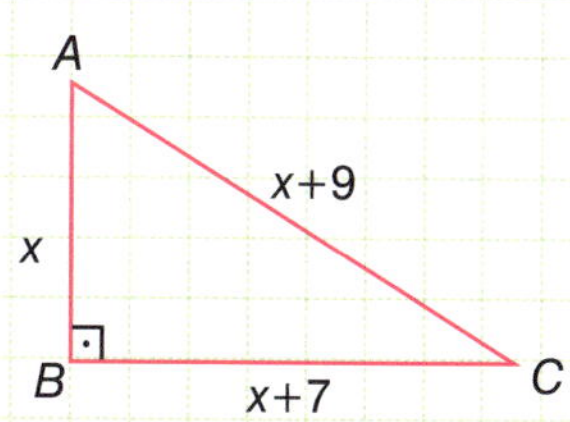

- 3–4–5 and scaling up

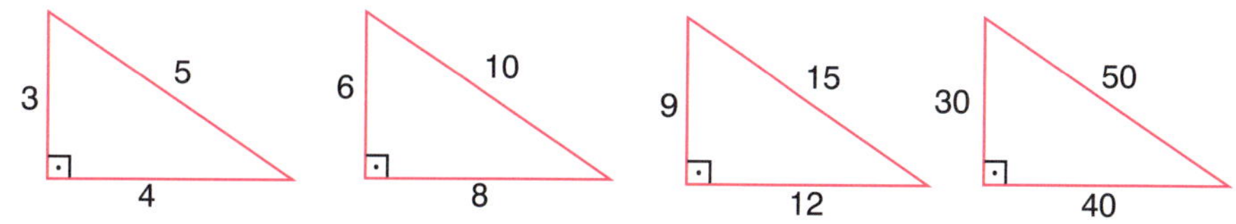

- 5–12–13 and scaling up

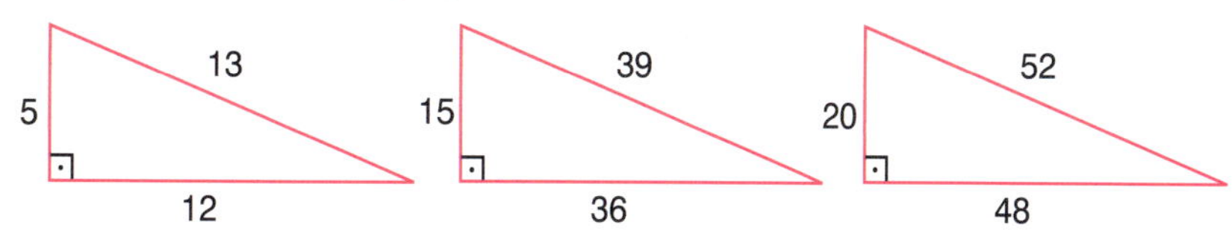

- 8–15–17 and scaling up

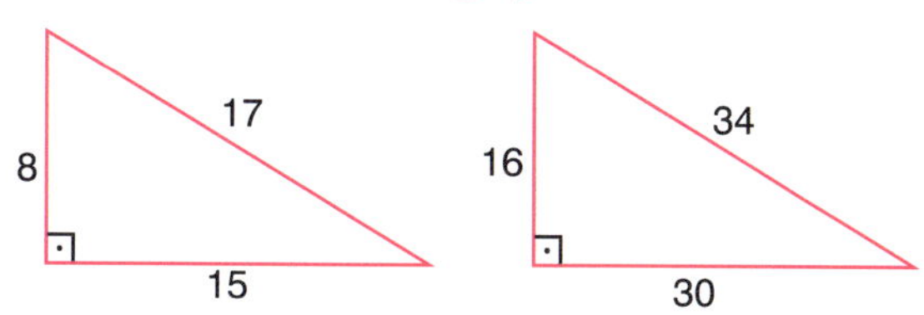

- 7–24–25 and scaling up

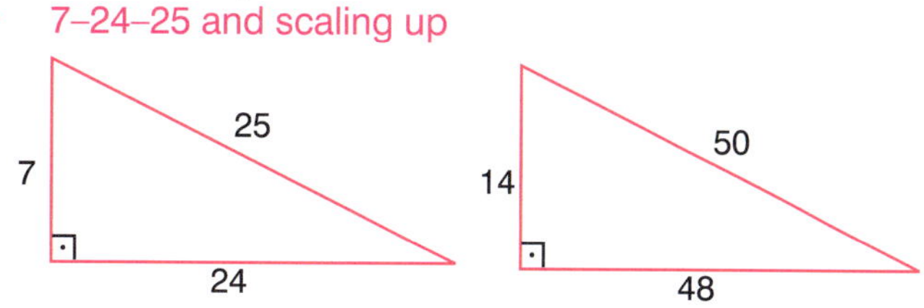

- 9–40–41 and scaling up

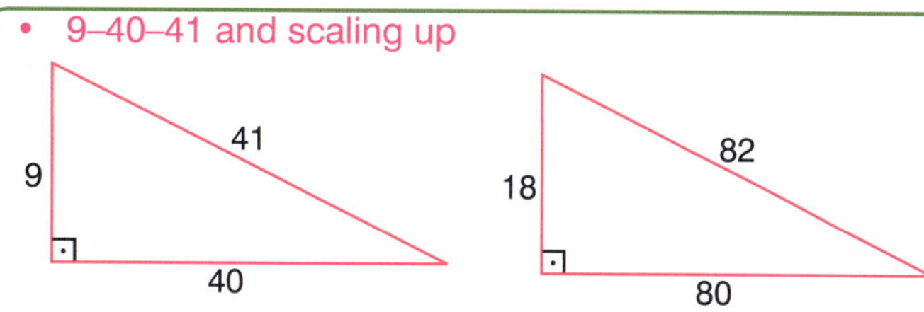

- 20–21–29 and scaling up

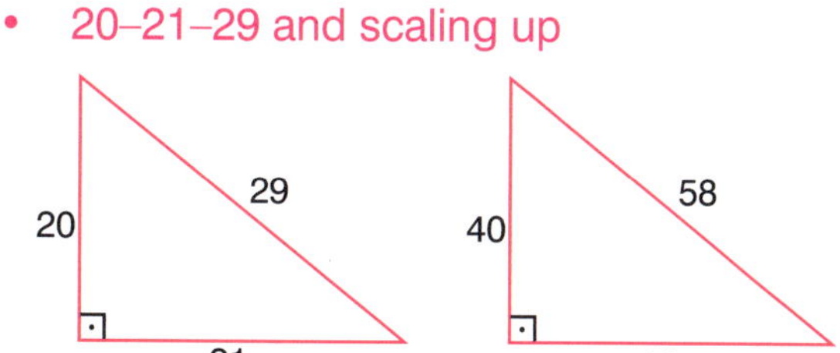

Example 2.2

In the figure,

$AB \perp AC$, and

$AD \perp BC$.

Find x and y.

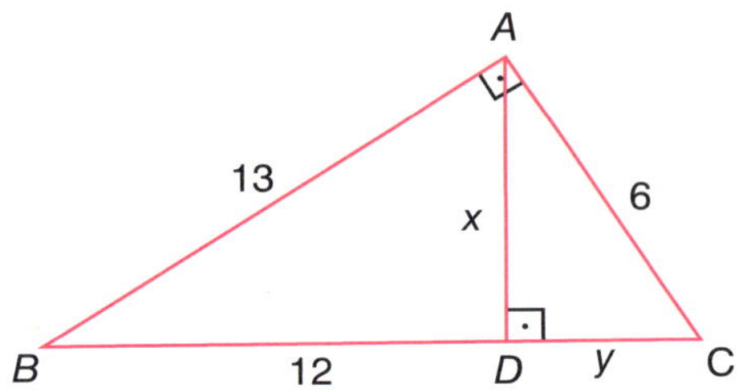

Solution 2.2

$13^2 = 12^2 + x^2$ (from ΔABD) and $13^2 - 12^2 = x^2$.

Then, we have $(13-12)(13+12) = x^2$, and $25 = x^2$. So, $x = 5$.

$6^2 = 5^2 + y^2$ (from ΔADC) and $36 - 25 = y^2$. So, $11 = y^2 \Rightarrow y = \sqrt{11}$.

Example 2.3

In the figure,

$AB \perp BC$, $AD = DB$,

$AC = 7$, and $DC = 5$.

Find DB.

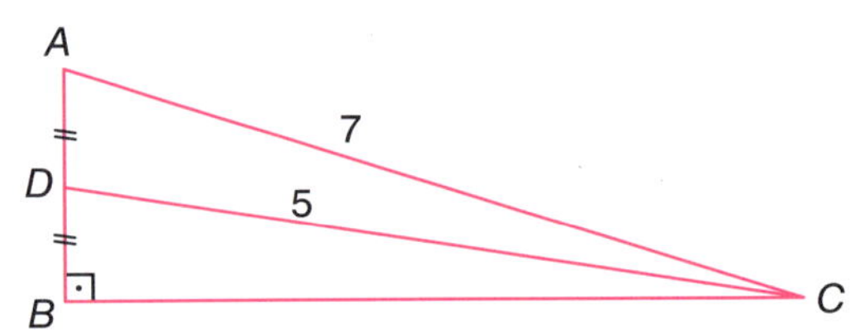

Solution 2.3

Let BC be a and $AD = DB = b$.

$a^2 + b^2 = 25$ (ΔDBC) (1)

$a^2 + 4b^2 = 49$ (ΔABC) (2)

Subtracting (1) from (2) gives

$3b^2 = 24 \Rightarrow b^2 = 8$. So, $b = 2\sqrt{2}$.

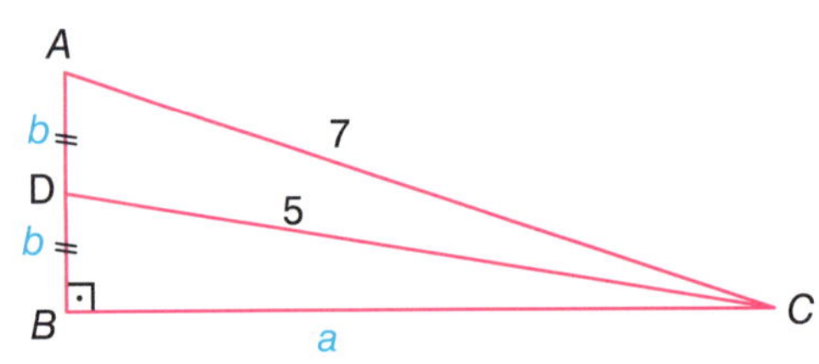

Example 2.4

In the figure,

$AD = AC$, $AB = 20$ cm, and

$BD = DC = 8$ cm,

Find $AD = AC$.

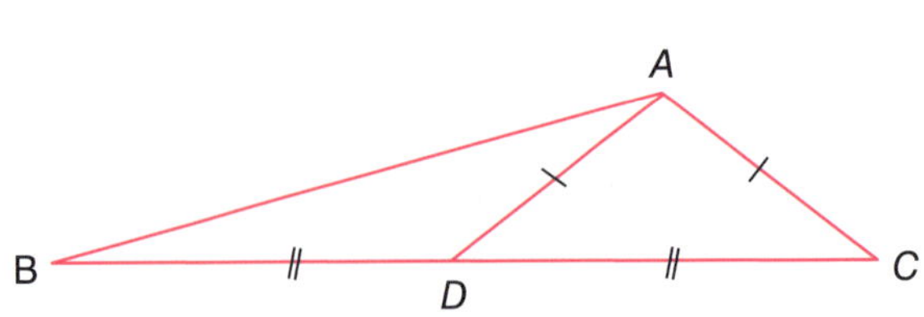

Try This

A triangle with sides lengths $4\sqrt{3}$, 1, and 7.

What is the angle opposite to the largest side?

Try This

In the figure,

$AD = DC = 4$, and

$AB = AC = 6$.

Find BD.

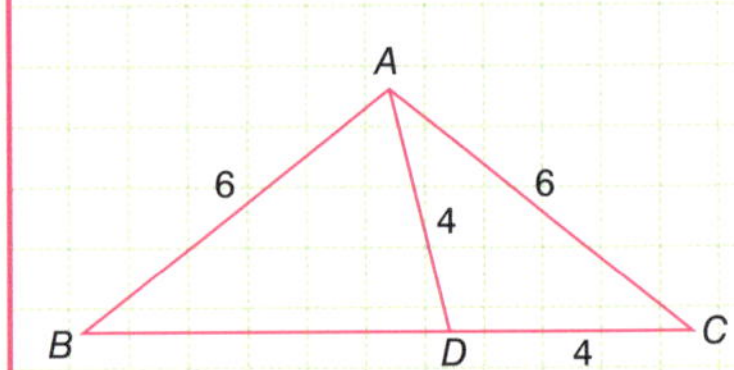

Proof

British Flag Theorem

A point P is chosen inside rectangle $ABCD$.

Prove that: $AP^2+CP^2 = BP^2+DP^2$.

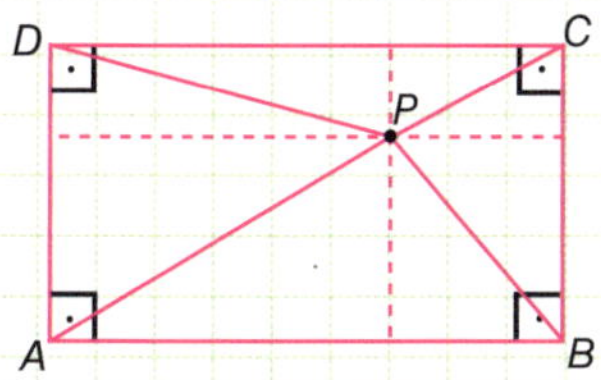

My Notes

Solution 2.4

$h^2+12^2 = 20^2$ (ΔABH)

$h^2 = 400-144$

$h^2 = 256$, so, $h = 16$.

Then, $16^2+4^2 = AC^2$ (ΔAHC)

$256+16 = AC^2 \Rightarrow AC = 4\sqrt{17}$

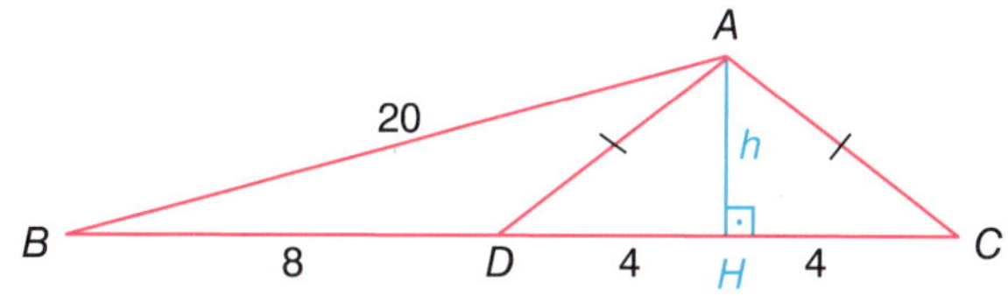

Example 2.5

In the figure,

AB = 1, BC = 7, and AC = $5\sqrt{2}$.

What is the area of ΔABC ?

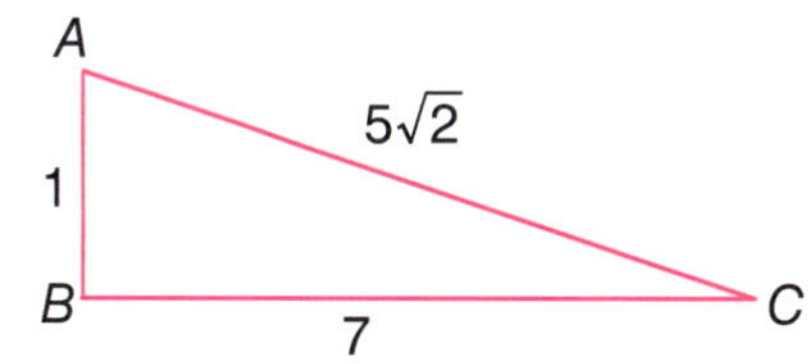

Solution 2.5

Observe that

$AB^2+BC^2 = AC^2$

$1^2+7^2 = (5\sqrt{2})^2$

$50 = 50$

Therefore, ΔABC is a right triangle, So, $m\angle B = 90°$.

The area of $\Delta ABC = \dfrac{1 \cdot 7}{2} = 3.5$.

Example 2.6

In the figure,

$a+b = 3c$ and

the area of ΔABC is 32 cm^2.

Find the value of c.

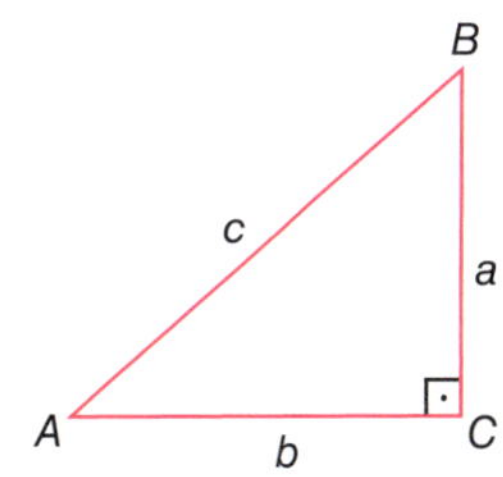

Solution 2.6

If $a^2+b^2 = c^2$, then $(a+b)^2-2ab = c^2$. $\left[(a+b)^2= a^2+2ab+b^2\right]$

The area of $\Delta ABC = \dfrac{1}{2}ab = 32$. So, $ab = 64$.

Substituting 64 for ab and $3c$ for $a+b$, we obtain

$(3c)^2-128 = c^2$

$8c^2= 128$. So, $c^2= 16 \Rightarrow c = 4$.

Example 2.7

In the figure,

$AB = 6$ cm, $BD = 2$ cm, and

$DC = 4$ cm.

Find $AD = AC$.

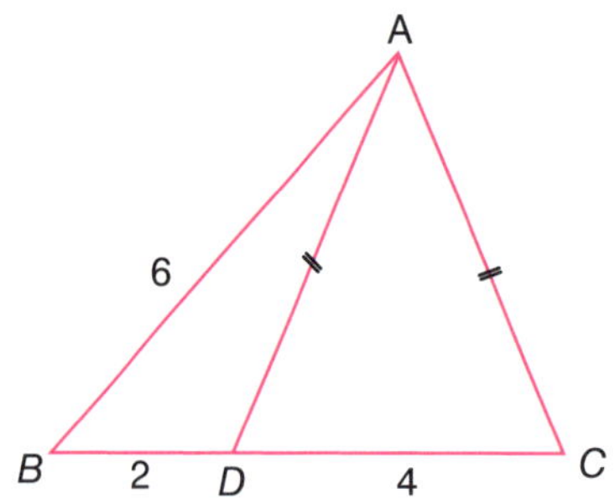

Solution 2.7

$6^2 = 4^2 + h^2$ (from ΔAHB)

$36 - 16 = h^2 \Rightarrow 20 = h^2$.

From ΔAHC,

$h^2 + 2^2 = AC^2$

$20 + 4 = AC^2 \Rightarrow 24 = AC^2$

$AC = \sqrt{24} \Rightarrow AC = 2\sqrt{6}$.

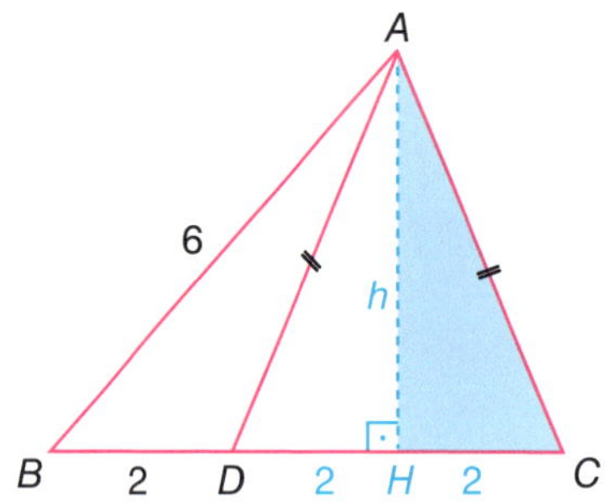

Example 2.8

In the figure,

$AB = 20$, $AC = 15$, and $DC = 18$.

If $m\angle ADB + m\angle ACB = 180°$ then find BD.

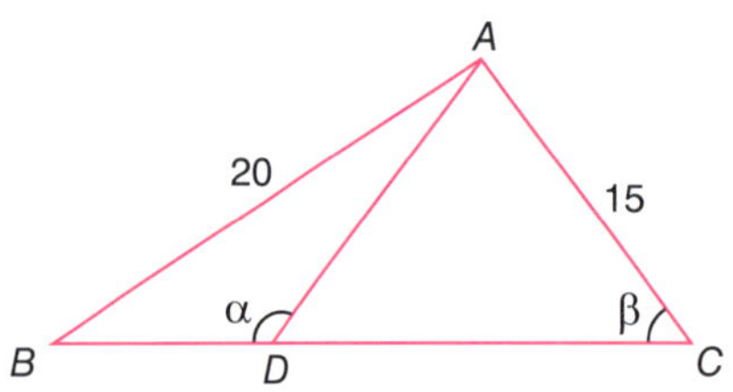

Solution 2.8

Step-1

By using given angle sum supplementary angles we get $m\angle ADC = \beta$ and therefore $AD = AC$.

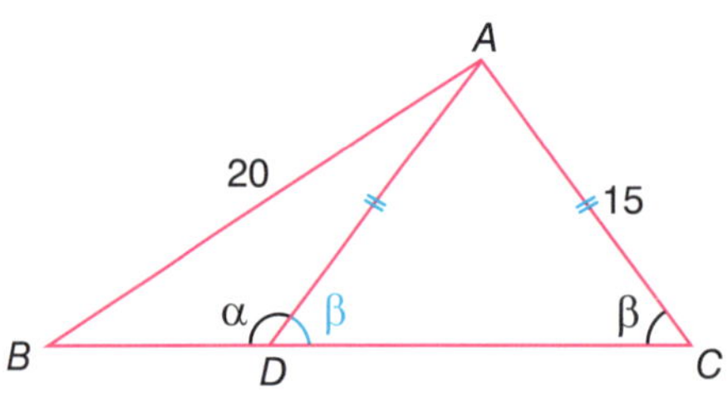

Step-2

By drawing a right triangle we obtained

$DH = HC = 9$, so $h = 12$. [12-9-15 triangle]

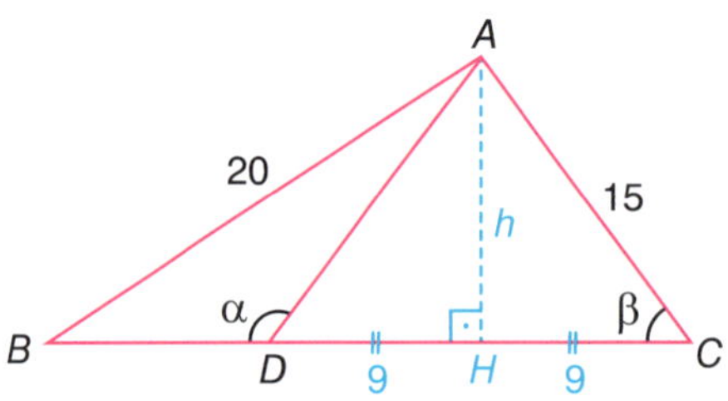

Step-3

$BH = 16$ (ΔAHB) [12-16-20 triangle]

Therefore $BD = 16 - 9 = 7$.

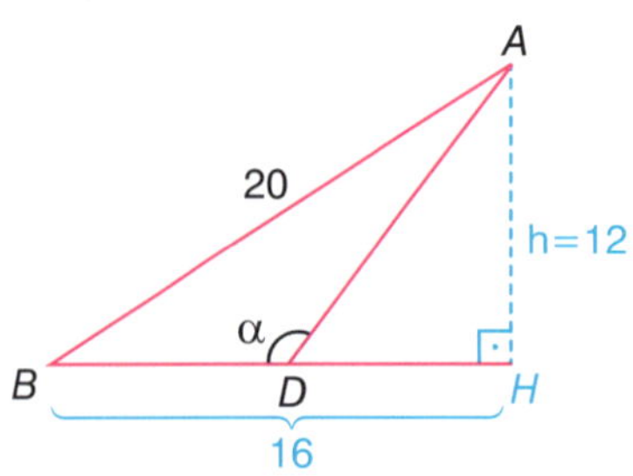

Catriona Agg @Cshearer41

Geometry puzzle

These two isosceles triangles have the same area.

What is the angle?

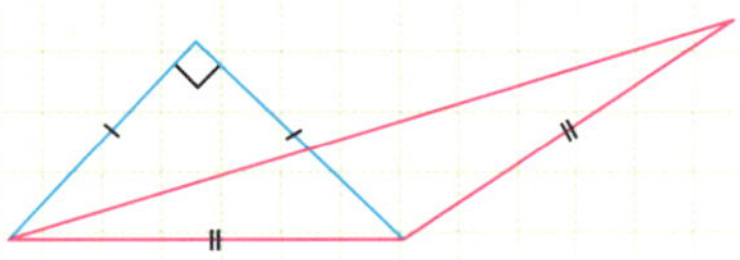

Try This

In the figure,

$BD = 4$, $DC = 8$, and $AC = 10$.

What is the shortest distance from point B to side AC?

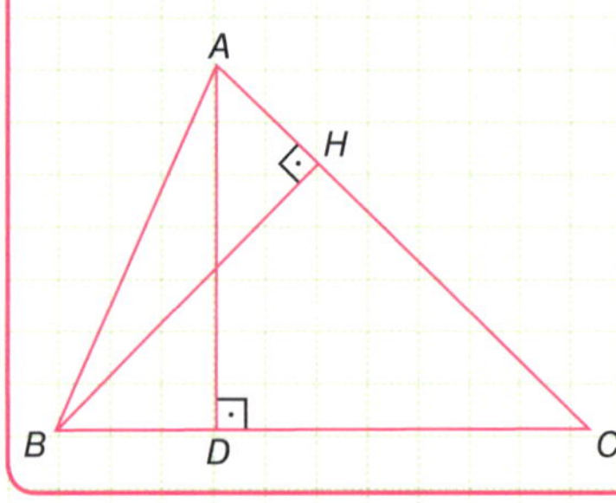

Try This

In the figure, point D is the mid-point of AB.

$AE = 3$ cm,

$EC = 7$ cm, and

$BC = 8$ cm.

Find DE.

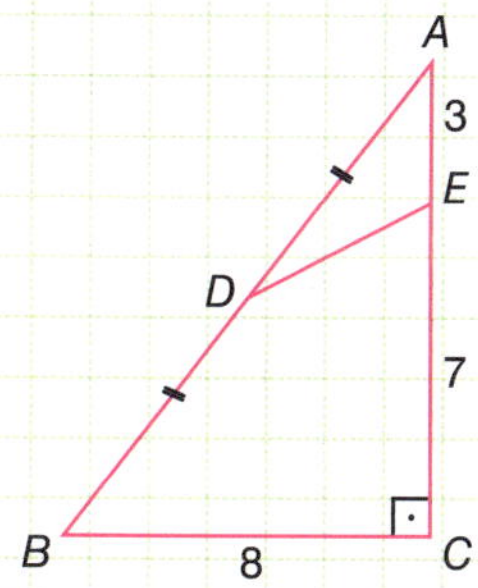

My Notes

Example 2.9

In rectangle $ABCD$, point P is chosen inside the rectangle so that

$AP = 10$, $BP = 16$, and $CP = 15$.

What is the length of DP?

Solution 2.9

From the British Flag Theorem,

$$AP^2 + CP^2 = BP^2 + DP^2$$

$$10^2 + 15^2 = 16^2 + DP^2 \Rightarrow DP^2 = 325 - 256$$

$$DP^2 = 69 \Rightarrow DP = \sqrt{69}.$$

Example 2.10

In the figure, $AB \perp BC$,

$BA = 3x - 1$,

$AC = 3x + 1$, and

$BC = x\sqrt{3}$.

Find the value of x.

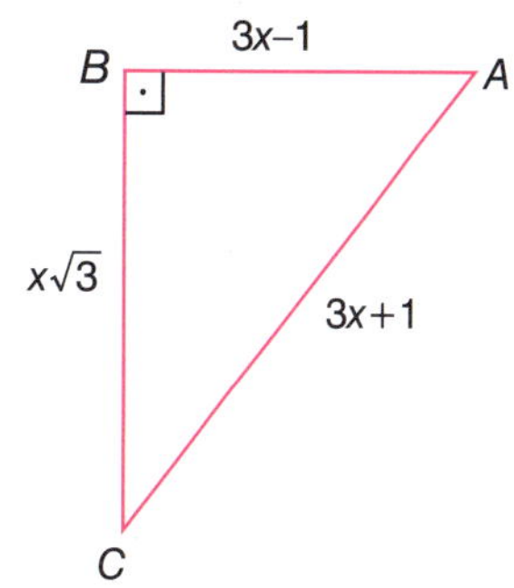

Solution 2.10

$$(3x-1)^2 + (x\sqrt{3})^2 = (3x+1)^2$$

$$9x^2 - 6x + 1 + 3x^2 = 9x^2 + 6x + 1$$

$$3x^2 = 12x \quad (x \neq 0)$$

$$3x = 12 \Rightarrow x = 4.$$

Example 2.11

In the figure,

$AB = 3$ cm, $BC = 13$ cm,

$CD = 10$ cm, $AD = x$, and

$m\angle BAD = m\angle ADC = 45°$.

Find the value of x.

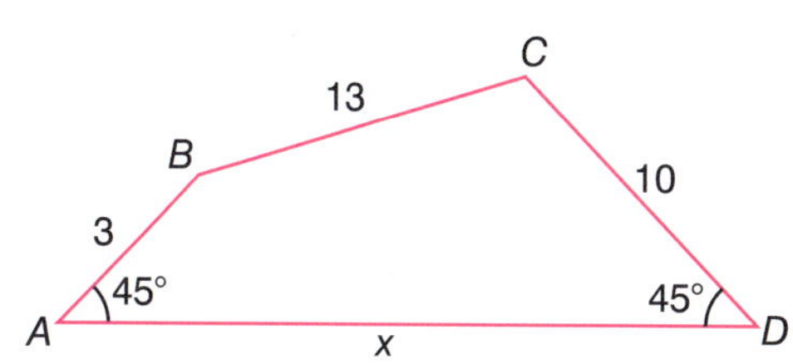

Solution 2.11

By extending AB and DC we obtain an isosceles right triangle.

If $EC = y$, then $EB = y + 7$.

The 5-12-13 triangle gives $y = 5$.

Therefore, $x = 15\sqrt{2}$ cm.

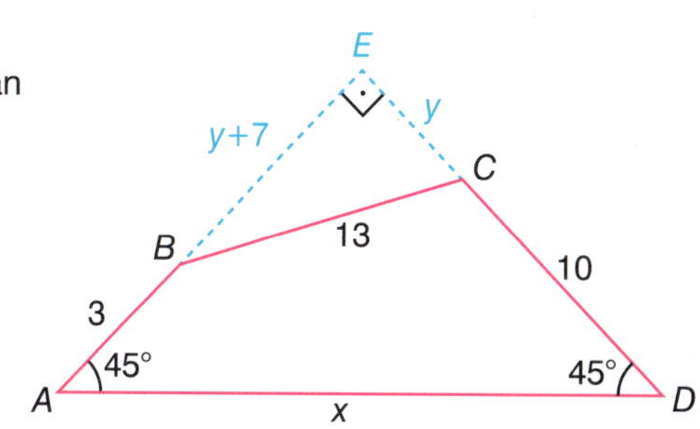

Example 2.12

In the figure,

$BA \perp AD$,

$AD = AB$,

$CB = 7$ cm, and

$BD = 16$ cm.

What is the length of AC?

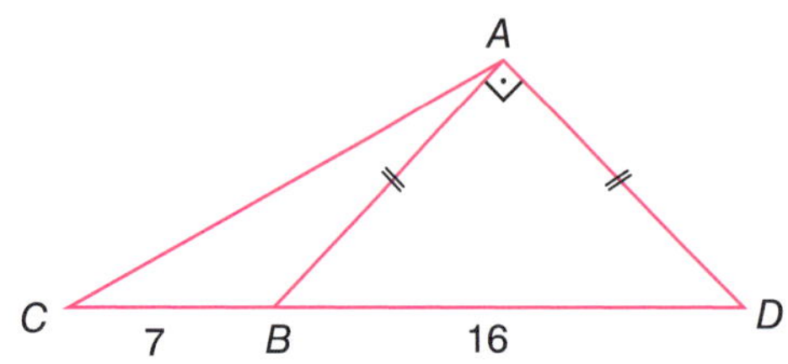

Solution 2.12

Draw altitude AH from A to BD. AH is a perpendicular to BD and $BH = HD = AH = 8$.

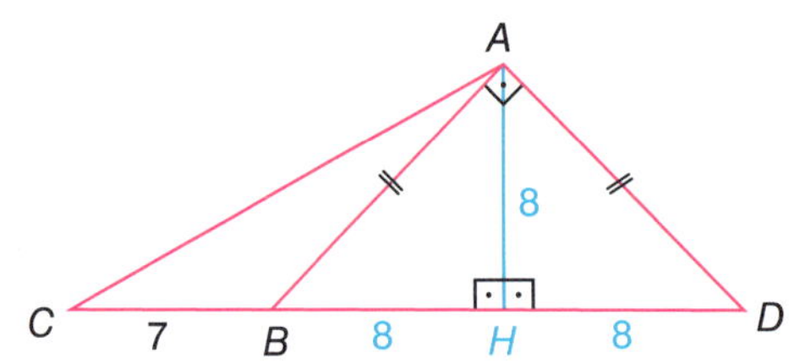

From ΔAHC, $8^2+15^2 = AC^2$

Therefore, $AC = \sqrt{8^2+15^2} = 17$.

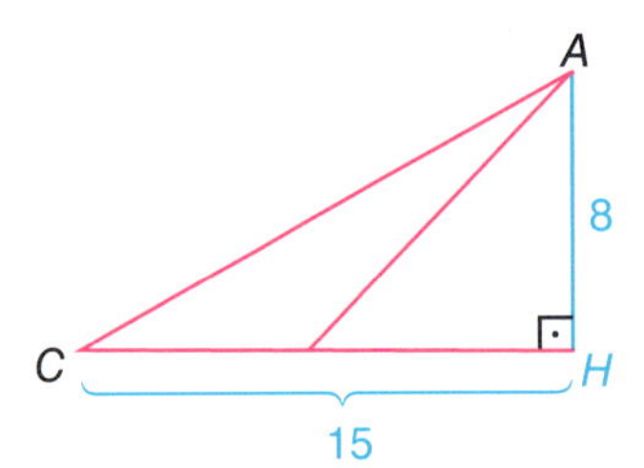

Example 2.13

$ABCD$ is a rectangle with

$AD = 10$ and

$DC = 21$.

If point E is the midpoint of AB, what is FC?

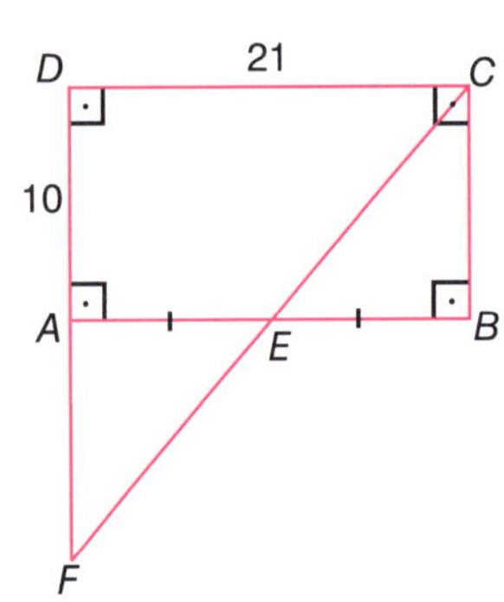

Solution 2.13

Note that triangles AFE and BCE are congruent.

Therefore, $BC = AF = 10$.

Hence, $DF = DA + AF = BC + AF = 20$.

From the triangle FDC, $FC = \sqrt{20^2+21^2} = 29$.

Catriona Agg @Cshearer41

Geometry puzzle

The three angles are equal.

What fraction of the rectangle is shaded?

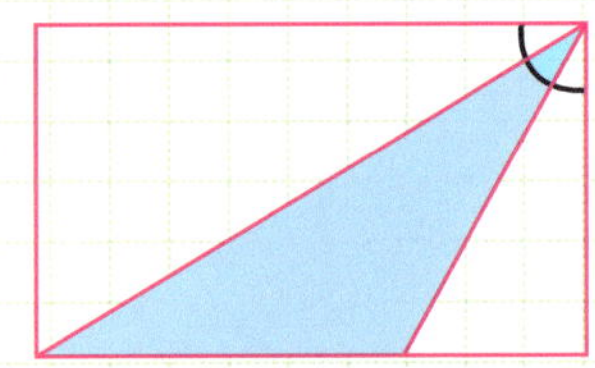

Try This

In the figure,

$AB \perp BC$, $BH \perp AC$,

$AB = 5$ cm, and $HD = 1$ cm.

What is the area of ADB?

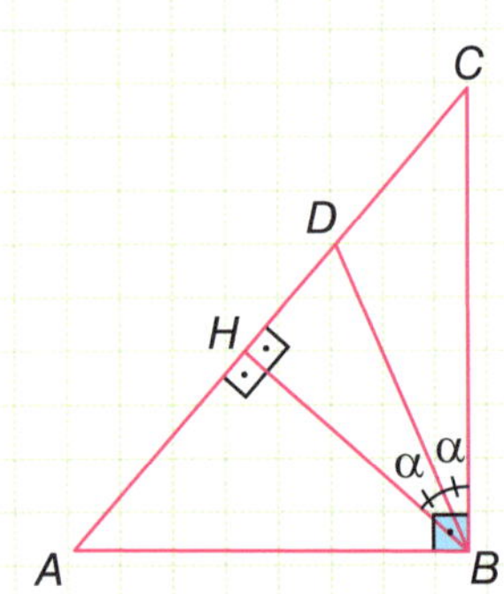

1. In the figure,

$AH \perp BC$,

$AB = 13$,

$HC = 16$, and

$BH = 5$.

Find AC.

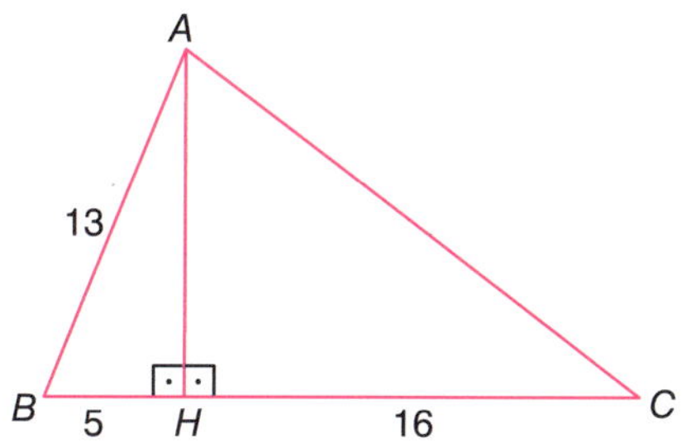

2. In the figure,

$AC \perp BC$,

$AD = 12$,

$DC = 8$, and

$BD = 17$.

Find AB.

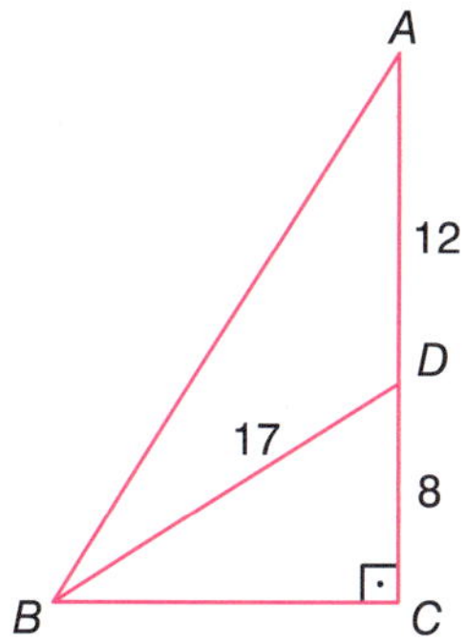

3. In the figure,

$m\angle BAC = 90°$,

$AD = EB = 3$,

$AE = DC = 2$,

$ED = x$, and

$BC = y$.

Find x^2+y^2.

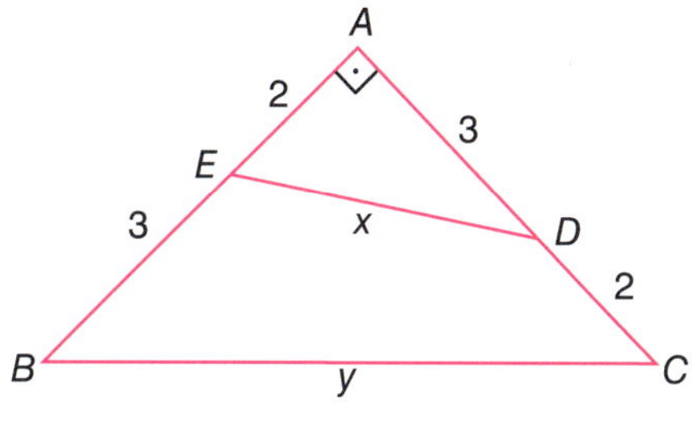

4. In the figure,

$m\angle B = 54°$,

$BH = HC$, and

$AB = DC$.

Find $m\angle BAC$.

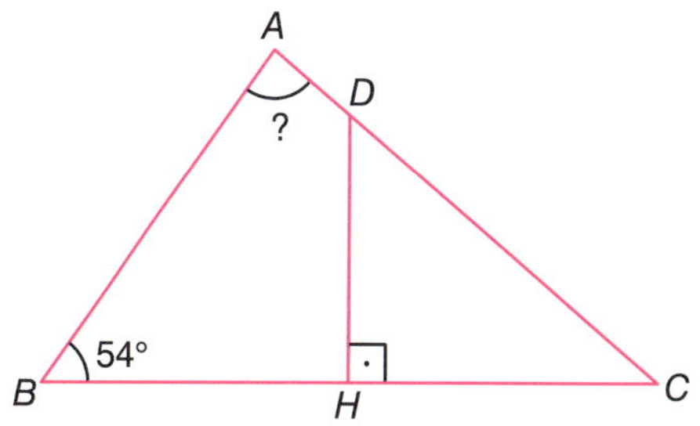

5. In the figure, $DE \perp AB$,

$m\angle C = 90°$,

$m\angle CAD = m\angle DAB$,

$CD = 6$ cm, and

$BD = 10$ cm.

Find AC.

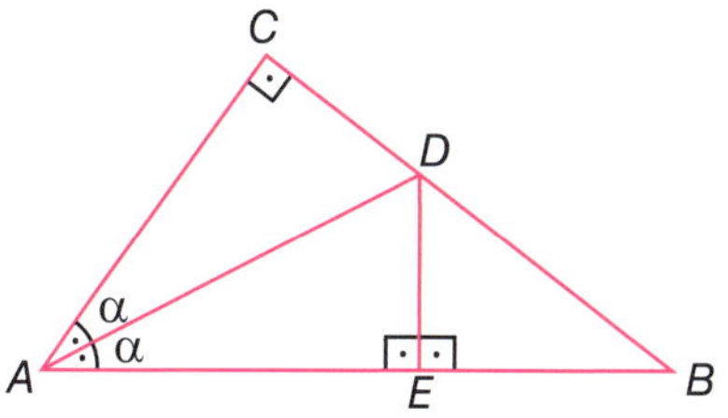

6. ΔABC is an equilateral triangle with

$AD = 8$ and

$DB = 2$.

Find DC.

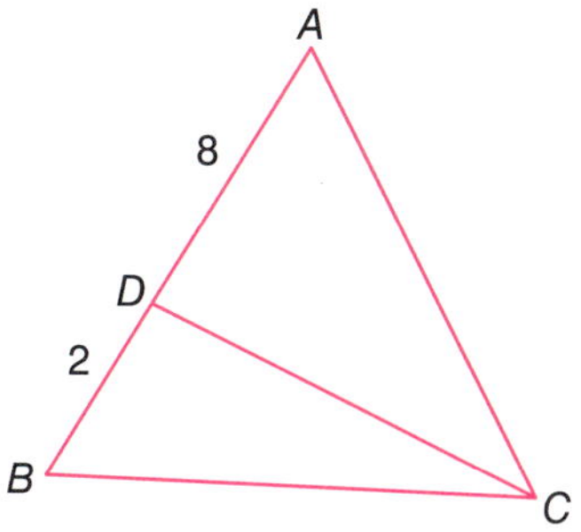

7.

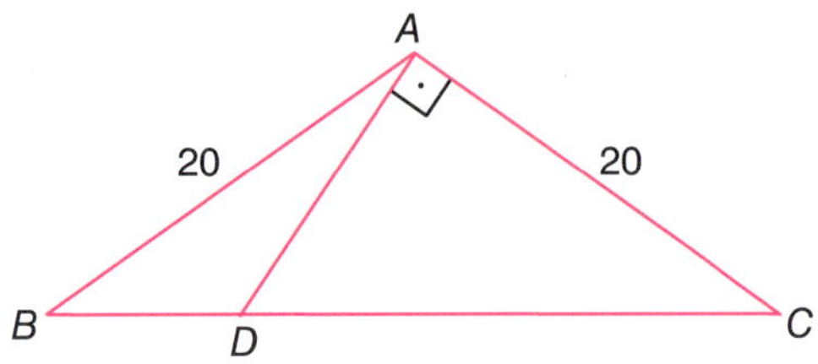

In the figure,

$AD \perp AC$, $AB = AC = 20$ and $BC = 32$

Find AD.

8. In the figure,

$AD \perp BD$,

$AB = 9$,

$AC = 7$, and

$BC = 5$.

Find CD.

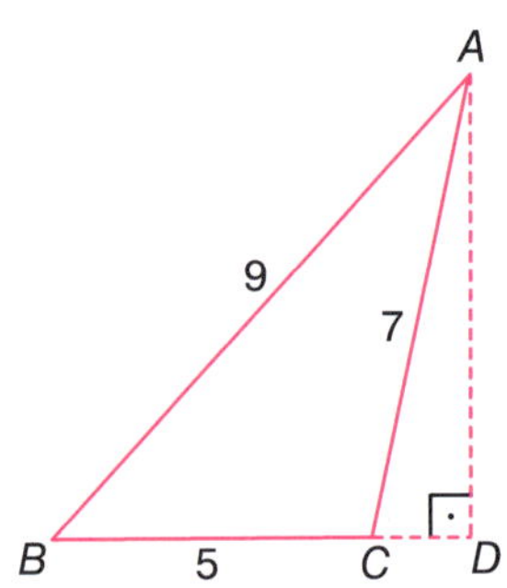

9.

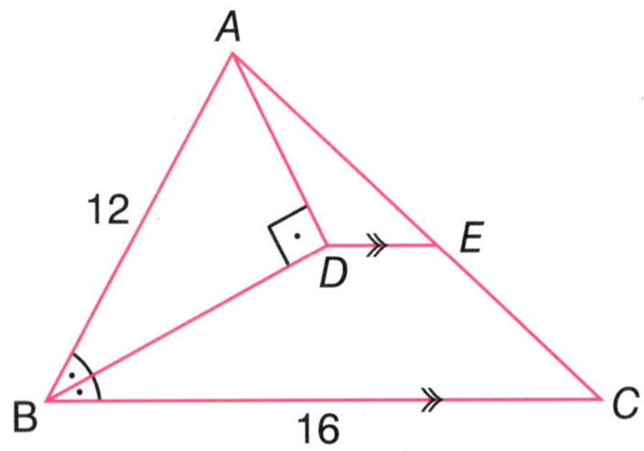

In the figure, BD bisects angle B.

Given $DE // BC$, $m\angle ADB = 90°$, $AB = 12$, and $BC = 16$.

Find the length of DE.

10. In the figure, ΔABC is an isosceles triangle,

$AB = AC$,

$m\angle BAC = 30°$, and

$2PE = PD = 4$.

Find AC.

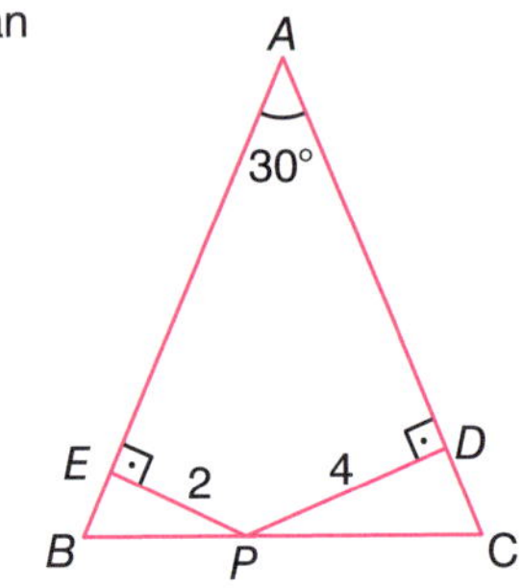

11.

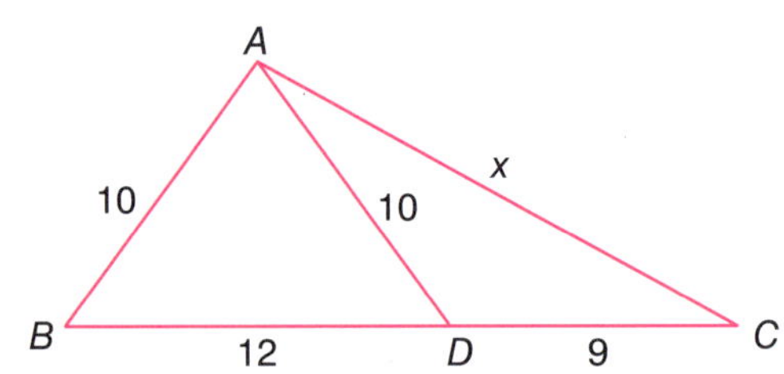

In the figure,

$AB = AD = 10$, $BD = 12$, $DC = 9$, and $AC = x$.

Find x.

12. In the figure,

$AB = AC$,

$AD = 4$,

$BD = 8$, and

$DC = 6$.

Find AB.

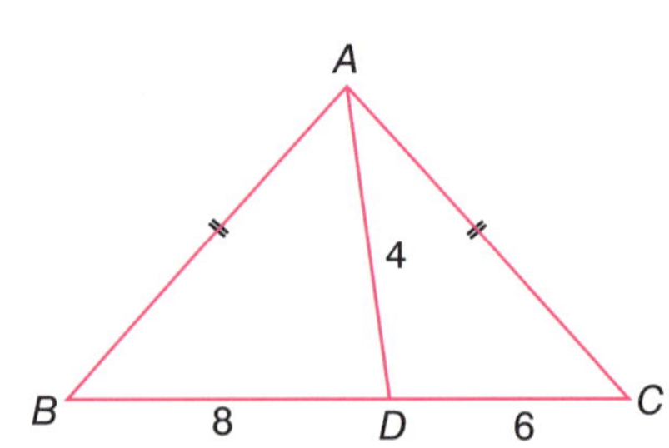

1.

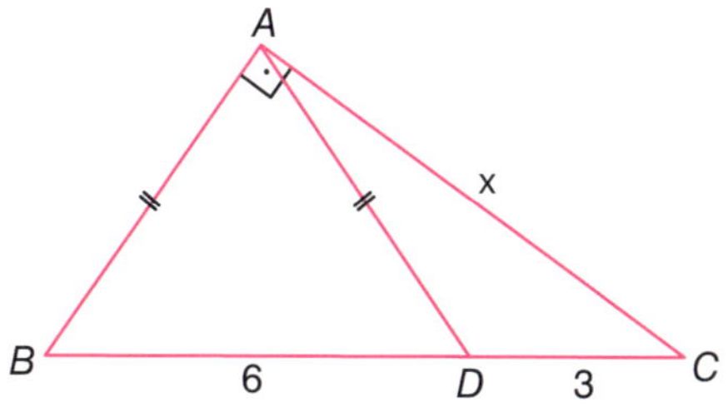

In the figure, $m\angle BAC = 90°$, $AB = AD$.

Find the value of x .

2. ABC is a equilateral triangle, $DP // BC$, $FP // AD$, and $PE // FC$.

$AD = 6$,

$FC = 10$, and

$BE = 8$.

Find perimeter of ΔABC.

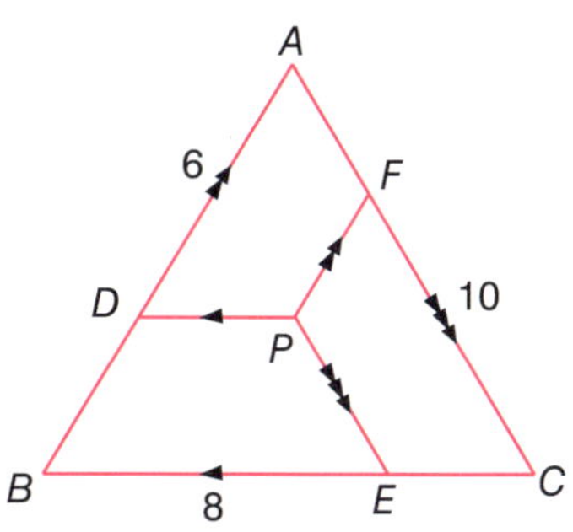

3.

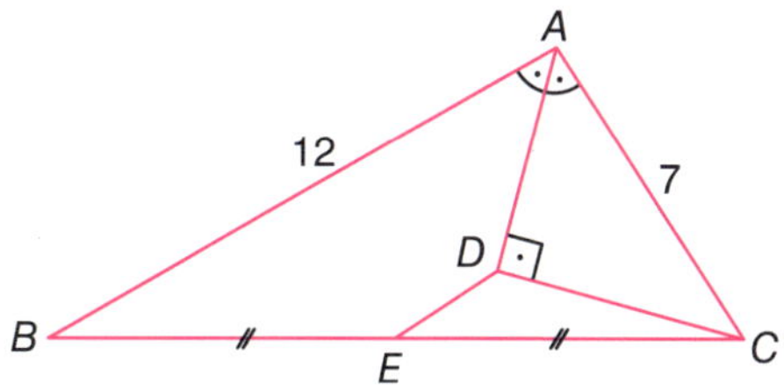

In the figure, $m\angle ADC = 90°$, $BE = EC$, and AD bisects the interior angle A.

Given $AB = 12$ and $AC = 7$, find the length of ED.

4. In the figure,

$m\angle DAC = 90°$,

$AD = AC$,

$BD = 3$ and

$DC = 18$.

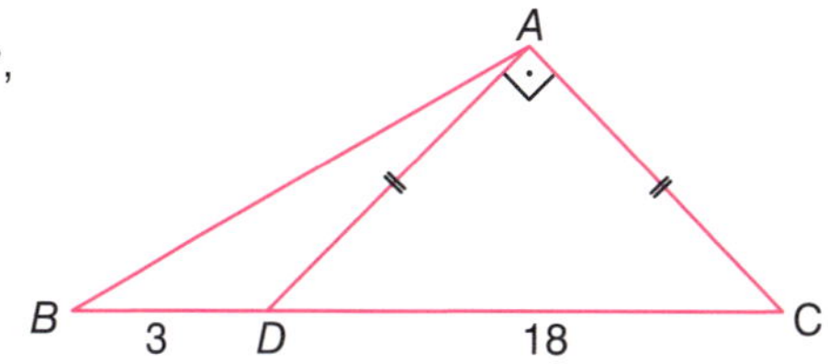

Find the length of AB.

5. In the figure,

$m\angle C = 90°$,

$m\angle CAD = m\angle DAB$,

$CD = 3$, and

$BD = 5$.

Find AC.

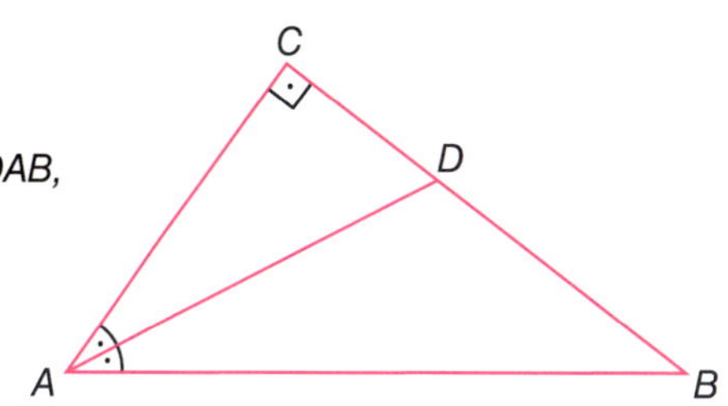

6. A right triangle has side lengths 3, 4 and 5.

Find the length of the shortest altitude.

7. ABC is an isosceles right triangle,

$AB = AC$,

$BH = 5$ cm,

$HC = 3$ cm are given.

Find the length $AD = x$.

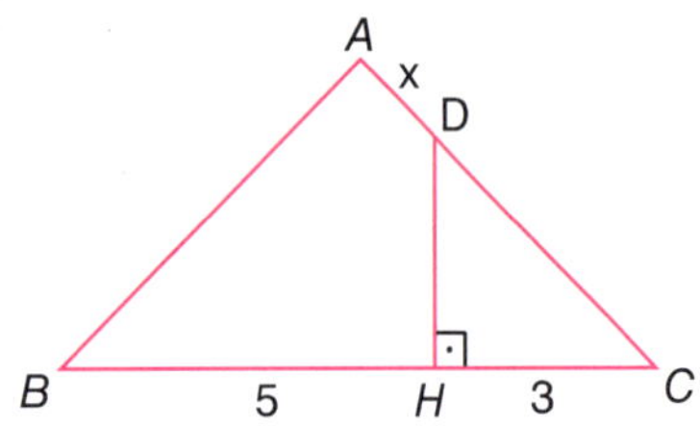

8. In ΔABC, $m\angle C = 90°$.

E is on the BC such that $AC = BE$.

D is on the AB such that $DE \perp BC$.

Given that $DE+BC = 1$, $BD = \frac{1}{2}$.

Find $m\angle B$ in degrees.

9. ABC is a right triangle with one leg 7 cm length. If the other leg and hypotenuse are positive integers find the perimeter of the triangle.

10. In ΔABC, $m\angle C = 90°$,

D is the midpoint of AC.

Prove that $AB^2+3BC^2 = 4BD^2$.

11. In ΔABC, $m\angle A = 90°$,

$AB = AC$, and D is the point on BC.

Prove that $BD^2+CD^2 = 2AD^2$

12. BAC is a right triangle with $m\angle BAC = 90°$,

$a+b+c = 28$, and

$a^2+b^2+c^2 = 288$

Find the area of the triangle.

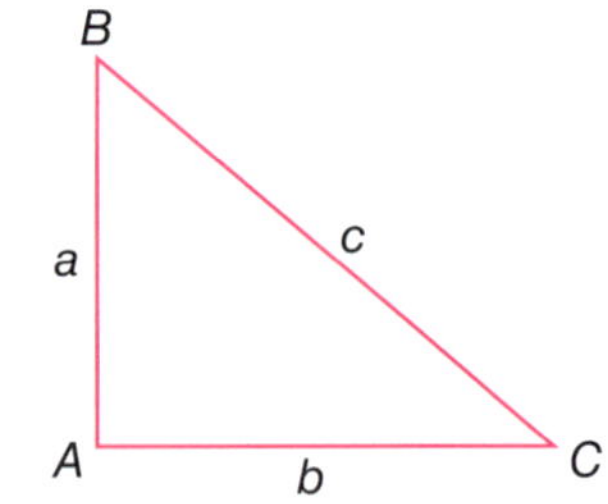

1. **(2003 AMC 8 # 6)**

Given the areas of the three squares in the figure, what is the area of the interior triangle?

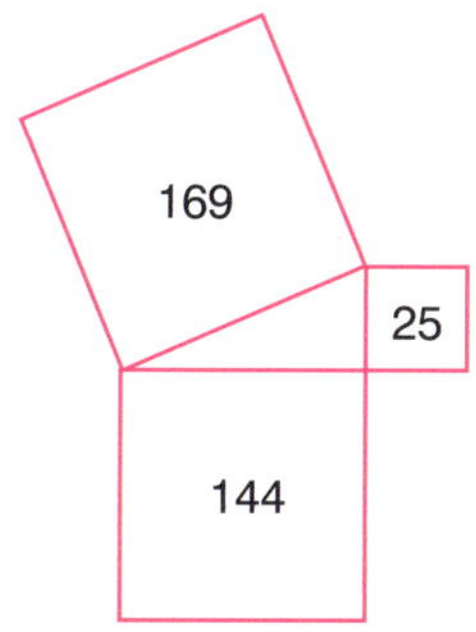

A) 13 B) 30 C) 60 D) 300 E) 1800

2. **(1999 AMC 8 # 14)**

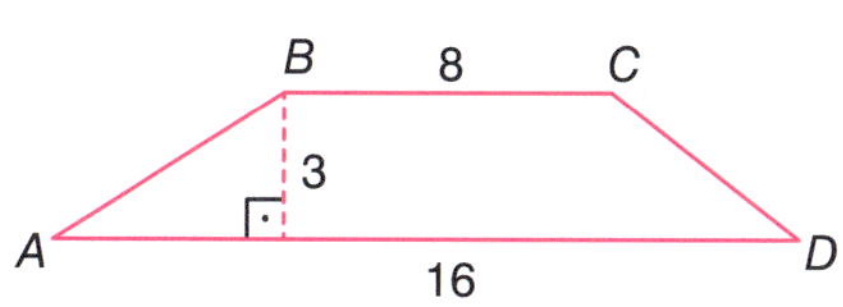

In trapezoid *ABCD*, the sides *AB* and *CD* are equal. The perimeter of *ABCD* is

A) 27 B) 30 C) 32 D) 34 E) 48

3. **(CWML- 2019-10 Meet 3 #17)**

Given ΔDEF with right angle at *E*, $m\angle D = 30°$, $EF = 12$, *X* is between *D* and *F*, and $\overline{EX} \perp \overline{DF}$.

What is the length *DX*?

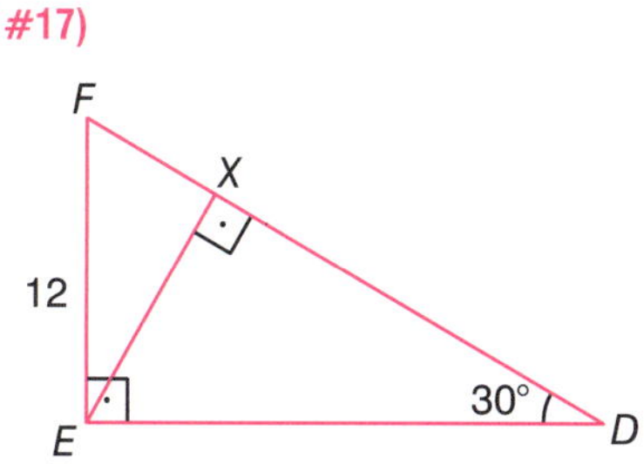

4. **(2018 AMC 10A #16)** Right triangle *ABC* has leg lengths $AB = 20$ and $BC = 21$.

Including $\overline{AB}$ and $\overline{BC}$, how many line segments with integer length can be drawn from vertex *B* to a point on hypotenuse $\overline{AC}$?

A) 5 B) 8 C) 12 D) 13 E) 15

5. **(2014 AMC 10A # 9)** The two legs of a right triangle, which are altitudes, have lengths $2\sqrt{3}$ and 6.

How long is the third altitude of the triangle?

A) 1 B) 2 C) 3 D) 4 E) 5

6. **(2012 AMC 10B # 12)** Point *B* is due east of point *A*. Point *C* is due north of point *B*. The distance between points *A* and *C* is $10\sqrt{2}$, and $\angle BAC = 45°$. Point *D* is 20 meters due north of point *C*.

The distance *AD* is between which two integers?

A) 30 and 31 B) 31 and 32 C) 32 and 33 D) 33 and 34 E) 34 and 35

7. **(2009 AMC 10B # 18)**

Rectangle $ABCD$ has $AB = 8$ and $BC = 6$. Point M is the midpoint of diagonal $\overline{AC}$, and E is on AB with $\overline{ME} \perp \overline{AC}$.

What is the area of ΔAME?

A) $\frac{65}{8}$ B) $\frac{35}{3}$ C) 9 D) $\frac{75}{8}$ E) $\frac{85}{8}$

8. **(2014 AMC 10B #21)** Trapezoid $ABCD$ has parallel sides $\overline{AB}$ of length 33 and $\overline{CD}$ of length 21. The other two sides are of lengths 10 and 14. The angles A and B are acute.

What is the length of the shorter diagonal of $ABCD$?

A) $10\sqrt{6}$ B) 25 C) $8\sqrt{10}$ D) $18\sqrt{2}$ E) 26

9. **(AlphaStar PiMC 2018 #5)** In triangle ABC with area 2, angle $\angle ABC = 90$ degrees.

What is the product of side lengths AB and BC?

10. **(2008 AMC 10A # 18)** A right triangle has perimeter 32 and area 20.

What is the length of its hypotenuse?

A) $\frac{57}{4}$ B) $\frac{59}{4}$ C) $\frac{61}{4}$ D) $\frac{63}{4}$ E) $\frac{64}{4}$

11. **(CWML 2016-17 Meet 3)**

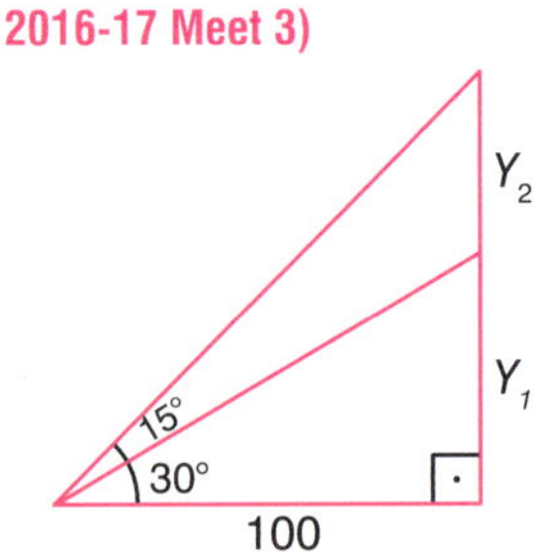

Which is nearest length Y_2?

A) 35 B) 37 C) 39 D) 41 E) 43

12. **(2002 AMC 12B # 20)** Let ΔXOY be a right-angled triangle with $m\angle XOY = 90°$. Let M and N be the midpoints of legs OX and OY, respectively

Given that $XN = 19$ and $YM = 22$, find XY.

A) 24 B) 26 C) 28 D) 30 E) 32

CHAPTER 3

Note

ABC is an isosceles right triangle.

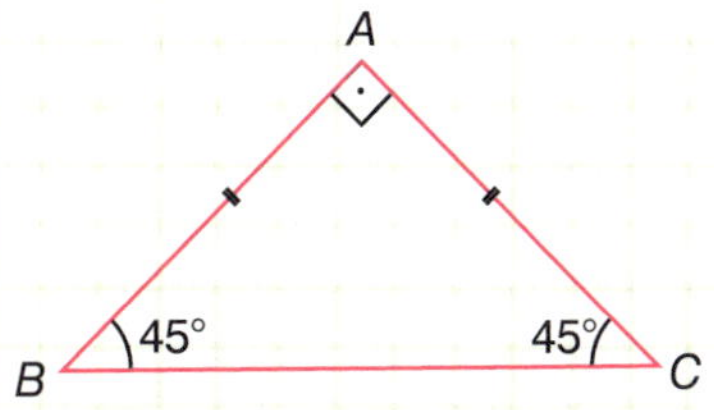

If we draw a perpendicular foot from point *A* to *BC*, then we create two 45°–45°–90° isosceles right triangles.

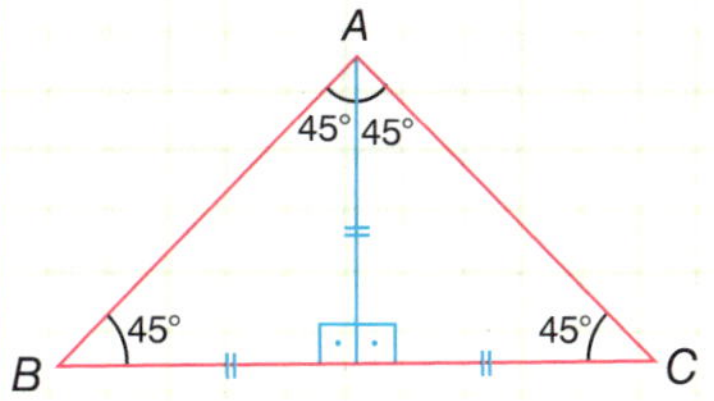

My Notes

Theorem 3.1

The ratio of the sides of an isosceles right triangle is $1 : 1 : \sqrt{2}$ or in general $a : a : a\sqrt{2}$.

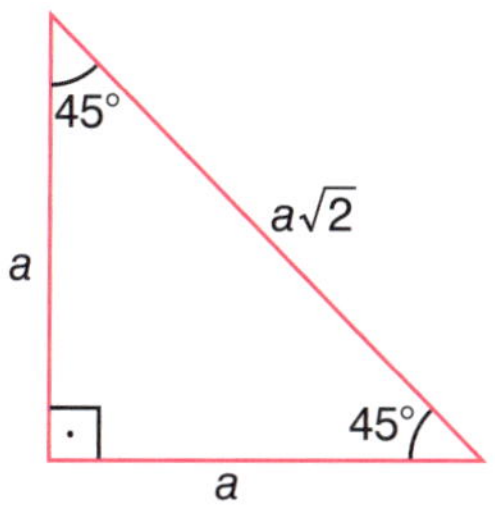

Example 3.1

What is *BC*?

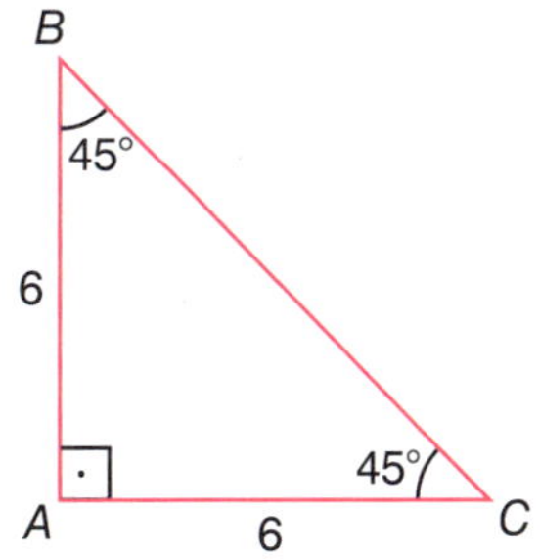

Solution 3.1

From the $a : a : a\sqrt{2}$ relationship the length of $BC = 6\sqrt{2}$.

Example 3.2

$AB = AC = ?$

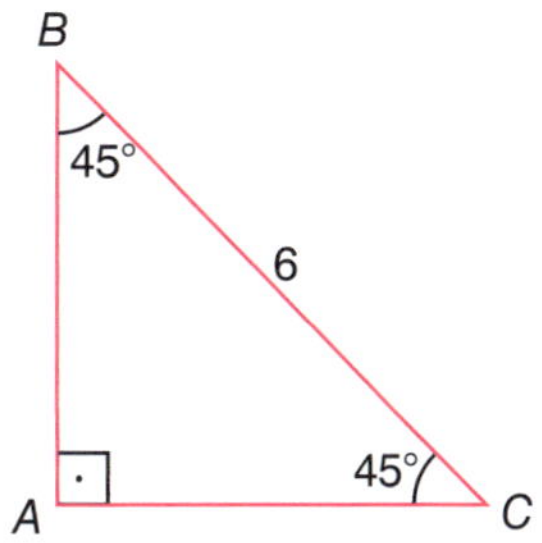

Solution 3.2

Let $AB = AC = a$.

From the isosceles right triangle ratio

$$BC = a\sqrt{2} = 6$$

$$a = \frac{6}{\sqrt{2}} = \frac{6\sqrt{2}}{2} = 3\sqrt{2}.$$

Example 3.3

$AB = AC = x = ?$

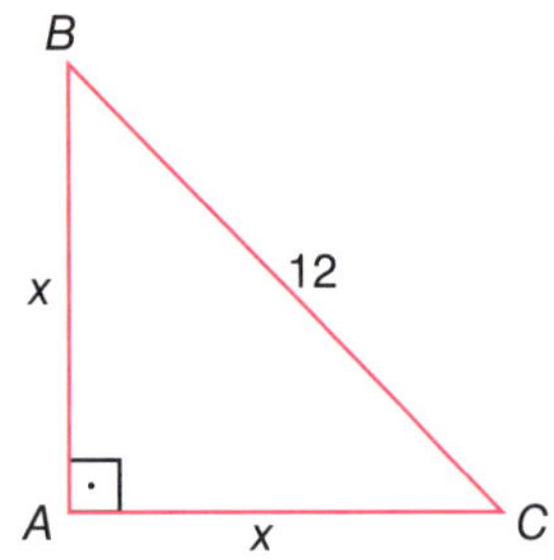

Try This

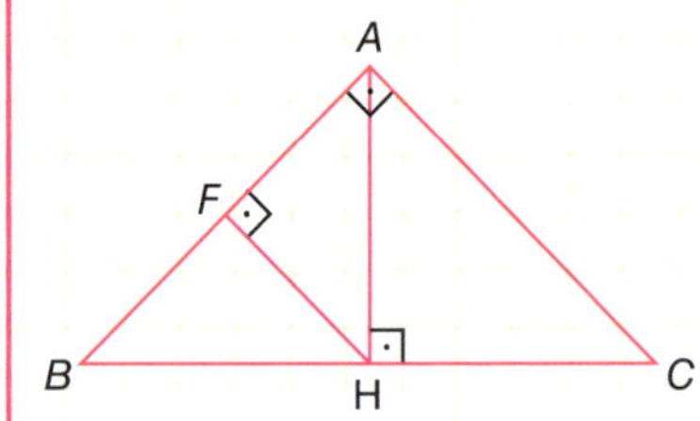

ABC is an isosceles right triangle with $BC = 3\sqrt{6}$.

If $AH \perp BC$ and $HF \perp AB$, then find FH.

Solution 3.3

From the 45°– 45°– 90° triangle,

$$x\sqrt{2} = 12$$

$$x = \frac{12}{\sqrt{2}} = \frac{12\sqrt{2}}{2} = 6\sqrt{2}.$$

Example 3.4

$AB = AC = x = ?$

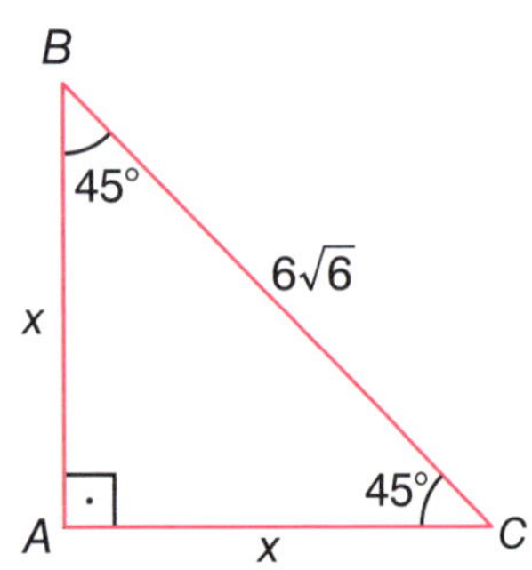

Solution 3.4

From the 90°– 45°– 45° triangle.

$$x\sqrt{2} = 6\sqrt{6}$$

$$x = \frac{6\sqrt{6}}{\sqrt{2}} = \frac{6\sqrt{3}\cdot\sqrt{2}}{\sqrt{2}} = 6\sqrt{3}.$$

My Notes

Theorem 3.2

In a 30°– 60°– 90° triangle, the side opposite of the 30° angle is half of hypotenuse, the side opposite of the 60° angle is $\sqrt{3}$ times the other leg.

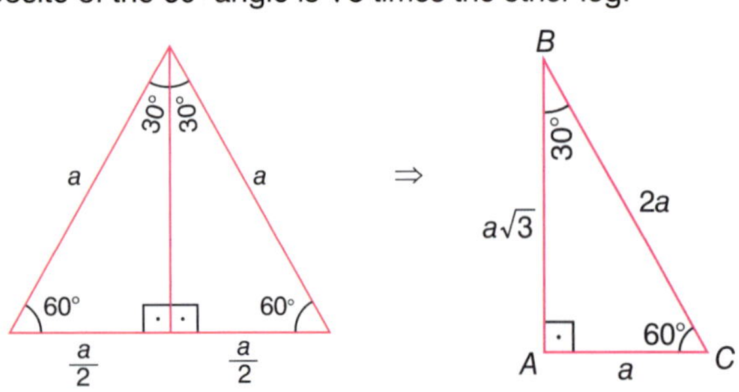

Note

ABC is an equilateral triangle with side length a. The length of the height of this triangle is $\frac{a}{2}\sqrt{3}$.

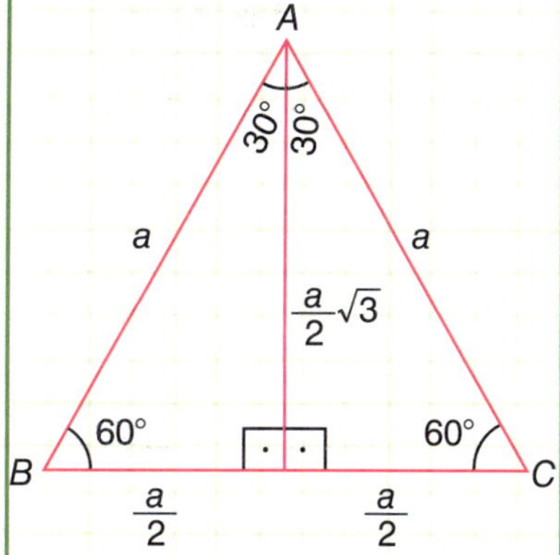

(30°– 30°– 120°)

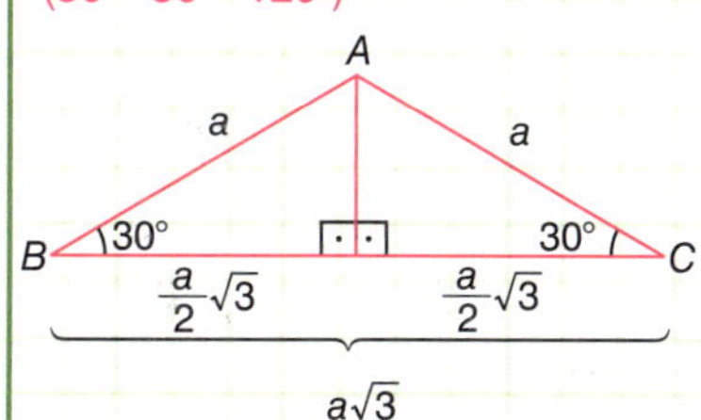

$BC = \sqrt{3}\,AB = \sqrt{3}\,AC$

My Notes

Example 3.5

Find AB and AC.

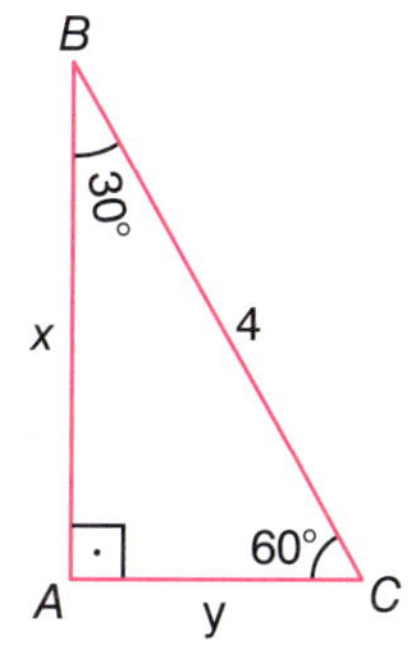

Solution 3.5

AC is across from the 30°angle, so, it is which is half of BC and $AC = 2$. AB is across from the 60° angle, so, it is $\sqrt{3}AC$ and $AB = 2\sqrt{3}$.

Example 3.6

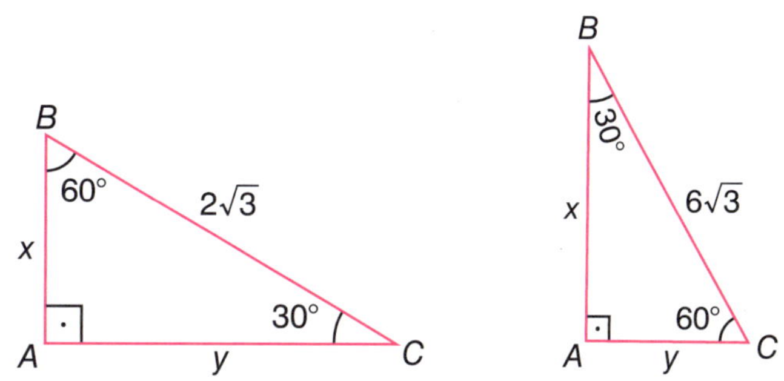

Solution 3.6

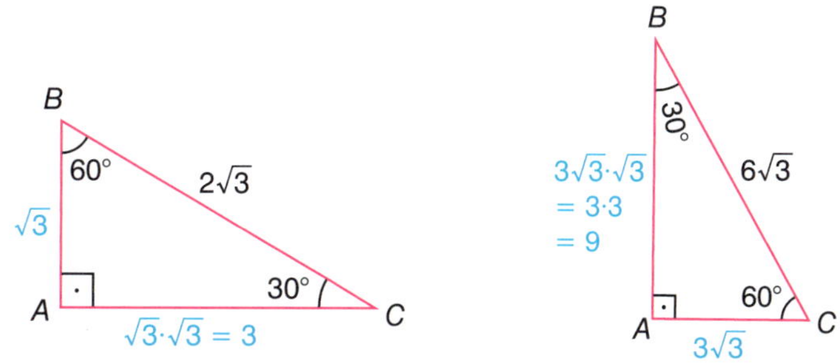

Example 3.7

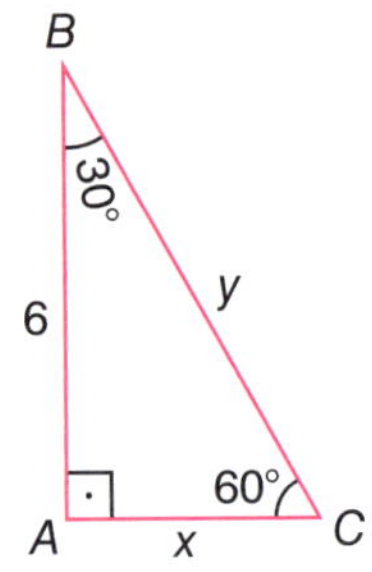

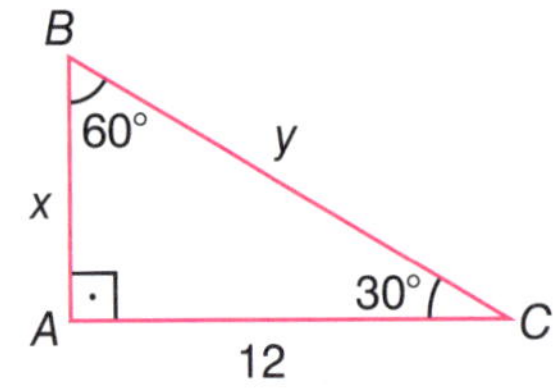

Try This

In the figure,

$m\angle A = 105°$,

$m\angle B = 60°$.

If $AC = 6\sqrt{6}$, then find BC.

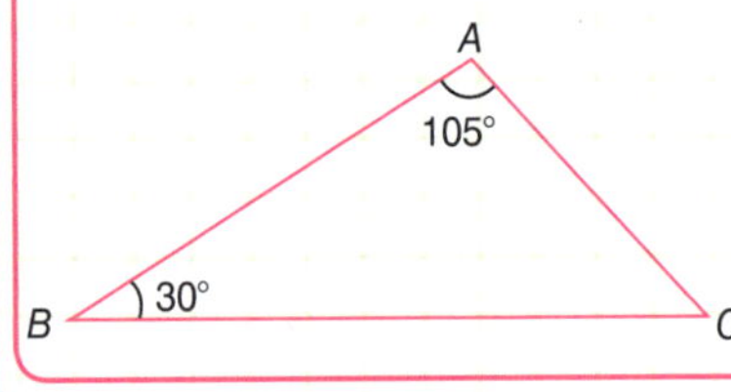

Solution 3.7

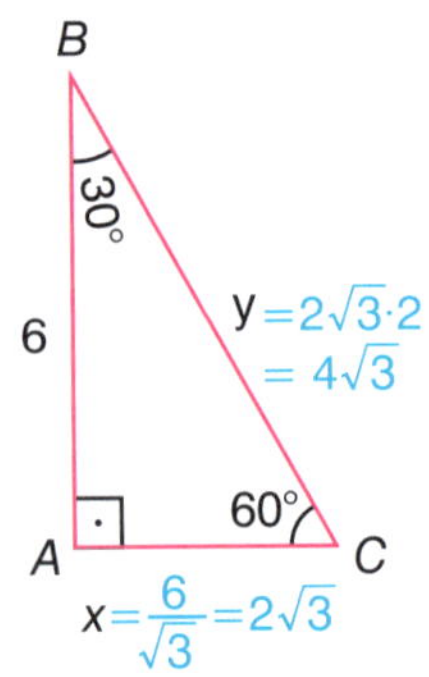

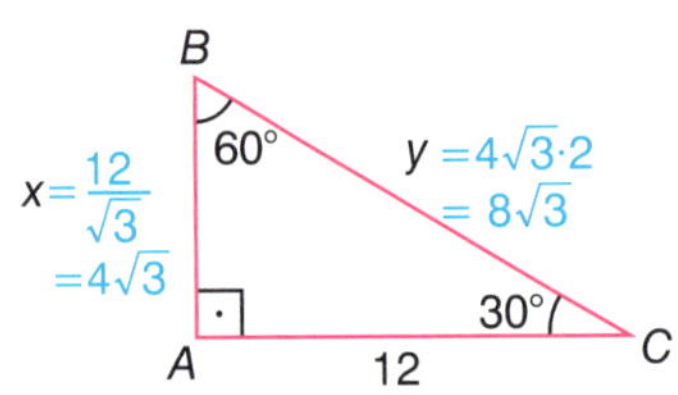

My Notes

Example 3.8

In the figure,

$AC = 6\sqrt{2}$, $BC = x$,

$m\angle A = 60°$, and $m\angle B = 45°$.

Find the value of x

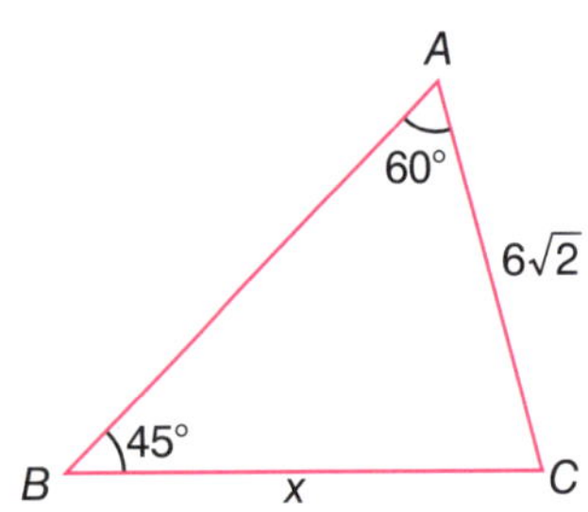

Solution 3.8

We create two special triangles by making a perpendicular foot from C to AB.

$$x = 3\sqrt{6}\cdot\sqrt{2}$$
$$= 3\sqrt{12}$$
$$= 6\sqrt{3}.$$

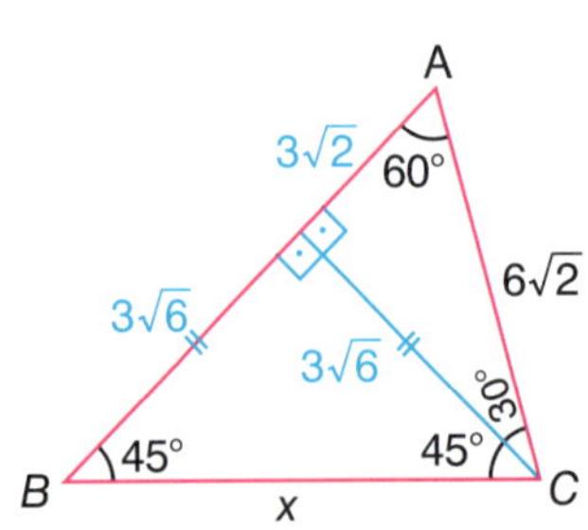

Try This

In the figure,

$MN \perp NK$, $ML \perp LK$,

$MN = 3\sqrt{3}$, $ML = 3$, and

$m\angle M = 150°$.

Find KL.

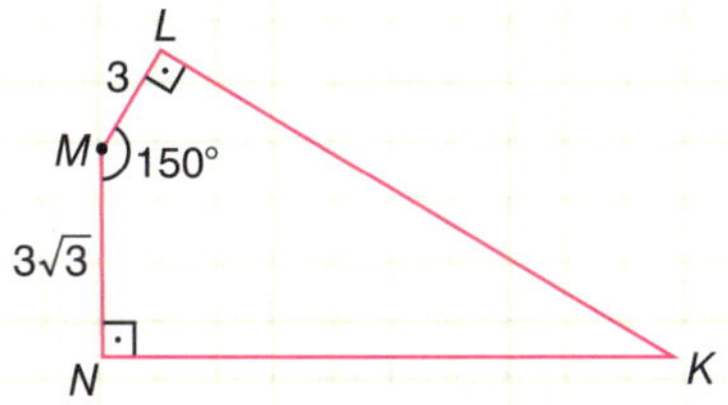

My Notes

Example 3.9

In the figure,

$BC = 9\sqrt{2}$,

$AC = x$,

$m\angle A = 60°$, and $m\angle B = 75°$.

Find the value of x.

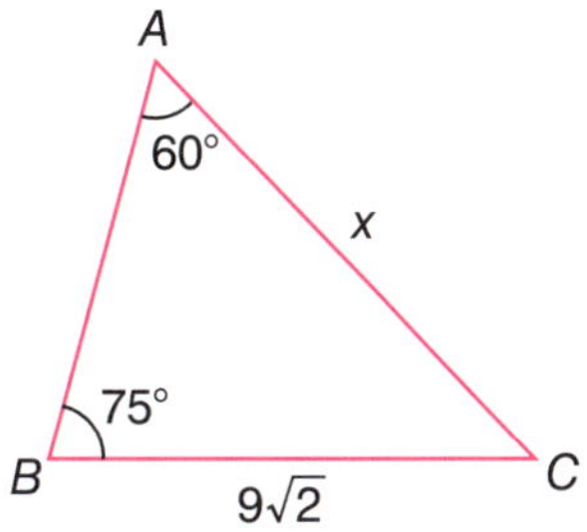

Solution 3.9

We can create two special triangles by drawing the altitude from B to AC. (See diagram)

Therefore, $x = 9+3\sqrt{3}$.

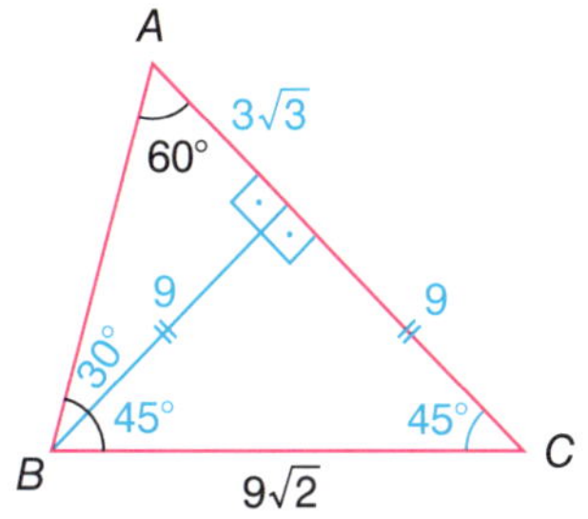

Example 3.10

In the figure, $MN \perp NK$, $ML \perp LK$,

$MN = 3\sqrt{2}$,

$ML = 2$, and

$m\angle M = 135°$.

Find the length of KL.

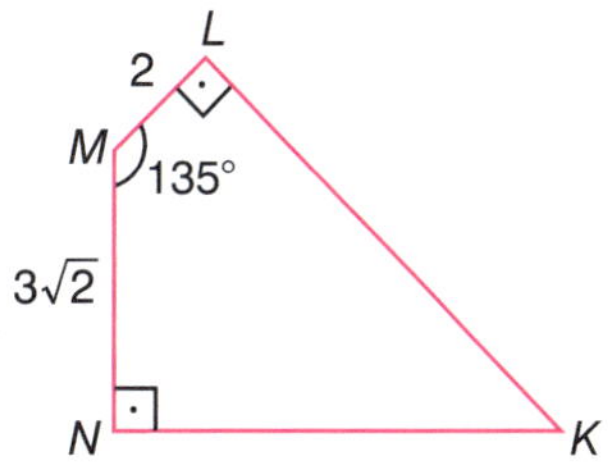

Solution 3.10

From $45°-45°-90°$ triangle,

$m\angle TML = m\angle LTM = 45°$,

$TK = 5\sqrt{2} \cdot \sqrt{2} = 10$,

$KL = 10-2 = 8$.

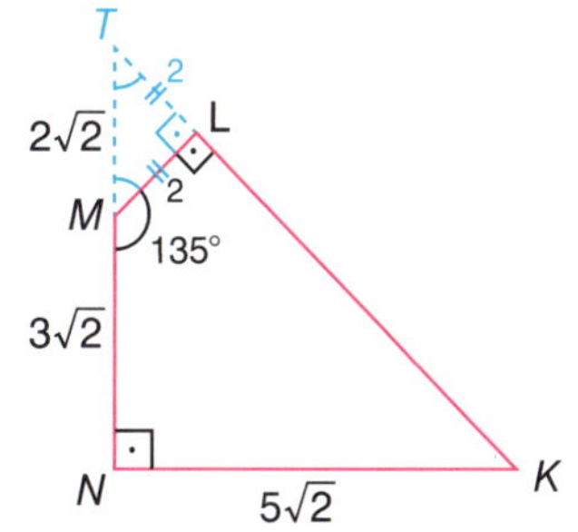

Example 3.11

In the figure,

$AD \perp DC$,

$AD = 7$, $BC = 4$, and

$m\angle A = m\angle B = 60°$.

Find the length of DC.

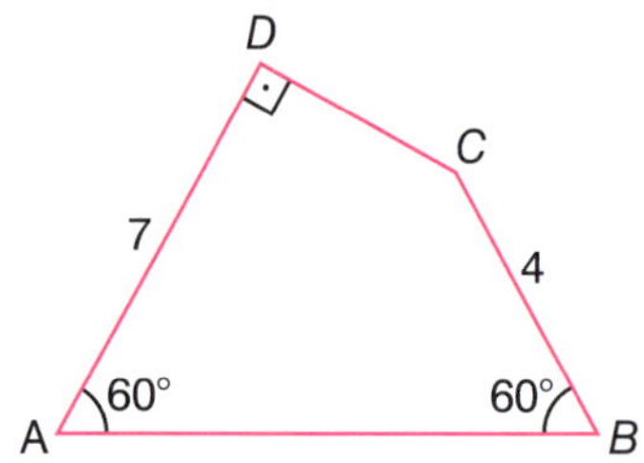

Try This

KLM is an isosceles right triangle with $m\angle KML = 90°$. Point D is on the hypotenuse KL such that

$DL = 2$ and

$DK = 4$.

Find DM.

Solution 3.11

$AE = 2 \cdot 7 = 14$

$CB = BE = 4$ ($DBCE$ is an isosceles triangle)

$DE = 7\sqrt{3}$ $(30°-60°-90°)$

Since $CE = 4\sqrt{3}$ $(30°-30°-120°)$

$DC = 7\sqrt{3} - 4\sqrt{3} = 3\sqrt{3}$.

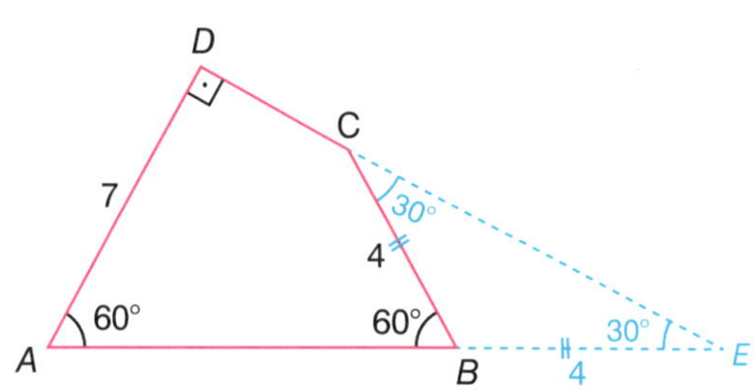

My Notes

Example 3.12

In triangle ABC,

$BC = 7\sqrt{2}$,

$m\angle BAC = 30°$, and $m\angle BCA = 45°$.

Find the length of the side AB.

Solution 3.12

We create two right triangle by drawing altitude from B to AC.

$BH = HC = 7$ $(45°-45°-90°)$

$AB = 2 \cdot 7 = 14$ $(30°-60°-90°)$

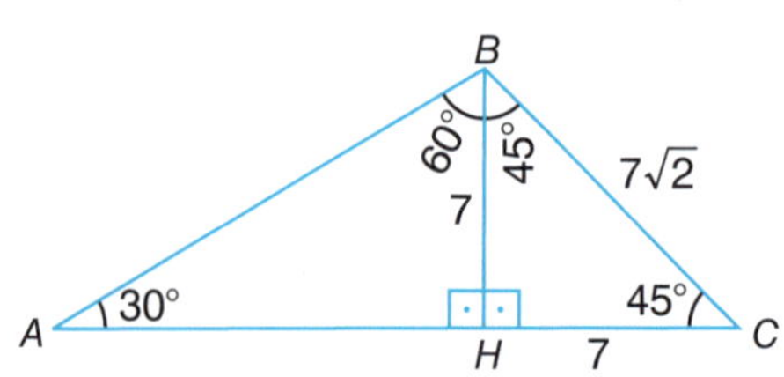

Note

If $m\angle A = 90°$ and BI and CI are angle bisectors, then $m\angle BIC = 135°$.

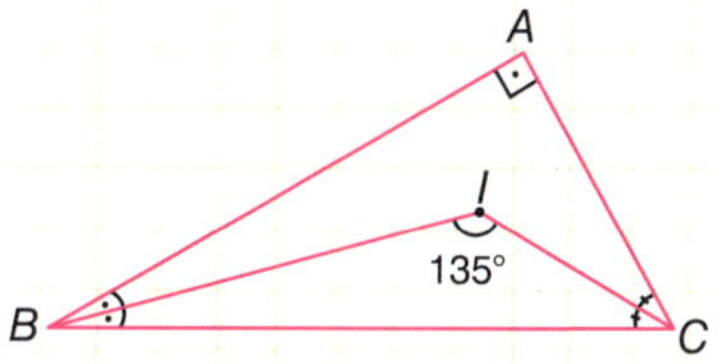

My Notes

Example 3.13

In the figure,

$m\angle B = 105°$,

$m\angle A = 30°$, and

$AB = 8$.

Find BC.

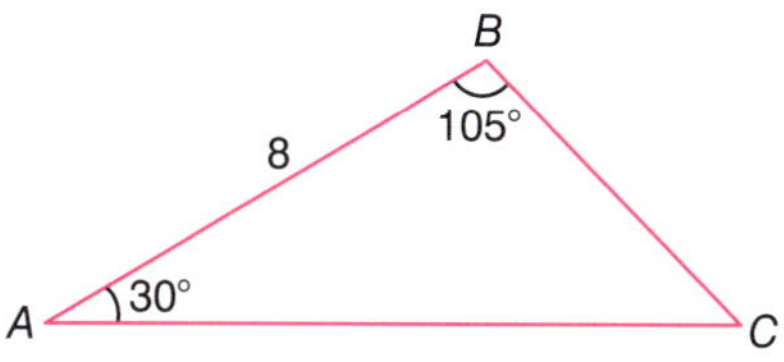

Solution 3.13

We create two special triangles

$(30°-60°-90°)$ and $(45°-45°-90°)$.

From ΔABH : $BH = 4$.

From ΔBHC : $|BC| = \sqrt{2} \cdot |BH| = 4\sqrt{2}$.

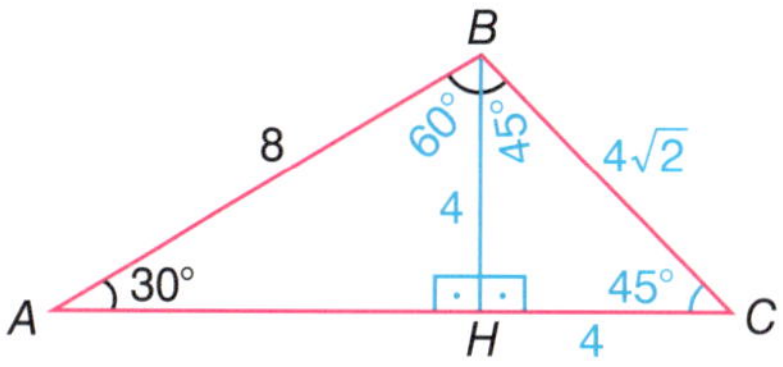

Example 3.14

In the figure BP and CP are their respective bisectors.

$AB \perp AC$, $BC = 5\sqrt{2}$, and $BP = 6$.

Find the length of CP.

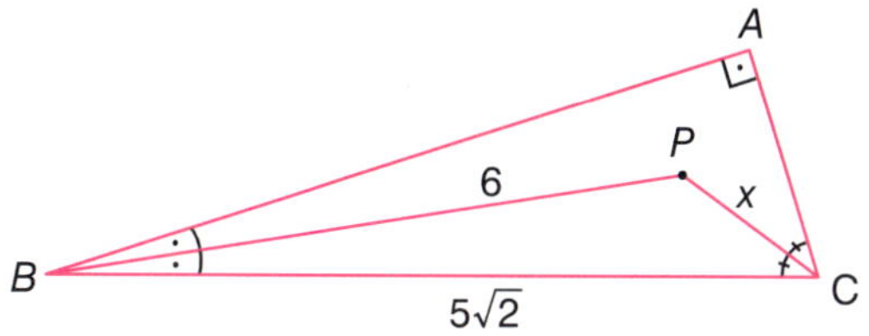

Solution 3.14

From Theorem (1.3)

$$m\angle BPC = 90° + \frac{90°}{2} = 135°$$

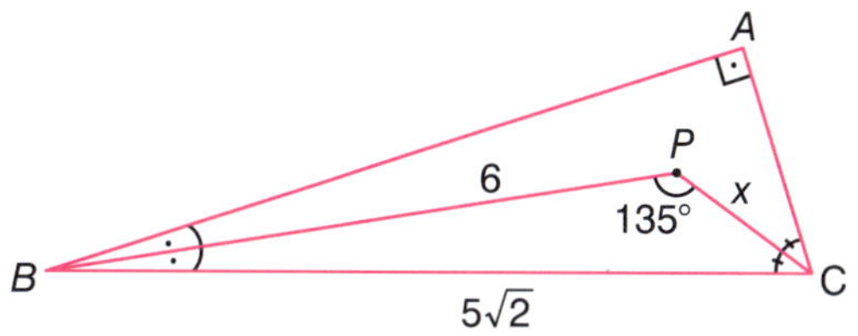

By using ΔBPC

$$PK = BK = 3\sqrt{2}$$

$$KC = x + 3\sqrt{2}$$

$$(KB)^2 + (KC)^2 = (BC)^2$$

$$(3\sqrt{2})^2 + (x + 3\sqrt{2})^2 = (5\sqrt{2})^2$$

$$(x + 3\sqrt{2})^2 = (5\sqrt{2})^2 - (3\sqrt{2})^2$$

$$(x + 3\sqrt{2})^2 = 32$$

$$x + 3\sqrt{2} = \sqrt{32}$$

$$x = 4\sqrt{2} - 3\sqrt{2} \Rightarrow x = \sqrt{2}.$$

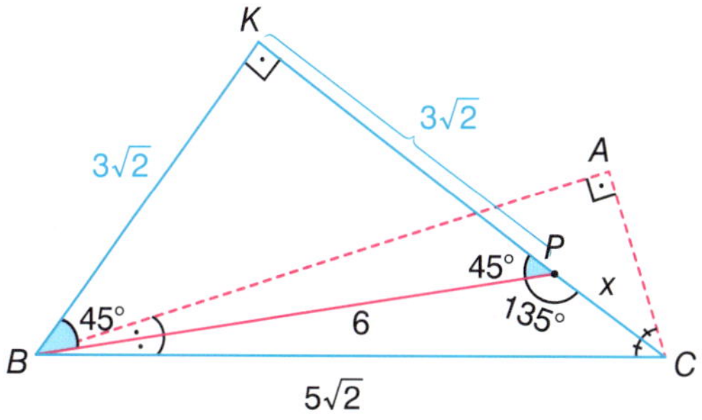

Example 3.15

In the figure,

$m\angle BAC = 150°$, and

$AB = AC = 2$.

Find the area of ΔABC.

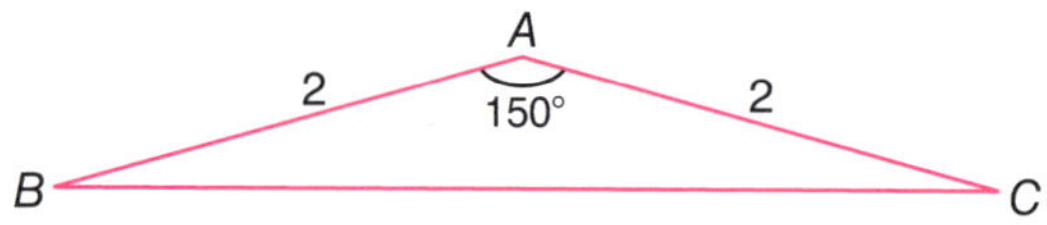

Solution 3.15

Extend BA to point H so that $AH \perp HC$. From 30°– 60°– 90° triangle (ΔAHC) $HC = 1$.

Therefore, the area of $\Delta ABC = \frac{2 \cdot 1}{2} = 1$.

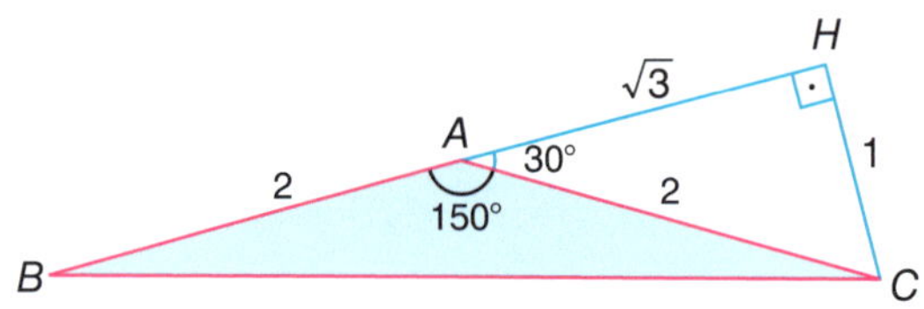

Try This

ABC is a right triangle.

$AD = DC$,

$\sqrt{3}\,BC = AB$, and

$AE = 2EB$

Find $m\angle AED$.

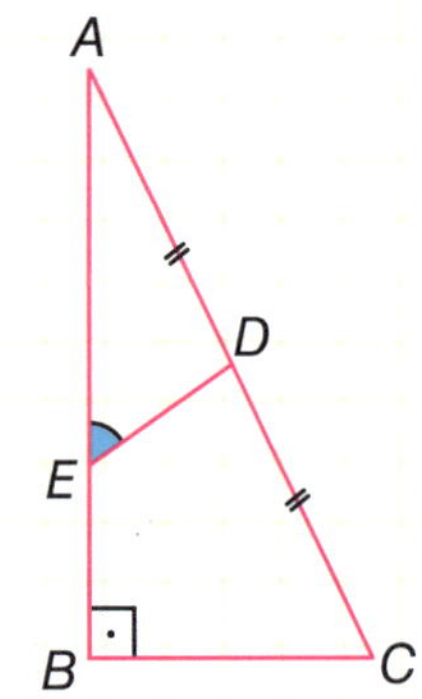

My Notes

Example 3.16

In the figure, $AB \perp BC$,

$AB = BC$,

$AC = 6$, and $m\angle ADB = 30°$.

Find the length of CD.

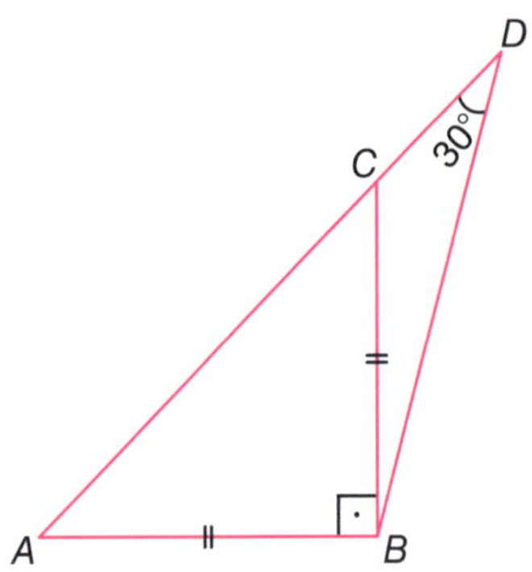

Solution 3.16

ΔABC is an isosceles triangle. Draw a perpendicular foot from B to AC.

$HB = AH = CH = 3$

$HD = 3\sqrt{3}$. (ΔDBH is 30°–60°–90°)

So, $CD = 3\sqrt{3} - 3$.

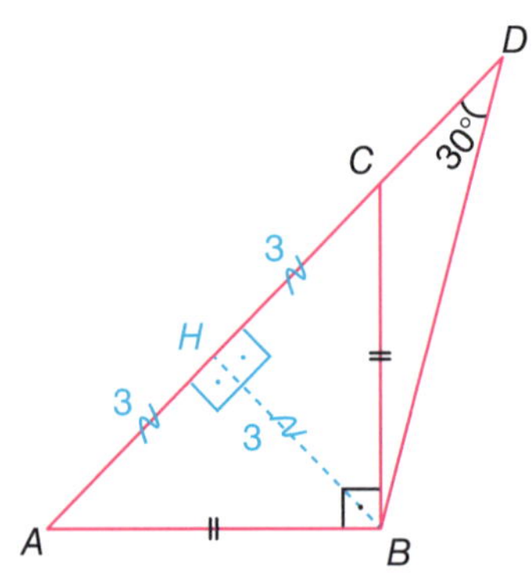

Try This

BAC is a right triangle,

$ED = AC$, $BD = DC$, and

$m\angle ACB = 63°$.

Find $m\angle EDC$.

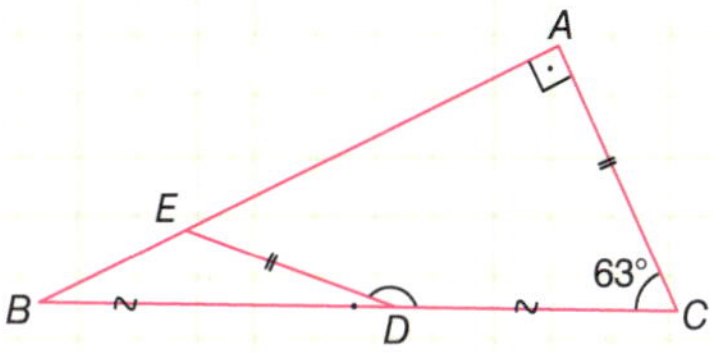

My Notes

Example 3.17

D is the midpoint of *BC*.

$BD = DC$,

$m\angle BAD = 2x$, and

$m\angle ACB = 3x$.

Find $m\angle ADC$.

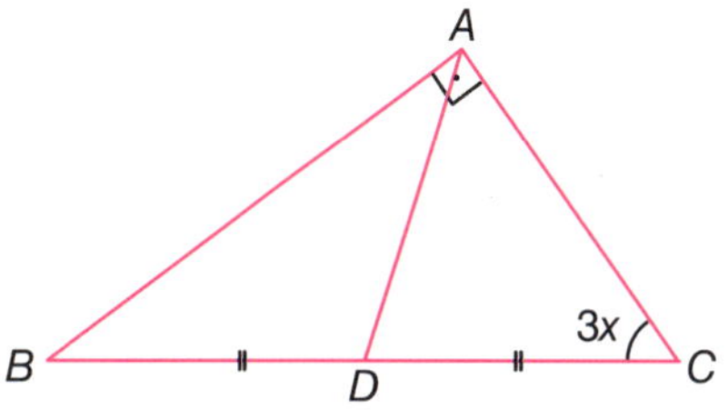

Solution 3.17

From 'the median of a right triangle,

$AD = BD = DC$.

From $\triangle ADC$,

$$4x+3x+3x = 180°$$

$$10x = 180°$$

$$x = 18°.$$

$m\angle ADC = 4x = 4\cdot 18° = 72°$.

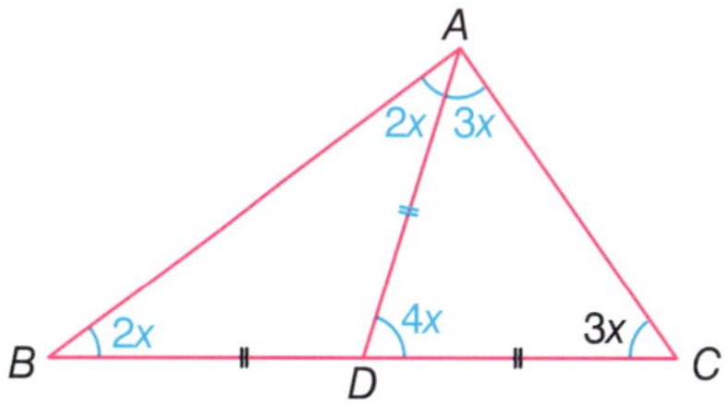

Example 3.18

D is the midpoint of *BC*, $CA \perp AB$,

$m\angle C = 60°$,

$AC = 6x-1$, and

$AD = 2x+3$.

What is *CB*?

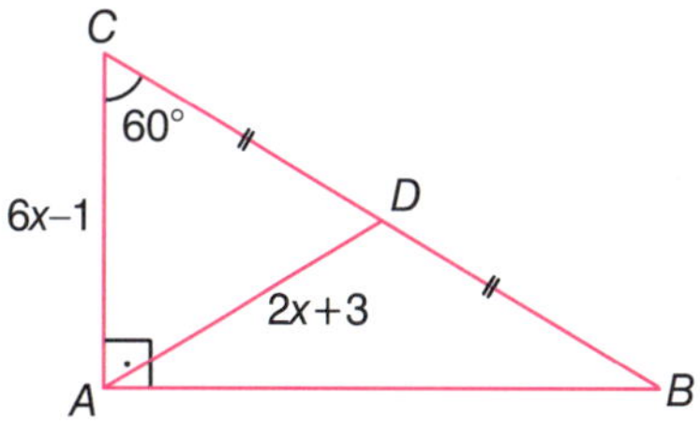

Solution 3.18

From the median of a right triangle,

$CD = DB = AD = 2x+3$

ADC is an equilateral triangle and

$$6x-1 = 2x+3$$

$$4x = 4 \Rightarrow x = 1.$$

So, $CB = (2x+3)+(2x+3) = 4x+6 \Rightarrow CB = 4+6 = 10$.

Theorem 3.3

In a $15°-75°-90°$ triangle, the length of the hypotenuse is 4 times the length of the altitude from to hypotenuse.

In the figure,

$BA \perp AC$, $AD \perp BC$. If $m\angle B = 15°$ and

$m\angle C = 75°$, then $BC = 4AD$.

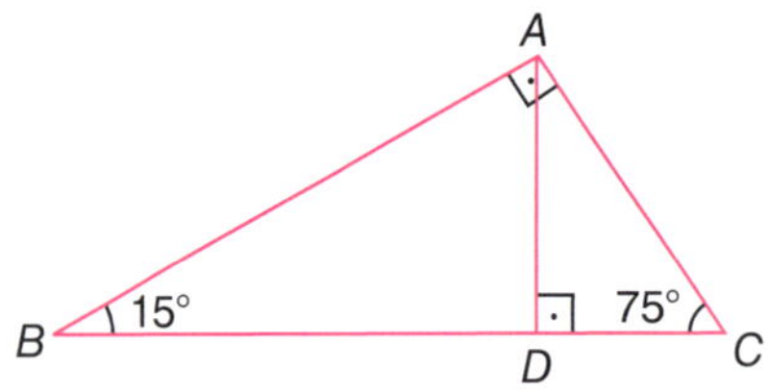

Proof 3.3

From the median of a right triangle
$BM = MC = AM$.

Suppose AD = a. Then $BM = MC = 2a$ (From $30°-60°-90°$ triangle).

The length of hypotenuse BC is

$2a+2a = 4a$ which is always 4 times of length of AD.

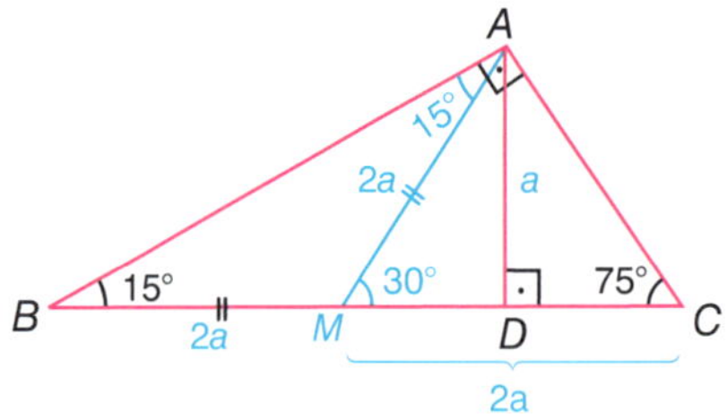

Theorem 3.4

In the figure, BAC is a right triangle,

$m\angle B = 15°$, and $m\angle BCA = 75°$.

If $AC = a$, then $AB = a(\sqrt{3}+2)$.

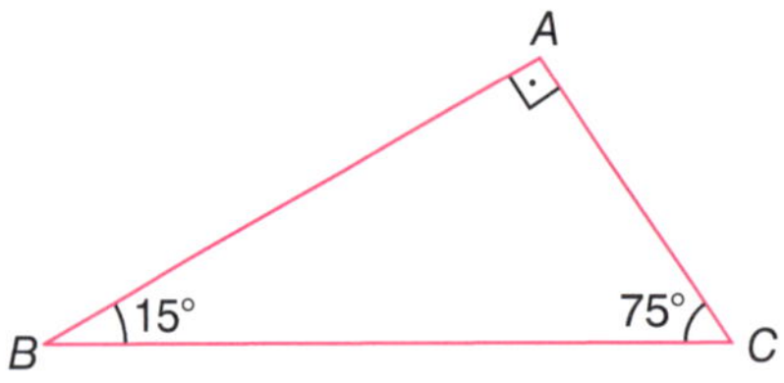

Proof 3.4

We divided angle C into 15° and 60° angles. ΔBKC isan isosceles triangle with 15° equal angles.

Let BK be $2a$ since BKC is an isosceles triangle then $KC = 2a$.
Then, $AC = a$ and $KA = a\sqrt{3}$. (From $30°-60°-90°$ triangle)

Therefore, AB $= 2a + a\sqrt{3} = a(2+\sqrt{3})$.

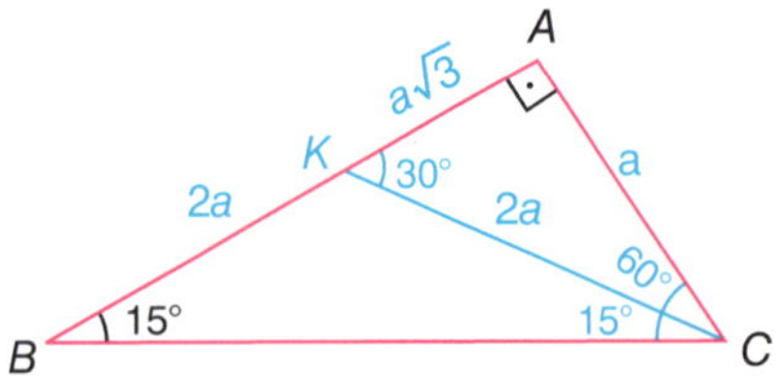

Try This

$ABCD$ is quadrilateral

$AB = AD = 3,$

$BC = 3\sqrt{2}$, and

$m\angle BAD = 90°$.

Find DC.

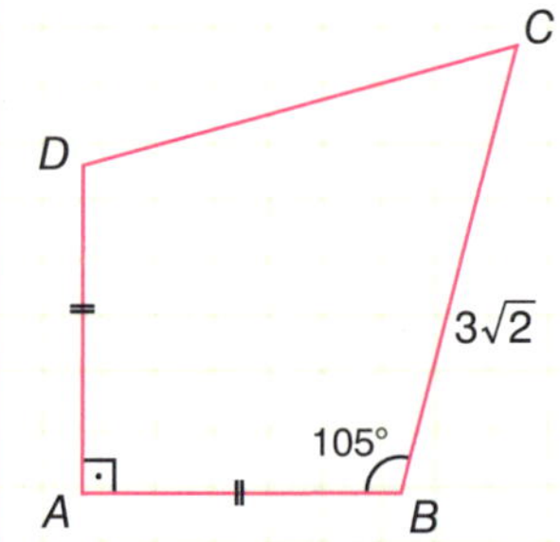

Try This

Find the height of a regular octagon with a side length of 8. Write your in simplest form.

Catriona Agg @Cshearer41
Geometry puzzle

The pink square has double the area of the blue square.

What is the angle?

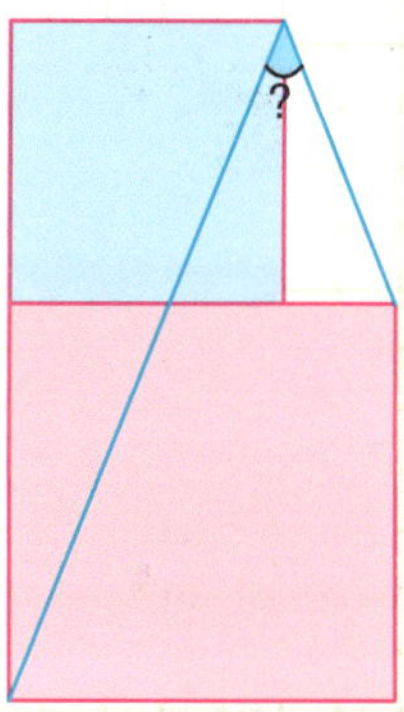

Try This

The area of triangle ΔABC is 8 unit^2.

Find *AH*.

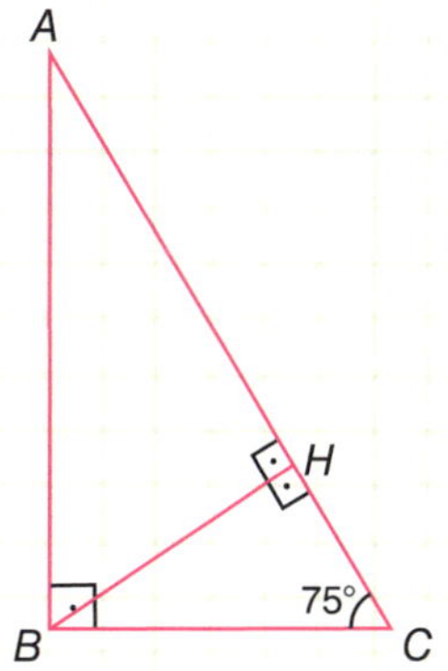

Example 3.19

ABC is a right triangle,

$m\angle B = 75°$,

$m\angle C = 15°$, and

$AB = 2-\sqrt{3}$.

Find the length of *AC*.

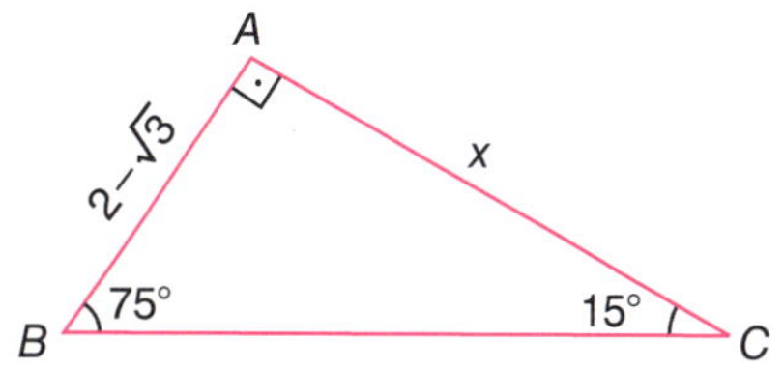

Solution 3.19

Using theorem 3.4,

$$AC = (2+\sqrt{3})\cdot(2-\sqrt{3})$$
$$= 2^2-\sqrt{3}^{\,2} = 1.$$

Example 3.20

BAC is a right triangle,

$AK \perp BC$, $AK = 1$,

$m\angle B = 75°$, and $m\angle C = 15°$.

Find the area of ΔABC.

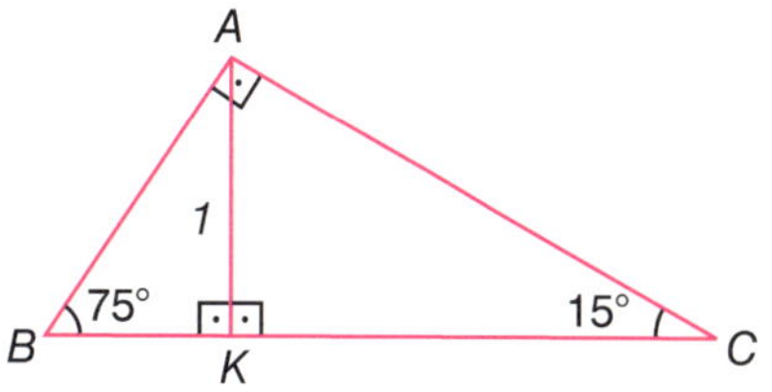

Solution 3.20

Using theorem 3.3,

Thus, $BC = 4AK$. But $AK = 1$, so $BC = 4$.

Therefore, area $\Delta ABC = \dfrac{4\cdot 1}{2} = 2.$

Example 3.21

In the figure,

$m\angle A = 15°$,

$m\angle B = 105°$, and

$BC = 2$.

Find *AC*.

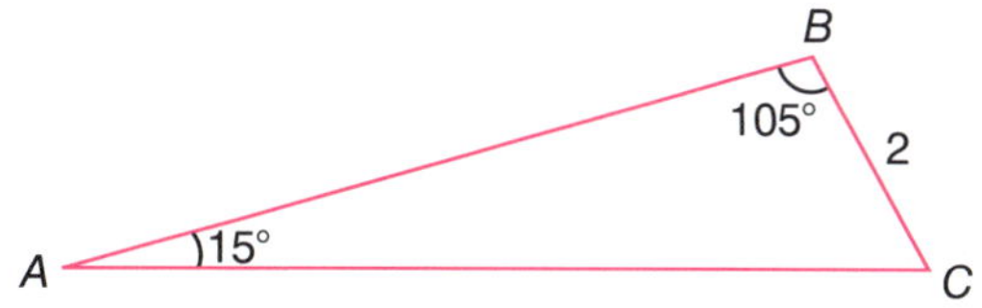

Solution 3.21

Using theorem 3.4,

$$AH = \sqrt{3}\cdot(2+\sqrt{3})$$
$$= 2\sqrt{3}+3$$

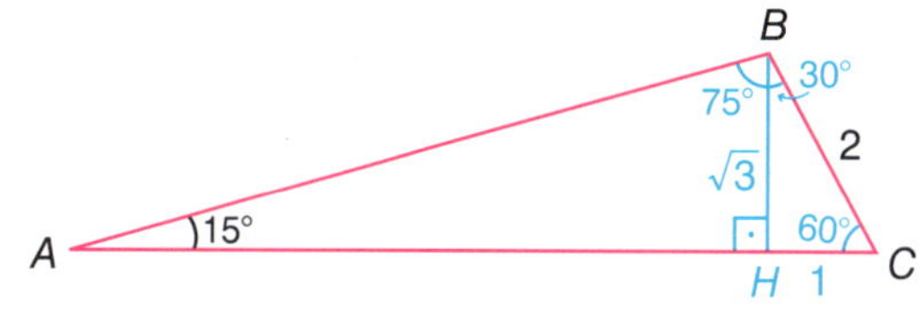

and also lasing ΔBHC $(30°-60°-90°)$ $HC = 1$.

Therefore, $AC = 2\sqrt{3}+3+1 \Rightarrow AC = 2\sqrt{3}+4$.

Example 3.22

In the figure,

$AB = 6+3\sqrt{3}$ and

$m\angle B = 15°$.

What is the length of AC?

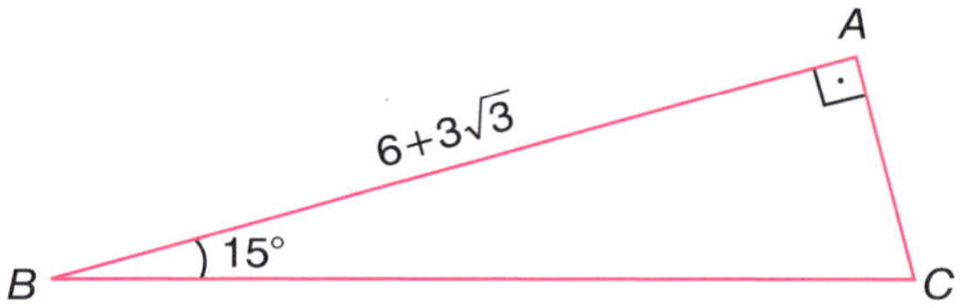

Solution 3.22

From $15°-75°-90°$ triangle (Theorem 3.3)

$$AC = \frac{6+3\sqrt{3}}{2+\sqrt{3}} = \frac{3(2+\sqrt{3})}{2+\sqrt{3}} = 3.$$

Theorem 3.5

In the figure,

$BA \perp AC$ and $AH \perp BC$.

If $AH = h$, then $BC = 2\sqrt{2}\,h$.

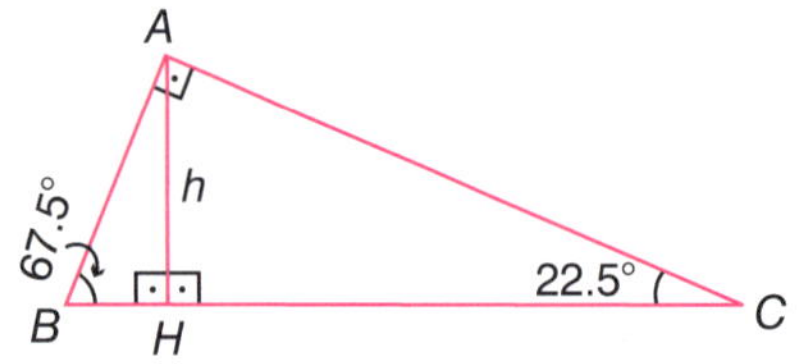

Proof 3.5

Very similar to $15°-75°-90°$ sides connection proof (Theorem 3.3), we create the median AM. Let AH be h; then, $AM = h\sqrt{2}$ (from $45°-45°-90°$ triangle). Since $AM = BM = MC$, $BC = h\sqrt{2} + h\sqrt{2} = 2h\sqrt{2}$.

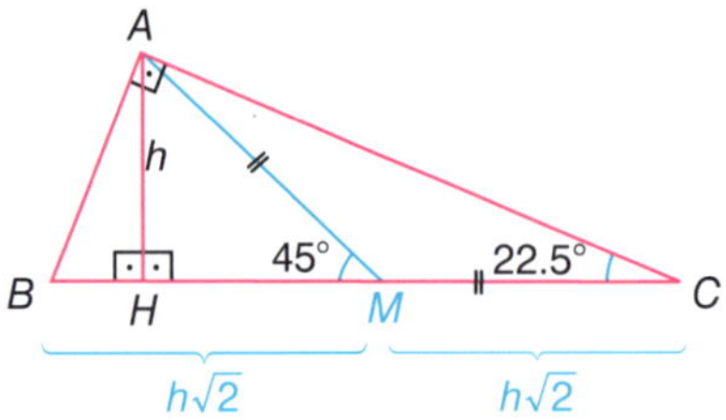

Theorem 3.6

In the figure,

$AB = a$,

$m\angle B = 67.5°$, and

$m\angle C = 22.5°$.

Then, $AC = a(\sqrt{2}+1)$.

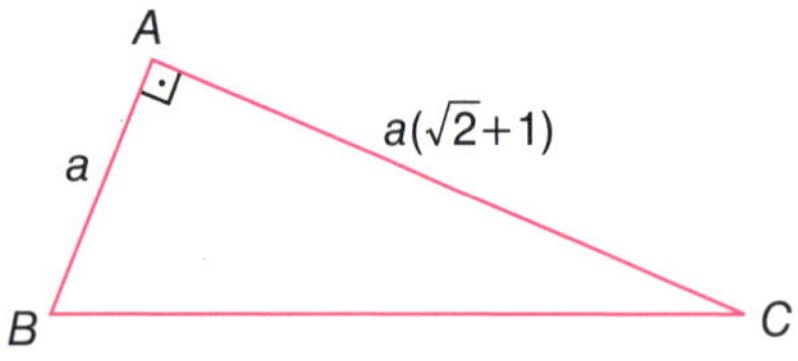

Proof 3.6

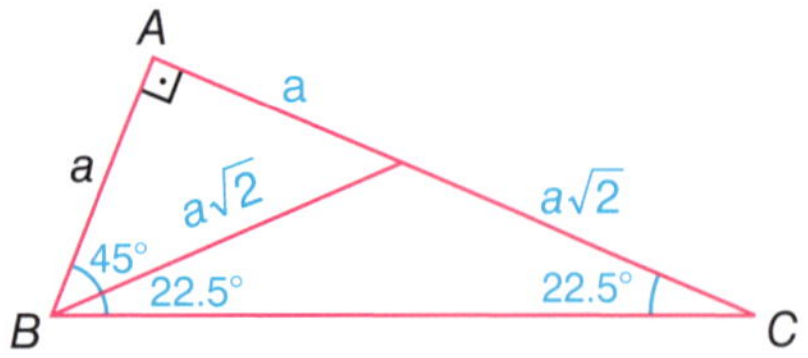

Catriona Agg @Cshearer41

Geometry puzzle

The point lies on the diagonal of the square so that $a + b = c$.

What is the angle?

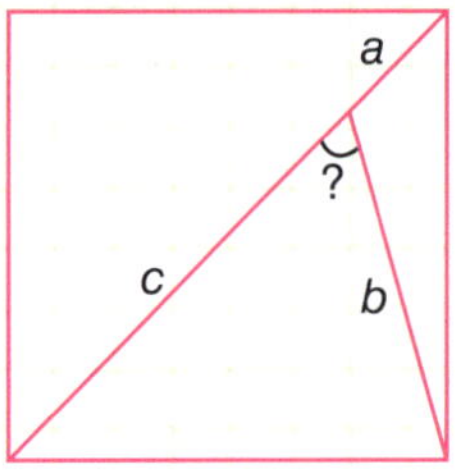

Try This

In the figure, if

$AB = \sqrt{2}\,AH$,

then find $m\angle HCB$.

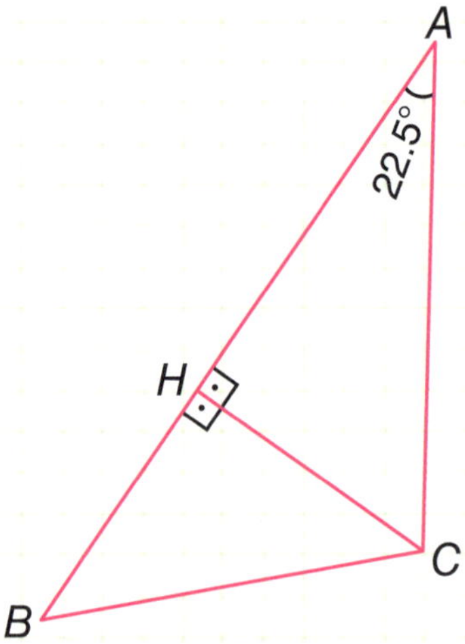

SPECIAL TRIANGLES- ANGLES TEST 1

1. In the figure,

$BC = 9\sqrt{2}$,

$m\angle BAC = 75°$, and

$m\angle ABC = 60°$.

Find AC.

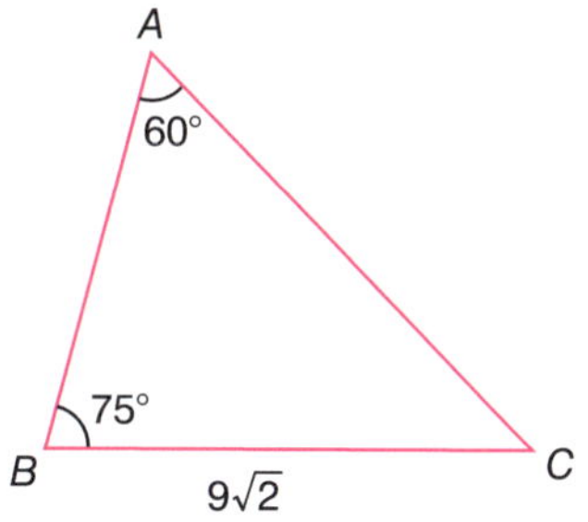

2. In the figure,

$BC = 12$,

$m\angle BAC = 90°$,

$m\angle ADC = 90°$ and

$m\angle ABC = 60°$.

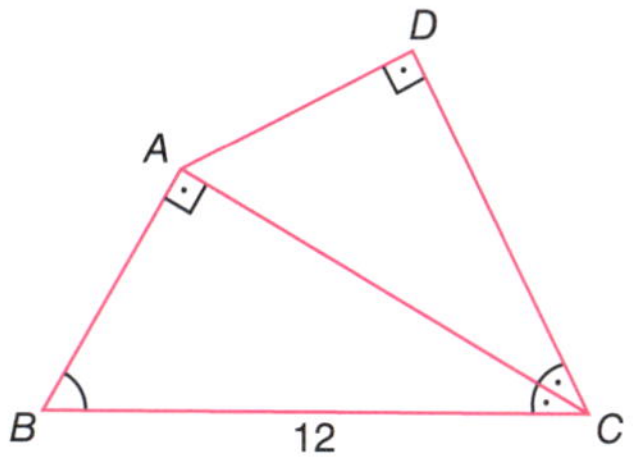

If AC is the angle bisector of $\angle C$, then find the length AD.

3. In the figure,

$m\angle BAC = 90°$,

$m\angle AHC = 90°$,

$m\angle B = 30°$, and

$HC = 1$ cm.

What is the length of BH?

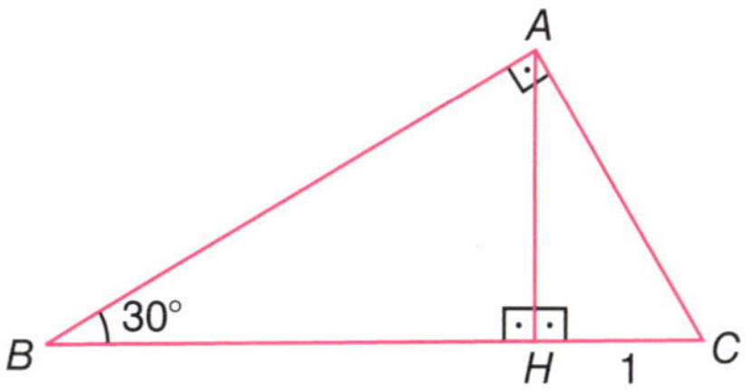

4. In the figure,

$m\angle ACB = 120°$,

$AC = 4$, and

$BC = 5$.

Find AB.

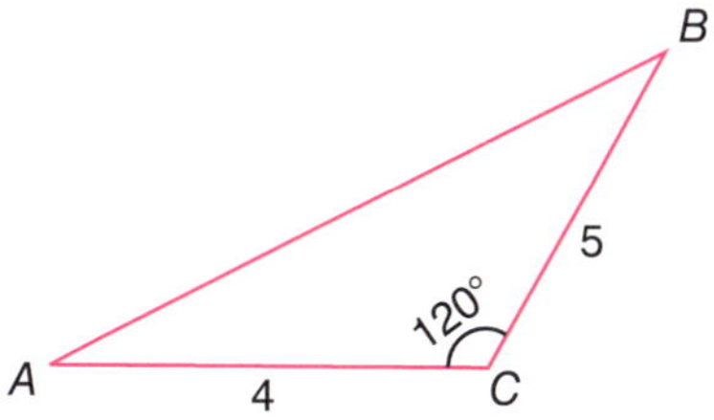

5. ΔABC is a right triangle with $m\angle A = 90°$, and AH is the altitude to the hypotenuse.

If $m\angle C = 30°$ and $BH = 2$ cm, find HC.

6. In the figure,

$AB = BC$,

$m\angle ADB = 30°$, and

$AC = 8$.

What is the length of CD?

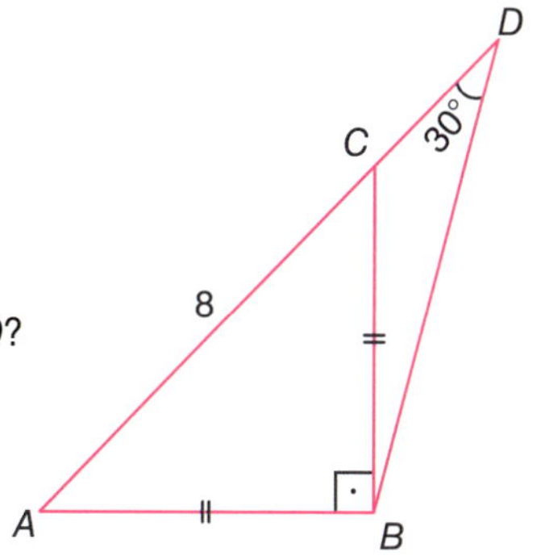

7. In the figure,

$AD \perp DC$,

$AD = 9$, $BC = 5$, and

$m\angle A = m\angle B = 60°$.

What is length of DC?

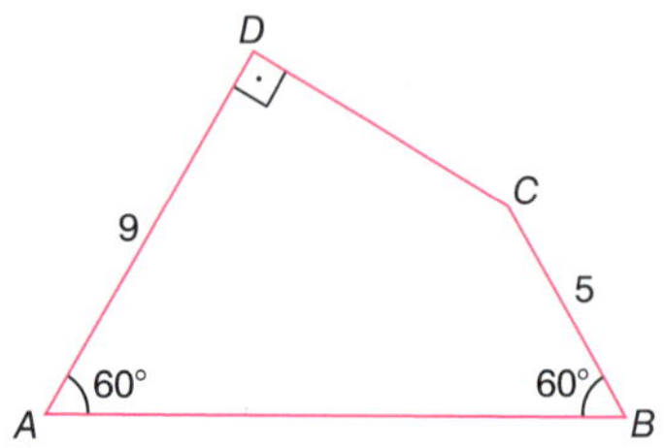

8. In the figure,

$m\angle C = 90°$,

$m\angle BAD = m\angle DAC$,

$BD = DA$, and

$DC = 3$ cm.

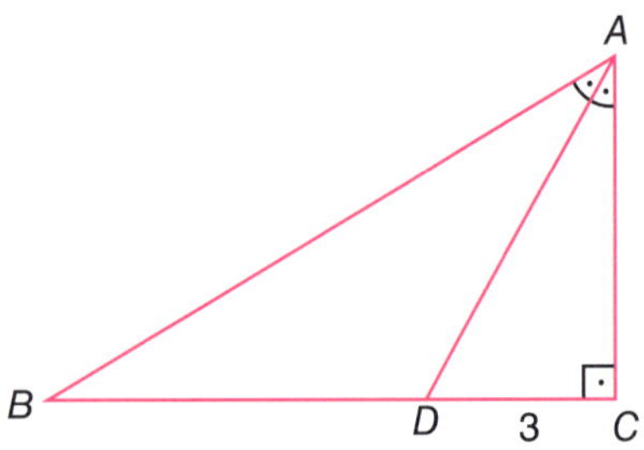

What is length of AB?

9. In the figure,

$AB = AC$,

$AH = HC$,

$m\angle A = 120°$, and

$PC = 8$ cm.

Find the length of BP.

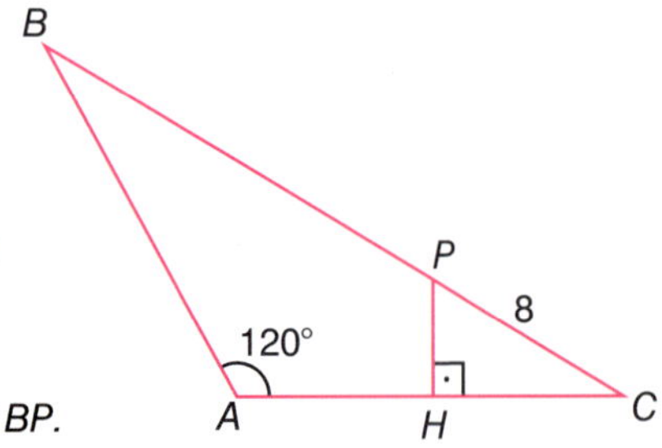

10. $ABCD$ a quadrilateral with $DA \perp AB$.

$m\angle ABC = 45°$,

$BC = 5\sqrt{2}$,

$AD = 10$, and

$AB = 17$.

Find DC.

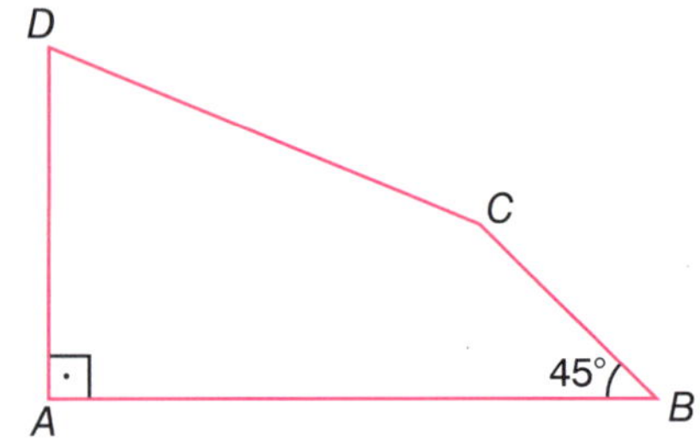

11. In the triangle DEF,

$m\angle DEF = 120°$,

$DE = 2\sqrt{5}$, and $DF = 8$.

Find the length of EF.

12. ABC is an isosceles triangle with

$AB = BC = 8$, and

$m\angle ABC = 150°$.

Find the area of ΔABC.

1.

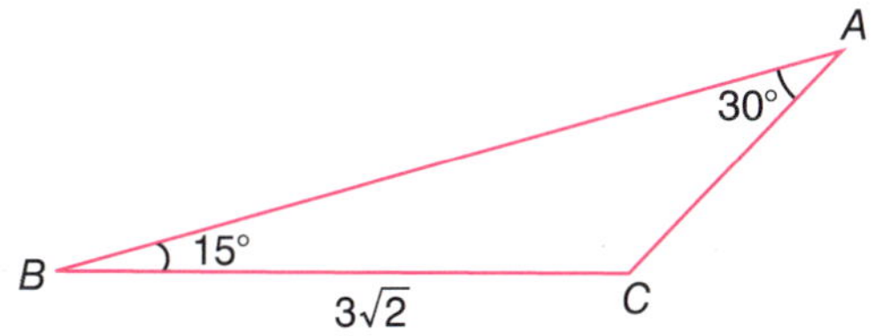

In the figure,

$m\angle ABC = 15°$, $m\angle BAC = 30°$, and $BC = 3\sqrt{2}$.

Find AC.

2. ABC is a right triangle ,

$m\angle A = 75°$,

$m\angle C = 15°$, and

AB $= 4-\sqrt{3}$.

Find the length of AC.

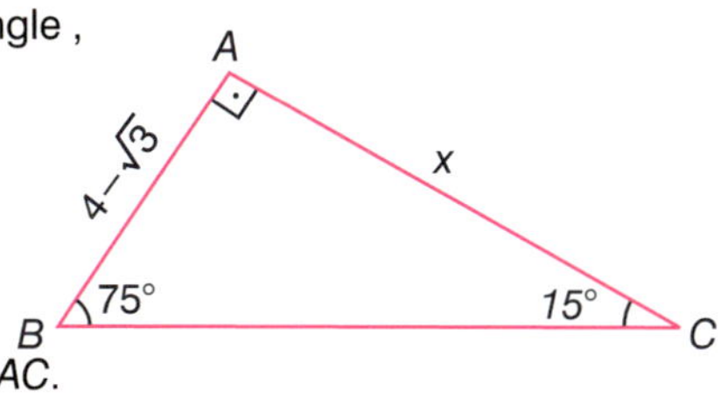

3. ABC is a right triangle,

$m\angle A = 67.5°$,

$m\angle C = 22.5°$, and

$AB = \sqrt{2}-1$.

Find the length of AC.

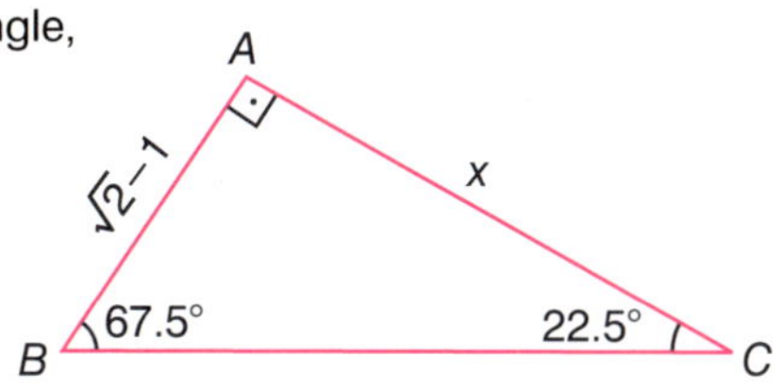

4. In the figure,

$m\angle ABC = m\angle BAD = 60°$,

$m\angle ADC = 150°$, and

$AB - AD = 6$ cm.

What is the length of DC?

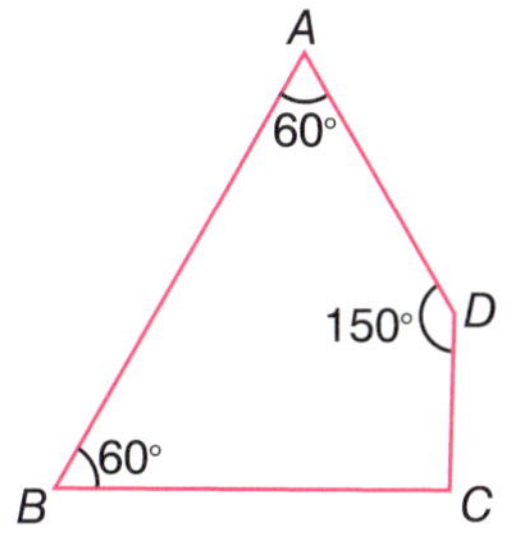

5.

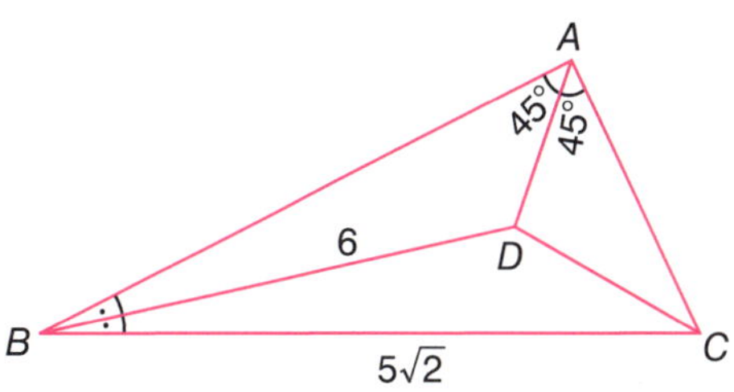

In the figure, BD and AD are angle bisectors of $\angle B$ and $\angle A$, respectively.

$m\angle BAD = m\angle CAD = 45°$, $BD = 6$, and $BC = 5\sqrt{2}$.

Find DC.

6.

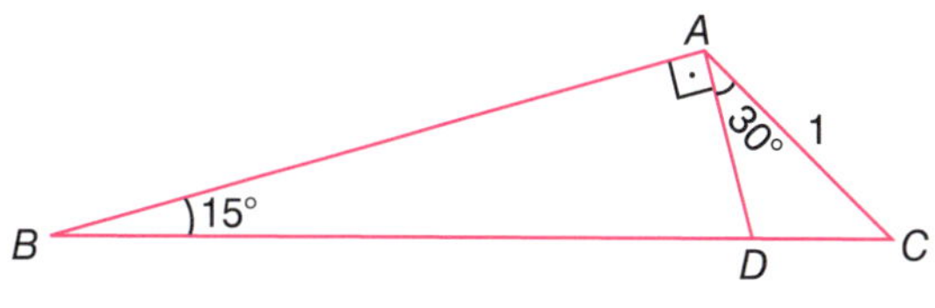

In the figure, $AB \perp AD$,

$m\angle DAC = 30°$, $m\angle ABC = 15°$, and $AC = 1$.

Find BD.

7.

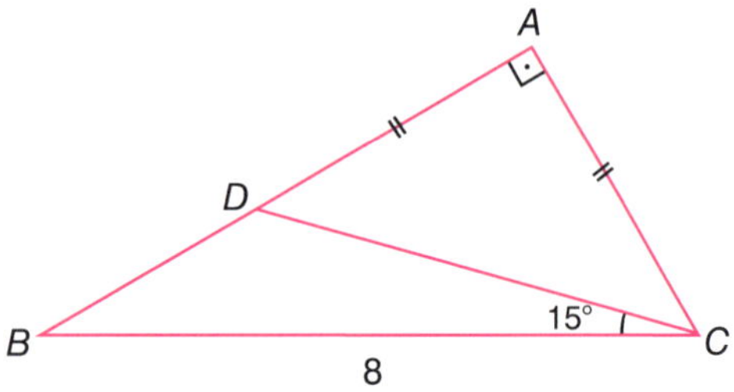

If ABC is a right triangle,

$m\angle A = 90°$, $m\angle BCD = 15°$, $AD = AC$, and $BC = 8$,

then find AB.

8. The area of the ABC is 32 unit^2. Find BH.

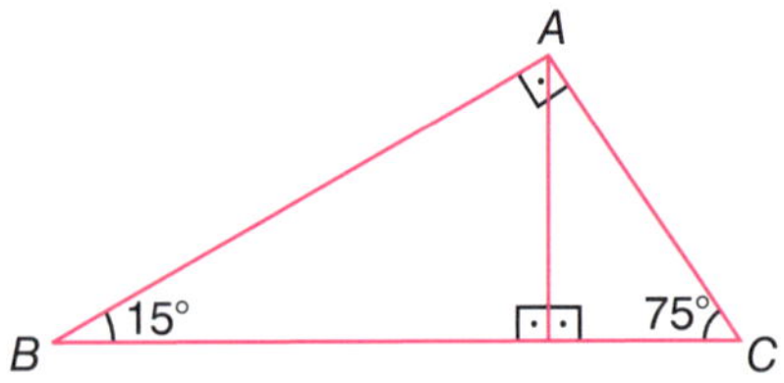

9. ABC is a right triangle, $AD = DC$, $3\ BC = AB$, and $AE = 2EB$. Find the $m\angle AED$.

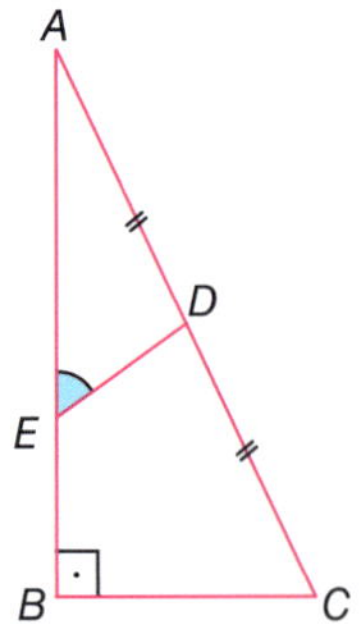

10. ABC is a right triangle with $m\angle A = 90°$ and acute angles 15° and 75°. If the shortest distance from point A to the hypotenuse is h then what is the length of the hypotenuse in terms of h?

11. Suppose $ABCDEF$ is a regular hexagon with side length $2\sqrt{3}$. Find the area of AEC?

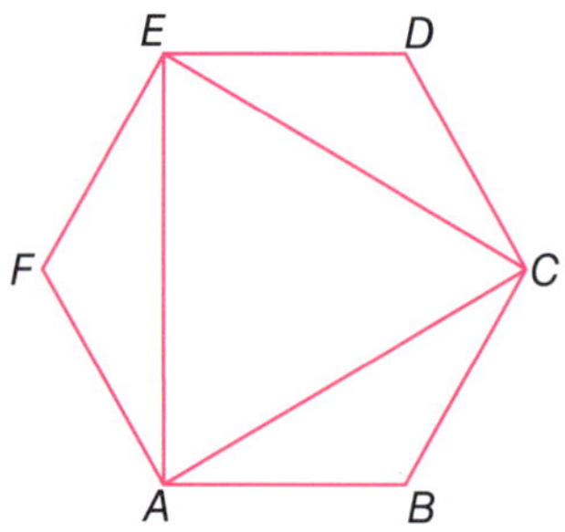

12. In the figure BHC is a right triangle, with $AB = 2HC$, and $m\angle BAC = 75°$.

Find $m\angle ABC$.

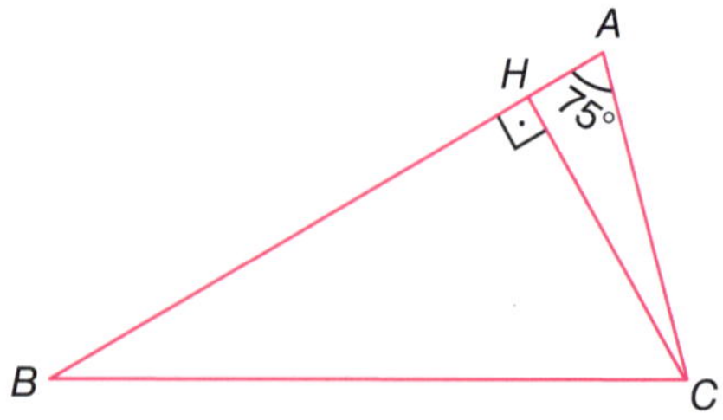

1. (2017 AMC 8 #16)

In the figure below, choose point D on $\overline{BC}$ so that ΔACD and ΔABD have equal perimeters.

What is the area of ΔABD?

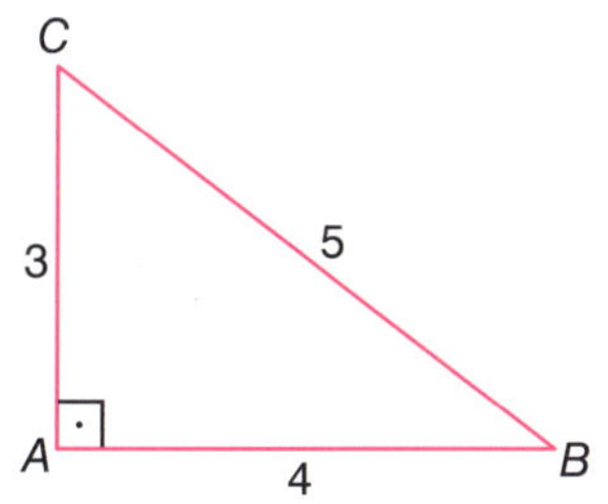

A) $\frac{3}{4}$ B) $\frac{3}{2}$ C) 2 D) $\frac{12}{5}$ E) $\frac{5}{2}$

2. (CWML 2014-15 Meet 12 # 1) What is the perimeter of an equilateral triangle with area $\sqrt{3}$ square units?

A) 2 units B) 3 units C) 6 units D) $2\sqrt{3}$ units E) $6\sqrt{3}$ units

3. (2014 AMC 10A # 9) The two legs of a right triangle, which are altitudes, have lengths $2\sqrt{3}$ and 6.

How long is the third altitude of the triangle?

A) 1 B) 2 C) 3 D) 4 E) 5

4. (CWML 2017-18 Meet 1 # 8)

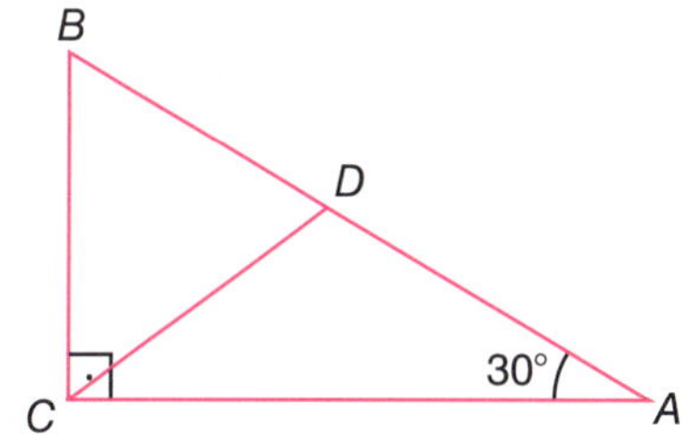

Triangle ABC is 30-60-90 shown.

If side BC is 5 units long and point D is the midpoint of AB find the exact length of segment CD.

5. (2005 AMC 8 #19)

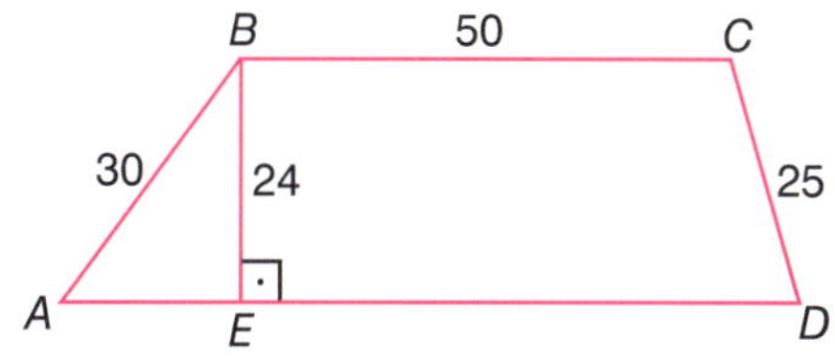

What is the perimeter of trapezoid $ABCD$?

A) 180 B) 188 C) 196 D) 200 E) 204

6. (CWML 2016-17 Meet 3 # 12)

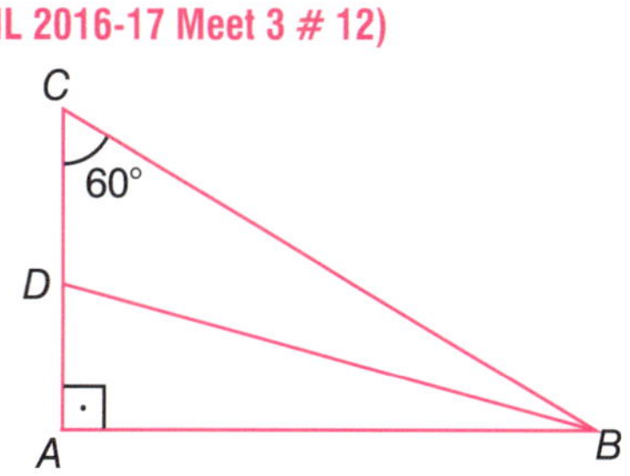

Triangle ABC is 30°-60°-90°, BD is angle bisector of $\angle ABC$, and side AC 4 units long as shown.

What is the length of line segment DC?

Give the exact, fully simplified answer.

7. **(2018 AMC 10A #16)** Right triangle ABC has leg lengths $AB = 20$ and $BC = 21$.

Including $\overline{AB}$ and $\overline{BC}$, how many line segments with integer length can be drawn from vertex B to a point on hypotenuse $\overline{AC}$?

A) 5 B) 8 C) 12 D) 13 E) 15

8. **(CWML 2017-18 Meet 3 # 12)** A triangle has side lengths 15, $5\sqrt{3}$ and $5\sqrt{3}$.

What is the measure of the largest angle in the triangle?

9. **(2012 AMC 10B # 12)** Point B is due east of point A. Point C is due north of point B. The distance between points A and C is $10\sqrt{2}$, and $m\angle ABC = 45°$. Point D is 20 meters due north of point C.

The distance AD is between which two integers?

A) 30 and 31 B) 31 and 32 C) 32 and 33 D) 33 and 34 E) 34 and 35

10. **(2001 AMC 12 #24)**

In ΔABC, $m\angle ABC = 45°$. Point D is on $\overline{BC}$ so that $2\,BD = CD$ and $m\angle DAB = 15°$.

Find $m\angle ACB$.

A) 54° B) 60° C) 72° D) 75° E) 90°

11. **(2006 AMC 10B #15)**

Rhombus $ABCD$ is similar to rhombus $BFDE$. The area of rhombus $ABCD$ is 24 and $\angle BAD = 60°$.

What is the area of rhombus $BFDE$?

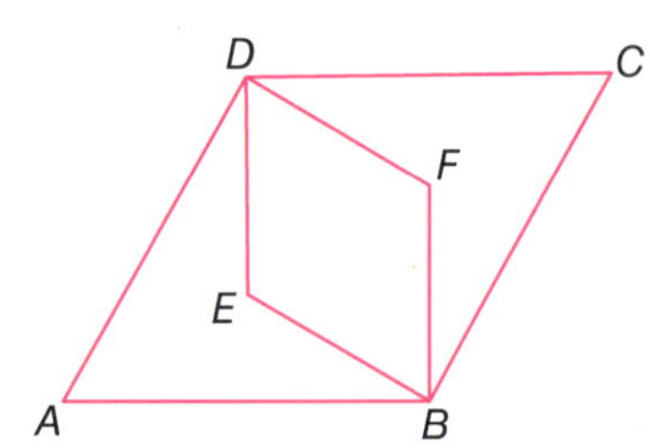

A) 6 B) $4\sqrt{3}$ C) 8 D) 9 E) $6\sqrt{3}$

12. **(MathCON 2019 Final Gr 11 #40)** Let P be a point inside of equilateral triangle ABC such that $m\angle APB = 150°$, $AP = 2\sqrt{3}$, and $BP = 2$.

Find PC.

CHAPTER 4

Notice and wonder

What do you notice and wonder?

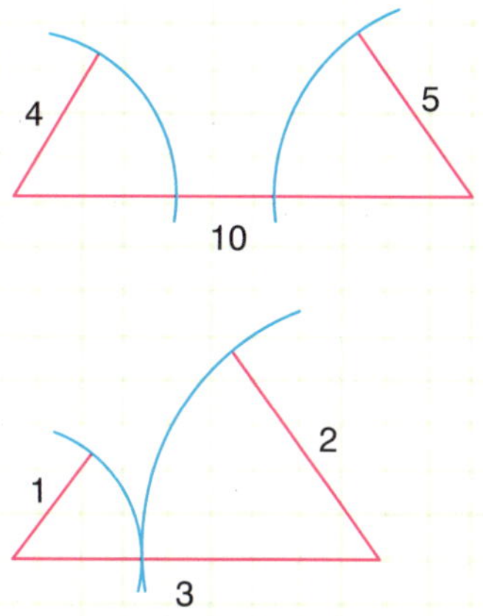

My Notes

Theorem 4.1

Angle -Side Connection Theorem

If two sides of a triangle have unequal lengths then the longer side is opposite the largest angle. For example, if a triangle has sides a, b, and c with a>b>c, then $m\angle A > m\angle B > m\angle C$.

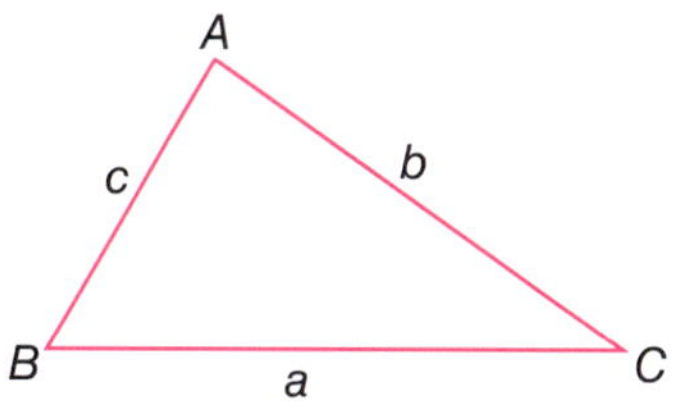

If $AB > AC$, then $m\angle C > m\angle B$.

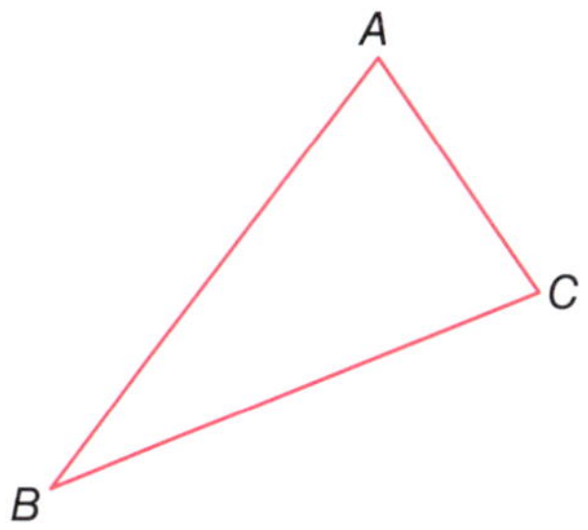

Proof 4.1

Let $AB>AC$. We can create an isosceles triangle ADC.

Let $m\angle B$ be α and $m\angle DCB$ be β with $\beta \neq 0$.
Then $m\angle ADC = m\angle ACD = \alpha+\beta$

Thus $m\angle B = \alpha$ and $m\angle C = \alpha+2\beta$.

Therefore, $m\angle C > m\angle B$ (since $\beta > 0$).

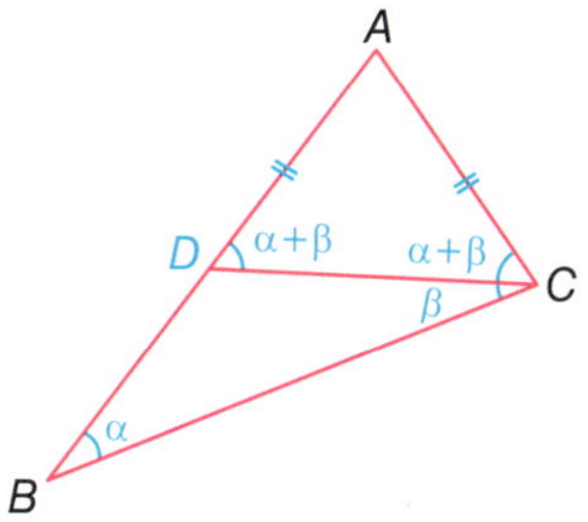

Exercise 4.1

Side lengths of a triangle ABC are given as 7, 5, and 3. By theorem 4.1, we can order the angles as follows: $m\angle A > m\angle B > m\angle C$.

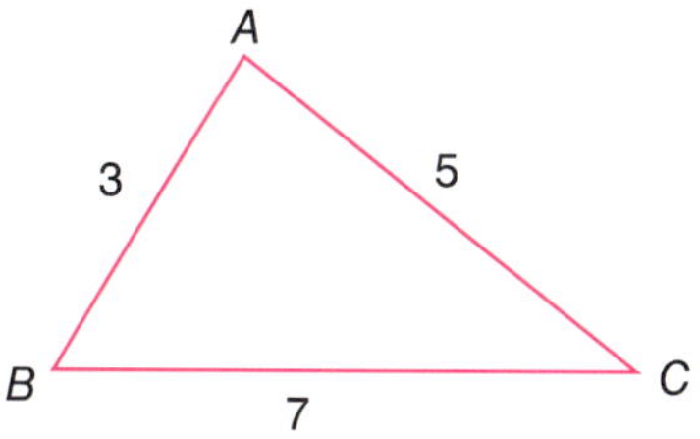

Example 4.1

Given that

$m\angle A = x+20°$,

$m\angle C = 2x-10°$, and

$m\angle B = 20°$,

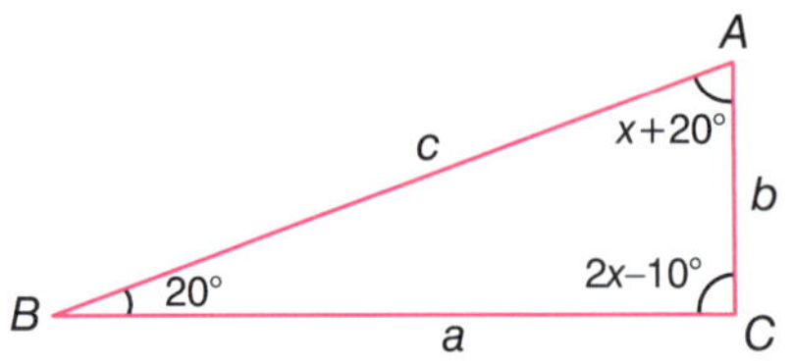

order the side lengths from the smallest to the largest.

Solution 4.1

$m\angle A+m\angle B+m\angle C = 180°$

$x+20° + 2x-10° + 20° = 180°$

$3x+30° = 180°$

$x = 50°$

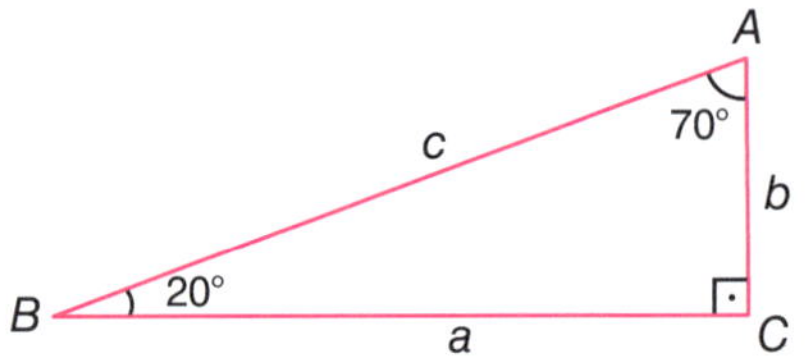

Therefore, $b < a < c$.

Example 4.2

If a, b, and c are the lengths of the sides of ΔABC, and h_a, h_b, and h_c are lengths of the altitudes, then prove that $a+b+c > h_a+h_b+h_c$.

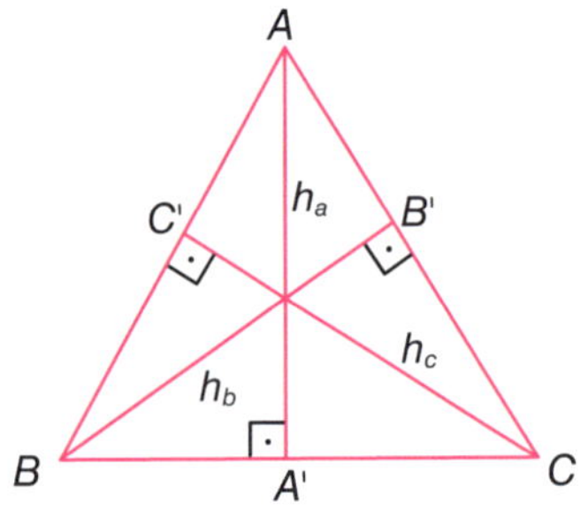

Solution 4.2

From $\Delta ABB'$ $c > h_b$ (1)

From $\Delta ACA'$ $b > h_a$ (2)

From $\Delta BCC'$ $a > h_c$ (3)

By adding the inequalities (1), (2), and (3) we get $a+b+c > h_a+h_b+h_c$.

Try This

Name the largest angle and the smallest angle of the triangles.

Largest:__

Smallest:__

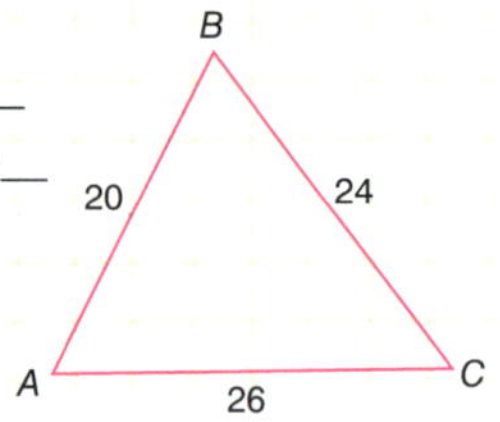

Largest:__

Smallest:__

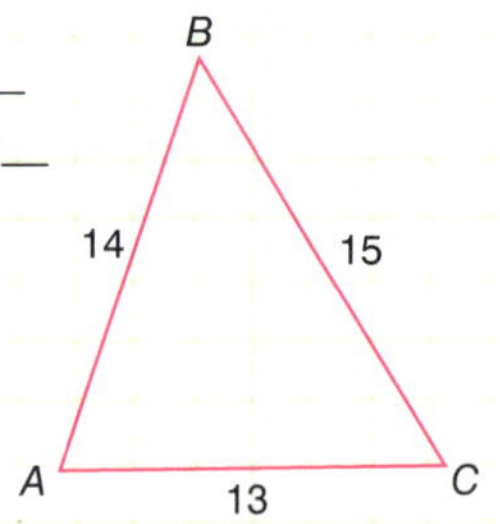

My Notes

Corollary

Let a, b, and c be positive numbers. Then there exists a triangle with sides lengths a, b, and c if and only if of the following hold.

(1) $a<b+c$, $b<a+c$, and $c<a+b$
(2) $|b-c| < a < b+c$
(3) $|a-c| < b < a+c$
(4) $|a-b| < c < a+b$

Corollary: An immediate consequence of a result already proved. Corollaries usually state more complicated theorems in a language simpler to use and apply.

My Notes

Example 4.3

Find the largest segment of the following figures.

Figure 1

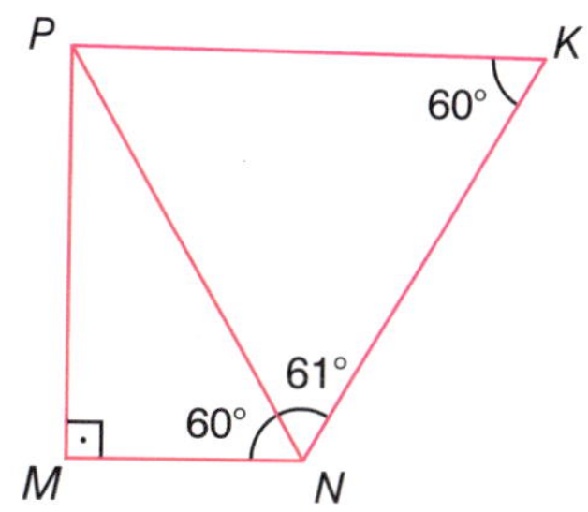

Figure 2

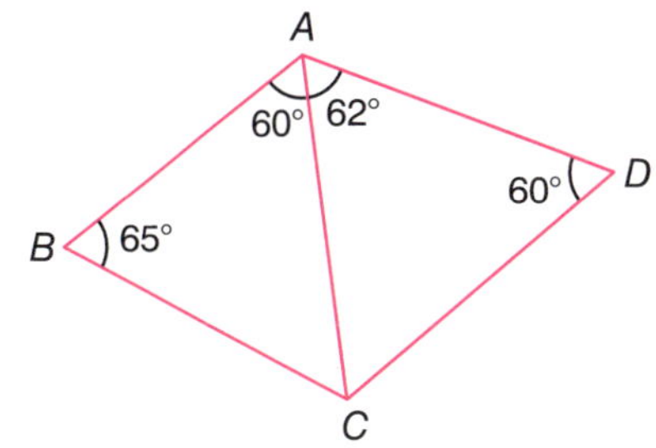

Solution 4.4

Figure 1

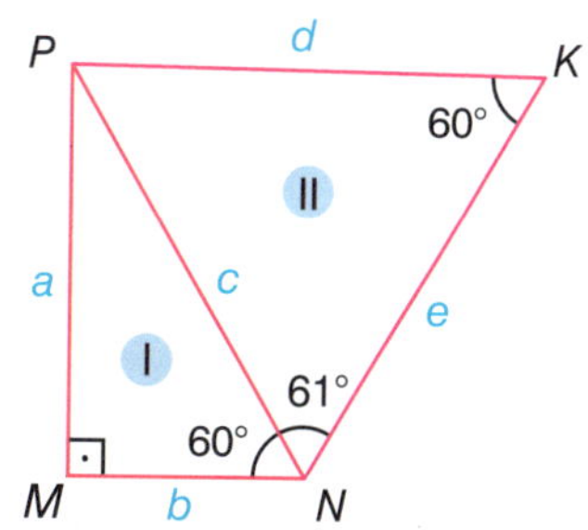

Figure 2

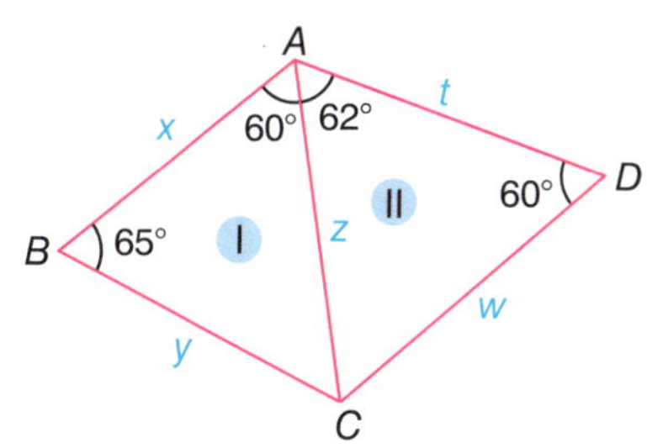

Figure 1 I) $c > a > b$

II) $d > c > e$, so the largest side is $d = PK$.

Figure 2 I) $z > y > x$

II) $w > z > t$, so the largest side is $w = DC$.

Theorem 4.2

The Triangle Inequality Theorem

Let a, b, c be any real numbers. Then the following are equivalent.

(1) $a<b+c$, $b<a+c$, and $c<a+b$
(2) $|b-c| < a < b+c$
(3) $|a-c| < b < a+c$
(4) $|a-b| < c < a+b$

Proof 4.2

Suppose (1) holds, $a<b+c$, $b<a+c$, and $c<a+b$ hold. Using the second and third inequality we get $b-c<a$ and $-a<b-c$, respectively. Combining both inequalities we get $-a<b-c<a$. Hence $|b-c|<a$. Combining this with $a<b+c$ we get $|b-c|<a<b+c$ which is (2). Parts (3) and (4) follow similarly. This shows (1) $\Rightarrow$ (2), (1) $\Rightarrow$ (3), and (1) $\Rightarrow$ (4). Now assume (2) holds. Since $|b-a|<a$, $-a<b-c<a$ which implies $c< a+b$ and $b<a+c$. Hence (1) holds. Similarly (3) $\Rightarrow$ (1) and (4) $\Rightarrow$ (1).

Exercise 4.2

I. 6, 7, 8 $\Rightarrow |6-7| < 8 < (6+7)$ or $|6-8| < 7 < (6+8)$ therefore possible,

II. 0.7, 0.4, 1 $\Rightarrow |0.7-0.4| < 1 < (0.7+0.4)$ or $|1-0.4| < 0.7 < (1+0.4)$ therefore possible.

III. $\frac{1}{2}, \frac{1}{3}, 1 \Rightarrow \left|\frac{1}{2}+\frac{1}{3}\right| < 1 < \left|\frac{1}{2}-\frac{1}{3}\right|$ this inequality does not hold, therefore impossible.

Example 4.4

In the figure,

$AB = BD = 5$,

$AC = 7$, $DC = 10$, and

$BC = x$.

Find all possible integer values of x.

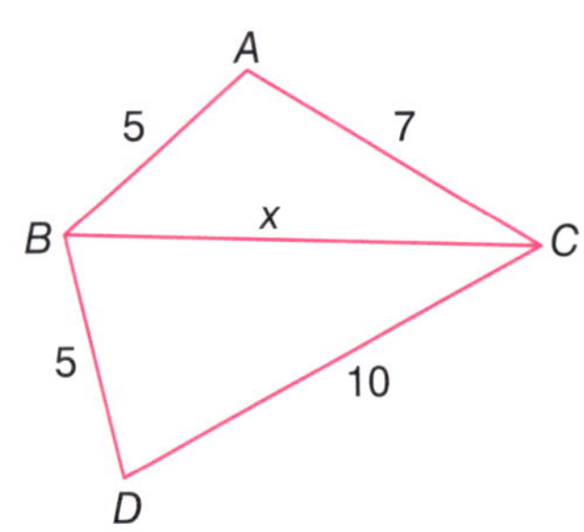

Solution 4.4

$5 < x < 15$ (From the lower triangle)

$2 < x < 12$ (From the upper triangle)

The intersection of the above intervals is $5 < x < 12$.

Therefore, all possible values of x are $\{6, 7, 8, 9, 10, 11\}$.

Example 4.5

In the figure,

$AB = 4$, $AC = 6$, $BD = 5$,

$DC = 3$, and $BC = x$.

Find the maximum integer value of x.

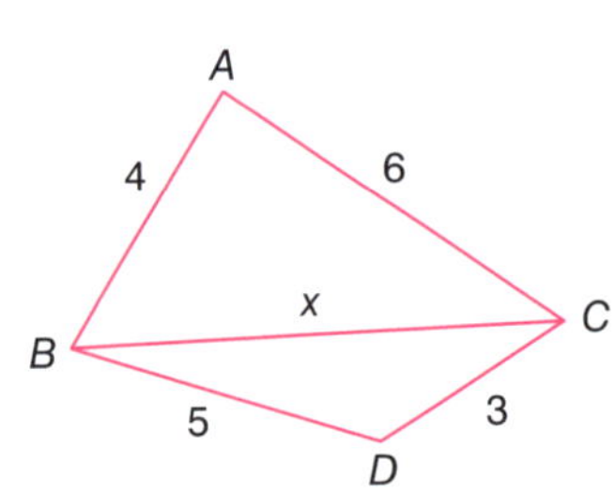

Solution 4.5

$2 < x < 10$ (From the upper triangle)

$2 < x < 8$ (From the lower triangle)

The intersection of the these intervals is $2 < x < 8$.

Therefore, all possible integer values are $\{3, 4, 5, 6, 7\}$ and the maximum value of x is 7.

Try This

In triangle ABC,

$AB = 6$ and $BC = 8$.

What is the sum of all possible integer values of AC?

My Notes

Prove This

Boomerang Inequality
If point D is inside ΔABC.

Prove that $AB+BC > AD+DC$.

My Notes

Theorem 4.3

In any triangle ABC with inside point P,

$$\frac{a+b+c}{2} < x+y+z < a+b+c.$$

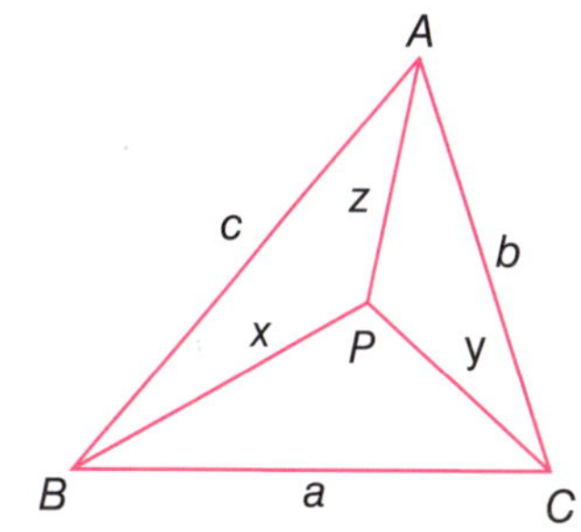

Proof 4.3

Consider triangle ABP. From the triangle inequality $c < x+z$.
Similarly from triangle APC and BPC, we get $b < y+z$ and $a < x+y$, respectively.
Adding the 3 inequalities $a+b+c < 2(x+y+z)$. Therefore, $\frac{a+b+c}{2} < x+y+z$.

$$x+z < a+b,\ y+z < c+a, \text{ and } y+x < c+b.$$ (Boomerang Inequality)

Then, $2(x+y+z) < 2(a+b+c) \Rightarrow (x+y+z) < (a+b+c)$.

Therefore, $\frac{a+b+c}{2} < x+y+z < a+b+c$.

Example 4.6

In triangle ABC, point P is inside ΔABC.

Given that,

$AB = 10$,

$AC = 8$, and

$BC = 9$,

find all possible integer values of $PB+PC$.

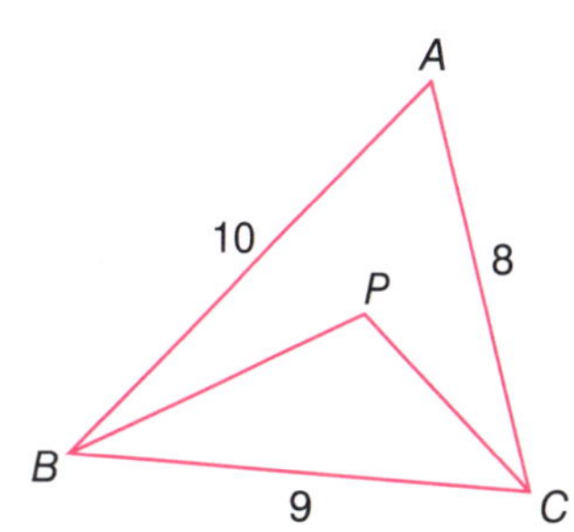

Solution 4.6

From ΔPBC, $9 < PB+PC$. (1)

From ΔABC, $PB+PC < 10+8$. (2)

By combining (1) and (2) we obtain $9 < PB+PC < 18$

Thus, all possible integers are

$$10, 11, 12, 13, 14, 15, 16, 17.$$

Example 4.7

In a triangle with integer side lengths, one side is 4 times as long as the second side and the length of the third side is 21.

What is the smallest possible perimeter of this triangle?

Solution 4.7

Let sides of a triangle ABC be s, $4s$, and 21.

By the triangle inequality theorem,

$$4s-s< 21 <4s+s$$

$$3s< 21 <5s \Rightarrow s< 7 < \frac{5s}{3}$$

Possible values of s are 5 and 6.

So, the smallest possible perimeter is $5+20+21 = 46$.

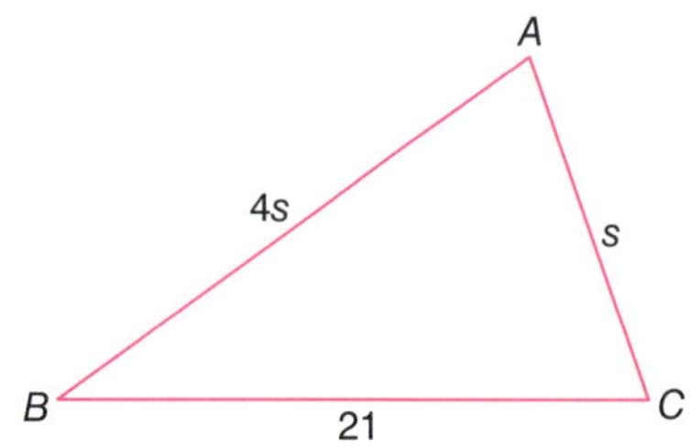

Example 4.8

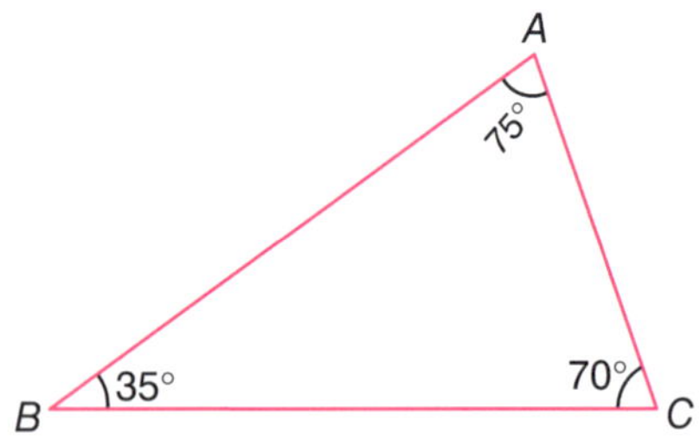

Figure 1

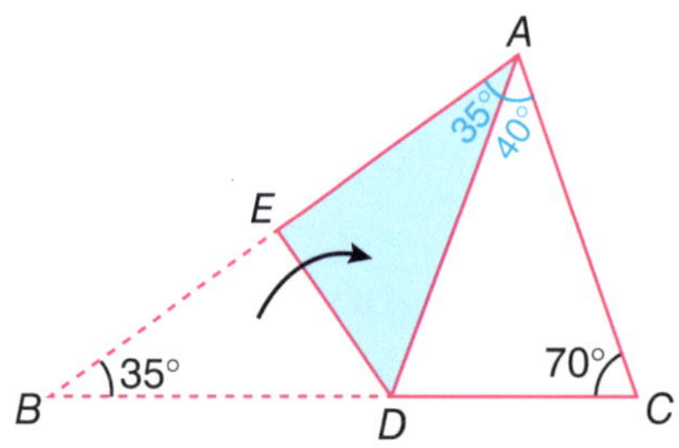

Figure 2

Two-sided triangle shape in figure [front side is not shaded and the back side is shaded] is folded by connecting point B to point A, as shown in figure 2.

What is the order of side AC, AE, and BD from the smallest to the largest?

Solution 4.8

$BD = AD = AC$

$BD = AD>AE$

So, $AE<BD = AC$.

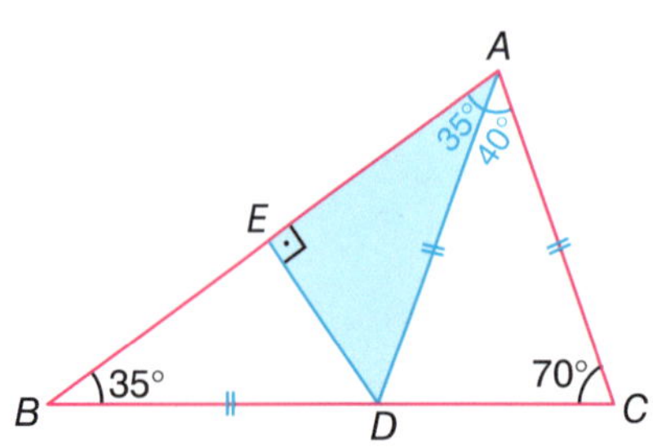

Prove This

Let P is a perimeter of a triangle and S is a sum of all three medians of the triangle prove that

(a) $S < P$

(b) $S > \frac{3}{4}P$.

My Notes

Try This

What is the integer length of AC?

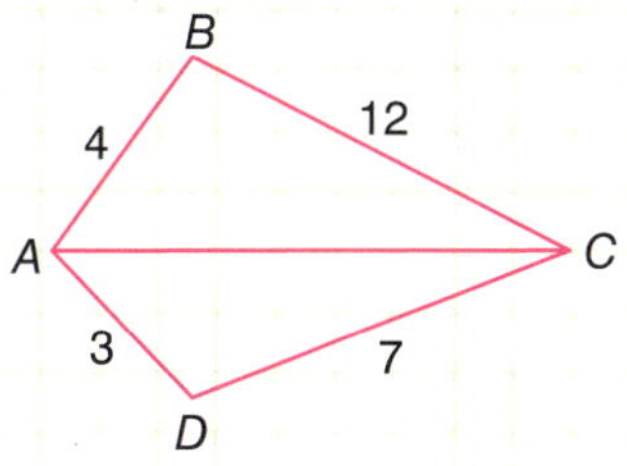

My Notes

Remark 4.1

In any triangle ABC with sides of a, b, and c and angles of $\angle A$, $\angle B$, and $\angle C$,

1. if $m\angle A = 90°$ then $\sqrt{b^2+c^2} = a < b+c$
2. if $m\angle A < 90°$ then $|b-c| < a < \sqrt{b^2+c^2}$
3. if $m\angle A > 90°$ then $\sqrt{b^2+c^2} < a < b+c$

Example 4.9

In the figure,

$AB = 6$,

$AC = 8$, and

$m\angle A > 90°$.

What are the possible integer values for BC?

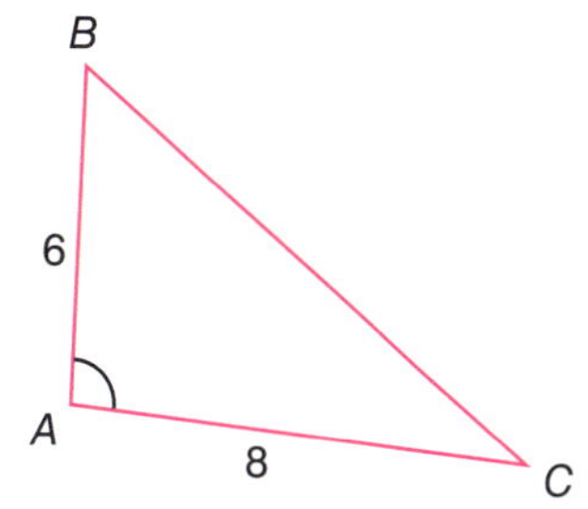

Solution 4.9

If $m\angle B = 90°$, then $BC = 10$ cm,

If $m\angle B > 90°$, then $BC > 10$ cm,

$2 < BC < 14$ (From the triangle inequality theorem)

By combining two intervals, we get $10 < BC < 14$ so, the possible values are 11, 12, and 13.

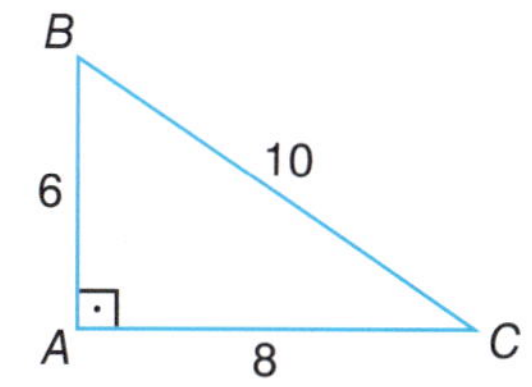

Example 4.10

ABC is a triangle with $m\angle ACB < 90°$.

What is the sum of all possible integer values of AB?

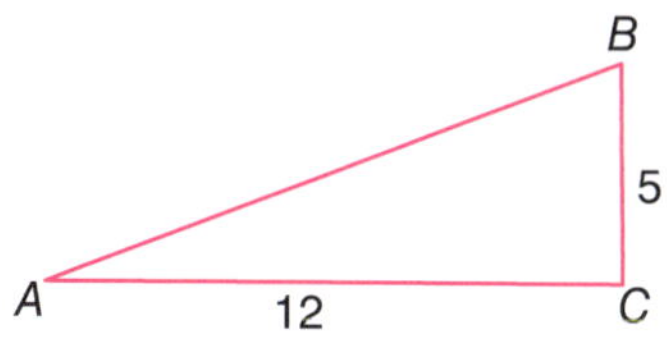

Solution 4.10

Since $m\angle ACB < 90°$, then $c < \sqrt{5^2+12^2} \Rightarrow c < 13$.

From the triangle inequality,

$12-5 < c < 12+5 \Rightarrow 7 < c < 17$.

Therefore, the sum is $8+9+10+11+12 = 50$.

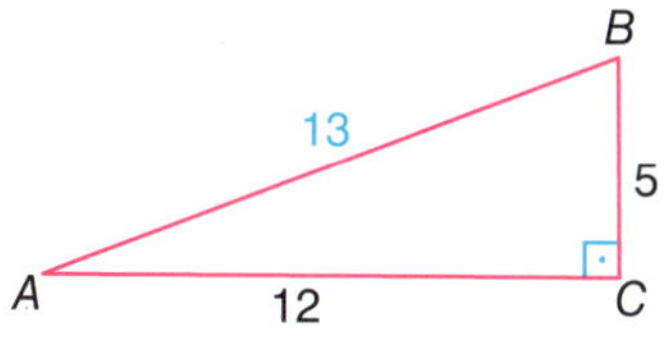

Example 4.11

In triangle ABC, $DE \perp CB$.

Given that

$CE = EB$,

$AB = 3$, and

$DC = 4$,

find all possible integer values of AD.

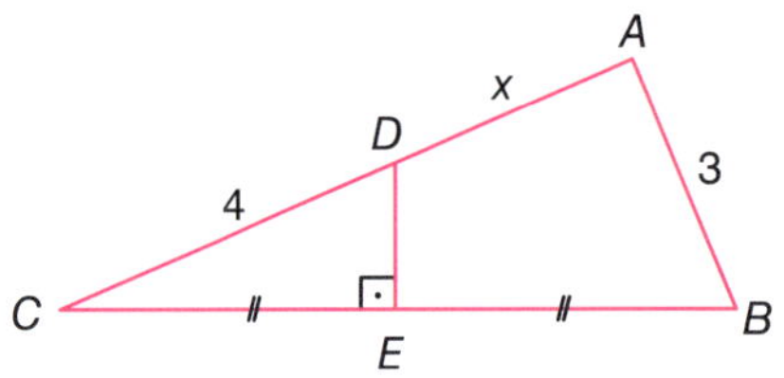

Solution 4.11

By connecting points B and D, we see that

$CD = DB = 4$.

Applying the triangle inequality to ΔABD we get,

$1 < x < 7$.

Therefore, all possible integer value of AD are 2,4,5, and 6.

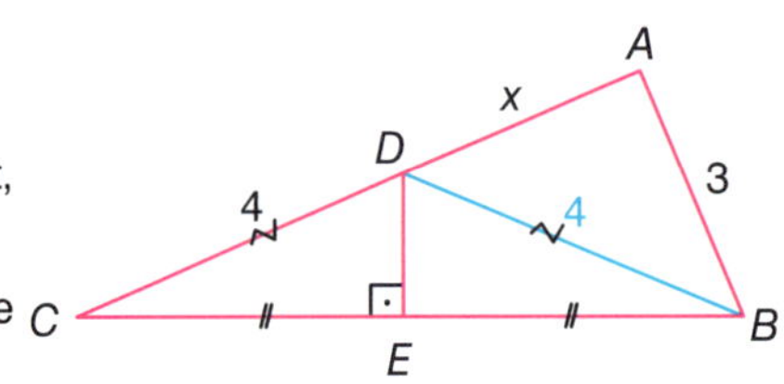

Example 4.12

In triangle ABC,

$\angle BAD = \angle ACB$,

$AB = 6$, $AD = 4$, and

$BD = x$.

Find all possible integer values of x.

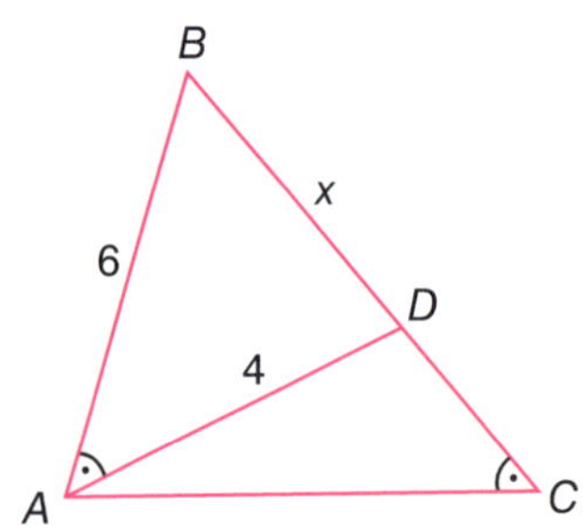

Solution 4.12

Let $\beta = m\angle BAD = m\angle ACB$ and let $\alpha = m\angle DAC$.

So, $m\angle ADB = \alpha + \beta = m\angle BAC$.

Applying the triangle inequality to ΔABD, we get,

$6 - 4 < x < 6 + 4 \Rightarrow 2 < x < 10$.

From angle-side connections ($\beta < \alpha + \beta$) so, $x < 6$.

Therefore, all possible integer values are 3,4, and 5.

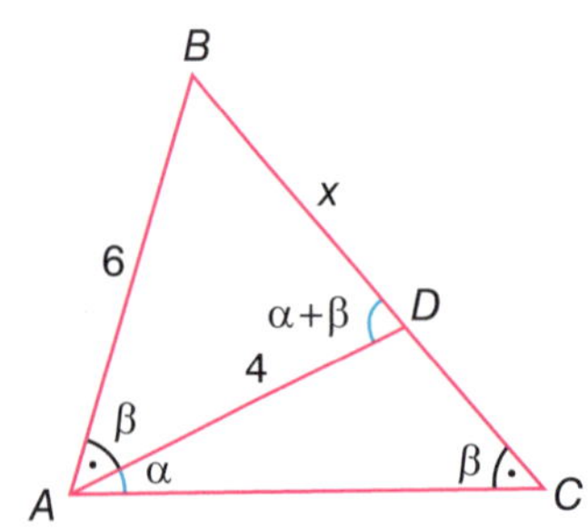

Catriona Agg @Cshearer41

Geometry puzzle

Three congruent triangles. The blue line bisects the top angle.

What angle does this line make with the horizontal?

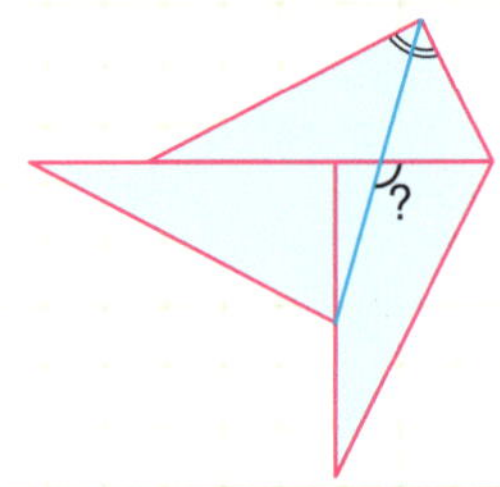

My Notes

1. Find the sum of all possible integer values of x?

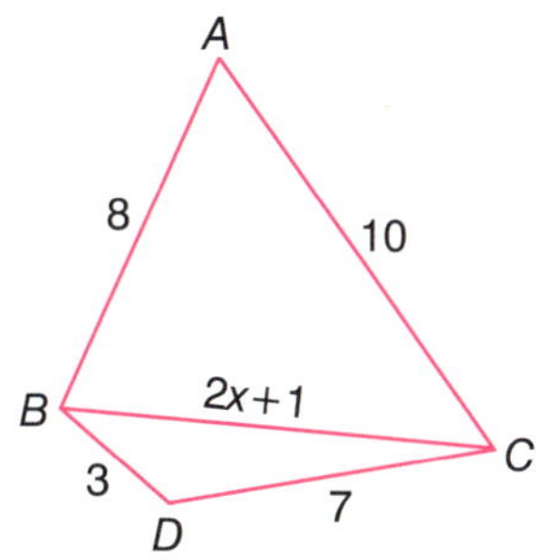

2. In the figure, a, b, and c are integers.

Find the smallest possible values of $a+b+c$?

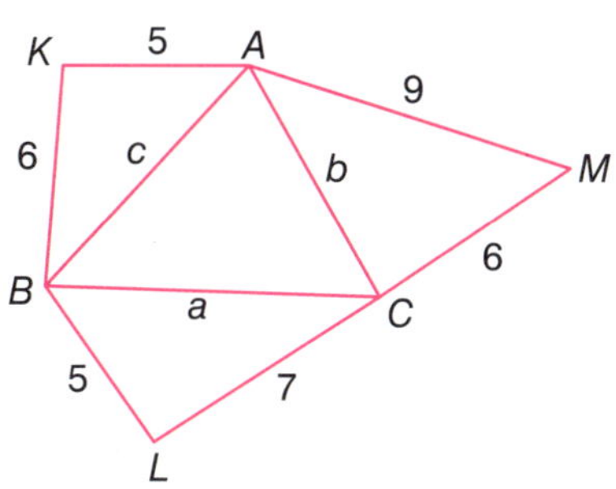

3. $m\angle BAC > 45°$,

$BD = BC$,

$AD = 8$ cm, and $AC = 16\sqrt{2}$.

What is the smallest integer value of the perimeter of $\triangle ABC$?

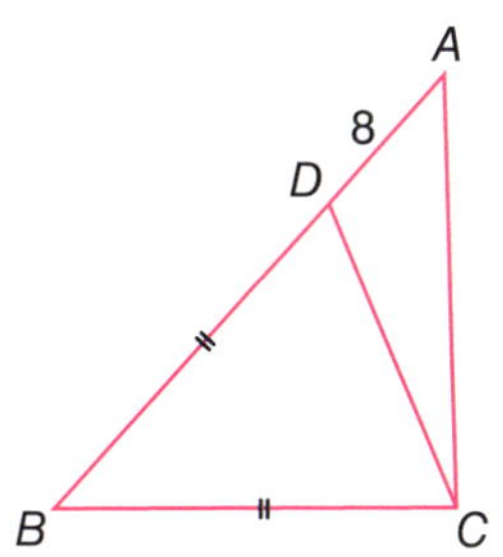

4. In the figure,

$AD = 5$ cm,

$AB = 12$ cm,

$BC = 15$ cm, and

$DC = 8$ cm

If $m\angle A > 90°$ and $m\angle C < 90°$,

find the sum of all possible integer values of BD?

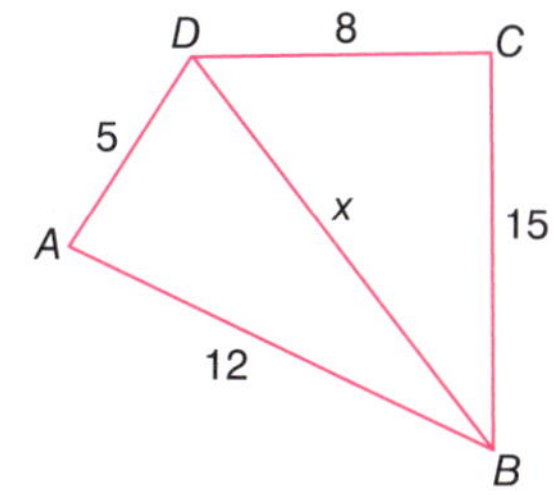

5. ABC is a triangle with $m\angle ACB < 90°$.

What is the largest integer values of AB?

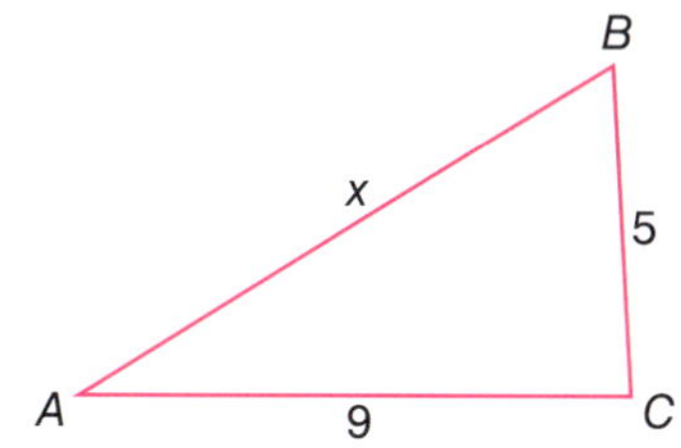

6. ABC is a triangle with $AD = DC$,

$AB = 3$, and

$BC = 9$.

What is the largest integer value of BD?

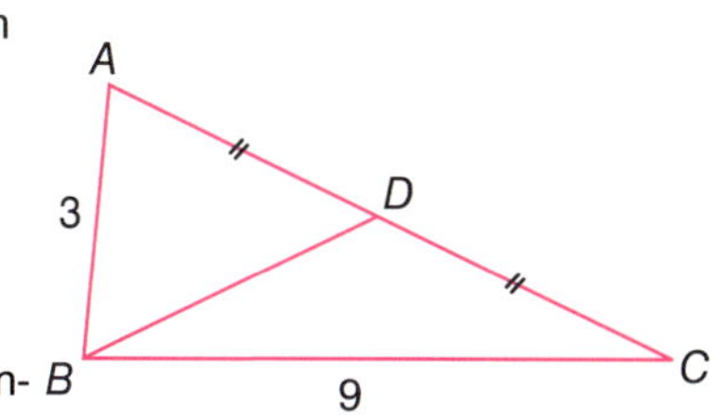

7. How many distinct isosceles triangles can be created with integer sides and perimeter of 200 units?

10. In the figure,

$m\angle C < 60°$,

$AC = 8$, and

$BC = 10$.

What is the greatest possible integer value of AB ?

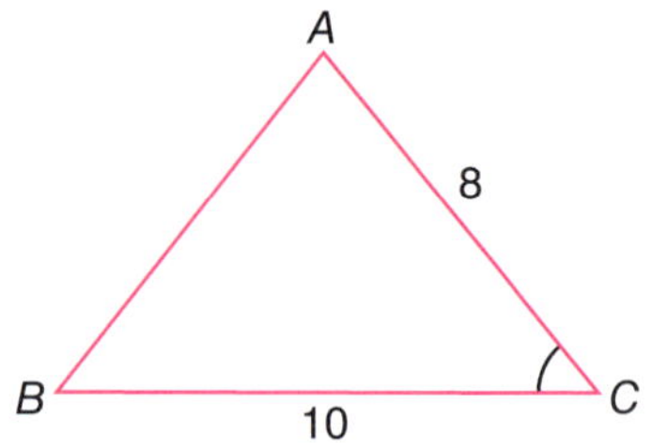

8. If $AC+BC = 10$, then what is the smallest possible integer value of AC.

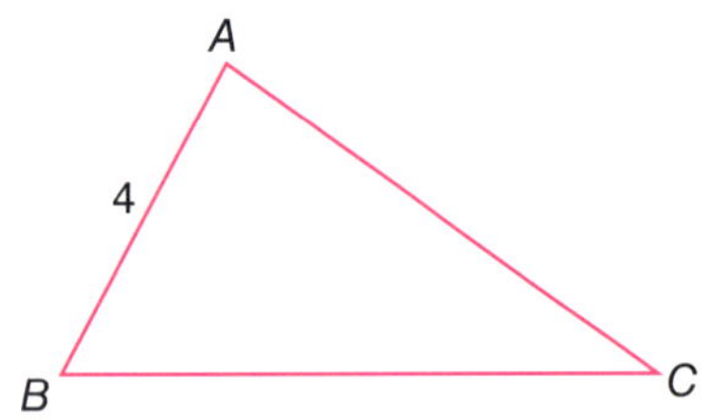

11. The sides of a triangle have lengths 8,14, and n where n is an integer. For how many n values is the triangle obtuse?

9. The side length of an equilateral triangle are $a+2b$, $3a-b$, and $5b-a$.

Which of the following **could not** be the values of a and b.

A) (12,8) B) $\left(\frac{9}{2},3\right)$ C) (10,6) D) (3,2) E) $\left(\frac{3}{2},1\right)$

12. The side lengths of a triangle are n, $3n$, and 32.
What is the sum all of the possible integer values of n?

1. According the figure, how many possible integer value that the perimeter of the triangle *ABC* have?

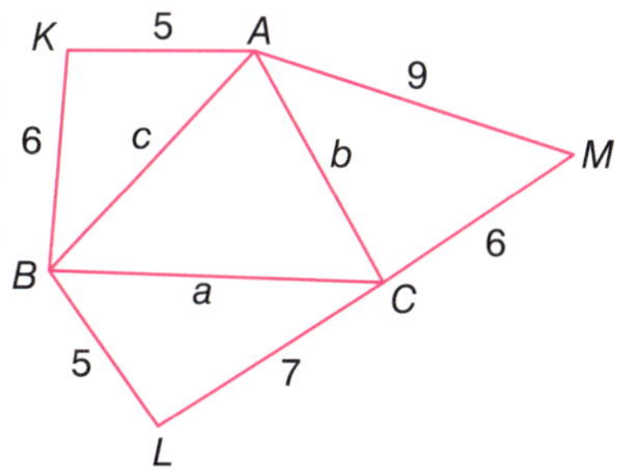

2. Find the greatest possible integer value of *m* in the figure, then find the smallest possible integer value of *n*.

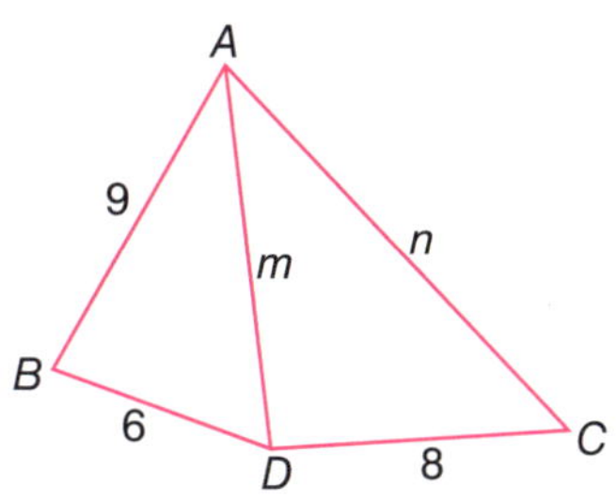

3. In a triangle *ABC*,

$m\angle A > 90°$,

$AB = 6$, and $AC = 8$.

Find the sum all the possible integer lengths of *BC*.

4. You have six sticks of length 1 cm, 2 cm, 3 cm, 11 cm, 12 cm, and 13 cm. You have to choose there of these sticks to from a triangle.

How many different choices of three sticks are there?

5. In a triangle *ABC*,

$AB = 8$ cm,

$BC = x$, and $AC = y$.

If $m\angle A > 90°$ and x and y are integers, then find the smallest possible value of $x+y$.

6. In the figure,

$AD = 6$ cm,

$AB = DC = 8$ cm, and

$BC = 15$ cm.

If $m\angle A > 90°$ and $m\angle C < 90°$, then find all possible integer values of *BD*.

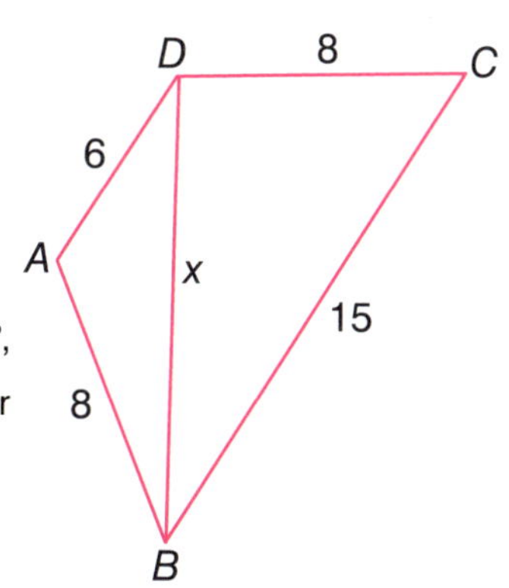

7. Suppose a, b, and c are integer sides of ΔABC with $b = c$. If $(a+b+c)\cdot(a+b-c) = 15$, then find the value of a.

8. On side of a triangle has length 6.3 and an other side 1.7. How many integer values are possible for the length of the third side?

9. For how many integers values of x is $2x$, $3x$, and $8-x$ form sides of an obtuse triangle?

10. In a triangle ABC,

$AB = 6$ cm

$BC = x$, and $AC = y$.

If $m\angle A > 90°$, where x and y are integer values.

What is the smallest possible value of $x+y$?

11. In the figure,

$AB = 2x$,

$AC = 13$, and

$BC = x+1$.

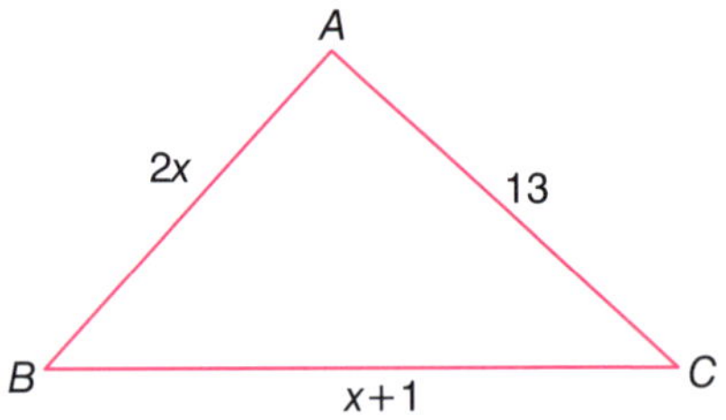

How many x integer values are there?

12. The sides of a triangle have lengths 7, 14, and c, where c is an integer. For how many values of c is the triangle acute?

1. **(CWML 2005-6 Meet 3 #3)** Three line segments are given with lengths x, $11-x$ and $x-4$.

Which one of the following indicates the set of all numbers x such that the three segments could be the sides of triangle?

A) $4<x<7$ B) $0<x<11$ C) $4<x<11$ D) $5<x<7$ E) $5<x<11$

2. **(CWML 2009-10 Meet 2 #2)** A triangle has two sides of length 15 feet and 21 feet.

If the length (in feet) of the third side is a whole number, what is the largest possible perimeter of the triangle?

3. **(CWML 2015-16 Meet 2 #10)**

In triangle ABC,

$AB = AC = 10$ cm

$BC = 12$ cm, and,

point P lies on side AB.

Find the minimum value for $PA+PC+PB$. *Given the answer in centimeters*

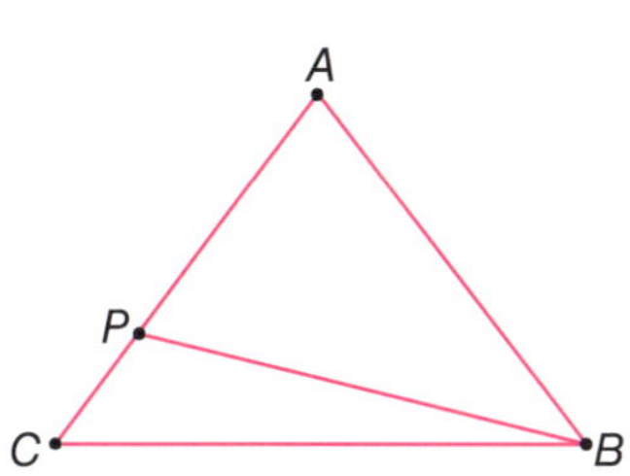

4. **(2006 AMC 10B #10)** In a triangle with integer side lengths, one side is three times as long as a second side, and the length of the third side is 15.

What is the greatest possible perimeter of the triangle?

A) 43 B) 44 C) 45 D) 46 E) 47

5. **(2009 AMC 12A #10)**

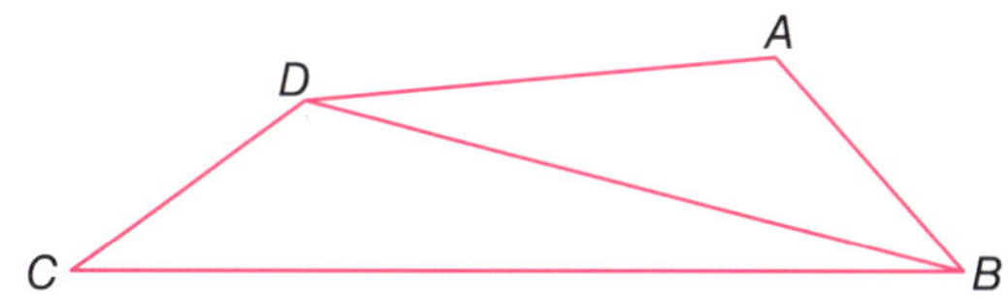

In quadrilateral $ABCD$, $AB = 5$, $BC = 17$, $CD = 15$, $DA = 9$, and BD is an integer.

What is BD?

A) 11 B) 12 C) 13 D) 14 E) 15

6. **(MathCON 2019 Finsl GR 8 # 16)** How many distinct isosceles triangles can be created with integer sides and perimeter of 200 units?

A) 49 B) 64 C) 81 D) 96 E) None of the preceding

7. (AlphaStar PiMC Fermat Ind. # 21) How many non-congruent triangles with integer side lengths have perimeter 15?

8. (MathCON 2019 Finsl GR 8 # 32) The sides of a triangle have lengths 7, 14, and c, where c is an integer.

For how many values of c is the triangle acute?

A) 2 B) 3 C)4 D) 5 E) 8

9. (AlphaStar PiMC Fermat Ind. # 19) P, Q, R are three similar triangles with areas proportional to 1,4, and 9, in some order.

If P has perimeter 12, what is the sum of all possible perimeters of Q?

A) 2 B) 3 C) 4 D) 5 E) 6

10. (2009 AMC 12B #13) Triangle ABC has $AB = 13$ and $AC = 15$, and the altitude to $\overline{BC}$ has length 12.

What is the sum of the two possible values of BC?

A) 15 B) 16 C) 17 D) 18 E) 19

11. (2010 AMC 12A #25) Two quadrilaterals are considered the same if one can be obtained from the other by a rotation and a translation.

How many different convex cyclic quadrilaterals are there with integer sides and perimeter equal to 32?

A) 560 B) 564 C) 568 D) 1498 E) 2255

12. (2006 AIME II #2) The lengths of the sides of a triangle with positive area are $\log_{10}12$, $\log_{10}75$, and $\log_{10}n$, where n is a positive integer

Find the number of possible values for n.

CHAPTER 5

Note

In similar triangles, the ratio of corresponding sides, medians, heights, and bisectors are equal to the similarity ratio.

My Notes

Theorem 5.1

Two triangles are called similar if their corresponding angles are equal and the ratio of corresponding sides is the same. This ratios is called the similarity ratio. If two triangles are similar to each other based on their two same angles, then this similarity is called Angle Angle Angle (A.A.A) similarity.

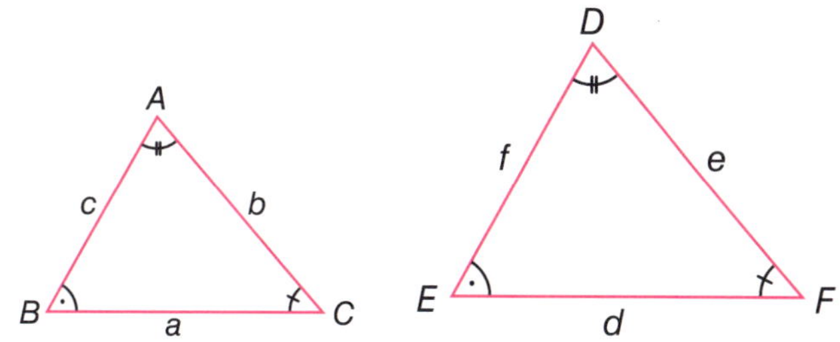

$\Delta ABC \sim \Delta DEF \quad \frac{a}{d} = \frac{b}{e} = \frac{c}{f} = k$ [*k* denotes the similarity ratio]

Example 5.1

In the figure, given that $m\angle BAC = m\angle DEC$,

$AB = 4$, $AC = 6$,

$DE = 8$, and $DC = 5$,

find $BC + CE$.

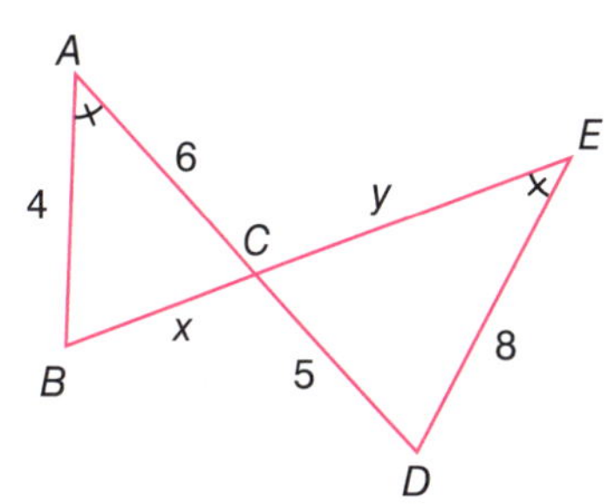

Solution 5.1

By the vertical angles,

$m\angle ACB = m\angle ECD$ and

$m\angle B = m\angle D$. Therefore, two triangles are similar(AAA).

We write the ratio

$$\frac{6}{y} = \frac{4}{8} = \frac{x}{5}$$

$y = 12$ and $x = 2.5$.

So, $y + x = 14.5$.

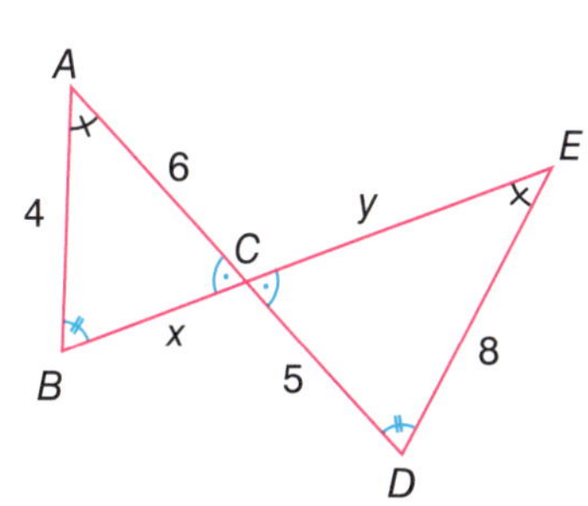

Example 5.2

In the figure, $m\angle EAB = m\angle ABC = 90°$ and AC and BE intersect at D.
Given that
$AB = 8$,
$BC = 4$, and
$AE = 6$,
find ED.

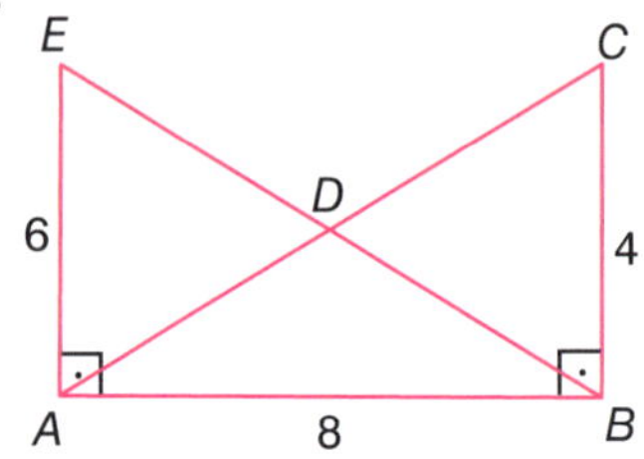

Solution 5.2

Notice that $AE // BC$ then $\Delta DBC \sim \Delta DEA$ [A.A.A./BOW-TIE].

$\frac{ED}{DB} = \frac{6}{4} = \frac{3}{2} \Rightarrow 2k+3k = 10$
($EB = 10$ by the Pythagorean Theorem)
$\Rightarrow k = 2$.

So, $ED = 3k = 6$.

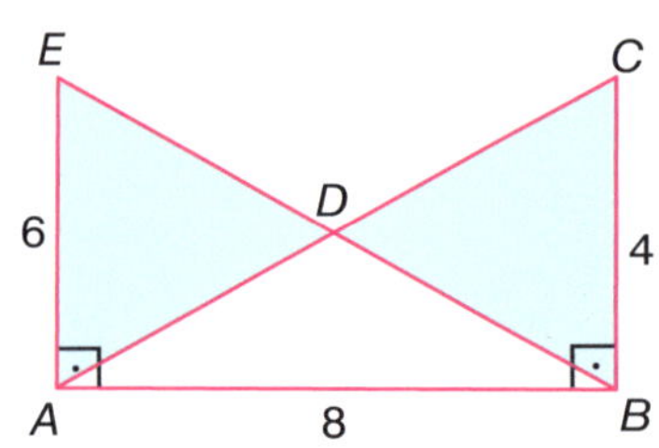

Example 5.3

In the figure,

$AB = 6$,

$AC = 8$,

$m\angle BCA = m\angle BCD = 90°$, and

BD is parallel to AC.

What is the length of BD?

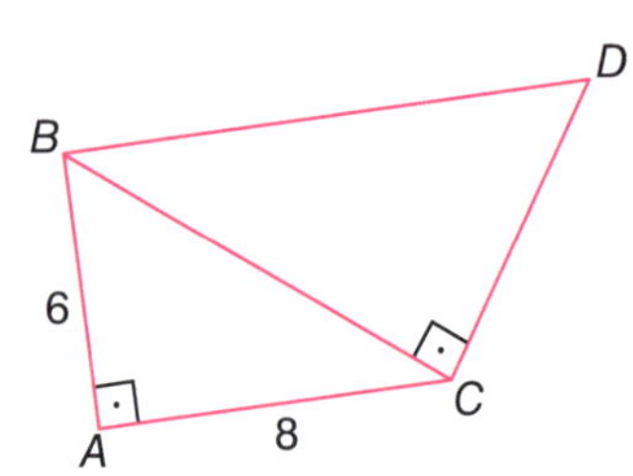

Solution 5.3

Using 6–8–10 triangle, we obtain $BC = 10$.

If $m\angle CAB = \alpha$, then $m\angle BCA = \alpha$.

If $m\angle ABC = \beta$, then $m\angle BDC = \beta$.

So, $\Delta ABC \sim \Delta CDB$, and the ratio

$\frac{10}{x} = \frac{8}{10} \Rightarrow x = 12.5$.

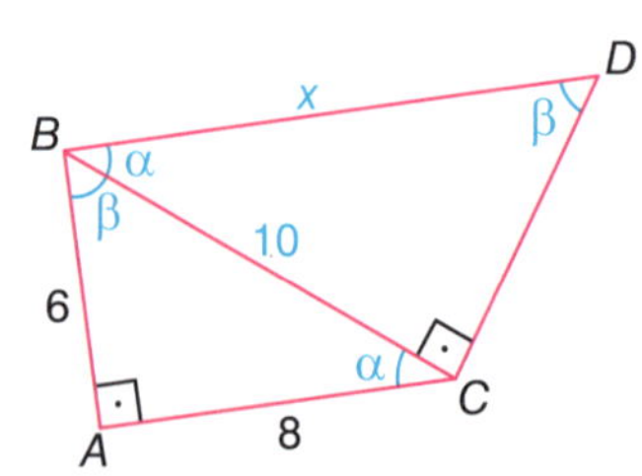

Try This

ΔABC has an inscribed rhombus $ADEF$.

If $AB = 12$ and $AC = 10$, find the side length of the rhombus.

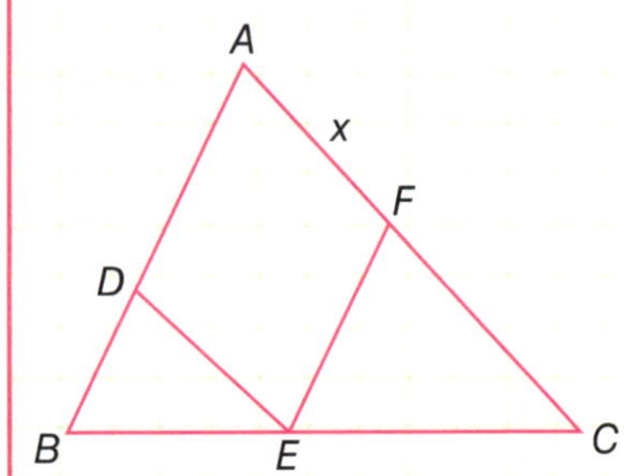

My Notes

My Notes

Example 5.4

In the figure, $BC // DE$,

$AB = m$, $BD = 7$,

$AC = 6$, and $CE = n$.

What is $m \cdot n$?

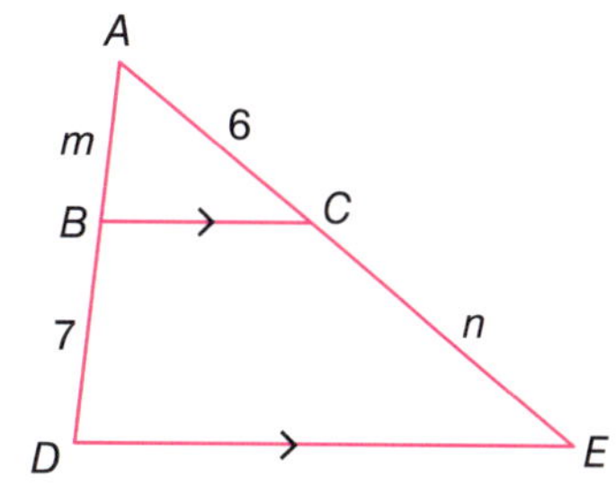

Solution 5.4

A is a common angle of ΔABC and ΔADC,

$m\angle ABC = m\angle D$, and $m\angle ACB = m\angle E$.

Therefore, $\Delta ABC \sim \Delta ADE$ (AAA similarity)

$$\frac{m}{7+m} = \frac{6}{6+n} \Rightarrow 6m+mn = 42+6m$$

$$\Rightarrow mn = 42.$$

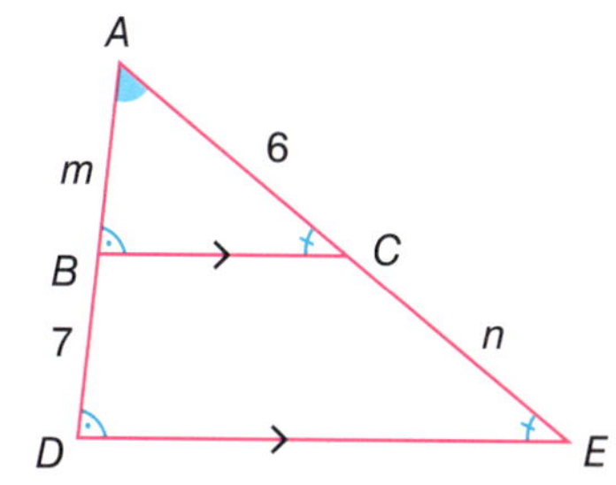

Example 5.5

In the figure, $BC // DE$,

$AB = 3$, $BD = 4$,

$AC = BC = 6$,

$CE = x$, and $DE = y$.

Find the value of $x+y$.

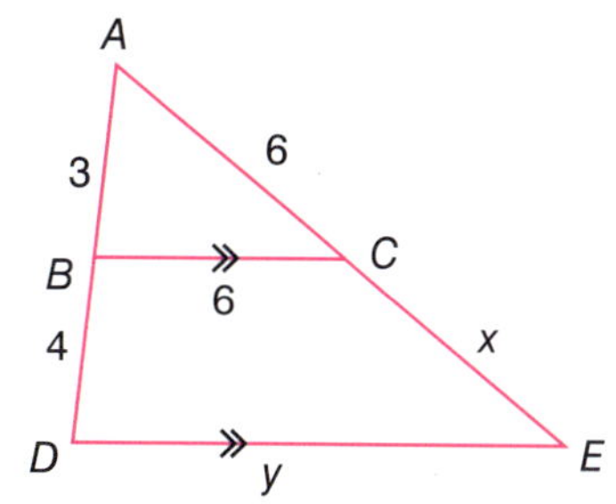

Solution 5.5

A is a common angle,

$m\angle ABC = m\angle D$, and $m\angle ACB = m\angle E$.

Therefore, ΔABC and ΔADE are similar (AAA).

$$\frac{3}{3+4} = \frac{3}{7} = \frac{6}{y} \Rightarrow y = 14$$

$$\frac{3}{7} = \frac{6}{6+x} \Rightarrow x = 8$$

$$x + y = 22.$$

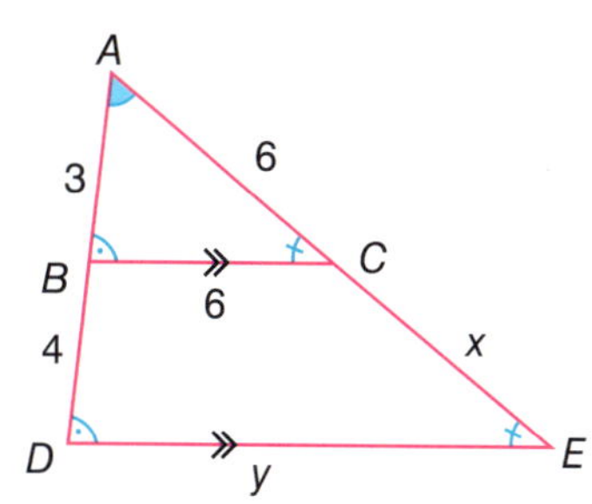

Example 5.6

In the figure, $DE//BC$ and

$$\frac{b-a}{b+a} = \frac{2}{5}.$$

What is AD?

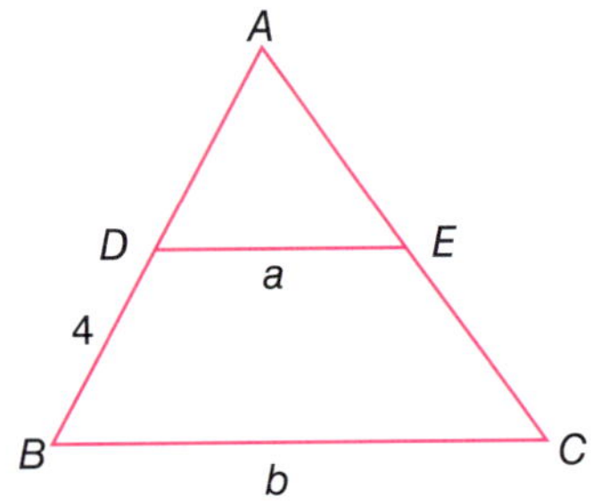

Solution 5.6

By cross multiplying, $\frac{b-a}{b+a} = \frac{2}{5} \Rightarrow 5b-5a = 2b+2a$

$$3b = 7a \Rightarrow \frac{a}{b} = \frac{3}{7}.$$

Let AD be x, then $\frac{3}{7} = \frac{x}{x+4} \Rightarrow 7x = 3x+12$

$4x = 12 \Rightarrow x = 3$. Therefore, $AD = 3$.

Example 5.7

In the figure, $m\angle ABD = m\angle BCA$,

$AB = 8$, $BC = 10$,

$AD = 4$, and $BD = x$.

What is BD?

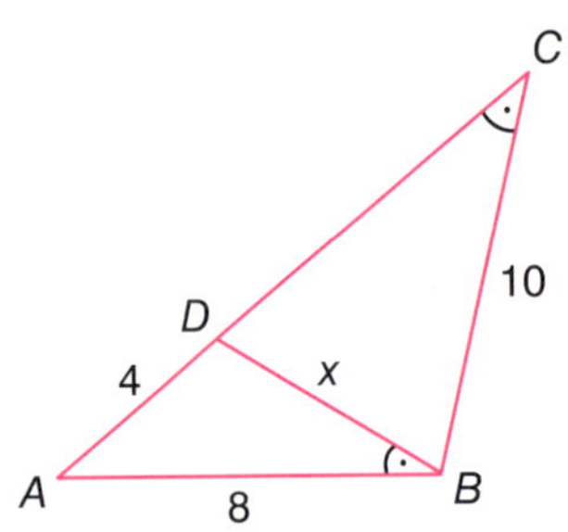

Solution 5.7

Since $\angle A$ is the common angle of ΔADB and ΔABC we can conclude that $m\angle ADB$ and $m\angle ABC$ are equal and denoted β.

So, $\Delta ADB \sim \Delta ACB$

$$\frac{4}{8} = \frac{x}{10} \Rightarrow x = 5$$

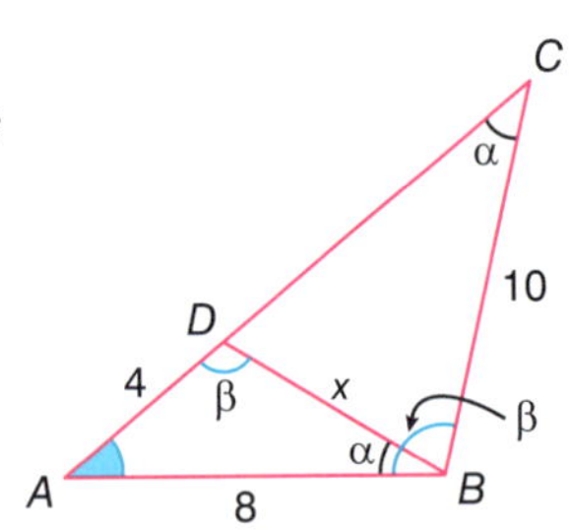

Remark 5.1

If a line is parallel to one side of a triangle and intersects the other two sides then it divides these two sides proportionally.

If $DE//BC$ then $\frac{a}{b} = \frac{x}{y}$.

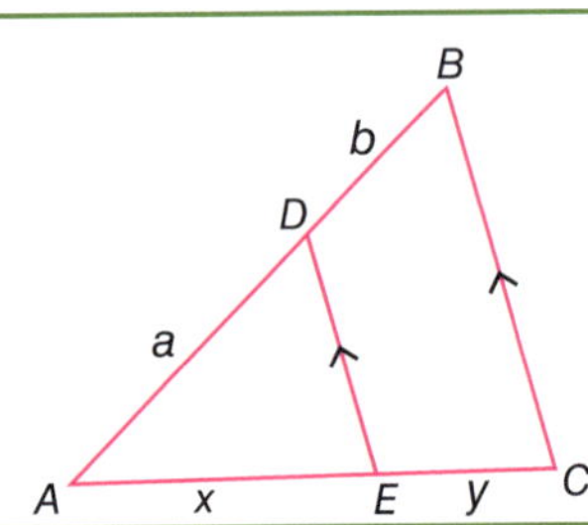

Exercise 5.1

If $DE//BC$, then, $\frac{AD}{DB} = \frac{AE}{AC}$

$$\frac{2}{3} = \frac{x}{6}$$

$x = 4$.

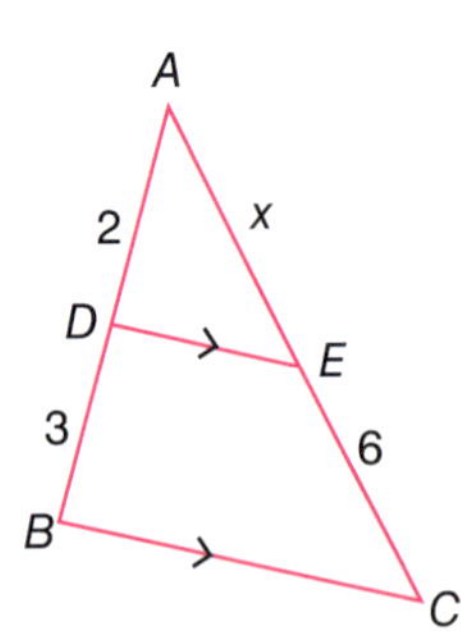

My Notes

Try This

ABCD is a parallelogram.

$AP = 4$,

$PK = 2$, and

$KE = x$.

What is x?

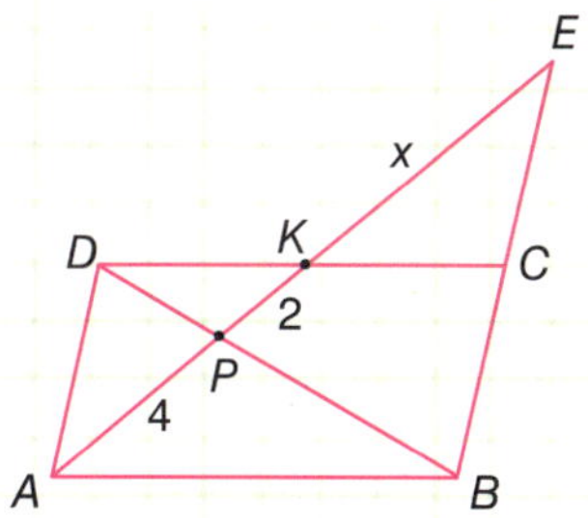

My Notes

Example 5.8

In the figure, $AB//CD$,

$AC = x+1$, $CE = x-2$,

$BD = x+5$, and $x = CE$.

Find x.

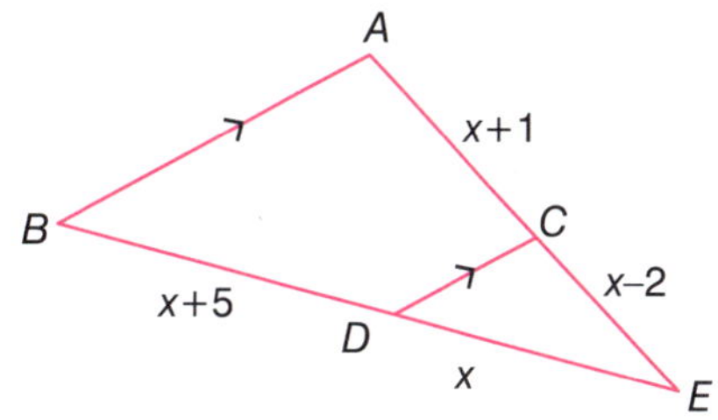

Solution 5.8

Method 1

E is a common angle and

$m\angle B = m\angle EDC$ and $m\angle A = m\angle ECD$.

Therefore, ΔEDC and ΔEBA are similar.

The similarity ratios:

$$\frac{ED}{EB} = \frac{EC}{EA} = \frac{x}{2x+5} = \frac{x-2}{2x-1}$$

$$2x^2-x = 2x^2+5x-4x-10$$

$$2x = 10 \Rightarrow x = 5.$$

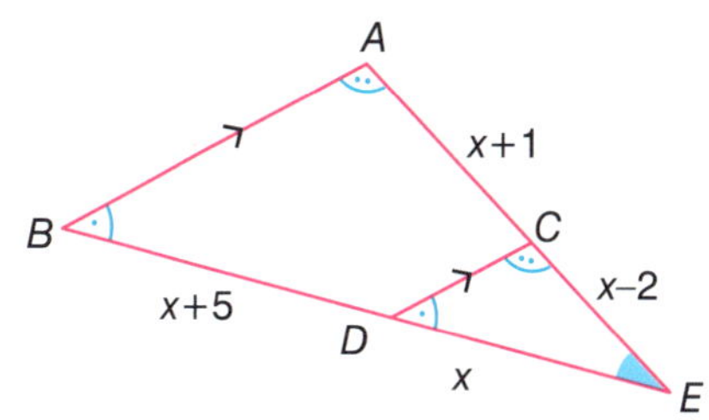

Method 2

By Remark 5.1

$$\frac{x}{x+5} = \frac{x-2}{x+1} \Rightarrow x^2+x = x^2+3x-10$$

$$2x = 10$$

$$x = 5.$$

Example 5.9

In the figure, $DE//BC$ and $FE//DC$,

$AF = 8$, $AE = 10$, $EC = 5$

$FD = x$, and $DB = y$.

What is $x+y$?

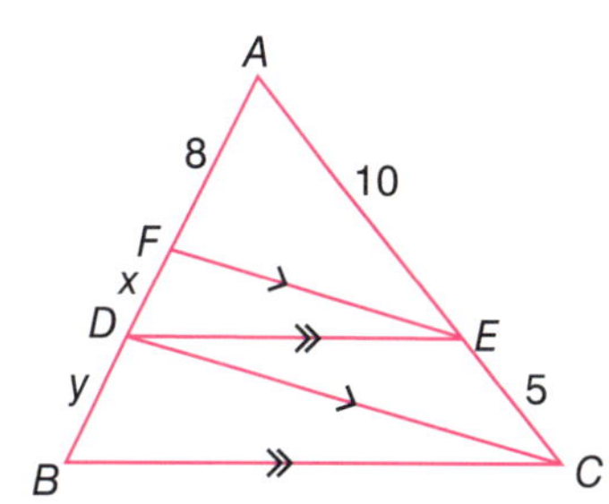

Solution 5.9

Since $FE//DC$ then

By remark 5.1, $\frac{8}{x} = \frac{10}{5} \Rightarrow x = 4$

Since $DE//BC$ then

By remark 5.1, $\frac{AD}{DB} = \frac{AE}{EC} \Rightarrow \frac{12}{y} = \frac{10}{5} \Rightarrow y = 6.$

So, $x+y = 10$.

Theorem 5.2

$$\frac{1}{y} = \frac{1}{x} + \frac{1}{z}$$

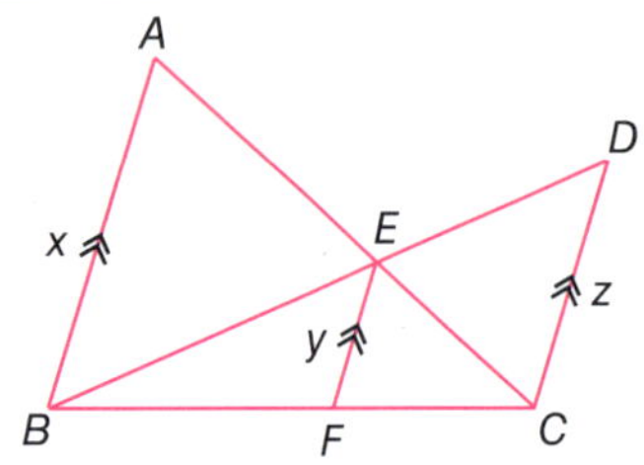

Proof 5.2

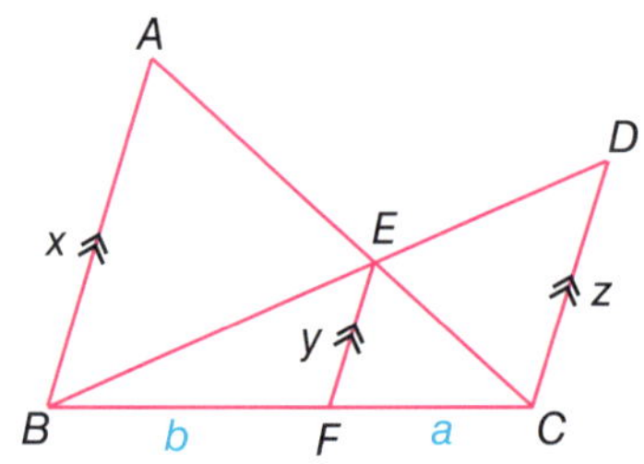

Since *EF//AB*, then *ABC* ~ *EFC*.

So, $\frac{y}{x} = \frac{a}{a+b}$ (1)

Since *EF//DC*, then *DCB* ~ *EFB*.

So, $\frac{y}{z} = \frac{b}{a+b}$ (2)

Add (1)+(2) $\frac{y}{x}+\frac{y}{z} = \frac{a}{a+b}+\frac{b}{a+b} \Rightarrow \frac{y}{x}+\frac{y}{z} = 1$

Thus, $\frac{1}{x}+\frac{1}{z} = \frac{1}{y}$.

Example 5.10

In the figure, *AB//CD//EF*
AB = *x*, *CD* = *m*, *EF* = *y*,
x+*y* = 15, and
x·*y* = 5.
What is *CD*?

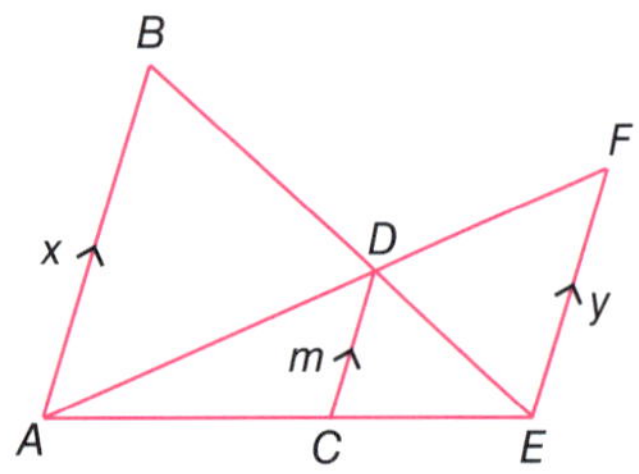

Solution 5.10

From Theorem 5.2

$$\frac{1}{m} = \frac{1}{x}+\frac{1}{y} = \frac{x+y}{x \cdot y} \Rightarrow \frac{1}{m} = \frac{15}{5} \Rightarrow \frac{1}{m} = 3. \text{ So, } m = \frac{1}{3}.$$

Theorem 5.3

S.A.S. Theorem with Example

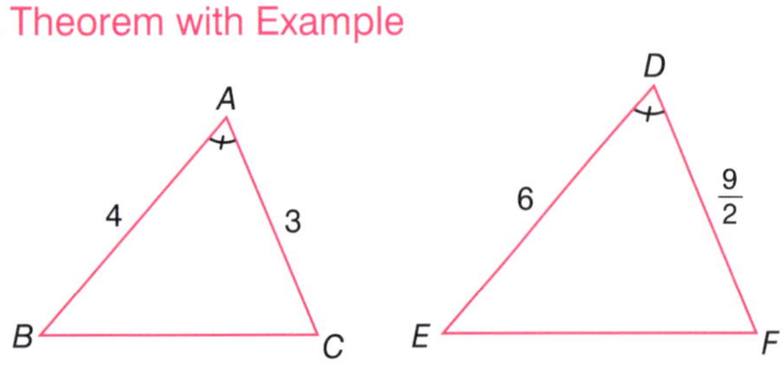

If two sides of triangles have the same ratio and the angles between them have same, then two triangles are similar.

If $\frac{AB}{DE} = \frac{AC}{DF} \Rightarrow \frac{4}{6} = \frac{3}{4.5}$ and $m\angle BAC = m\angle EDF$, then $\Delta ABC \sim \Delta DEF$

(S.A.S Similarity)

Prove This

ABCD is a parallelogram and *DB* is a diagonal. Prove that

$a^2 = b(b+c)$.

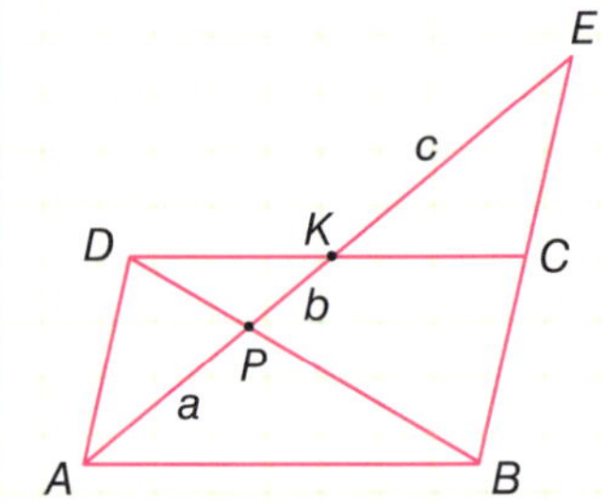

My Notes

Try This

ABC is a equilateral triangle and point *E* and *D* are on *AB* and *AC*, respectively.

Given that,

$AE = CD$ and

$m\angle DBC = 20°$,

find $m\angle AEC$.

My Notes

Example 5.11

In the figure,

$AB = 5, AC = 4, CE = 11,$

$BD = 7,$ and $DE = 10.$

What is *BC*?

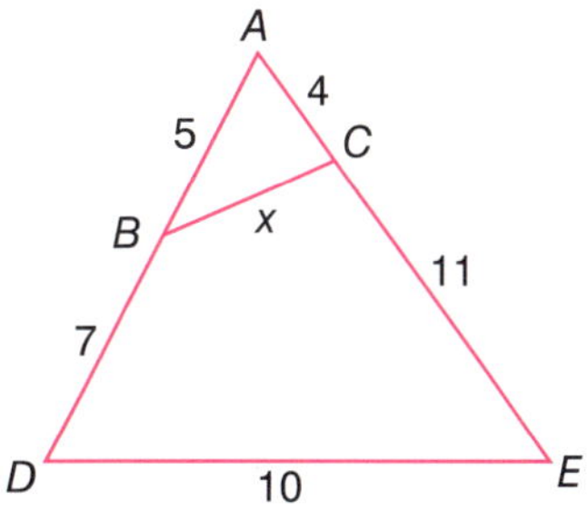

Solution 5.11

From S.A.S similarity

$\Delta BAC \sim \Delta EAD$ since

$\frac{4}{12} = \frac{5}{15}$

Therefore, $\frac{x}{10} = \frac{1}{3}$.

So, $x = \frac{10}{3}$.

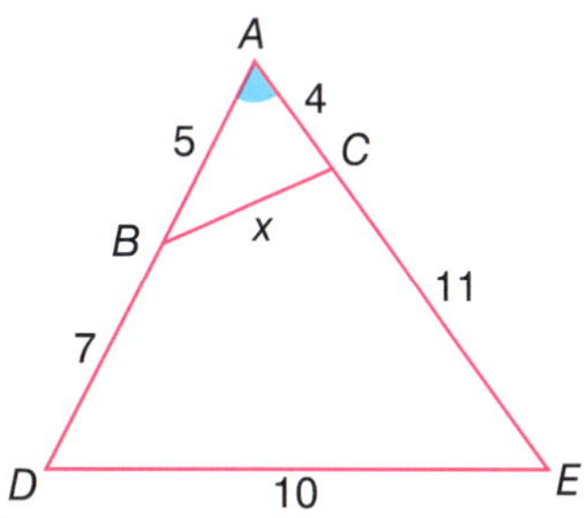

Example 5.12

In the figure, $AB // CD$

$AB = 4, AC = 8,$

$BC = 6,$ and $CD = 9.$

What is *BD*?

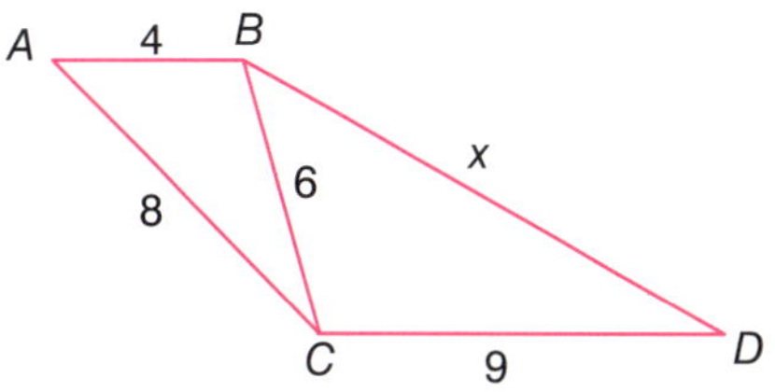

Solution 5.12

Since

$\frac{AB}{BC} = \frac{BC}{CD} = \frac{2}{3}$ and $m\angle ABC = m\angle BCD.$

Therefore, $\Delta ABC \sim \Delta BCD.$

(S.A.S similarity)

So, $\frac{2}{3} = \frac{8}{x} \Rightarrow x = 12.$

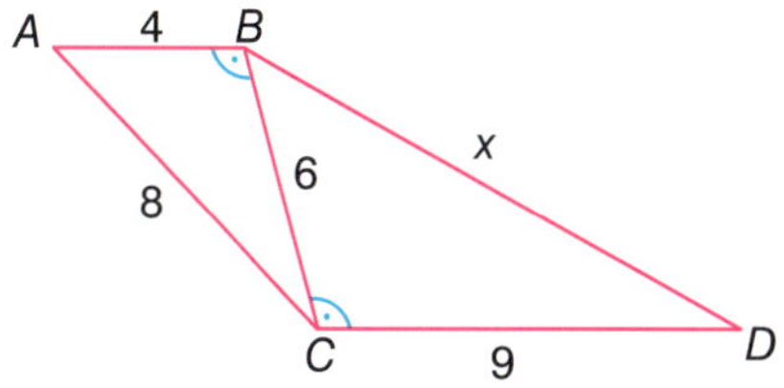

Example 5.13

In the figure, DB and EC intersect at A.

$AB = 12$ cm, $BC = 18$ cm,

$AC = 15$ cm, $AE = 8$ cm, and

$AD = 10$ cm.

What is the length of ED in cm?

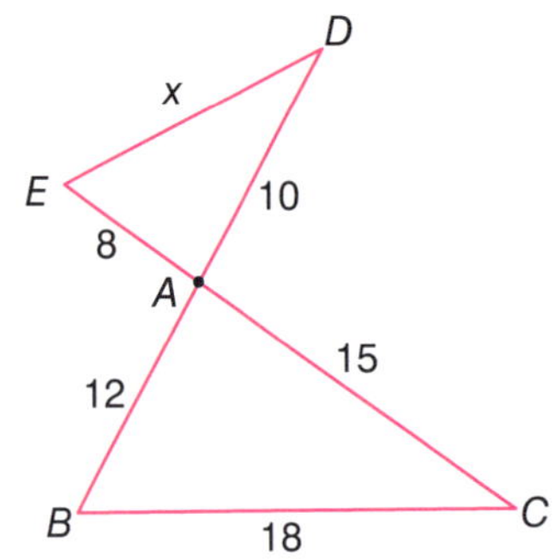

Solution 5.13

$\frac{AE}{AB} = \frac{AD}{AC} = \frac{2}{3}$ and $m\angle EAD = m\angle BAC$ (S.A.S similarity)

Then, $\frac{ED}{BC} = \frac{x}{18} = \frac{2}{3} \Rightarrow x = 12$ cm

Try This

In the figure, $\angle DAB = \angle CBD$, $AD = 6$, and $BC = 4$.

What is DC?

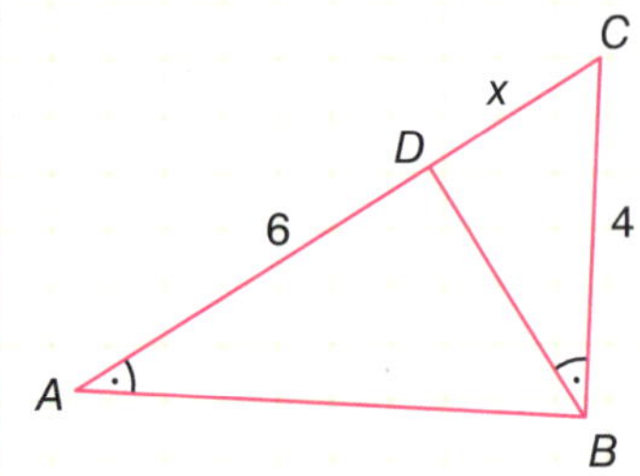

My Notes

Theorem 5.4

S.S.S. Theorem with Example

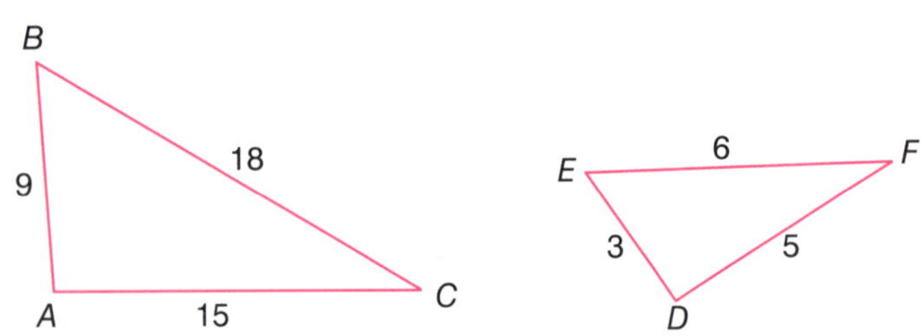

If $\frac{AB}{ED} = \frac{9}{3} = \frac{AC}{DF} = \frac{15}{5} = \frac{BC}{EF} = \frac{18}{6}$, then $\Delta ABC \sim \Delta DEF$ (S.S.S. similarity)

Theorem 5.5

Theorem of the Right Triangle

$h^2 = m{\cdot}n$

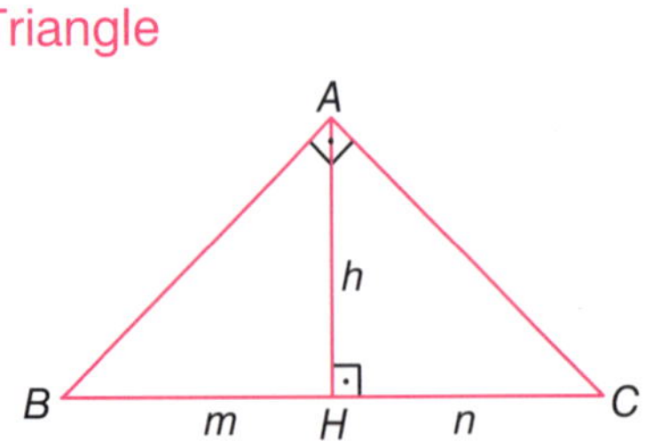

Challenge

In the figure, D is on AB. AC and BF intersect K, AC and DF intersect E, and $DF // BC$.

$DE = 4$ cm,

$EF = 6$ cm, and

$BC = 12$ cm.

Find $\frac{AE}{KC}$.

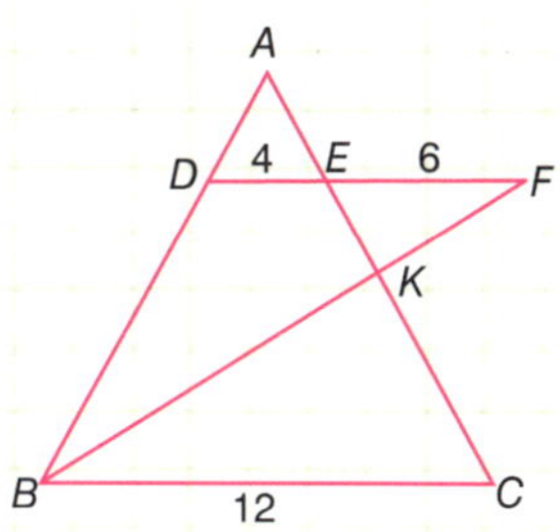

My Notes

Proof 5.5

$\Delta AHB \sim \Delta CHA$

$$\frac{m}{h} = \frac{h}{n} \Rightarrow h^2 = m \cdot n.$$

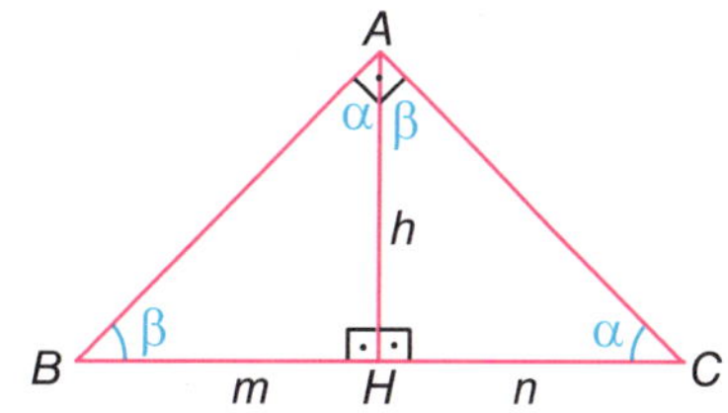

Theorem 5.6

$$k^2 = m \cdot (m+n)$$

$$l^2 = n \cdot (n+m)$$

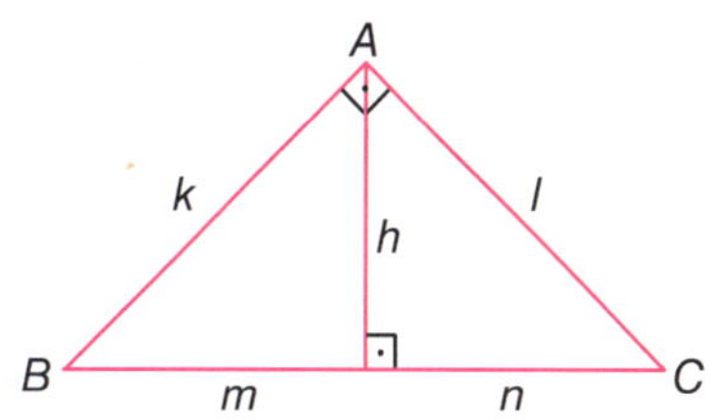

Proof 5.6

$\Delta AHB \sim \Delta CAB$

$$\frac{k}{m+n} = \frac{m}{k} \Rightarrow k^2 = m \cdot (m+n)$$

Similarly,

$\Delta CHA \sim \Delta CAB$

$$\frac{l}{m+n} = \frac{n}{l} \Rightarrow l^2 = n \cdot (m+n).$$

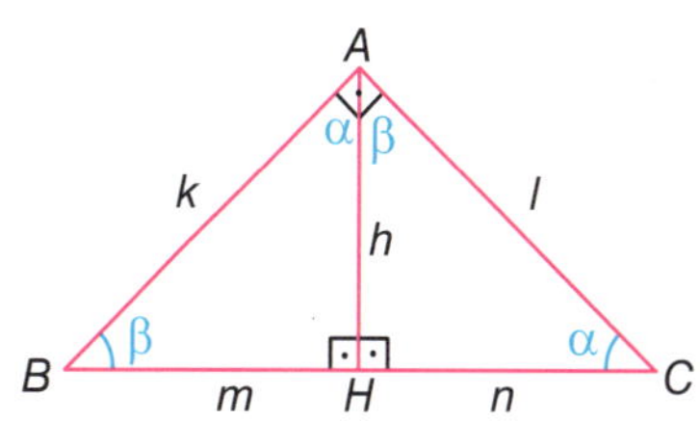

Theorem 5.7

If $AB = c$, $AC = b$, and

$BC = a$, then $c \cdot b = a \cdot h$.

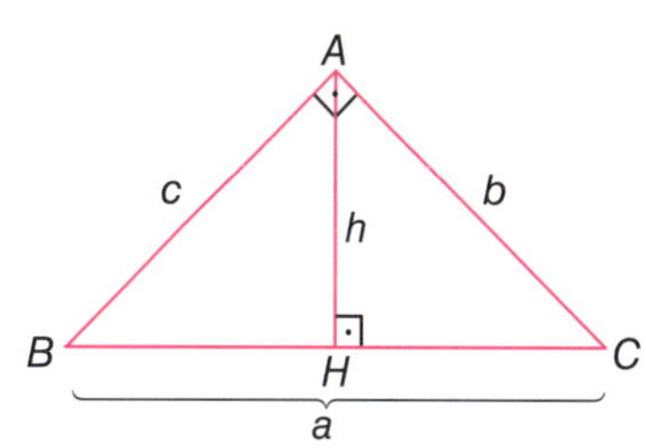

Proof 5.7

From the triangle area formula.

$$\text{Area } \Delta ABC = \frac{c \cdot b}{2} = \frac{a \cdot h}{2} \Rightarrow c \cdot b = a \cdot h.$$

Theorem 5.8

Inverse Pythagorean Theorem(IPT)

$$\frac{1}{b^2}+\frac{1}{c^2}=\frac{1}{h^2}.$$

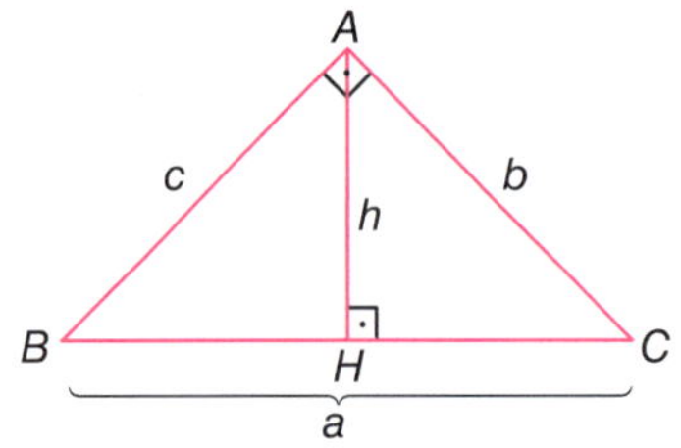

Proof 5.8

Starting from the left side of the equation,

$$\frac{1}{b^2}+\frac{1}{c^2}=\frac{b^2+c^2}{(b\cdot c)^2}$$

$$=\frac{a^2}{b^2\cdot c^2}=\frac{a^2}{a^2\cdot h^2}=\frac{1}{h^2} \quad \left(b^2\cdot c^2=a^2\cdot h^2 \text{ Theorem 5.7}\right)$$

Therefore, $\frac{1}{b^2}+\frac{1}{c^2}=\frac{1}{h^2}$.

Example 5.14

In the figure, $BA\perp AC$ and $AH\perp BC$.

$AC=8$, $BH=12$,

$HC=x$, and $AB=y$.

Find x, y, and h.

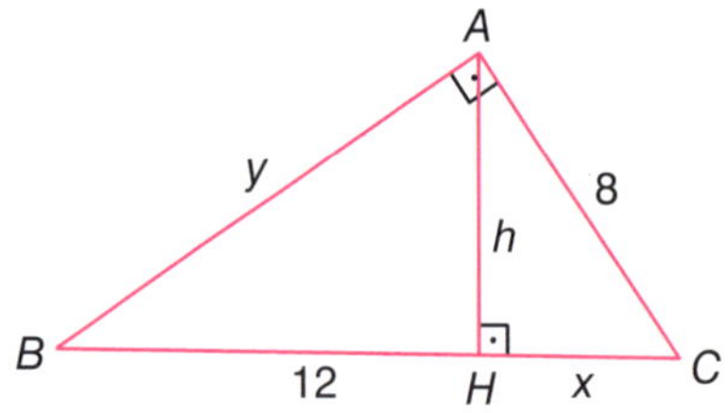

Solution 5.14

By Theorem 5.6, $8^2=x(12+x)\Rightarrow 64=x(12+x)$

$$x=4$$

By Theorem 5.6, $y^2=12(12+4)\Rightarrow y^2=12\cdot16$

$$y=\sqrt{12\cdot16}$$

$$y=8\sqrt{3}$$

By Theorem 5.5, $h^2=12\cdot x\Rightarrow h^2=12\cdot4$

$$h=\sqrt{48}$$

$$h=4\sqrt{3}.$$

Prove This

For any triangle ABC, three medians must intersect at one common point G, and each median is partitioned by G as a ratio 2 to 1.

My Notes

Catriona Agg @Cshearer41

Geometry puzzle

The areas of the three squares are given.

What's the area of the blue triangle?

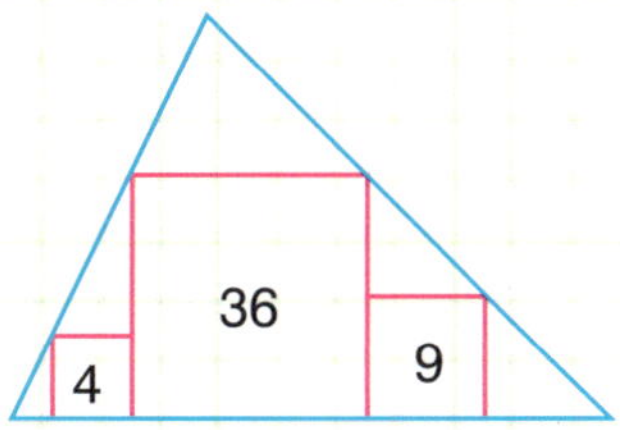

My Notes

Example 5.15

In the figure, $BA \perp AC$, $AH \perp BC$ and

$AC = 3AB$.

What is the ratio of $\frac{x}{y}$?

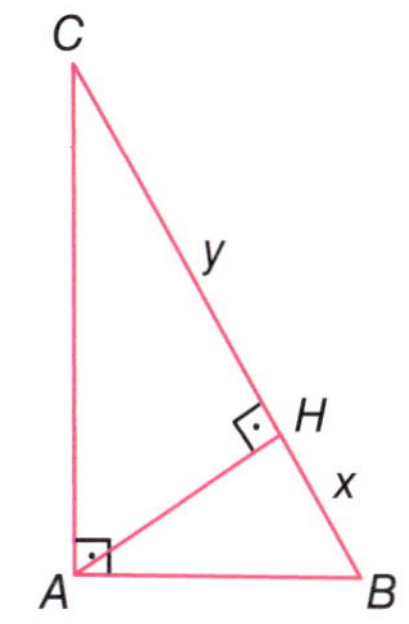

Solution 5.15

We write two equations from $\triangle ABC$ by using theorem 5.6.

$$c^2 = x \cdot (x+y)$$

$$9c^2 = y \cdot (y+x)$$

Ratio of two equations gives us $\frac{x}{y}$

$$\frac{c^2}{9c^2} = \frac{x(x+y)}{y(x+y)}$$

$$\frac{1}{9} = \frac{x}{y}.$$

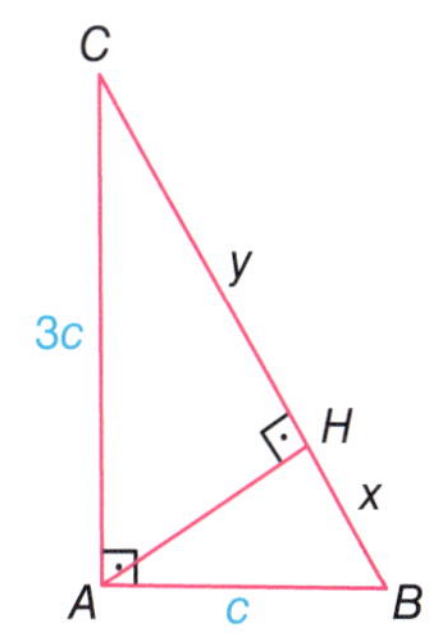

Example 5.16

In the figure, CB is an angle bisector of $\angle DCA$, $BA \perp AC$, and $DB \perp BC$.

If $AC = 4$, $DC = 9$, and

$BC = x$, then find x.

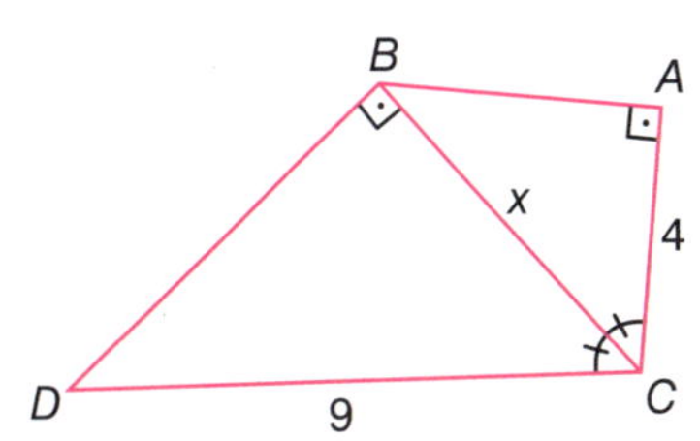

Solution 5.16

Solution 1: Let BE be h.

$h^2 = 5 \cdot 4 = 20$ (Theorem 5.5)

$x^2 = h^2 + 4^2 = 20 + 16 = 36$ ($m\angle BEC$)

$x^2 = 36 \Rightarrow x = 6$.

Solution 2: By Theorem 5.4

$x^2 = 4\,(4+5) = 36$. So, $x = 6$.

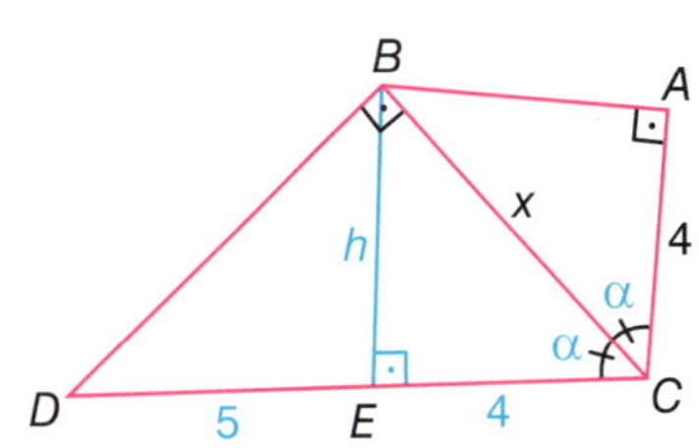

Example 5.17

In the figure, $BA \perp AC$,

$AB = AH$,

$BH = 6$, and

$HC = 3$.

Find the value of x.

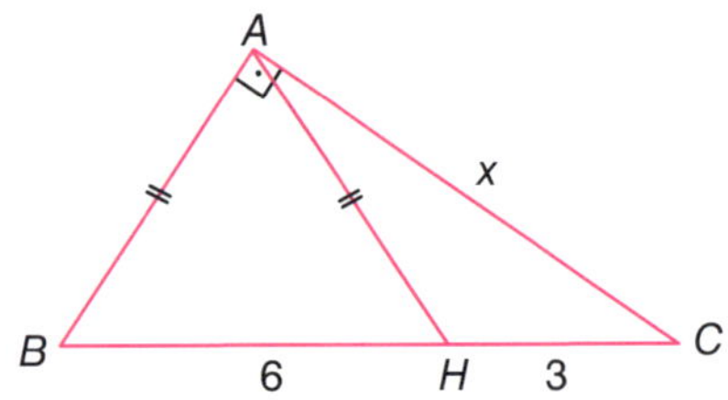

Solution 5.17

Always make a perpendicular foot when we have an isosceles triangle.

$x^2 = TC \cdot BC$ (Theorem 5.6)

$x^2 = 6 \cdot (6+3)$

$x^2 = 6 \cdot 9$

$x = 3\sqrt{6}$.

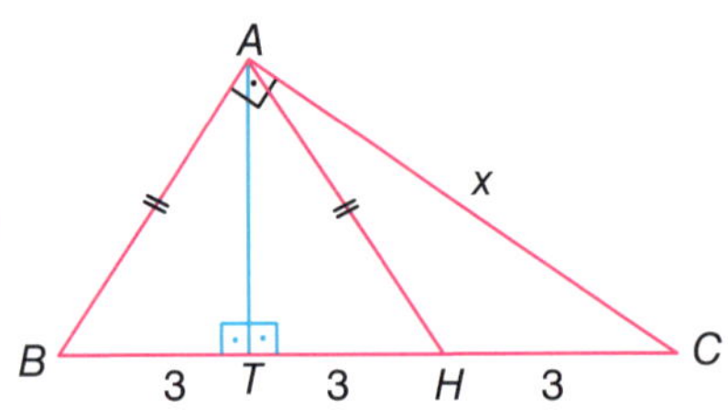

Example 5.18

In the figure, $AB \perp AC$ and $AH \perp BC$.

$AB = 4$,

$AC = 6$, and

$AH = x$.

Find the value of x.

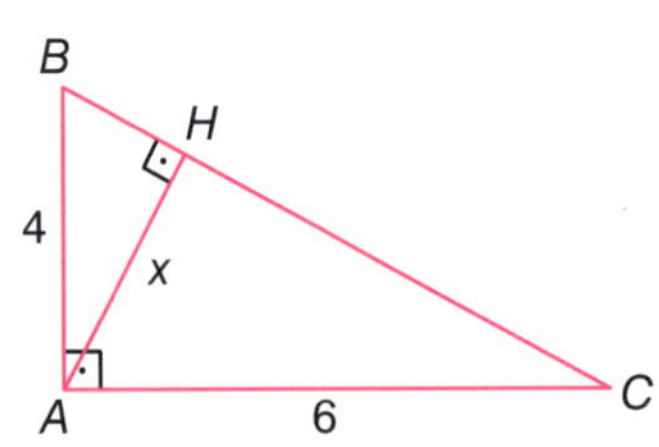

Solution 5.18

From theorem 5.6: (Inverse Pythagorean Theorem - IPT).

$$\frac{1}{x^2} = \frac{1}{4^2} + \frac{1}{6^2}$$

$$\frac{1}{x^2} = \frac{1}{16} + \frac{1}{36}$$

$$\frac{1}{x^2} = \frac{36+16}{16 \cdot 36}$$

$$x = \frac{4 \cdot 6}{\sqrt{52}}$$

$$x = \frac{24}{2\sqrt{13}} = \frac{12\sqrt{13}}{13}.$$

$$\text{So, } x = \frac{12\sqrt{13}}{13}.$$

Catriona Agg @Cshearer41

Geometry puzzle

These squares are the same size.

What is the angle?

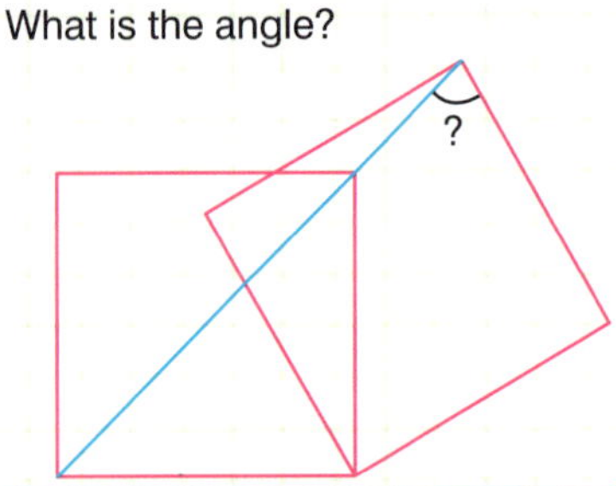

My Notes

1. In a triangle ABC, P and R the midpoints of AB and BC. $AC = 3x-1$ and $PR = x+4$ are given.

What is PR?

2. In the figure, $m\angle A = m\angle D$.

Find $x+y$.

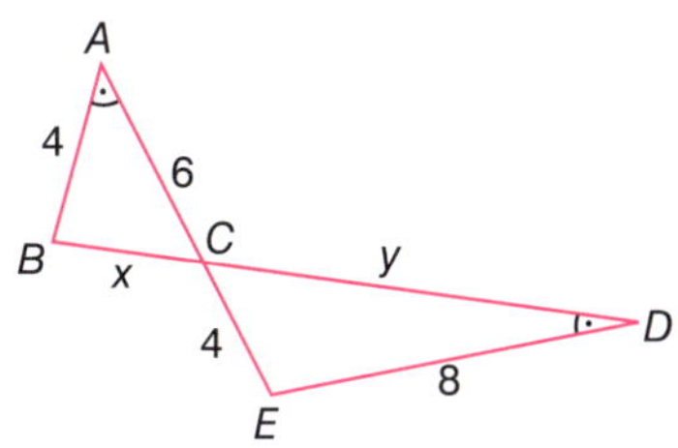

3.

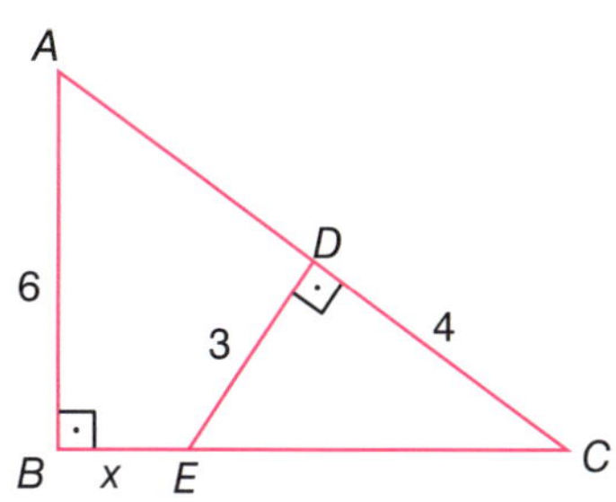

In the figure, $m\angle ABC = m\angle EDC = 90°$,

$AB = 6$, $DE = 3$, and $DC = 4$.

Find $BE = x$.

4. ABC is a triangle with $m\angle B = 90°$, $AB = 8$, and $BC = 6$. Points M and N are on sides AB and BC, respectively, such that $MN // AC$ and $AM = 3$.

What is MN?

5. In the figure, BC // DE.

What is AD?

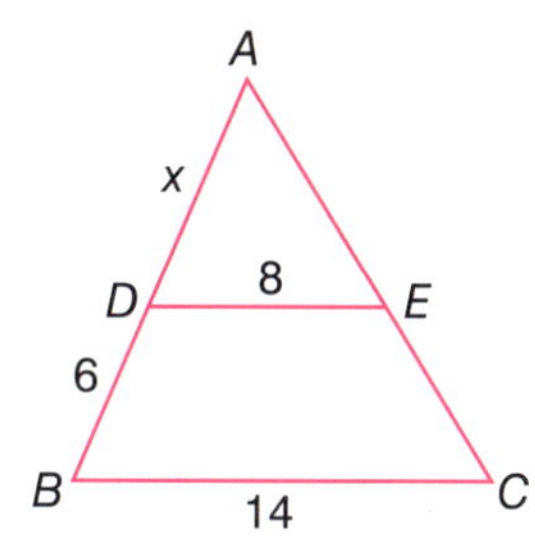

6. In the figure, BC // DE.

Find $x + y$.

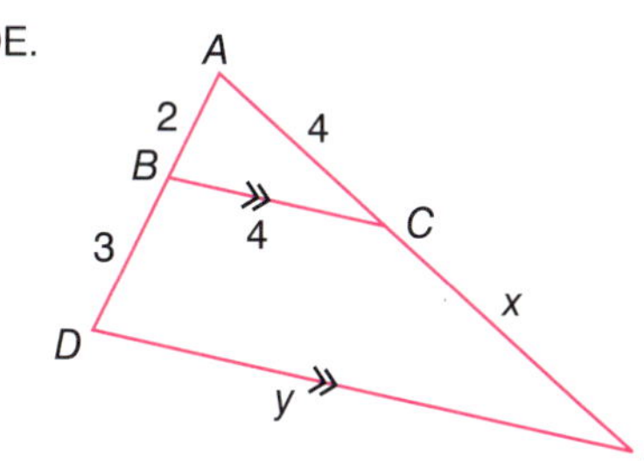

7. The ratio of the side lengths of the similar triangles is 5:4 and perimeter of the greater triangle is 30 cm.

What is length of the shortest side of the smaller triangle, if its side lengths are consecutive even numbers?

8. In the figure,

$m\angle BCD = m\angle CAB$,

$BD = 4$,

$BC = 6$, and

$AC = 8$.

What is the length of CD?

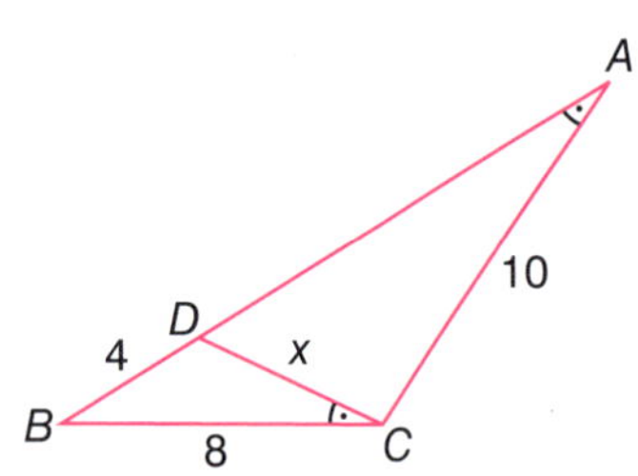

9. In the figure,

$m\angle BEA = m\angle CDA = 90°$,

$AE = 4$,

$EC = 5$, and

$DB = 3AD$.

Find AD.

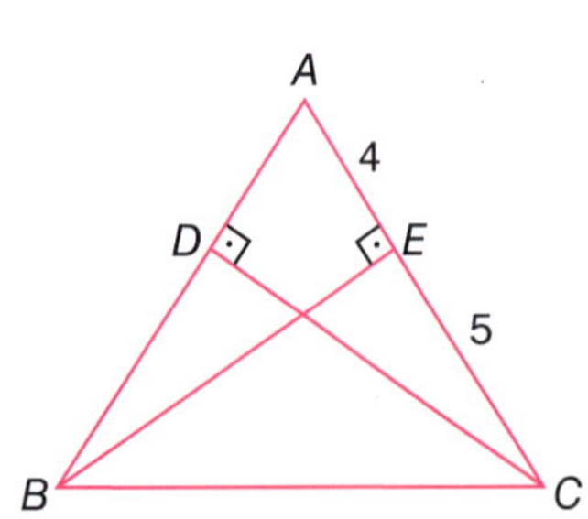

10. In the figure,

$PE \parallel AC$, $PH \parallel AB$,

$AB = AC = 13$ cm, and

$PE = 4$ cm.

Find the length of $\frac{PH}{AB}$.

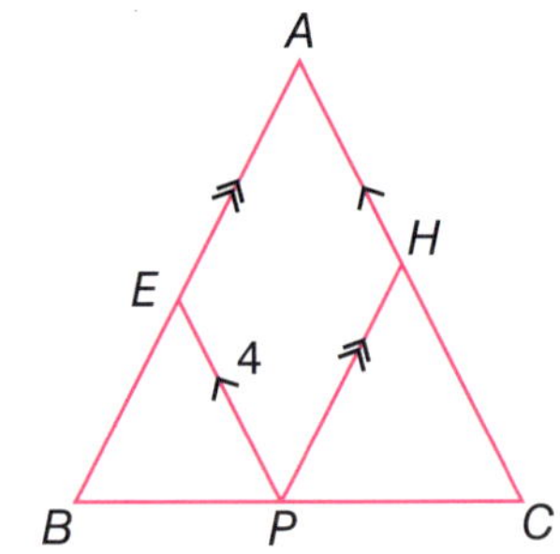

11. In the figure,

$AC \perp AB$ and $AH \perp BC$.

If $AC = 6$ cm and $AB = 8$ cm, what is the length of AH?

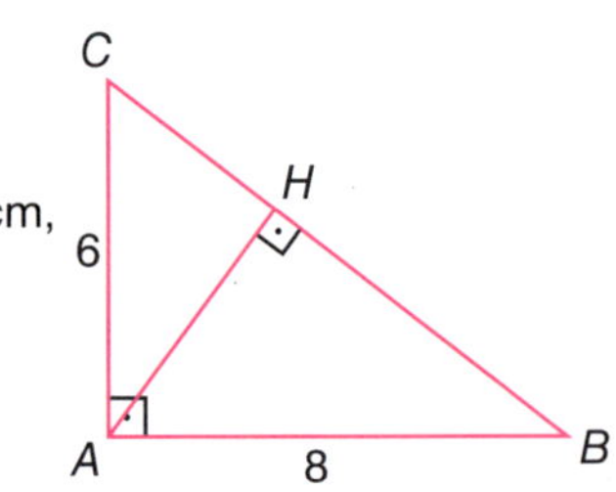

12. In the figure,

$BC \parallel EF$ and $CD \parallel FG$.

Find $AD = x$.

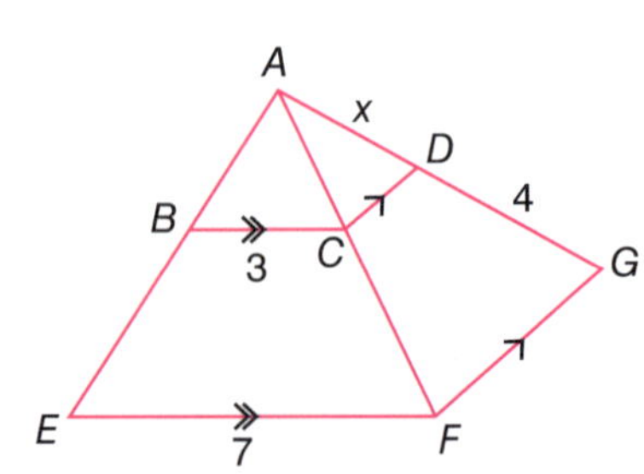

1. In the figure,

   $m\angle ABC = m\angle ADE$,

   $AE = 3$, $AD = 2$, and

   $EB = 2$.

   Find DC.

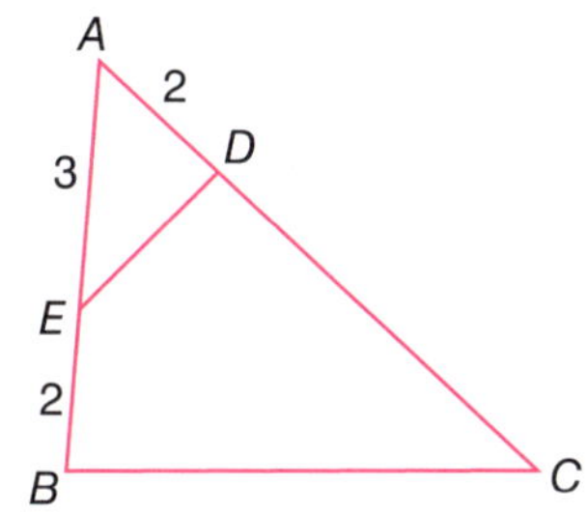

2. In triangle ABC,

   $CA \perp AB$, $AC = 2AB$,

   $BH = x$, and $HC = y$.

   Find the ratio of $\frac{x}{y}$.

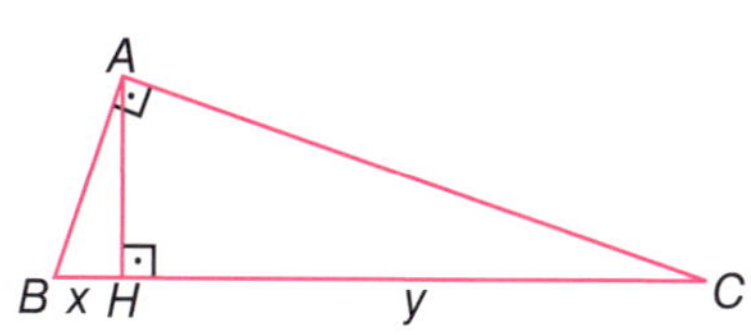

3. In the figure,

   $AH = 2\sqrt{5}$ and

   $m^2+n^2 = 41$.

   What is $b \cdot c$?

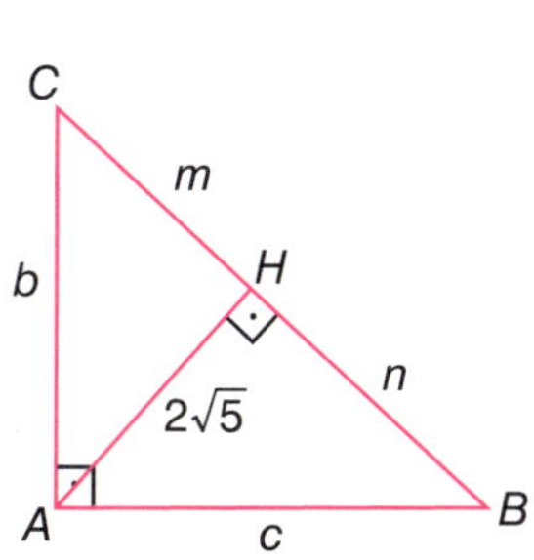

4. In the figure,

   $AB // EF // CD$,

   $AB = 6$, $BF = 2$, and

   $FC = 1$.

   Find $x+y$.

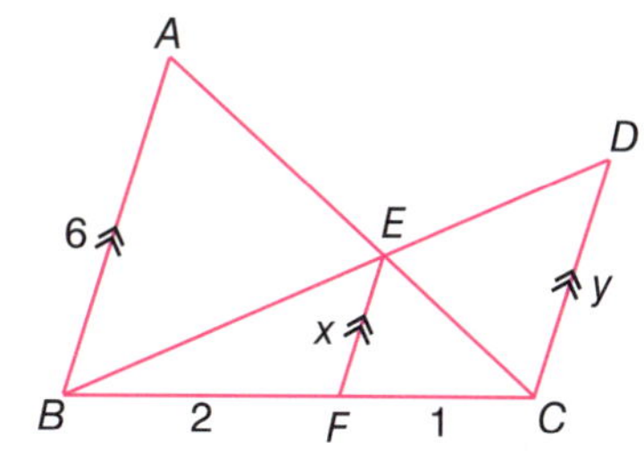

5. ABC is a triangle with

   $AB = 15$, and $BC = 30$.

   If $BDEF$ is a square, then find the length of BF.

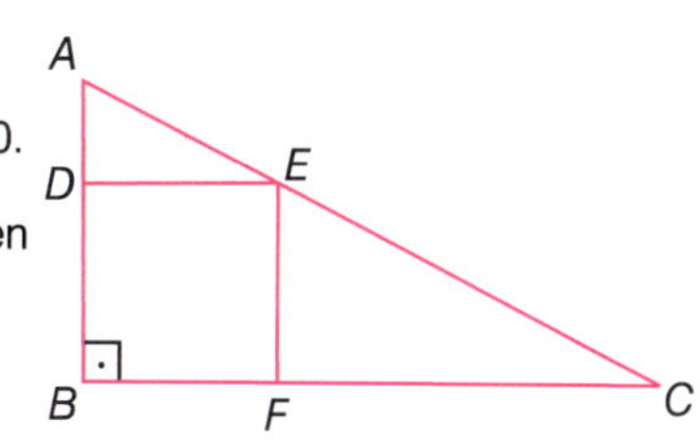

6. In the figure,

   DC//EF//AB,

   $DE = 10$, $EA = 12$, and

   $FB = 9$.

   What is the length of CF?

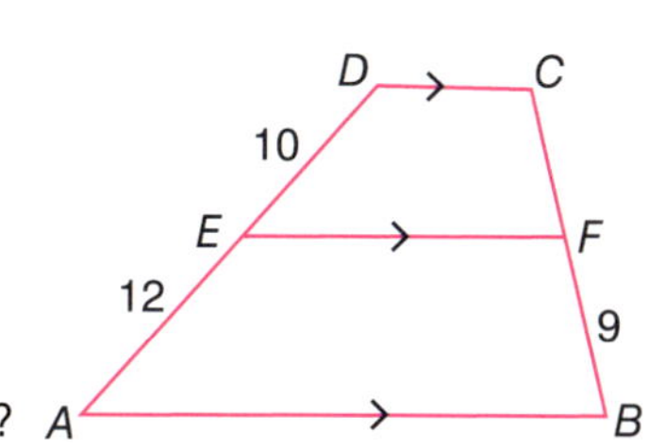

7.

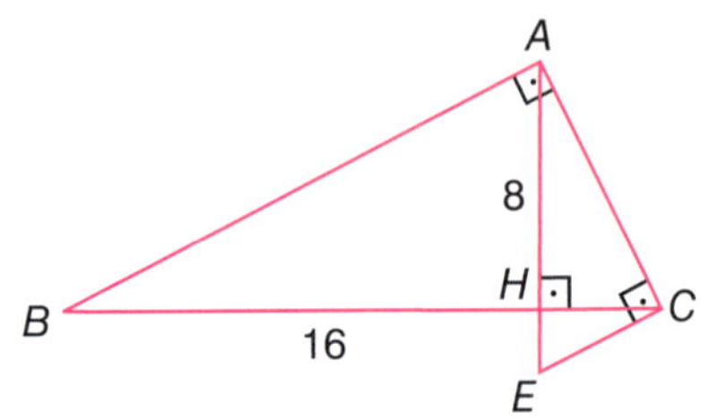

In the figure, $m\angle BAC = m\angle ACE = 90°$ and $AH \perp BC$.

If $BH = 16$ and $AH = 8$, then what is the length of CE?

10. In the figure, $DC//AB$,

$AB = 9$, $AD = 12$,

$BD = 6$, and $DC = 4$.

What is the value of x?

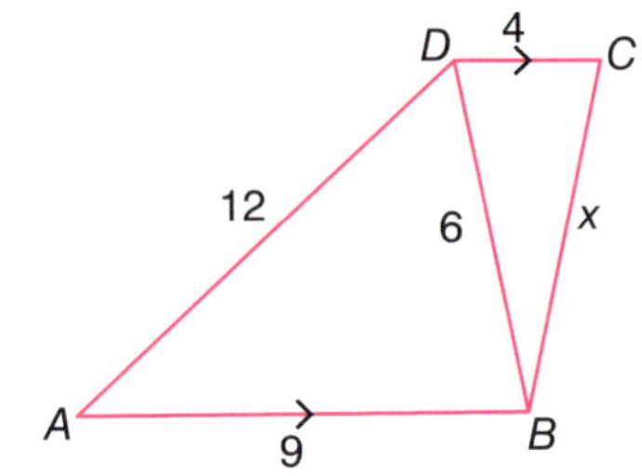

8. In the figure, given that

$AD = DE = 4$, $BD = 5$,

$AF = 3$, $FC = 9$, and $BF = 8$.

What is EC?

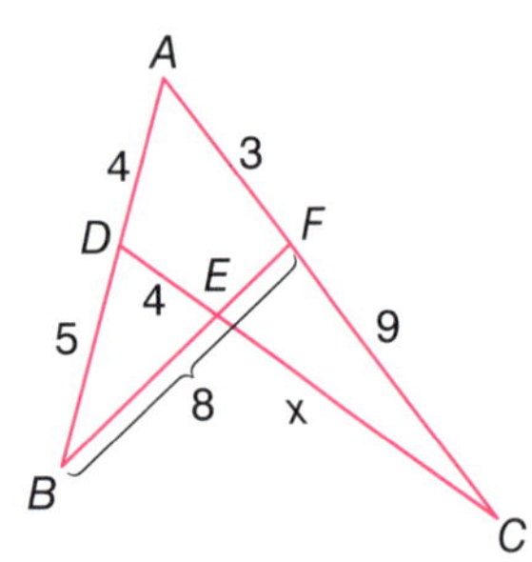

11. Suppose x, y and z are real numbers such that

$x^2 + y^2 = 9$,

$y^2 + z^2 = 16$, and

$y^2 = x \cdot z$.

What is $x+y+z$?

9. In the figure, $DE//BC$, and EF is the angle bisector of $m\angle DEC$.

If $AE = 4$ and $DE = EC = 6$, then find the length of BF.

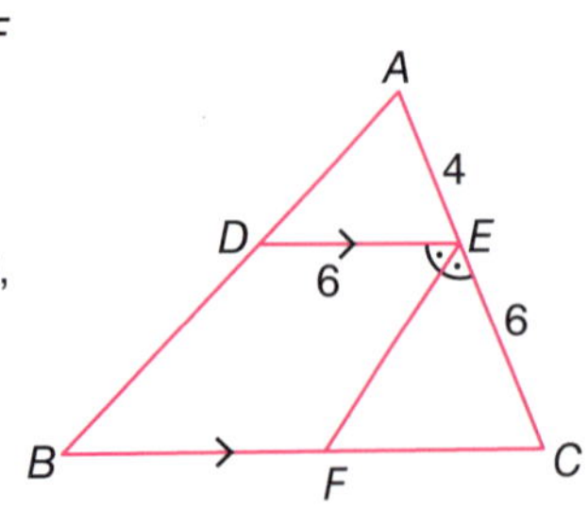

12. In the figure,

$CE = 3ED$,

$BF = FC$, and $KF = 6$.

Find AK.

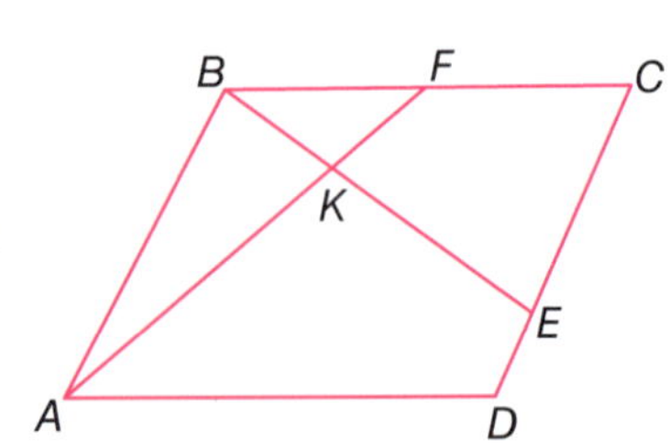

1. (2011 AMC 10B #9)

The area of ΔEBD is one third of the area of $3-4-5$ ΔABC. Segment DE is perpendicular to segment AB

What is BD?

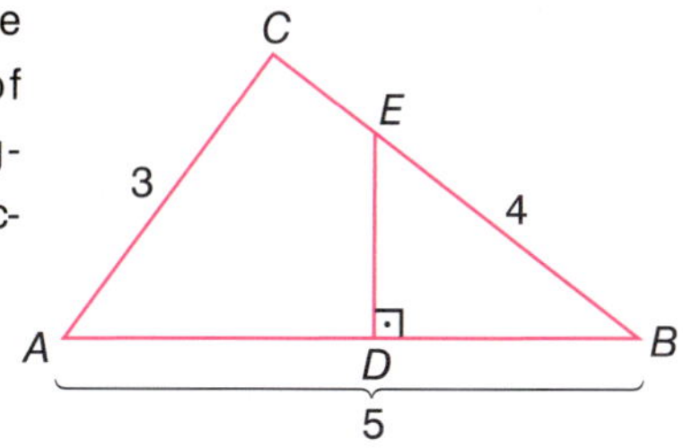

A) $\frac{4}{3}$ B) $\sqrt{5}$ C) $\frac{9}{4}$ D) $\frac{4\sqrt{3}}{3}$ E) $\frac{5}{2}$

2. (CWML 2017-18 Meet 3 #9)

Given the figure to the right, find the numerical value of the length of $\overline{DA}$ in order for $\overline{DE}$ to be parallel to $\overline{AB}$.

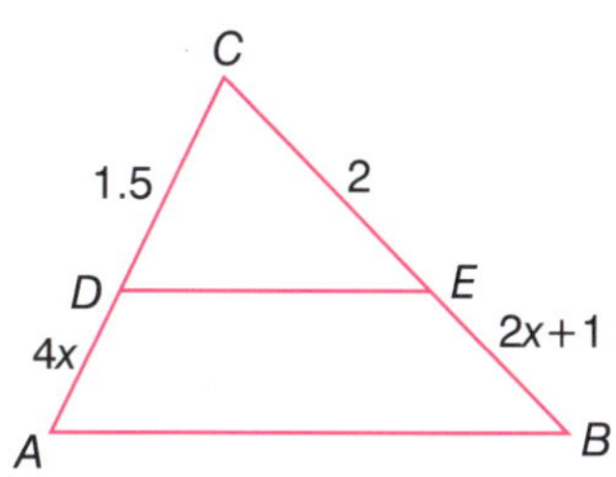

3. (2009 AMC 10A #10)

Triangle ABC has a right angle at B. Point D is the foot of the altitude from B, $AD = 3$, and $DC = 4$.

What is the area of ΔABC?

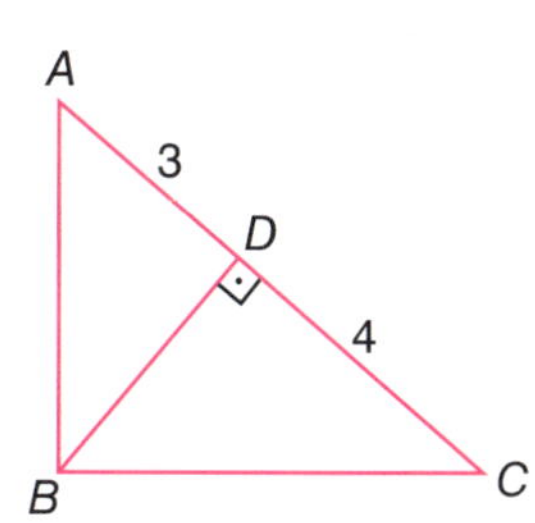

A) $4\sqrt{3}$ B) $7\sqrt{3}$ C) 21 D) $14\sqrt{3}$ E) 42

4. (CWML 2016-17 Meet 3 #9)

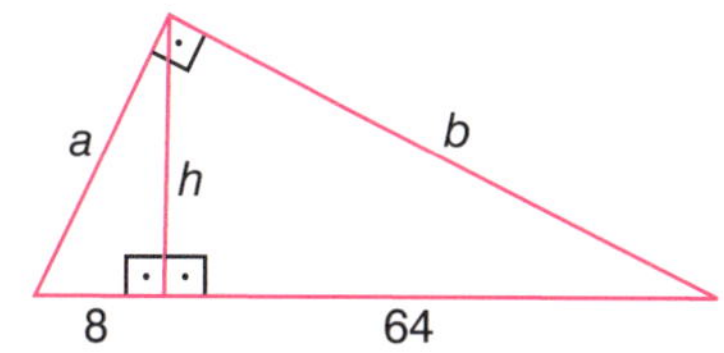

Find the exact length a in the given triangle.

5. (2004 AMC 12A #8)

In the overlapping triangles ΔABC and ΔABE sharing common side AB, $m\angle EAB$ and $m\angle ABC$ are right angles, $AB = 4$, $BC = 6$, $AE = 8$, and $\overline{AC}$ and $\overline{BE}$ intersect at D.

What is the difference between the areas of ΔADE and ΔBDC?

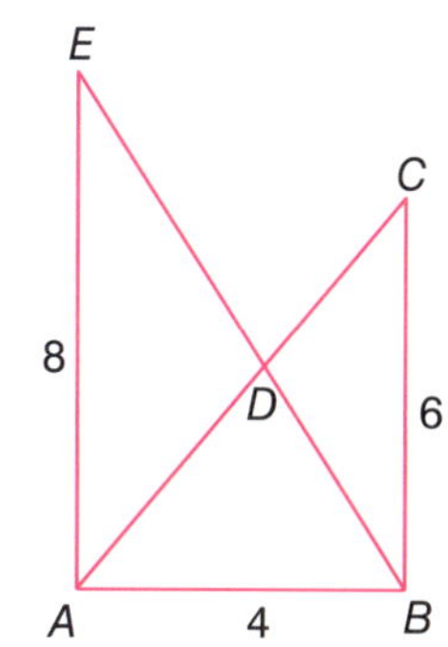

A) 2 B) 4 C) 5 D) 8 E) 9

6. (2013 AMC 10A #12)

In ΔABC, $AB = AC = 28$ and $BC = 20$. Points D, E, and F are on sides $\overline{AB}$, $\overline{BC}$, and $\overline{AC}$, respectively, $\overline{AC}$ such that $\overline{DE}$ and $\overline{EF}$ are parallel to $\overline{AC}$ and $\overline{AB}$, respectively.

What is the perimeter of parallelogram $ADEF$?

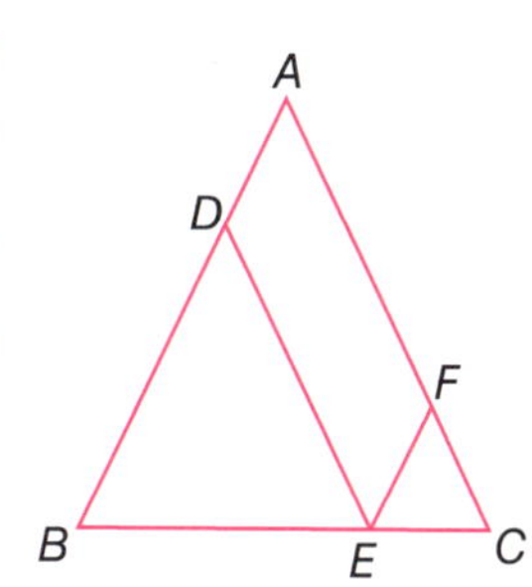

A) 48 B) 52 C) 56 D) 60 E) 72

7. (2019 AMC 10B #16) In $\triangle ABC$ with a right angle at C, point D lies in the interior of $\overline{AB}$ and point E lies in the interior of $\overline{BC}$ so that $AC = CD$, $DE = EB$ and the ratio $\frac{AC}{DE} = \frac{4}{3}$.

What is the ratio $\frac{AD}{DB}$?

A) $\frac{2}{3}$ B) $\frac{2}{\sqrt{5}}$ C) $\frac{1}{1}$ D) $\frac{3}{\sqrt{5}}$ E) $\frac{3}{2}$

8. (2013 AMC 12B #19) In triangle ABC, $AB = 13$, $BC = 14$, and $CA = 15$. Distinct points D, E, and F lie on segments $\overline{BC}$, $\overline{CA}$, and $\overline{DE}$, respectively, such that $\overline{AD} \perp \overline{BC}$, $\overline{DE} \perp \overline{AC}$, and $\overline{AF} \perp \overline{BF}$. The length of segment $\overline{DF}$ can be written as $\frac{m}{n}$, where m and n are relatively prime positive integers.

What is $m+n$?

A) 18 B) 21 C)24 D) 27 E) 30

9. (CWML 2008-9 Meet 2 #13)

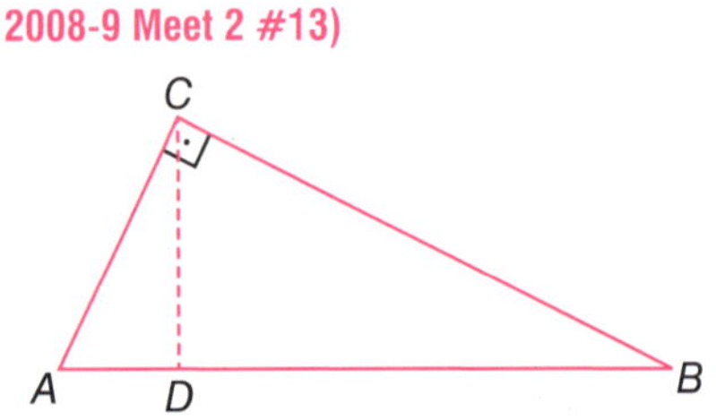

Triangle ABC is a right triangle with right angle at C, D is between A and B, $\overline{CD} \perp \overline{BA}$. Then we can prove

i) $\frac{BD}{CD} = \frac{CD}{DA}$

ii) $\frac{BD}{BC} = \frac{DA}{AC}$

iii) $\frac{CD}{BC} = \frac{DA}{AC}$

10. (2018 AMC 12A #11) A paper triangle with sides of lengths 3, 4, and 5 inches, as shown, is folded so that point A falls on point B.

What is the length in inches of the crease?

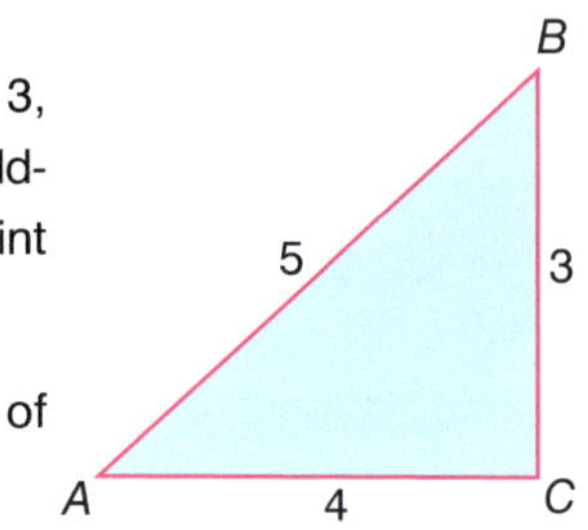

A) $1+\frac{1}{2}\sqrt{2}$ B) $\sqrt{3}$ C) $\frac{7}{4}$ D) $\frac{15}{8}$ E) 2

11. (2007 AMC 10B #21) Right $\triangle ABC$ has $AB = 3$, $BC = 4$ and $AC = 5$. Square $XYZW$ is inscribed in $\triangle ABC$ with X and Y on $\overline{AC}$, W on $\overline{AB}$, and Z on $\overline{BC}$.

What is the side length of the square?

A) $\frac{3}{2}$ B) $\frac{60}{37}$ C) $\frac{12}{7}$ D) $\frac{23}{13}$ E) 2

12. (1998 AIME # 6)

Let $ABCD$ be a parallelogram. Extend $\overline{DA}$ through A to a point P, and let $\overline{PC}$ meet $\overline{AB}$ at Q and $\overline{DB}$ at R.

Given that $PQ = 735$ and $QR = 112$, find RC.

CHAPTER 6

Note

Step 1: Center O, radius less than OA.

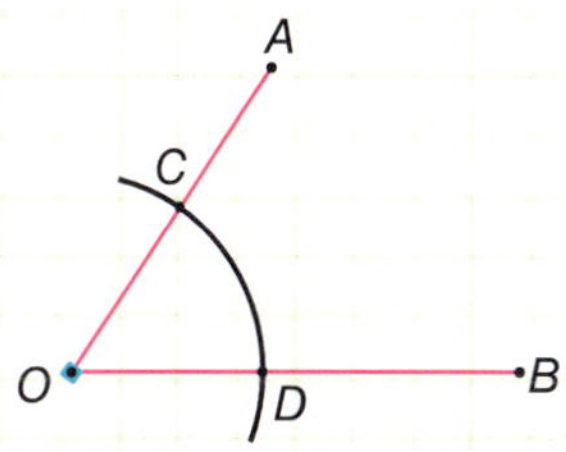

Step 2: Center C, radius R.

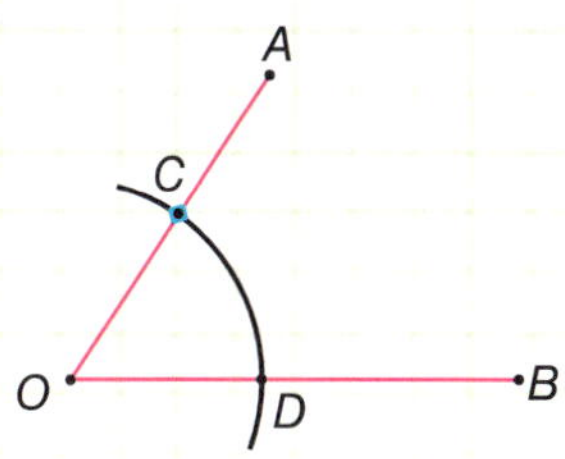

Step 3: Center D, radius R.

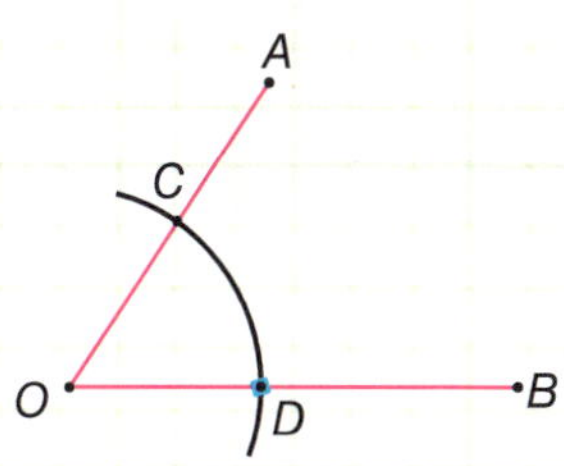

Step 4: Center O to intersected arcs.

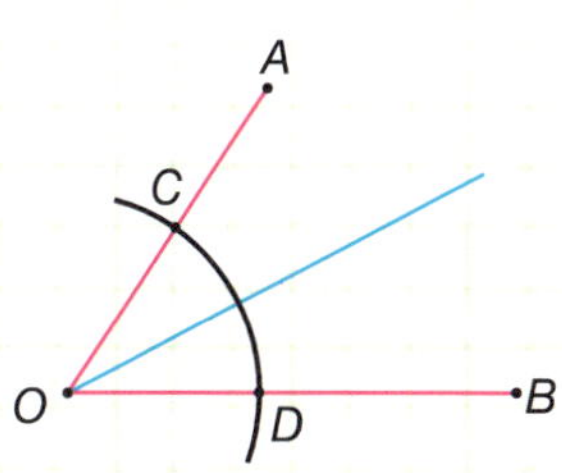

Angle Bisector Construction

Theorem 6.1

The Interior Angle Bisector Theorem

The bisector of an interior angle of a triangle divides the opposite side in the same ratio as the sides. Let ABC be a triangle, and let N be the point on BC such that $m\angle BAN = m\angle CAN$.

Then, $\frac{AB}{AC} = \frac{BN}{NC}$ or $\frac{AB}{BN} = \frac{AC}{NC}$.

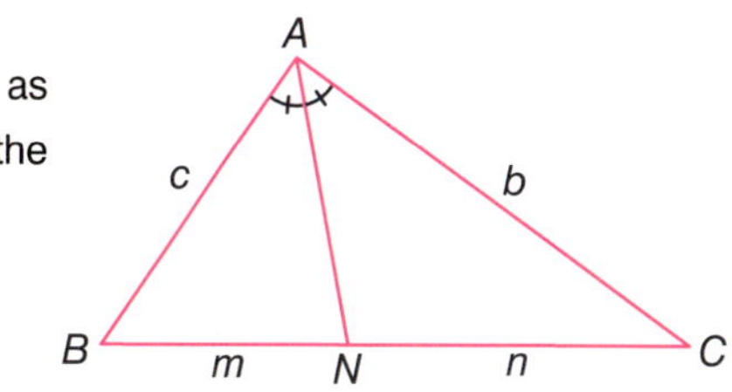

Proof 6.1

Let CT be parallel to AB and let $m\angle ATC$ be α. Since two segments are parallel each other, then $m\angle NAC$ and $m\angle ANB$ become α.

Thus, $AC = TC = b$ and

$\Delta ANB \sim \Delta TNC$ (similarity AAA – BOW TIE)

$$\frac{m}{n} = \frac{c}{b} \text{ or } \frac{c}{m} = \frac{b}{n}.$$

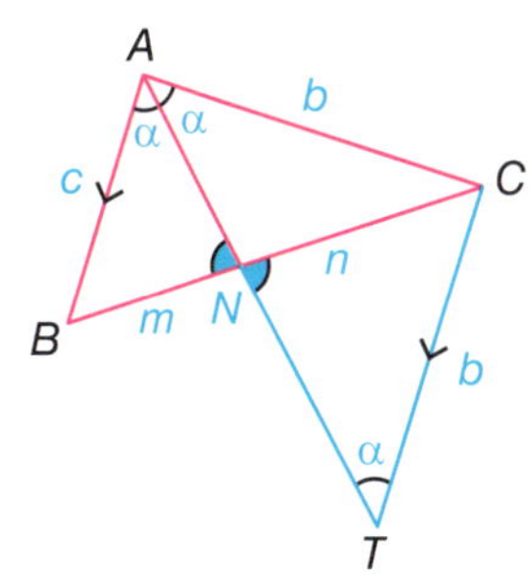

Notes 6.1

In triangle ABC with AN the angle bisector of $m\angle BAC$ the following ratios are always true.

$$\frac{m}{n} = \frac{c}{b} \text{ or } \frac{c}{m} = \frac{b}{n}$$

or $c \cdot n = b \cdot m$

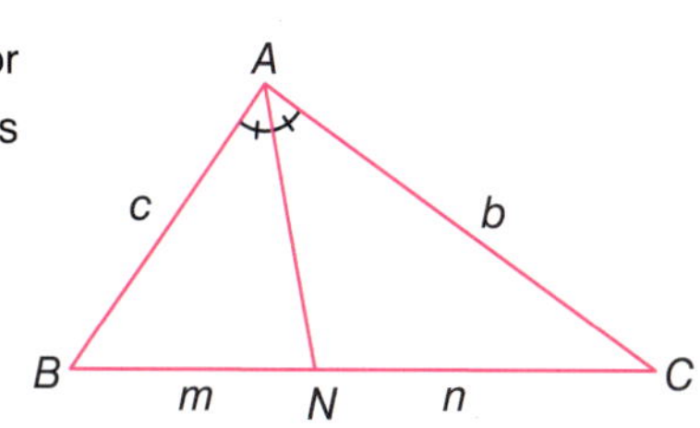

Exercise 6.1

If two sides are given as 5 and 7 units, then BN and NC can be written as $5k$ and $7k$.

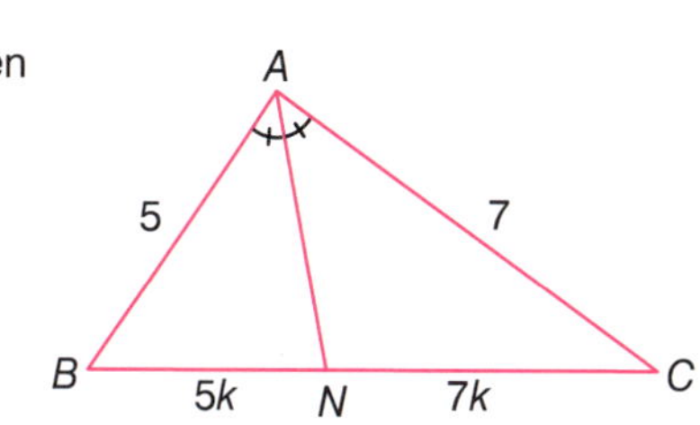

Exercise 6.2

If AN bisects angle $\angle BAC$, then

$$\frac{12}{8} = \frac{6}{x}$$

or $12 \cdot x = 8 \cdot 6$

So, $x = 4$.

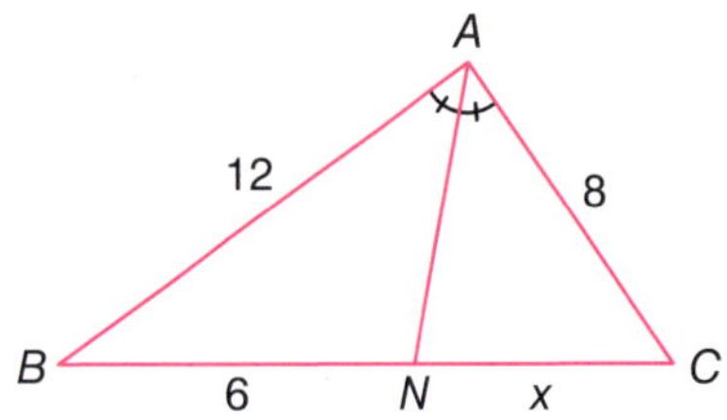

Example 6.1

In the figure, AN is the angle bisector of $\angle BAC$.

Given that $AB = 5$, $AC = 6$, and $BC = 22$, find NC.

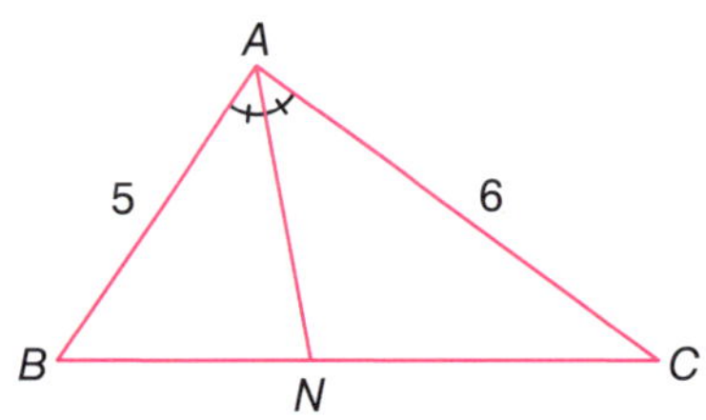

Solution 6.1

Let BN be $5k$ and NC be $6k$.

So, $BN + NC = BC$ and $5k + 6k = 22$

$$11k = 22$$

$$k = 2$$

$NC = 6k = 6 \cdot 2 = 12$.

Theorem 6.2

Exterior Angle-Bisector Theorem

In triangle ABC, Let AN be the angle bisector of the exterior angle at A as shown. Then,

$$\frac{AB}{BN} = \frac{AC}{CN}.$$

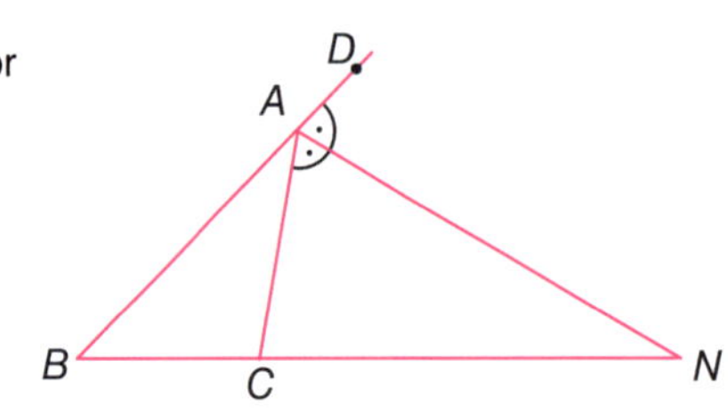

Proof 6.2

Let CT be the line segment parallel to AB, where T is on AN and C is on BN as shown.

Then $m\angle DAT = m\angle TAC = m\angle ATC = \beta$.

Thus, $AC = TC = b$.

From $TNC \sim ANB$ (AAA similarity)

$$\frac{b}{c} = \frac{n}{m+n} \Rightarrow \frac{b}{n} = \frac{c}{m+n}.$$

Therefore, $\dfrac{AB}{BN} = \dfrac{AC}{CN}$.

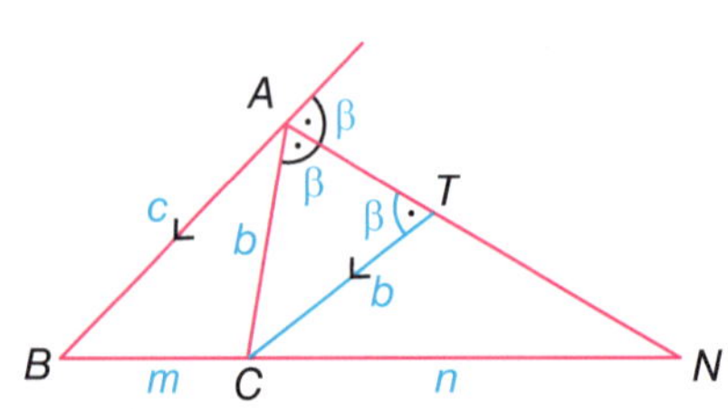

Note

Angle Trisection

Angle trisection is the division of an *arbitrary* angle into three equal angles. It was one of the three geometric problems of antiquity for which solutions using on1y compass and straightedge were sought. The problem was algebraically proved impossible by Wantzel (1836).

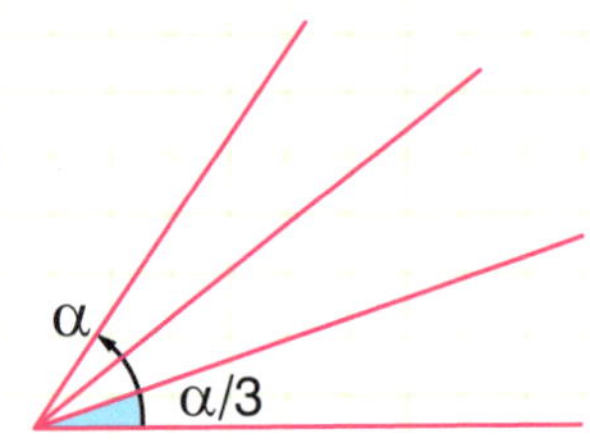

Trisection

My Notes

Try This

Triangle ABC has a right angle at B, and AD is the bisector of $\angle BAC$.

If $AB = 5$ and $BC = 12$, then find BD.

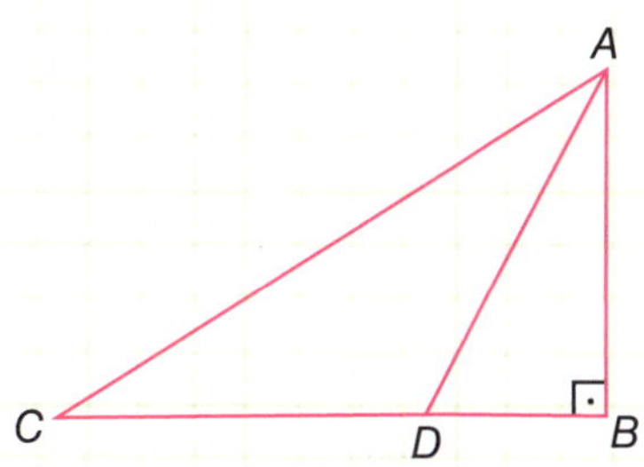

My Notes

Notes 6.2

If two sides of $\triangle ABC$ are given as 4 and 6 units,then CN (close to 4) and BN (close to AB) can be written as $4k$ and $6k$, respectively.

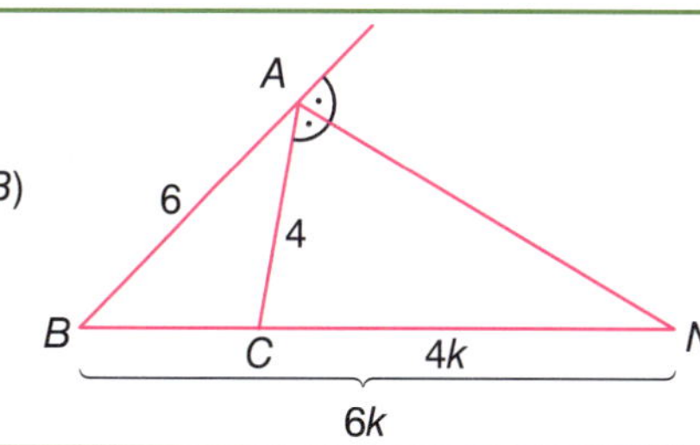

Example 6.2

In the figure,

$AN = 4$,

$BN = 6$, and

$AC = NC$.

Find AC.

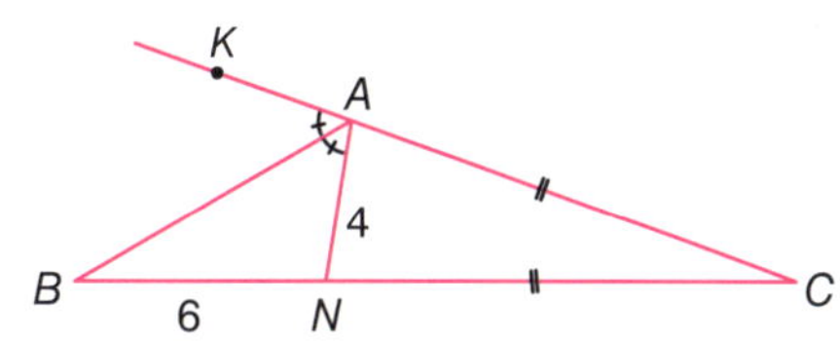

Solution 6.2

Let $x = AC = NC$. From the Exterior Angle Bisector Theorem,

$$\frac{4}{6} = \frac{x}{x+6} \Rightarrow \frac{2}{3} = \frac{x}{x+6} \Rightarrow 3x = 12 + 2x$$

$$x = 12.$$

Example 6.3

In the figure,

$AD = 5$ and

$DB = 4$.

Find the area of $\triangle ABC$.

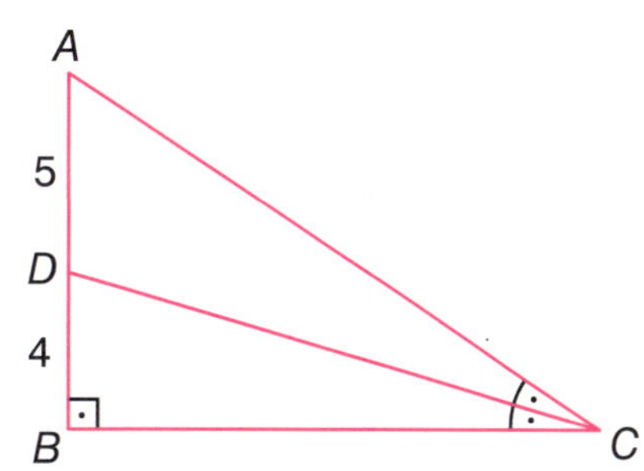

Solution 6.3

From the Interior Angle Bisector Theorem,

we can write $BC = 4k$ and $AC = 5k$.

$$(5k)^2 = (4k)^2+9^2$$

$$25k^2 = 16k^2+81$$

$$9k^2 = 81$$

$$k^2 = 9$$

$$k = 3 \Rightarrow BC = 4k = 4\cdot 3 = 12$$

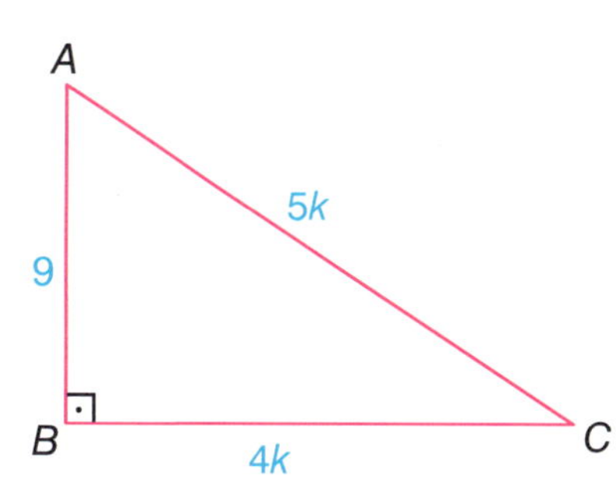

$$\text{Area } \triangle ABC = \frac{9\cdot 12}{2} = 54 \text{ unit}^2.$$

Example 6.4

In the figure $m\angle CAB = 2m\angle ABC$.

Given that $AC = 4$ cm, $AB = 5$ cm, and the bisector length from A to BC is $\frac{10}{3}$.

Find the length of BC.

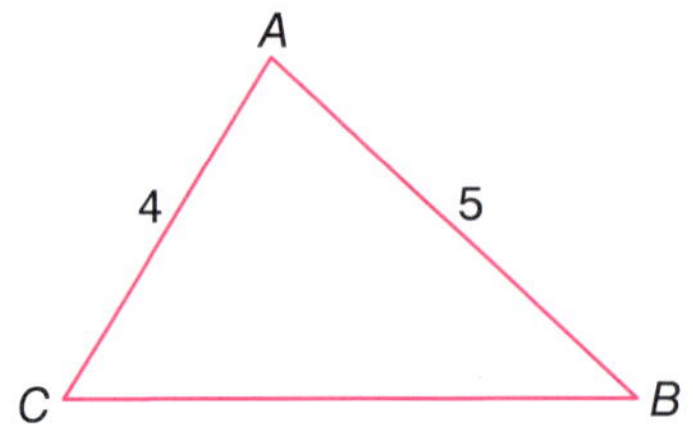

Solution 6.4

We replace α for $m\angle ABC$ and 2α for $m\angle CAB$.

Let AN be the angle bisector of $m\angle CAB$.

Triangle ANB is isosceles, and $AN = NB = 5k$.

So, $5k = \frac{10}{3} \Rightarrow k = \frac{2}{3}$

Therefore, $BC = 9k = 9 \cdot \frac{2}{3} \Rightarrow BC = 6$

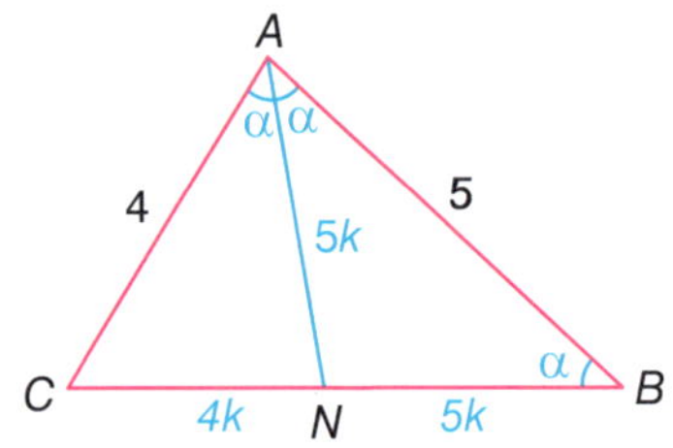

Remark 6.1

Let AD be on the angle bisector of $\angle CAB$ as shown. If G is any point on the bisector line such that $EG \perp AE$ and $GF \perp AF$ where E is on line AC and F is on line AF, then $EG = GF$.

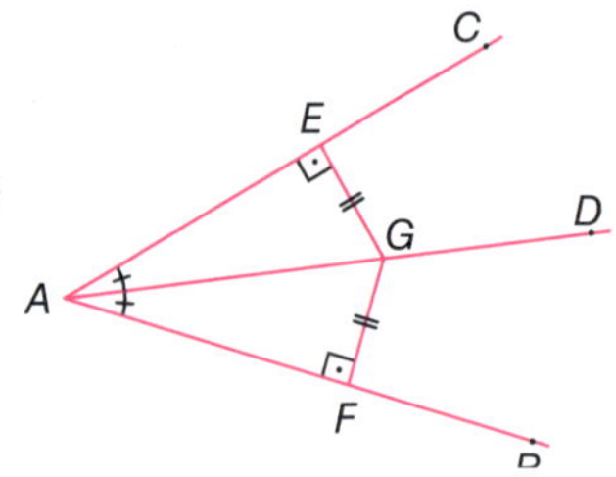

Example 6.5

AC is the angle bisector of $\angle BAD$.

$AD = 13$, $BC = 9$, and $AB = 25$.

Find DC.

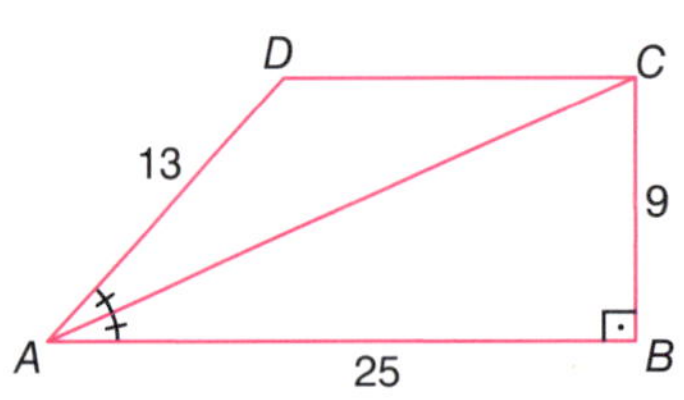

Solution 6.5

We extend the segment of AD to point H.

From Remark 6.1, $HC = CB = 9$ and $AH = AB = 25$. Therefore, $DH = 12$. So, $DC = 15$ (9–12–15 triangle).

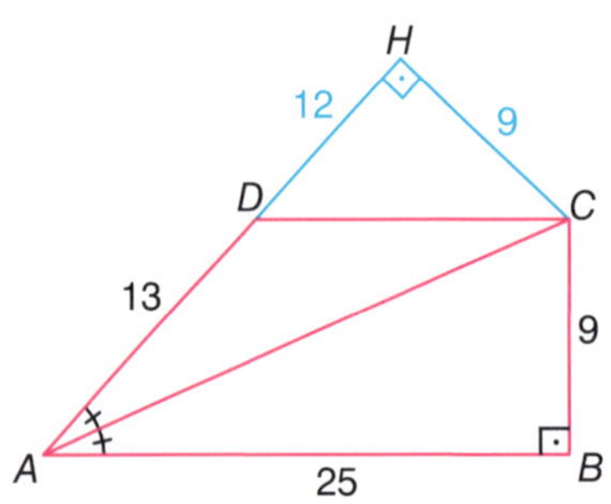

Prove this

ABC is the angle bisector of $m\angle BAD$ and AD is the angle bisector of $m\angle CAF$.

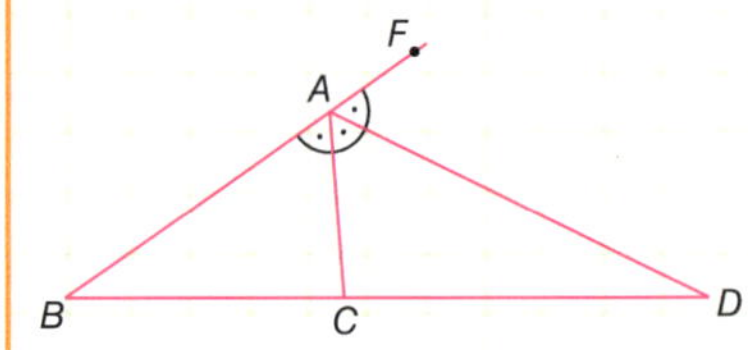

Prove that $\frac{1}{AC} = \frac{1}{AB} + \frac{1}{AD}$.

My Notes

Try This

Three side lengths of a triangle are consecutive integers. The largest angle of the triangle is two times of the smallest angle.

What is the length of the largest side?

My Notes

Example 6.6

In the figure,

$BD = 17$,

$BA = 8$, and

$BC = 10$

If BC is the angle bisector of $\angle ACB$, then find DC.

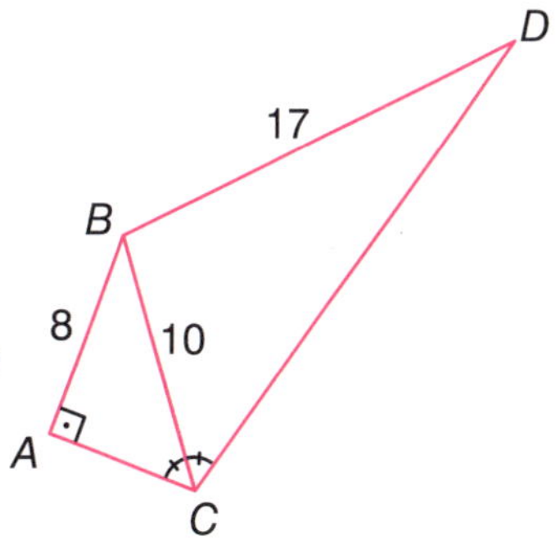

Solution 6.6

By drawing a perpendicular foot from point B to CD, (Remark 6.1) the following lengths connections give the value of DC.

$BH = AB = 6$ and

$HD = 15$ (8–15–17 triangle)

Therefore, DC = 6+15 = 21.

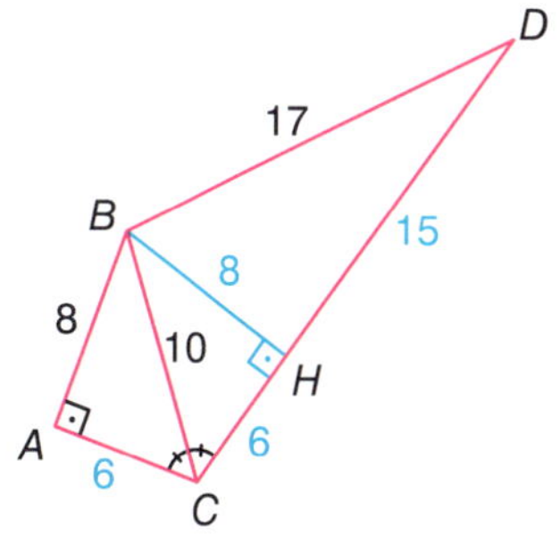

Example 6.7

AC is the angle bisector of $\angle BAD$,

$m\angle ADC = 90°$,

$m\angle ABC = 60°$, and

$BC = 12$,

find AD.

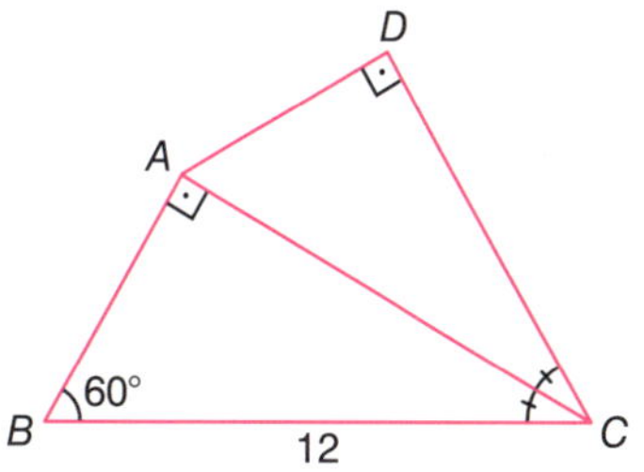

Solution 6.7

By using remark 6.1 and drawing a perpendicular foot from point A to BC.

We obtain AD = AH = $3\sqrt{3}$. (30°– 60°– 90°)

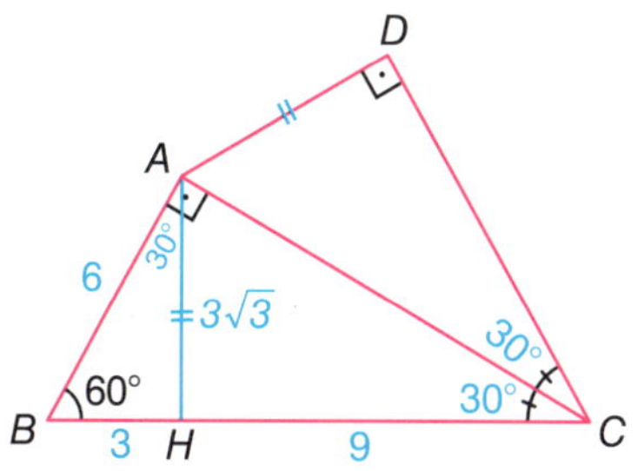

Example 6.8

AN and AM are the angle bisectors of $m\angle CAK$ and $m\angle BAC$, respectively.

If $MC = 3$ and $CN = 9$, then find BM.

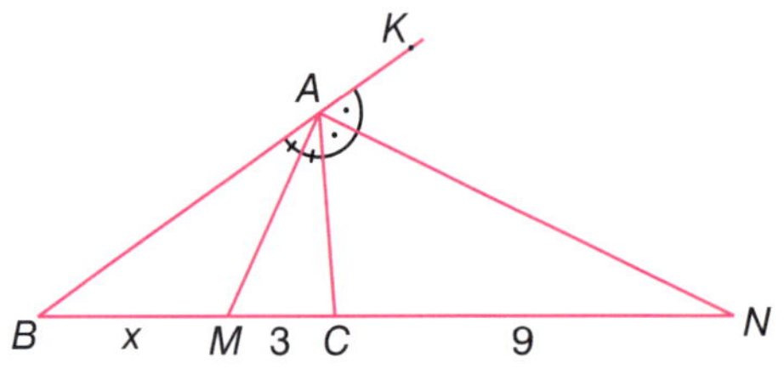

Solution 6.8

From the Interior Angle Bisector Theorem,

$AC = 3k$ and $AB = xk$.

From the Exterior Angle Bisector Theorem,

$$\frac{3k}{9} = \frac{xk}{12+x} \Rightarrow \frac{1}{3} = \frac{x}{12+x}$$

$$2x = 12$$

$$x = 6.$$

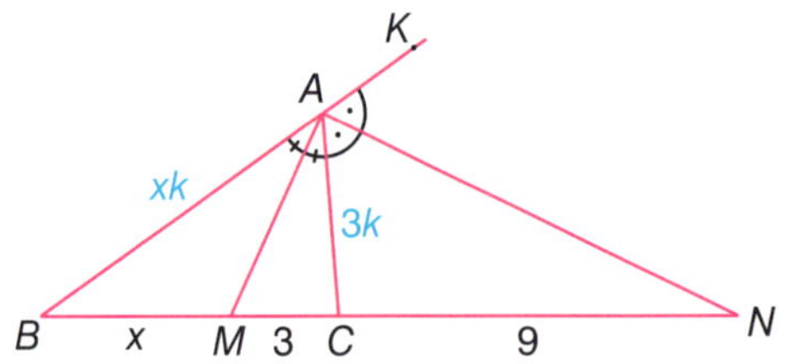

Example 6.9

In ΔABC,

$AB = 8$, $BC = 15$, and $CA = 12$.

Point N lies on BC and AN bisects $\angle BAC$.

Point E lies on AC and BE bisects $\angle ABC$.

Two bisectors intersect at F. What is the ratio of $\frac{AF}{FN}$?

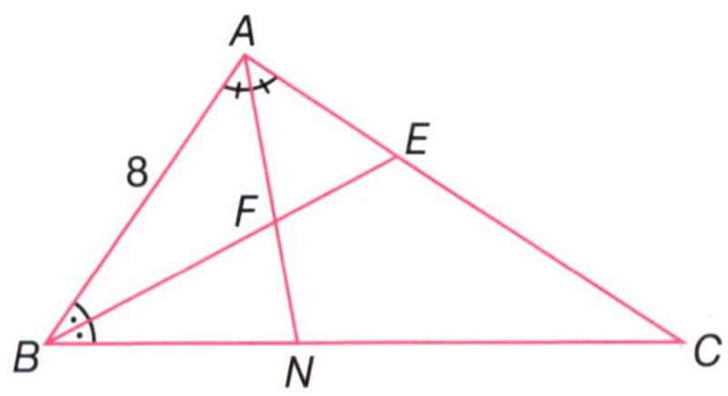

Solution 6.9

Using the angle bisector AN to $\angle ABC$

$BN = 8k$ and $NC = 12k$

$$8k + 12k = 15$$

$$20k = 15 \Rightarrow k = \frac{3}{4}.$$

So, $BN = 6$ and $NC = 9$.

Using the angle bisector BF to ΔABN,

We get $\frac{AF}{FN} = \frac{8}{6} \Rightarrow \frac{AF}{FN} = \frac{4}{3}.$

Try This

In the figure, $m\angle C = 2m\angle DAC$ and AD is the angle bisector of $\angle MAC$. Given that, $AB = 8$,

$AC = 6$, and $AM = 5$,

find DC.

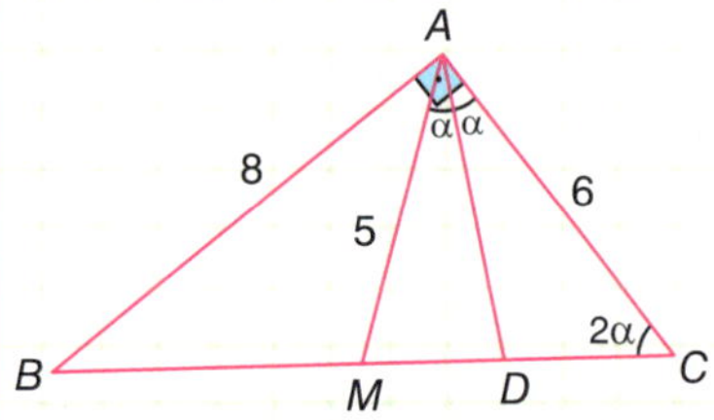

My Notes

Challenge

ABC is a right triangle with $m\angle B = 90°$. AB and DC are bisectors of $m\angle DAC$ and $m\angle ACB$.

If $2AB = DC$, then find $m\angle ADC = \alpha$.

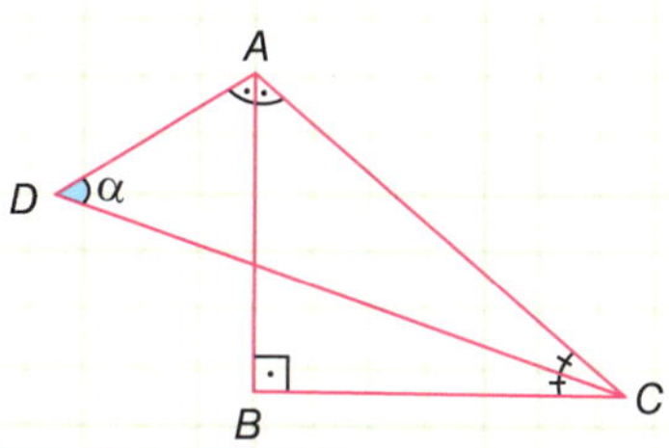

My Notes

Example 6.10

A bisector of a right angle of a right triangle ABC divides into its hypotenuse the ratio of 1 to 3.

Find the area of ΔABC.

Solution 6.10

Let x and y be the lengths of AB and AC, respectively.

From the Angle Bisector Theorem, $y = 3x$.

From the Pythagorean Theorem,

$$x^2 + y^2 = (1+3)^2 = 16$$

$$x^2 + 9x^2 = 16 \quad (y = 3x)$$

$$10x^2 = 16 \Rightarrow x^2 = \frac{8}{5}.$$

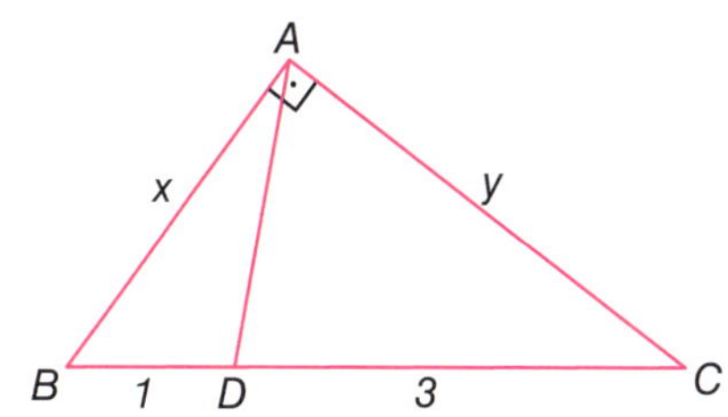

The area of the right triangle is

$$\text{Area } \Delta ABC = \frac{x \cdot y}{2} = \frac{x \cdot 3x}{2} = \frac{3}{2}x^2 = \frac{3}{2}\frac{8}{5} = \frac{12}{5}.$$

Theorem 6.3

The Length of Angle Bisector

In ΔABC let N be a point on BC such that AN is the angle bisector of $\angle BAC$. Let x, m, and n be the lengths of AN, BN, and NC, respectively.

$$AN^2 = AB \cdot AC - BN \cdot NC$$

$$x^2 = c \cdot b - m \cdot n$$

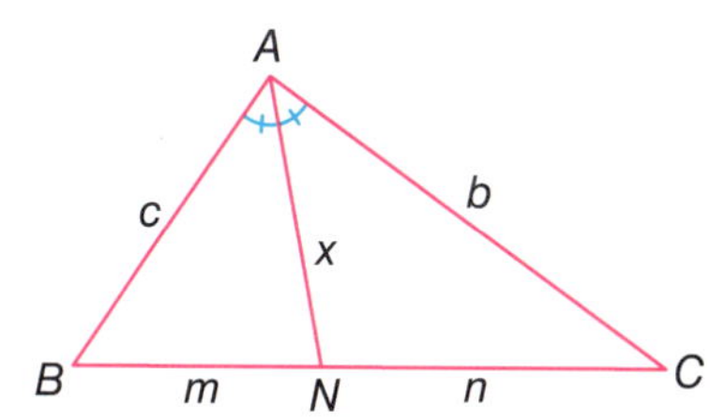

Note: Proof 6.3 will be provided in chapter 7.

Exercise 6.2

To find $AN = x$, we first need to find NC.

By using the Angle Bisector Theorem,

$$\frac{6}{4} = \frac{3}{NC} \Rightarrow NC = 2.$$

By the Length of Angle bisector theorem,

$$x^2 = 6 \cdot 3 - 4 \cdot 2$$

$$x^2 = 18 - 8 \Rightarrow x = \sqrt{10}.$$

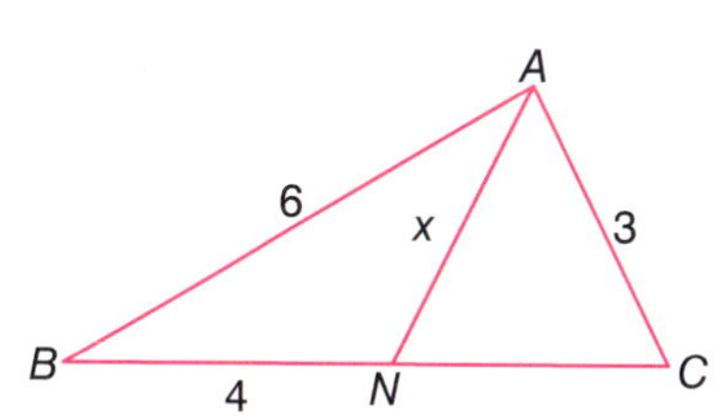

Example 6.11

ABC is a triangle with

$AB = 6$, $AC = 8$, and $BC = 7$.

Find the length of the angle bisector from point A to *BC*.

Solution 6.11

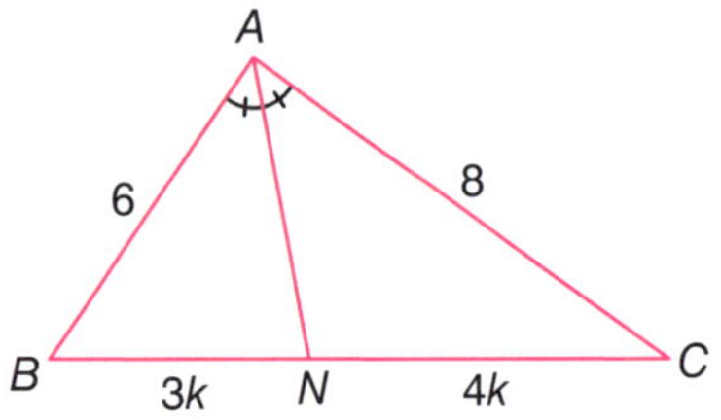

In the figure, $AB = 6$ and $AC = 8$ are given.
From the Interior Angle Theorem,

$$BN+NC = 3k+4k = 7$$

$$BN = 3 \text{ and } NC = 4.$$

From the Angle Bisector Length Theorem.

$$AN^2 = 6{\cdot}8-4{\cdot}3 = 36.$$

Therefore, $AN = 6$.

Example 6.12

In a triangle *ABC*,

$$m\angle ABC = 2m\angle BCA,$$

$$BC = 7, \text{ and } AB = 9.$$

Find *AC*.

Solution 6.12

By creating the angle bisector *BN* from point *B* to *AC*,

$BN = 7k$ and $AN = 9k$.

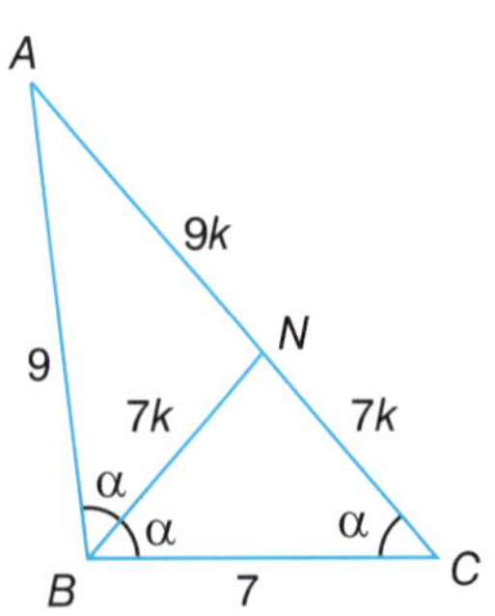

From the Angle Bisector Theorem,

$$(7k)^2 = 9{\cdot}7 - 9k{\cdot}7k$$

$$49k^2 = 63 - 63k^2$$

$$49k^2 + 63k^2 = 63$$

$$112k^2 = 63.$$

So, $k = \frac{3}{4}$.

Therefore, $AC = 16k = 16 \cdot \frac{3}{4} = 12$.

Catriona Agg @Cshearer41

Geometry puzzle

This pattern is made from three right angled triangles.

What is the area of pink region?

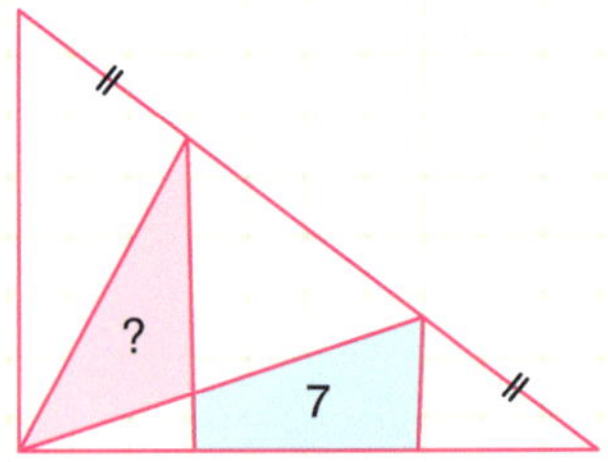

My Notes

Catriona Agg @Cshearer41
Geometry puzzle

What is the shaded area?

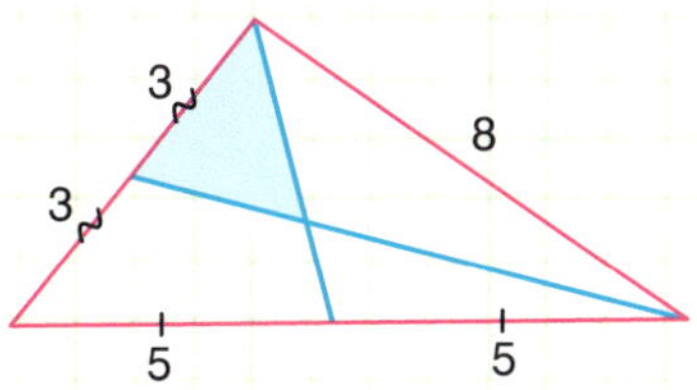

My Notes

Example 6.13

AF is the angle bisector of $\angle BAC$ and DE is the angle bisector of $\angle BDC$.

Given that $AB = 8$, $DC = 16$, and $BD = AC = 12$, find $\frac{FE}{EC}$.

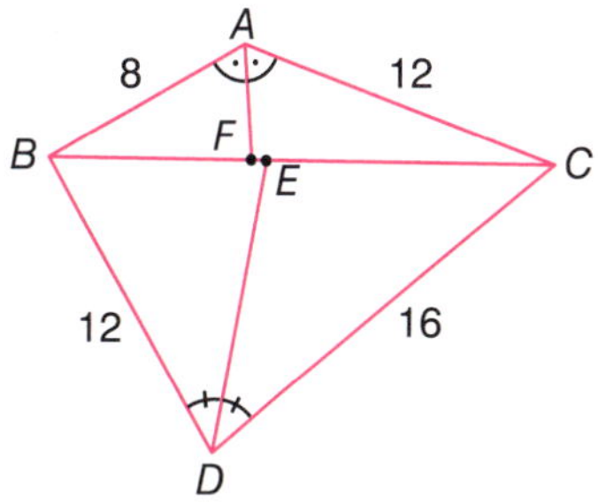

Solution 6.13

From ΔBDC,

$EC = 4k$ and $BE = 3k$.

Let $FE = x$ and $BF = 3k - x$.

From ΔABC (Angle Bisector Theorem 6.1)

$$\frac{3}{12} = \frac{2}{3} = \frac{3k-x}{4k+x} \Rightarrow 8k+2x = 9k-3x$$

$$5x = k.$$

So, $FE = x = \frac{k}{5}$ and $EC = 4k$.

Therefore, the ratio of $\frac{FE}{EC} = \frac{\frac{k}{5}}{4k} = \frac{k}{5} \cdot \frac{1}{4k} = \frac{1}{20}$.

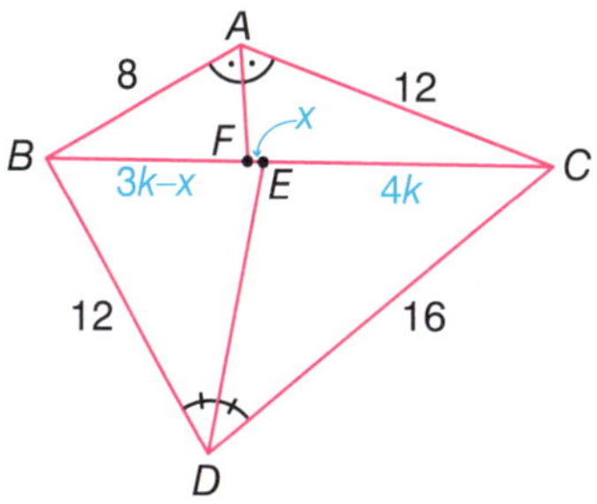

Example 6.14

AN and AM are angle bisectors of $\angle BAC$ and $\angle CAK$, respectably.

Given that, $AN = 6$, $AM = 8$, and $BN = 4$, find NC.

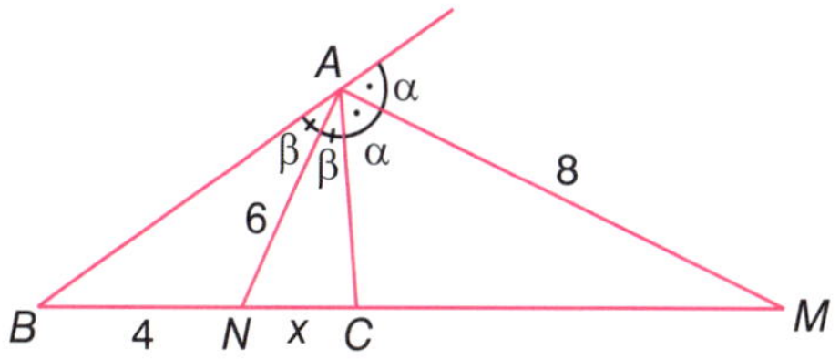

Solution 6.14

$2\alpha + 2\beta = 180^\circ \Rightarrow \alpha + \beta = 90^\circ$.

So, $m\angle NAM = 90^\circ$ and $NM = 10$.

Let $AC = xk$ and $AB = 4k$.

Then, the Exterior Aangle Bisector Theorem

$$\frac{4k}{14} = \frac{xk}{10-x} \Rightarrow \frac{2}{7} = \frac{x}{10-x} \Rightarrow 7x = 20-2x$$

$$9x = 20.$$

Therefore, $x = \frac{20}{9}$.

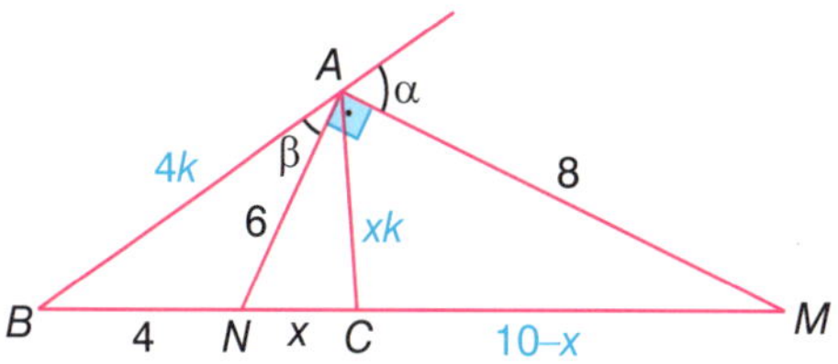

Remark 6.2

In a triangle all angle bisectors intersect at one point. This point is the center of the circle inscribed into the triangle.

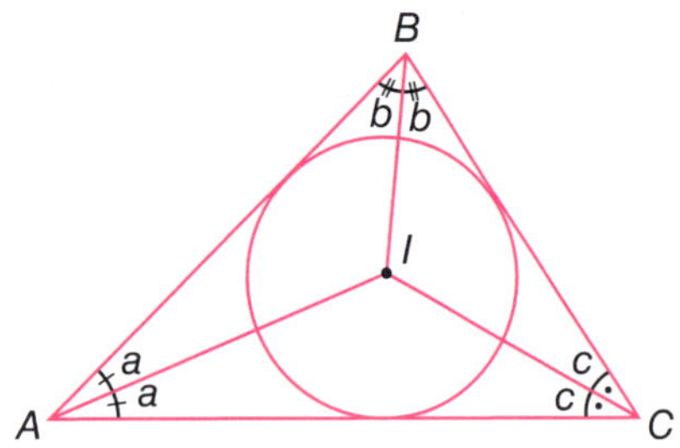

I : The incenter

Example 6.15

ABC is a triangle and the length of the bisector of $\angle CAB$ is 10.

Given that,

$AC = 12$ and $AB = 15$,

find CB.

Solution 6.15

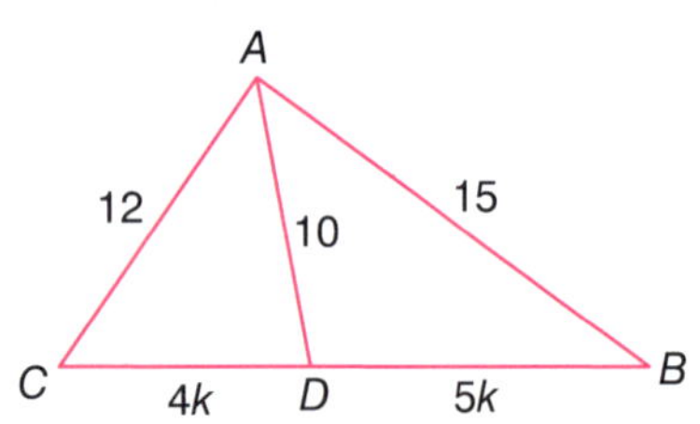

By the Angle Bisector Theorem,

$$\frac{CD}{DB} = \frac{12}{15} = \frac{4}{5}.$$

So, $CD = 4k$ and $DB = 5k$. From the Angle Bisector Length Theorem,

$$(AD)^2 = AB \cdot AC - CD \cdot DB$$

$$10^2 = 15 \cdot 12 - 20k^2$$

$$20k^2 = 180 - 100$$

$$k^2 = 4 \Rightarrow k = 2.$$

Therefore, $CB = 9k = 9 \cdot 2 = 18$.

Challenge

If $\frac{bc}{ad} = 2$, then

find the value of $\frac{\alpha}{\beta}$.

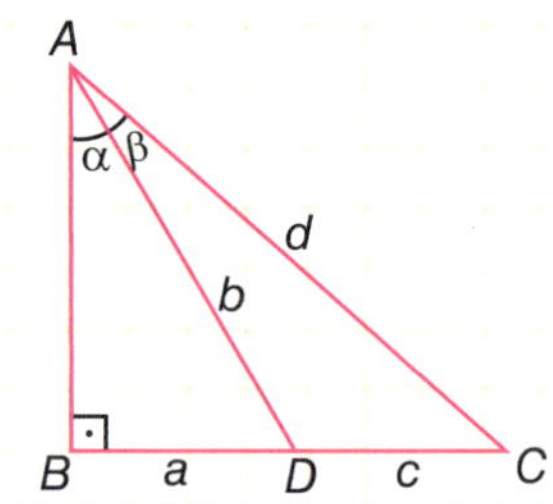

My Notes

1. In the figure, AD is the bisector of $\angle BAK$.

   Find $AC = BC = x$.

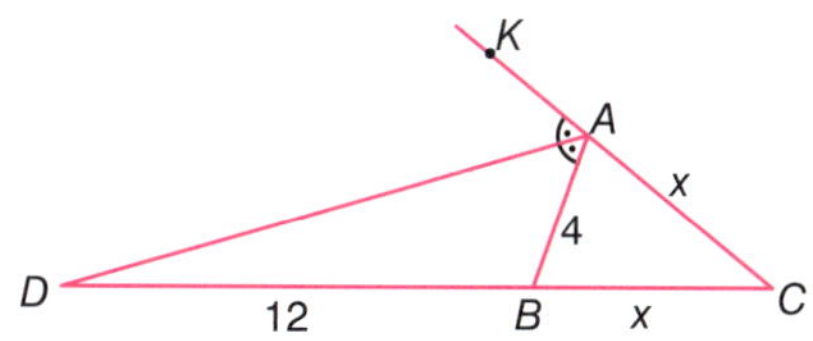

2. In the figure, AD is the bisector of angle $\angle BAC$.

   Given that $AB = 6$, $BD = 3$, and $DC = 2$, find AD.

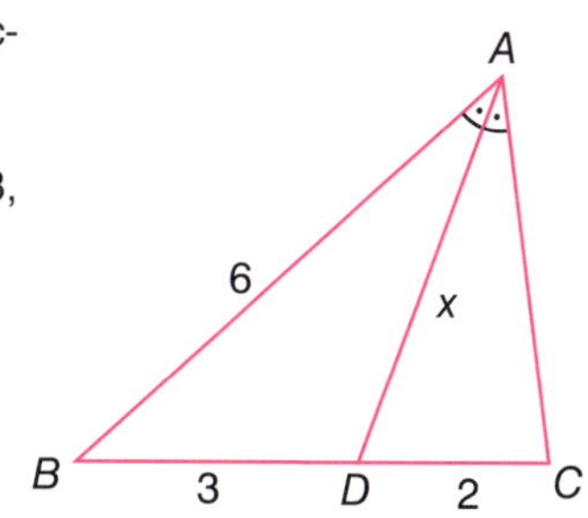

3. In the figure,

   $BC = DC$,

   $m\angle ACB = m\angle ACD$,

   $AB = 2x-2$, and $AD = x+1$.

   Find x.

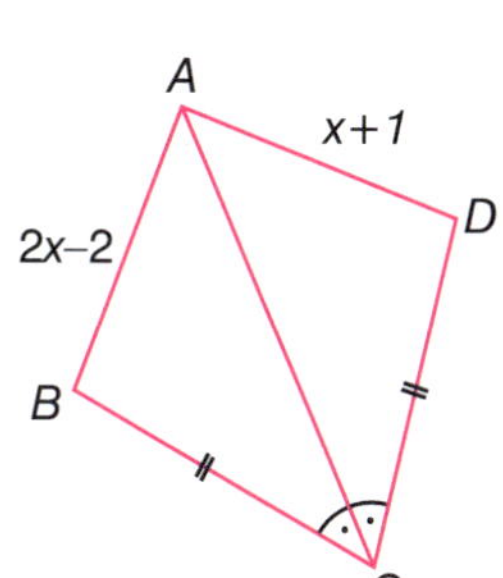

4.

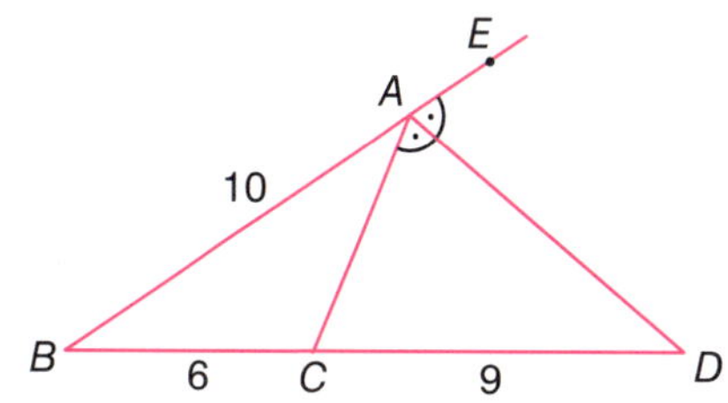

AD is the bisector of $\angle CAE$, $AB = 10$, $BC = 6$, and $CD = 9$

Find AC.

5.

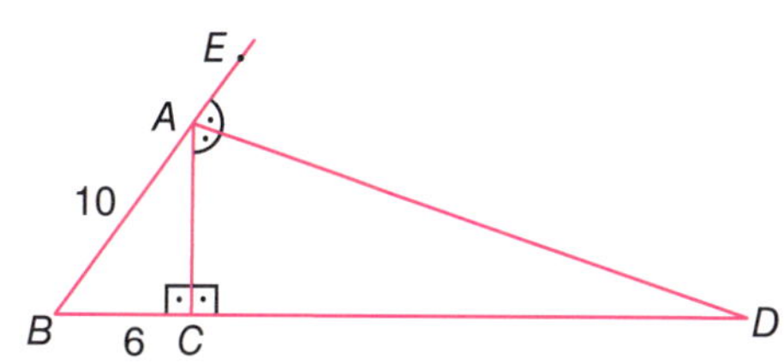

In the figure, $m\angle CAD = m\angle DAE$ and $AC \perp BD$.

If $AB = 10$, $BC = 6$, then find CD.

6. ABC is a right triangle, CD is the bisector of $\angle ACB$ and $AB \perp BC$.

   If $DB = 2$ and $AD = 3$, find AC.

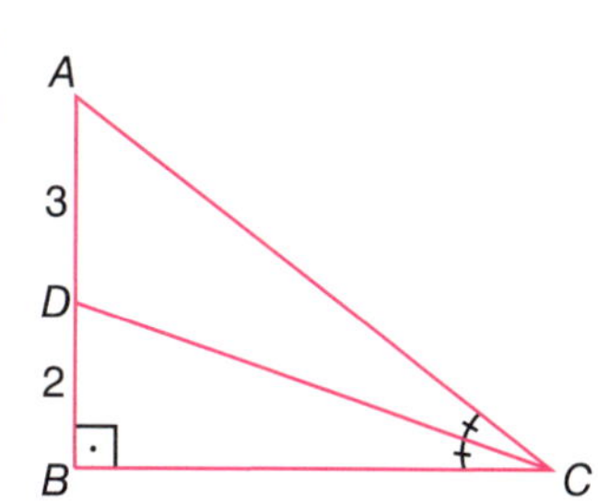

7. In a triangle ABC, point D is on BC, and AD bisects $\angle A$.

If $AB = 6$ cm, $BD = 3$ cm, and $DC = 2$ cm, then find the length of AD.

8. In a triangle ABC, D lies on the side BC and AD is the angle bisector of $\angle A$.

If $AC = BD$, $AB = 9$ cm, and $DC = 4$ cm, the find the length of BD.

9. In the figure,

$m\angle C = 90°$,

$m\angle BAD = m\angle DAC$,

$BD = DA$, and

$DC = 3$ cm.

Find the length of BD.

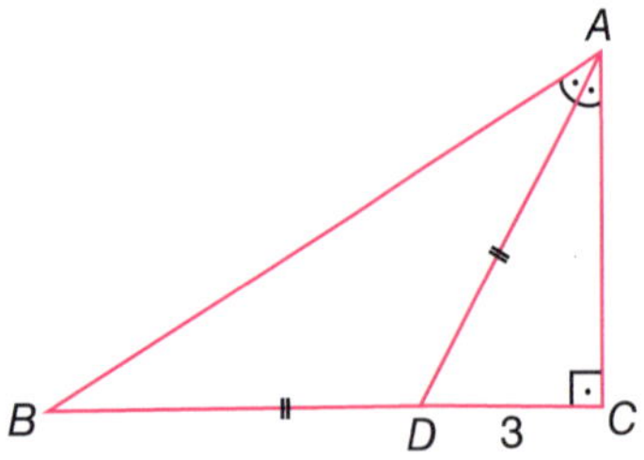

10. In the figure, AN and BE are the angle bisectors of $\angle A$ and $\angle B$, respectively.

If $AB = 4$, $AC = 6$, and $BC = 5$, then find the length of BE.

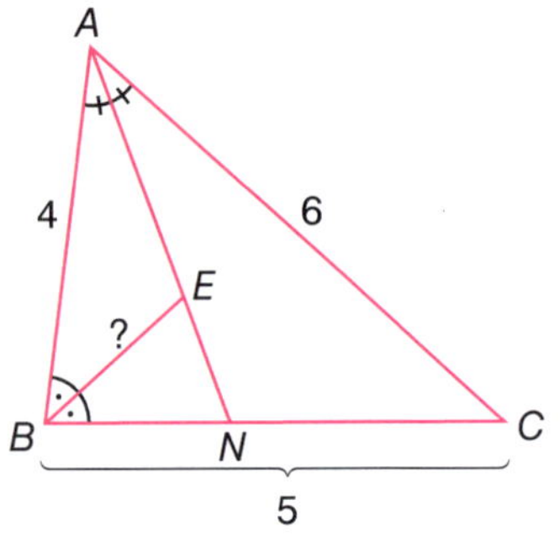

11.

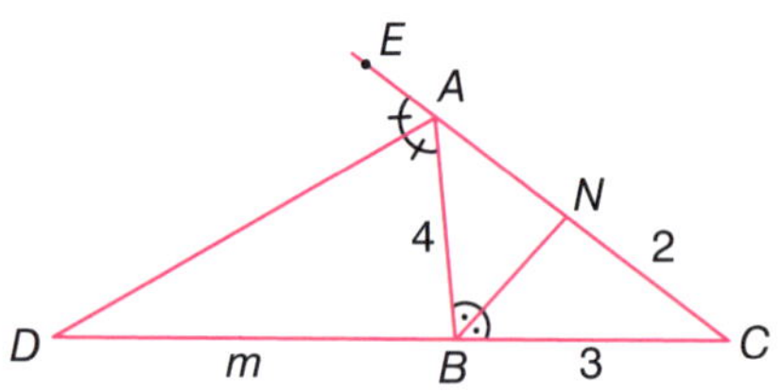

In the figure, AD is the angle bisector of $\angle BAE$ and BN is the angle bisector of $\angle ABC$.

Given that $DB = m$, $AB = 4$, $BC = 3$, and $NC = 2$, find the value of m.

12. In the figure, $AB \perp BC$,

$m\angle BCA = m\angle ACD$,

$AC = 10$,

$BC = 5$, and

$AD = 13$.

Find CD.

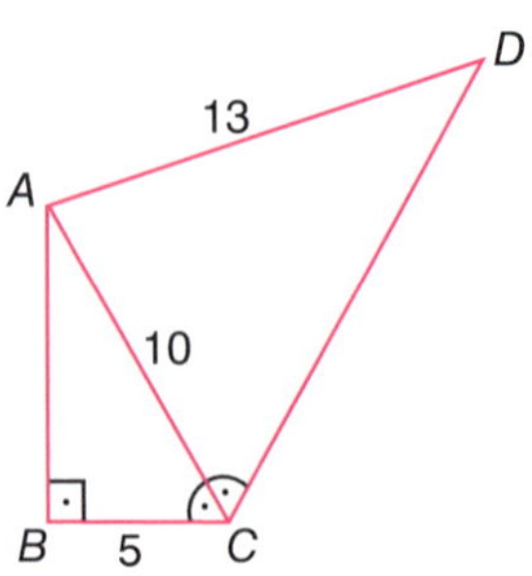

1.

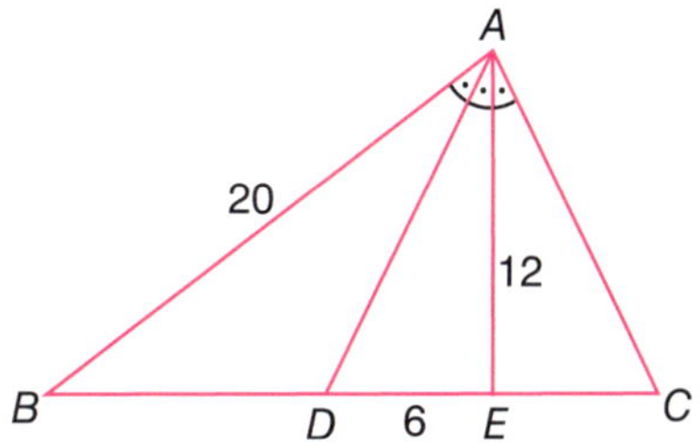

In the figure, $AB = 20$ cm, $AE = 12$ cm, and $DE = 6$ cm.

If $m\angle BAD = m\angle DAE = m\angle EAC$, then find BC.

2. In the figure,

$m\angle CAB = 2m\angle ABC$,

$AC = 4$ cm, and

$AB = 5$ cm.

What is BC?

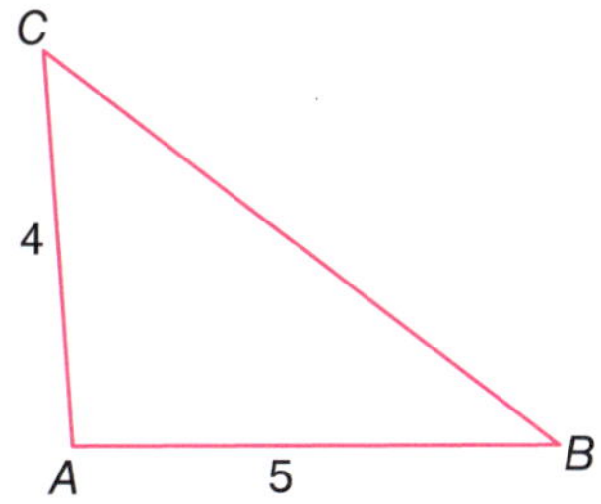

3.

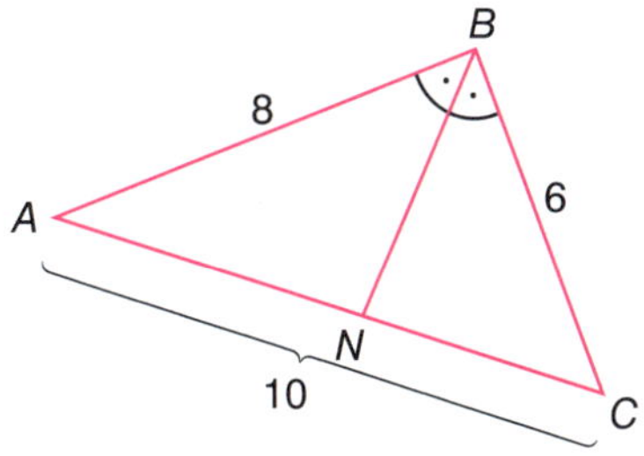

In the figure, $AB = 8$ cm, $BC = 6$ cm, and $AC = 10$ cm.

What is the length of the bisector of $\angle ABC$.

4. In the figure, AD and BE are the angle bisectors of $\angle BAC$ and $\angle ABC$, respectively.

If $\frac{AE}{ED} = \frac{3}{2}$ and $AC = 12$, then find DC.

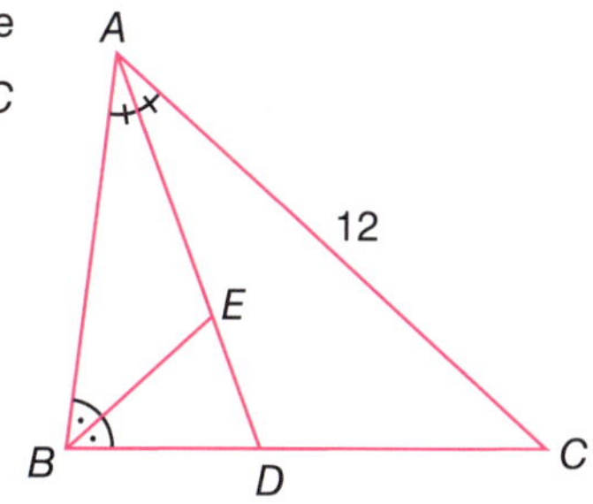

5. In the figure, $AB \perp BC$,

$m\angle BAC = m\angle CAD$,

$\frac{\text{Area } \Delta ABC}{\text{Area } \Delta ACD} = \frac{4}{3}$, and

$AD = BC = 6$ cm.

What is AC?

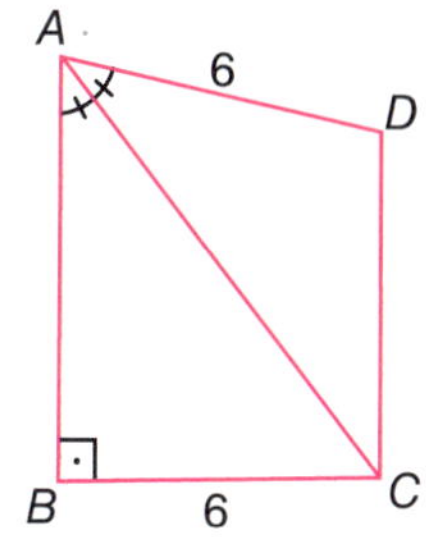

6. In the figure, AK is a bisector of $\angle CAB$.

$AC \perp BC$, $AD \perp BD$, and

$AB = 10$, $BC = 6$.

Find AD.

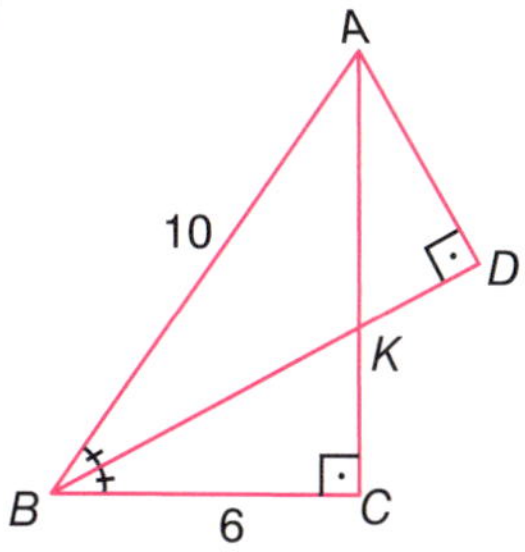

7.

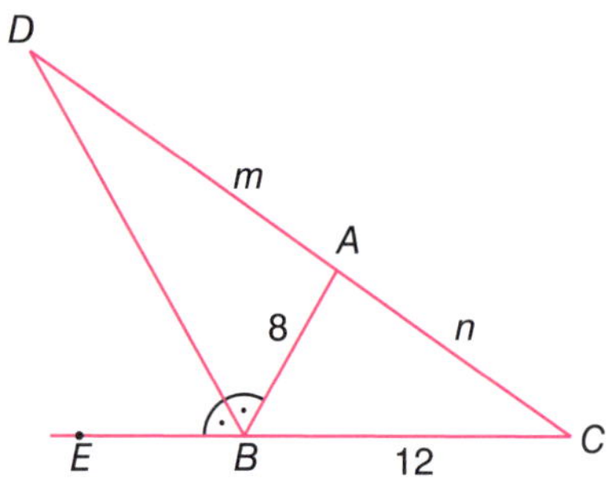

BD is the angle bisector of $\angle ABE$.

$AB = 8$, $BC = 12$, $DA = m$, and $AC = n$.

What is the value of $\frac{(2m+n)}{m}$?

8. In the figure, $AC \perp BC$ and $CH \perp AB$. DC is the angle bisector of $\angle HCA$, $HD = 1$, and $BC = 5$.

What is DC?

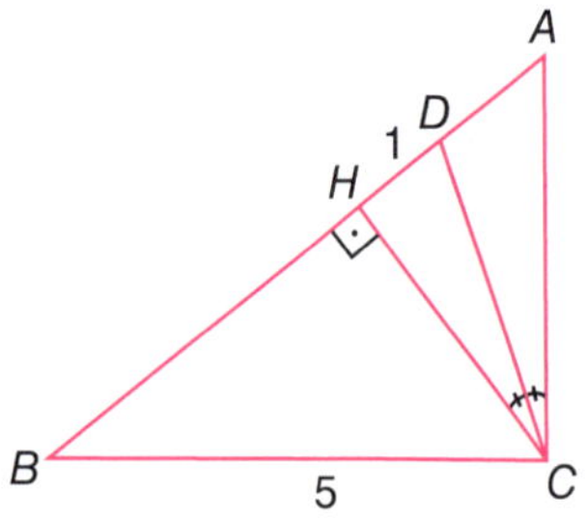

9. BD is the angle bisector of $\angle ABC$.

Given that $BD = 3\sqrt{5}$,

$AB = 8$, and

$DC = \frac{3}{2}$,

find $AD + BC$.

10. In the figure,

$AC \perp BC$ and

$m\angle DAC = m\angle BAC$.

If $AD = 14$, $AB = 29$, and $BC = 8$, then find DC.

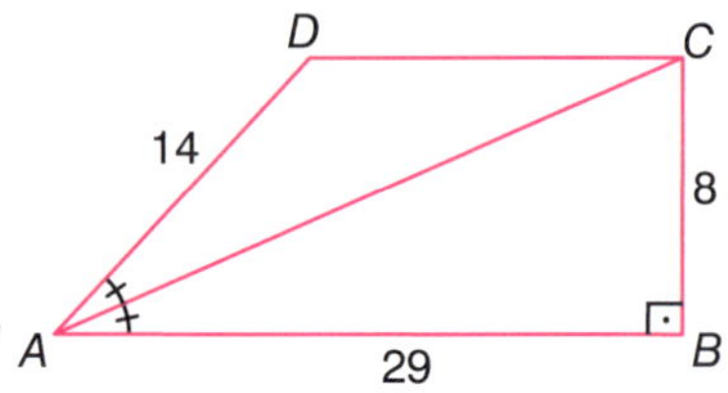

11.

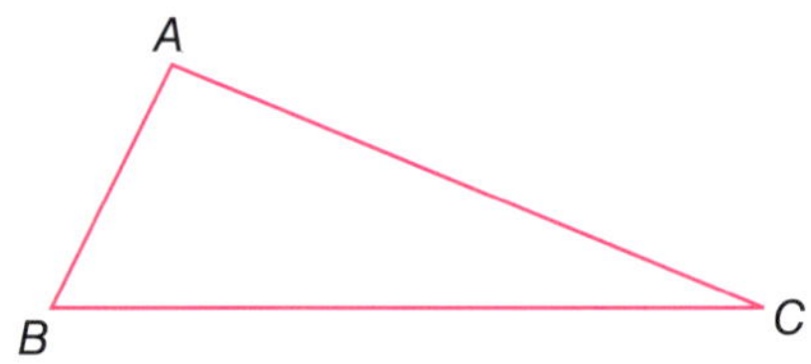

In a triangle ABC, $\angle B$ is an acute angle and $m\angle ABC = 2m\angle ACB$.

Given that $AB = 10$, and $BC = 22$, find AC.

12. In triangle ABC,

$AB = 12$,

$BD = 4\sqrt{6}$, and

$DC = 3$.

If BD is the angle bisector of $\angle ABC$, then find the perimeter of the triangle ABC.

1. (MathCON 2019 Finals Gr 7 #12)

In the figure at the right AN is a angle bisector.

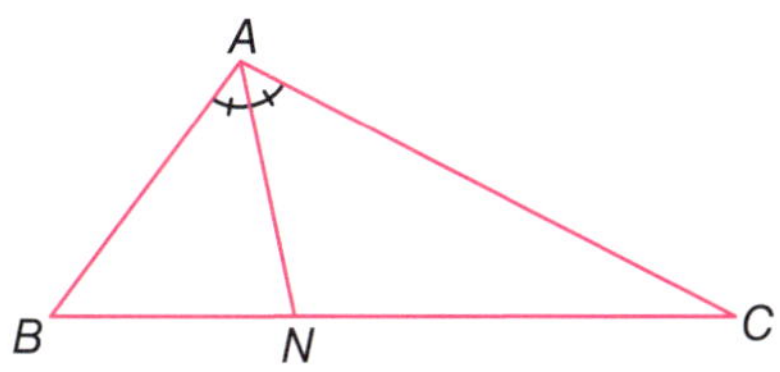

$m\angle ANB = 80°$ and $m\angle B = 2m\angle C$. Find $m\angle B$.

A) 20° B) 25° C) 30° D) 35° E) 40°

2. (CWML 2001/02 Meet 1 #15)

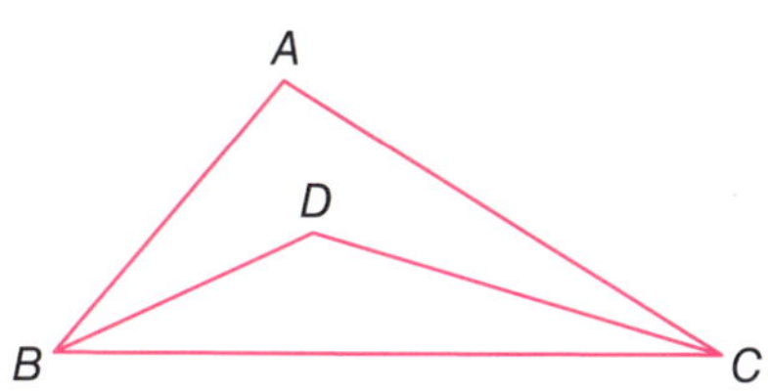

In the triangle ABC, the measure of angle A is 100° and the bisectors of angles B and C intersect at point D.

Find the exact degree measure of angle BDC.

3. (CWML 2003/04 Meet 1 # 14) Point S lies on segment RT and point V is chosen so that angle RSV measures 39°.

If segment US bisects angle RSV and segment WS bisects angle VST, what is the degree measure of angle USW?

4. (Alpha PiMC 2019 Final #23)

In the triangle ABC, points M and N are on AB and AC respectively such that $BC = CM = MN = NA = 10$ cm. CM is interior angle bisector of triangle ABC.

Find the largest integer less than or equal to the length of BN.

5. (MathCON 2017 Online #40)

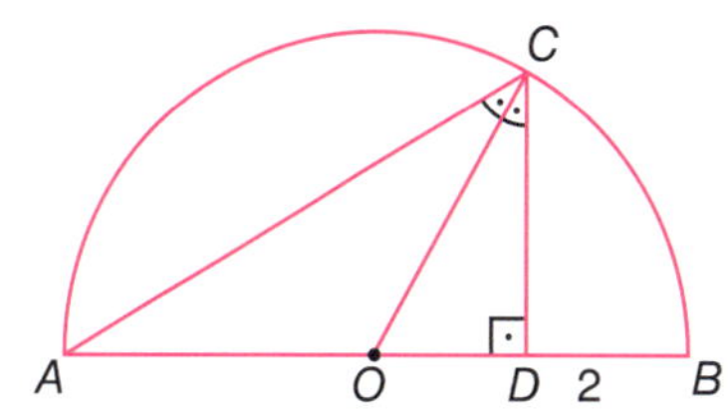

O is the center of the semicircle, where $m\angle ACO = m\angle OCD$.

If CD is perpendicular to AB and $DB = 2$, then what is the value of AB?

A) 4 B) 5 C) 8 D) 10 E) 12

6. (2009 AMC 10B #20)

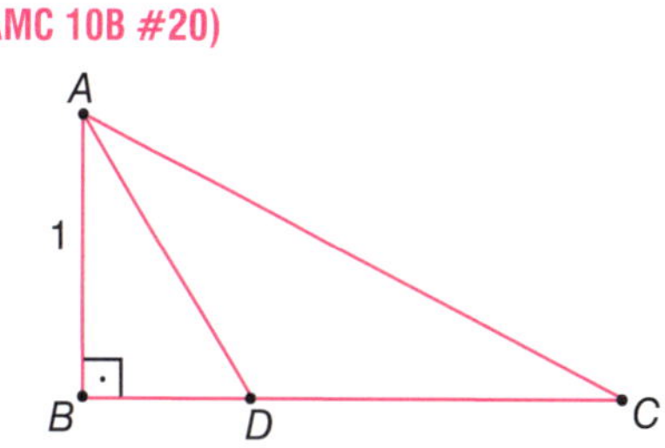

Triangle ABC has a right angle at B, $AB = 1$, and $BC = 2$. The bisector of $m\angle BAC$ meets $\overline{BC}$ at D.

What is BD?

A) $\frac{\sqrt{3}-1}{2}$ B) $\frac{\sqrt{5}-1}{2}$ C) $\frac{\sqrt{5}+1}{2}$ D) $\frac{\sqrt{6}+\sqrt{2}}{2}$ E) $2\sqrt{3}-1$

7. **(2002 AMC 12A #23)** In triangle ABC, side AC and the perpendicular bisector of BC meet in point D, and BD bisects $\angle ABC$.

If $AD = 9$ and $DC = 7$, what is the area of triangle ABD?

A) 14 B) 21 C) 28 D) $14\sqrt{5}$ E) $28\sqrt{5}$

8. **(2001 AMC 12 #24)** In ΔABC, $m\angle ABC = 45°$. Point D is on $\overline{BC}$ so that $2BD = CD$ and $m\angle DAB = 15°$.

Find $\angle ACB$.

A) 54° B) 60° C) 72° D) 75° E) 90°

9. **(2016 AMC 12B #17)**

In ΔABC shown in the figure, $AB = 7$, $BC = 8$, $CA = 9$, and $\overline{AH}$ is an altitude. Points D and E lie on sides $\overline{AC}$ and $\overline{AB}$, respectively, so that $\overline{BD}$ and $\overline{CE}$ are angle bisectors, intersecting $\overline{AH}$ at Q and P, respectively.

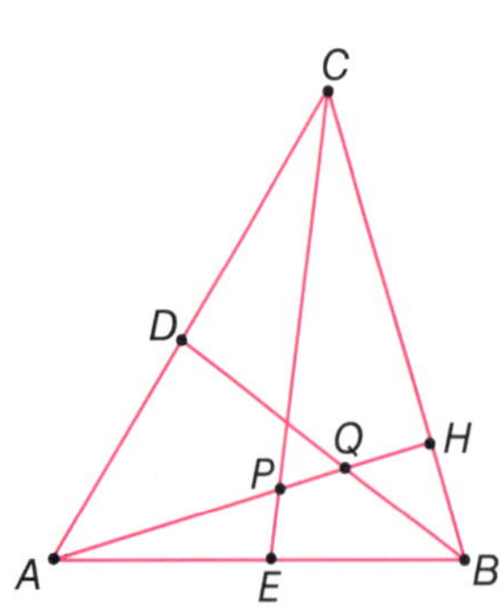

What is PQ?

A) 1 B) $\frac{5}{8}\sqrt{3}$ C) $\frac{4}{5}\sqrt{2}$ D) $\frac{8}{15}\sqrt{5}$ E) $\frac{6}{5}$

10. **(2004 AMC 10B #24)** In triangle ABC we have $AB = 7$, $AC = 8$, $BC = 9$. Point D is on the circumscribed circle of the triangle so that AD bisects angle BAC.

What is the value of $\frac{AD}{CD}$?

A) $\frac{9}{8}$ B) $\frac{5}{3}$ C) 2 D) $\frac{17}{7}$ E) $\frac{5}{2}$

11. **(2018 AMC 10A #24)** Triangle ABC with $AB = 50$ and $AC = 10$ has area 120. Let D be the midpoint of $\overline{AB}$, and let E be the midpoint of AC. The angle bisector of $\angle BAC$ intersects $\overline{DE}$ and $\overline{BC}$ at F and G, respectively.

What is the area of quadrilateral $FDBG$?

A) 60 B) 65 C) 70 D) 75 E) 80

12. **(2009 AIME I # 5)** Triangle ABC has $AC = 450$ and $BC = 300$. Points K and L are located on $\overline{AC}$ and $\overline{AB}$ respectively so that $AK = CK$, and $\overline{CL}$ is the angle bisector of angle C. Let P be the point of intersection of $\overline{BK}$ and $\overline{CL}$, and let M be the point on line BK for which K is the midpoint of PM.

If $AM = 180$, find LP.

CHAPTER 7

Note

A Centroid of a triangle is the point where the three medians meet. The Centroid is also called the center of the gravity of the triangle.

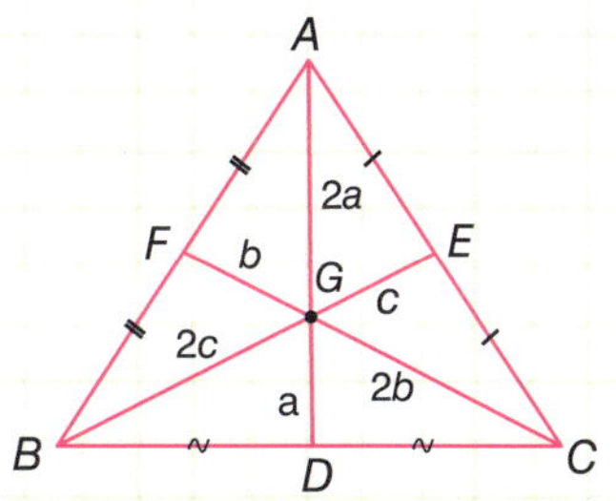

The Centroid divides each median in the ratio of 2:1.

My Notes

Triangle Length Theorems

Theorem 7.1

Stewart's Theorem

In ΔABC, with point D on BC, let m, n, and d, be the lengths of BD, DC, and AD, respectively. Then the following holds.

$$d^2 = \frac{b^2m + c^2n}{m+n} - mn.$$

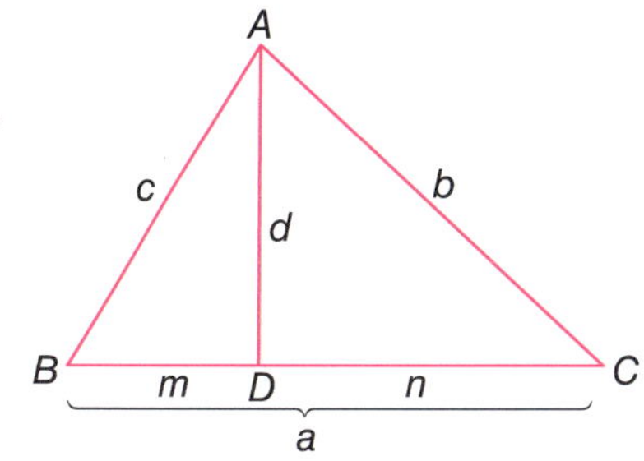

This formula can also be memorized by using mnemonic

man + *dad* = *bmb* + *cnc*

(a *man* and his *dad* put a *bomb* in the *sink*)

Theorem 7.2

The Interior Angle Bisector Length Theorem

If AN is the angle bisector of $\angle BAC$, the following holds:

$$d^2 = cb - mn.$$

$$d = \sqrt{cb - mn}.$$

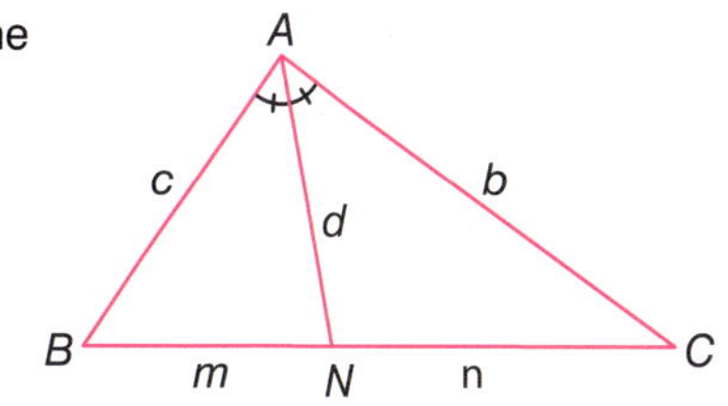

Proof 7.2

From Stewart's Theorem,

$$d^2 = \frac{b^2m + c^2n}{m+n} - mn$$

From the Angle Bisector Theorem

$$\frac{c}{m} = \frac{b}{n} \Rightarrow bm = cn$$

$$d^2 = \frac{b^2m + c^2n}{m+n} - mn = \frac{b^2m + c \cdot cn}{m+n} - mn \quad \text{[Replace } bm \text{ for } cn\text{]}$$

$$= \frac{b^2m + cbm}{m+n} - mn = \frac{b(bm + cm)}{m+n} - mn$$

$$= \frac{b(cn + cm)}{m+n} - mn = \frac{bc(n+m)}{m+n} - mn$$

Therefore, $d^2 = bc - mn.$

Example 7.1

In the figure, AN is the angle bisector of $\angle ABC$

$AB = 6$,

$BN = 3$, and

$NC = 2$.

What is the value of x?

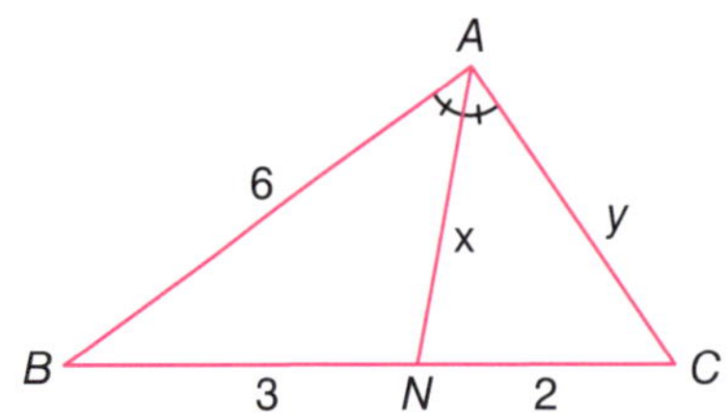

Solution 7.1

From the Interior Angle Theorem.

$$\frac{6}{3} = \frac{y}{2} \Rightarrow y = 4.$$

From Theorem 7.2 $x^2 = 6{\cdot}y - 3{\cdot}2$

$$x^2 = 6{\cdot}4 - 3{\cdot}2 = 18$$

$$x = \sqrt{18} = 3\sqrt{2}.$$

Try This

In the figure,

$DE = EC = 4$, $AD = 10$,

$DB = 6$, and $BE = 8$.

What is the length of AC?

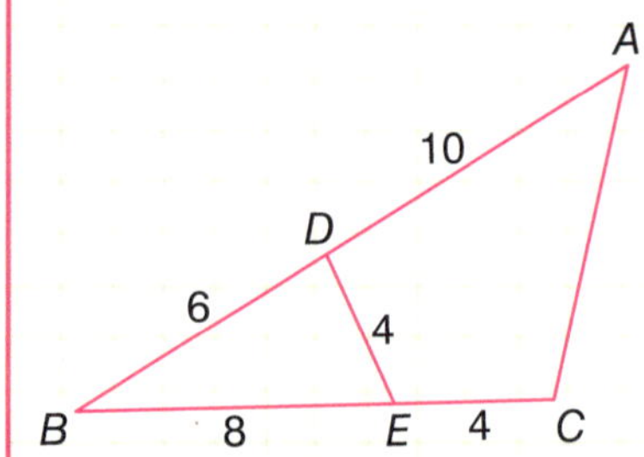

My Notes

Theorem 7.3

The Median Theorem

If AN is a median of ΔABC, then the following relationship holds:

$$2d^2 = b^2 + c^2 - \frac{a^2}{2}.$$

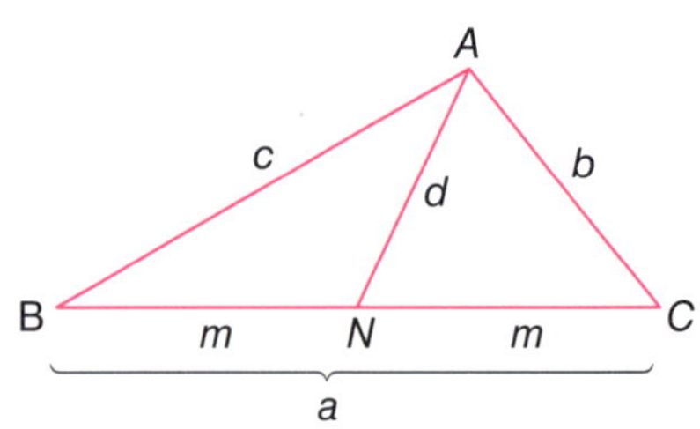

Proof 7.3

Apply Stewart's Theorem to ΔABC.

We get $d^2 = \dfrac{b^2m + c^2m}{m+m} - mm = \dfrac{m(b^2+c^2)}{2m} - m^2$

$$d^2 = \frac{b^2+c^2}{2} - m^2 \qquad \left[a = 2m \Rightarrow a^2 = 4m. \text{ So, } \frac{a^2}{2} = 2m^2\right]$$

We obtain $2d^2 = b^2 + c^2 - 2m^2$.

By replacing $\dfrac{a^2}{2}$ for $2m^2$, we get $2d^2 = b^2 + c^2 - \dfrac{a^2}{2}$.

Prove This

The Exterior Angle Bisector Length Theorem.

$$b^2 = (m+n)\cdot n - c\cdot d$$

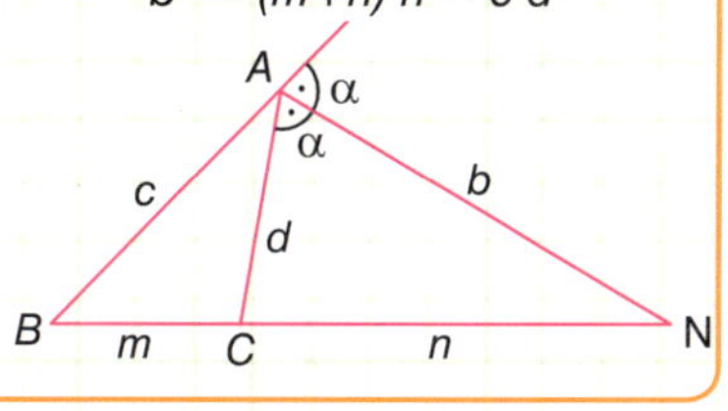

My Notes

Example 7.2

In the figure, AD is a median of ΔABC.

Given that

$AB = 7$,

$AC = 5$, and

$BD = DC = 2$,

find the length of AD?

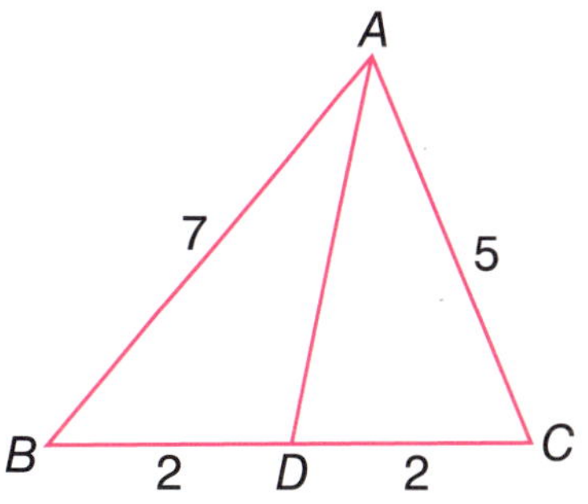

Solution 7.2

From the Median Theorem.

$$2AD^2 = 7^2+5^2-\frac{4^2}{2} = 49+25-8 = 66$$

$$2AD^2 = 66 \Rightarrow AD^2 = 33.$$

Therefore, $AD = \sqrt{33}$.

Remark 7.1

If m_a, m_b, and m_c represent the length of the medians, then we can write the following three median formulas.

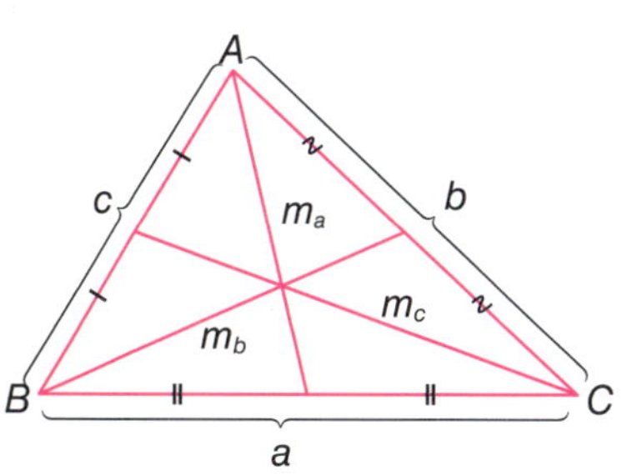

$$2m_a^2 = b^2 + c^2 - \frac{a^2}{2}$$

$$2m_b^2 = a^2 + c^2 - \frac{b^2}{2}$$

$$2m_c^2 = a^2 + b^2 - \frac{c^2}{2}$$

We add them together.

$$2(m_a^2+m_b^2+m_c^2) = \frac{3}{2}(a^2+b^2+c^2).$$

We obtain, $m_a^2+m_b^2+m_c^2 = \frac{3}{4}(a^2+b^2+c^2)$.

My Notes

Example 7.3

In the figure,

$AD = 2BD$,

$DC = 3BD$,

$AB = 2$, and $AC = 4$.

What is BC?

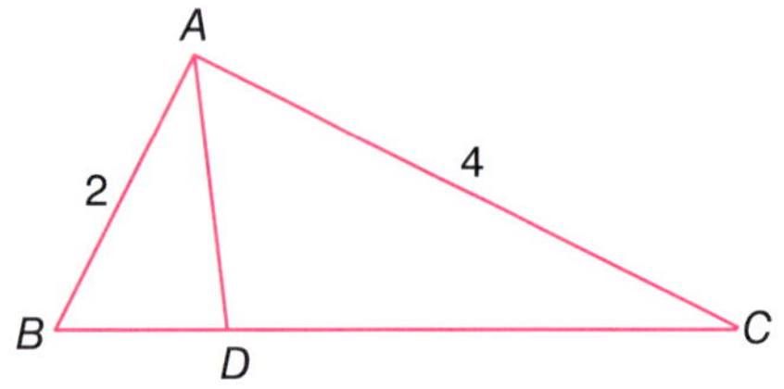

Solution 7.3

Let a be the length of BD. Then $AD = 2a$ and $DC = 3a$. By Stewart's Theorem

$$(2a)^2 = \frac{2^2(3a)+4^2a}{(a+3a)} - a\cdot 3a$$

$$4a^2 = \frac{12a+16a}{4a} - 3a^2$$

$$7a^2 = \frac{28a}{4a}$$

$$7a^2 = 7 \Rightarrow a = 1.$$

Therefore, $BC = 4a = 4\cdot 1 = 4$.

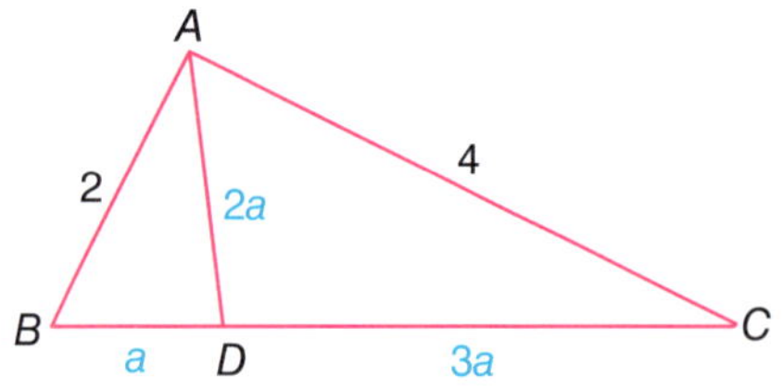

Example 7.4

In the figure,

$AD = DC = x$,

$AE = 5$, $EB = 1$,

$BC = 4$, and $ED = 2$.

What is AD^2+DB^2?

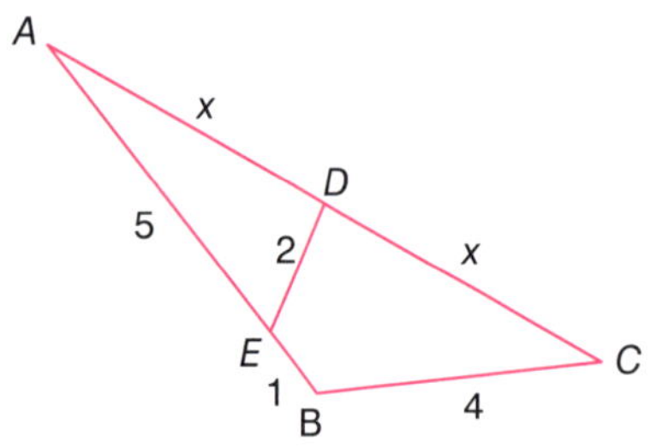

Solution 7.4

Let $DB = y$, we apply the Median Theorem ΔABC where DB is the median.

$$2y^2 = 6^2+4^2-\frac{(2x)^2}{2} \left(2DB^2 = AB^2+BC^2-\frac{AC^2}{2}\right)$$

$$2y^2 = 36+16-2x^2$$

$$2x^2+2y^2 = 52 \Rightarrow x^2+y^2 = 26.$$

Therefore, $AD^2+BD^2 = 26$.

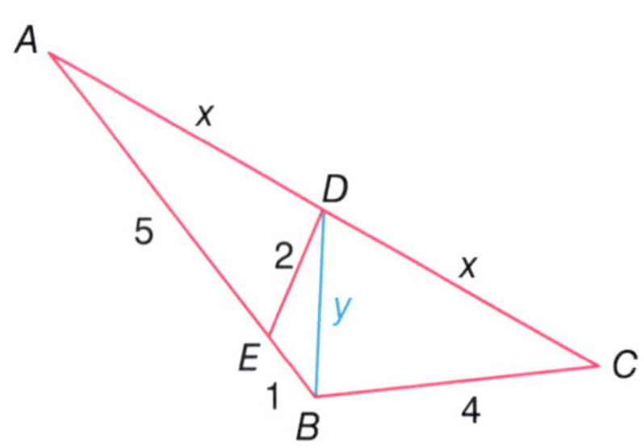

Ceva's Theorem

Tommaso Ceva was an Italian Mathematician. He discovered a beautiful theorem that was named after him.

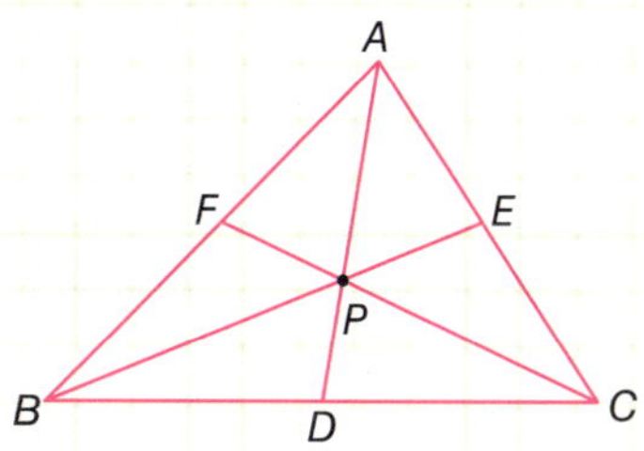

Given a triangle with a point P inside the triangle, extend lines *AP, BP, CP* to *D, E,* and *F,* respectively.

Ceva's Theorem states.

$$\frac{AF}{FB} \cdot \frac{BD}{DC} \cdot \frac{CE}{EA} = 1.$$

My Notes

Example 7.5

In the figure,

$AE = 5$, $AF = 3$, and

$FB = BD = DC = 2$

Find EC .

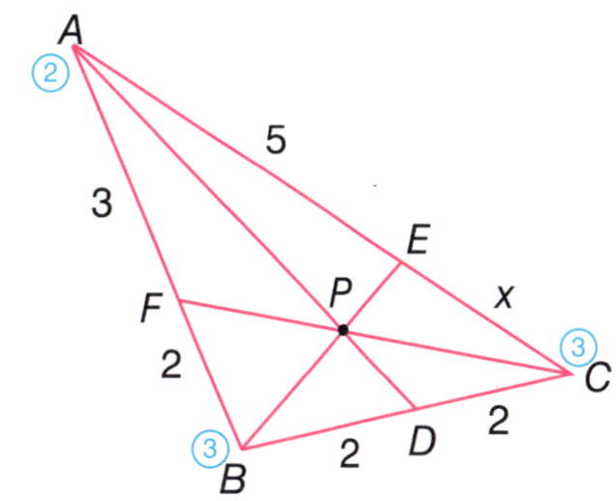

Solution 7.5

By using Ceva's Theorem.

$$\frac{2}{2} \cdot \frac{x}{5} \cdot \frac{3}{2} = 1 \quad \text{(Counterclockwise)}$$

$$\frac{3x}{10} = 1$$

$$x = \frac{10}{3}.$$

Example 7.6

In the figure,

$AF = 4$,

$FB = DC = 6$, and

$EC = 8$.

Find $\frac{BD}{AE}$.

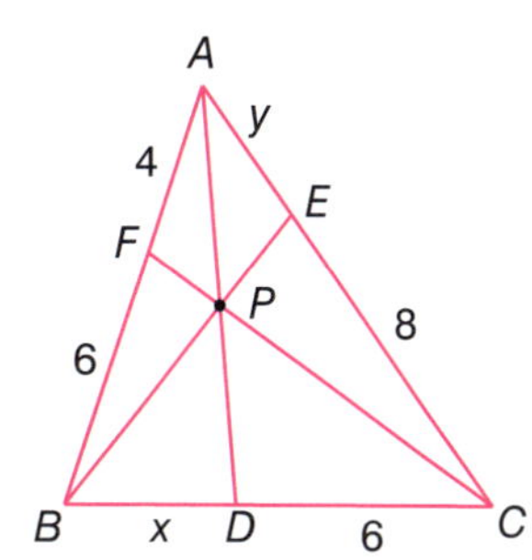

Solution 7.6

By using Ceva's Theorem.

$$\frac{x}{6} \cdot \frac{8}{y} \cdot \frac{4}{6} = 1$$

$$\frac{32x}{36y} = 1$$

$$32x = 36y \Rightarrow \frac{x}{y} = \frac{36}{32} = \frac{9}{8}.$$

My Notes

Theorem 7.4

Menelaus' Theorem

$$\frac{x}{x+y} \cdot \frac{a}{b} \cdot \frac{c}{d} = 1$$

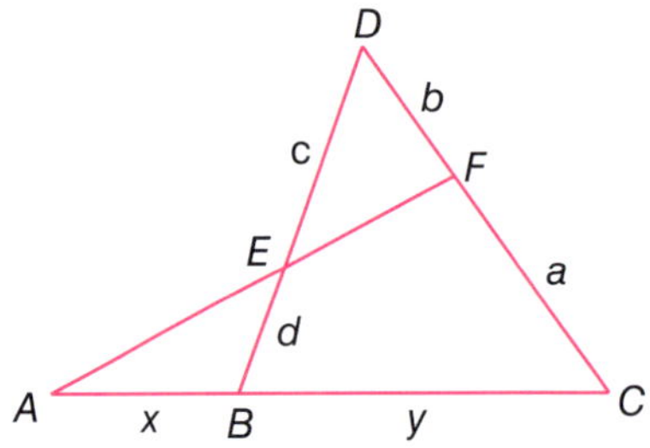

Proof 7.4

We create an auxiliary line (helping line) from point B to AF that is parallel to DC. Let GB be k.

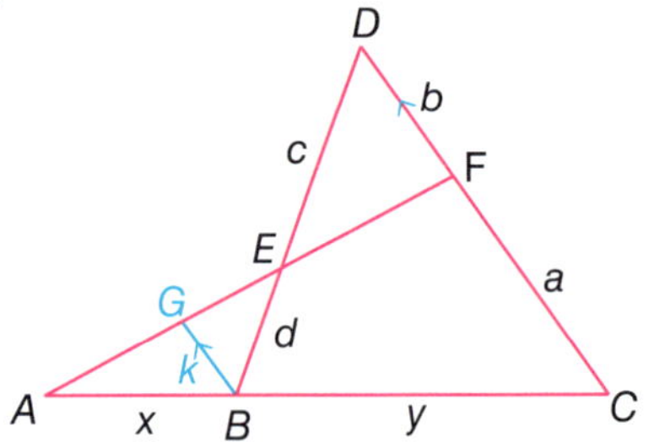

$\Delta ABG \sim \Delta AFC \quad \frac{x}{x+y} = \frac{k}{a}$ (1).

$\Delta BGE \sim \Delta DFE \quad \frac{d}{c} = \frac{k}{b} \Rightarrow k = \frac{d \cdot b}{c}$ (2).

Substitute $k = \frac{db}{c}$ into equation (1)

$$\frac{x}{x+y} = \frac{\frac{d \cdot b}{c}}{a} \Rightarrow \frac{x}{x+y} = \frac{d \cdot b}{a \cdot c}.$$

Then, we can write $\frac{x}{x+y} \cdot \frac{a \cdot c}{b \cdot d} = 1$

Therefore, $\frac{x}{x+y} \cdot \frac{a}{b} \cdot \frac{c}{d} = 1.$

Example 7.7

In the figure,

$AC = 22$, $AF = FD$,

$BD = 8$, and $DC = 6$.

What is EC?

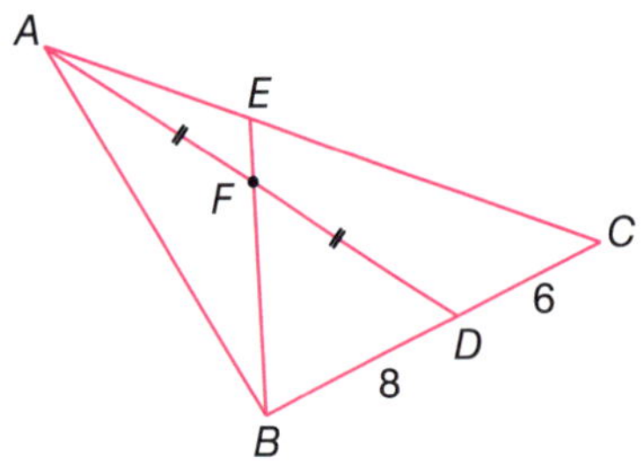

Solution 7.7

What point (what vertex) should we start with? Point B or A?

If we apply the theorem from vertex A, then we write $\frac{AE}{AC} \cdot \frac{6}{8} \cdot \frac{FB}{EF} = 1$

But, we don't no know the ratio of $\frac{FB}{EF}$. So, vertex A does not help. Let's use vertex B and apply the Theorem from vertex B.

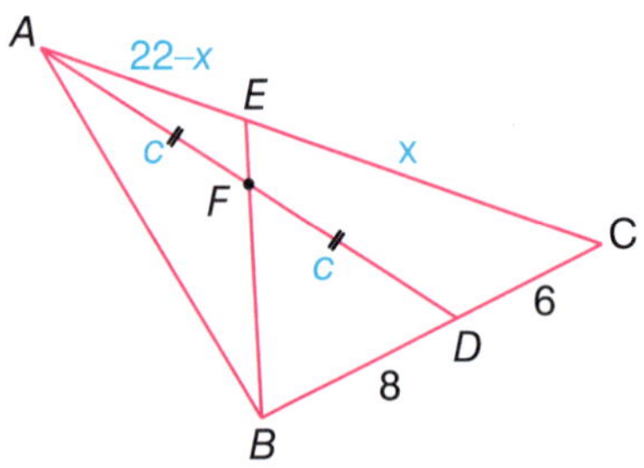

So, $\frac{8}{8+6} \cdot \frac{x}{22-x} \cdot \frac{c}{c} = 1 \Rightarrow x = 14$. So, $EC = 14$.

My Notes

Example 7.8

In the figure,

$EC = 2AE$,

$BK = KE$, and

$AF = 6$.

Find the length of FB.

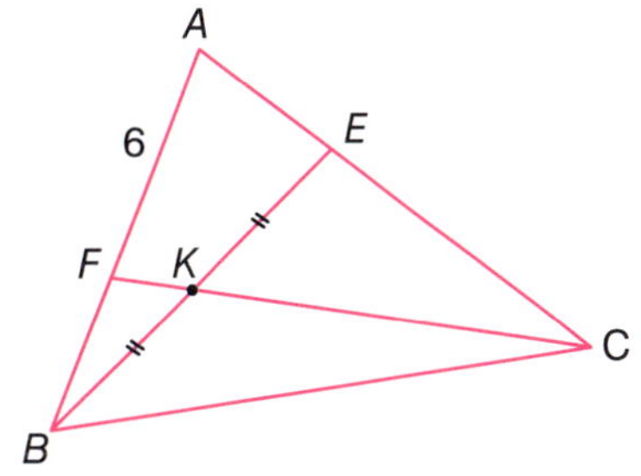

Solution 7.8

Should we start with point B or C?

Let's use point C.

$$\frac{2a}{3a} \cdot \frac{6}{FB} \cdot \frac{c}{c} = 1$$

$$\frac{2}{3} \cdot \frac{6}{FB} = 1 \Rightarrow 3FB = 12.$$

So, $FB = 4$.

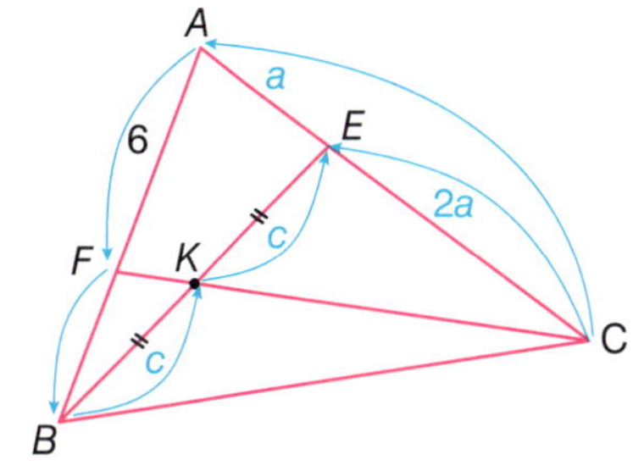

Example 7.9

In the figure,

$ED = 2AE$,

$BD = DC$, and

$AB = AC = 15$.

Find FC.

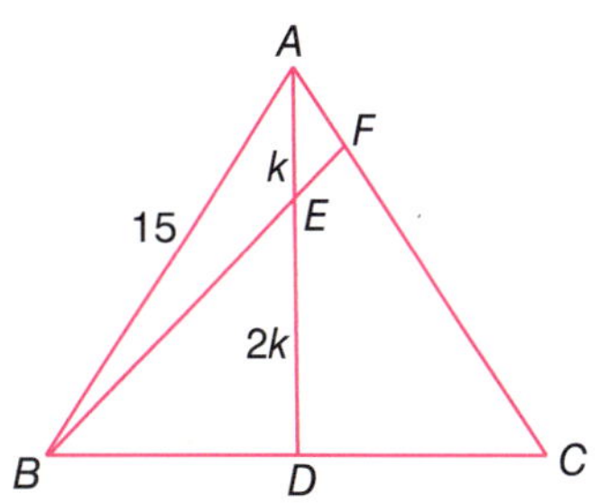

Solution 7.9

Let's apply Menelaus' Theorem from vertex B with counterclockwise.

So, $\frac{a}{2a} \cdot \frac{x}{15-x} \cdot \frac{k}{2k} = 1 \Rightarrow \frac{x}{2(15-x)2} = 1.$

$$x = 60-4x$$

$$5x = 60$$

$$x = 12.$$

Therefore, $FC = 12$.

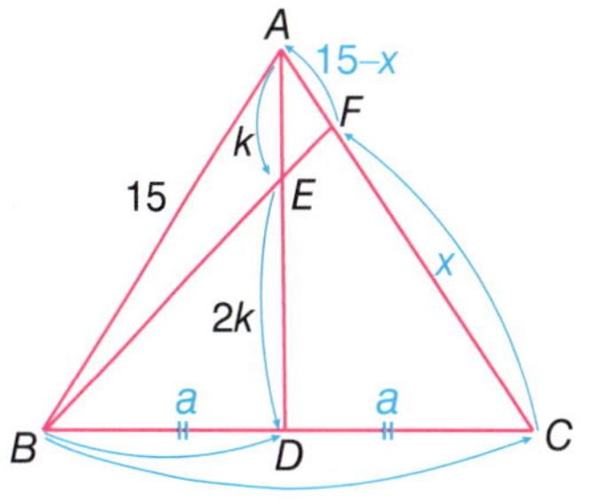

Example 7.10

In the figure,

$AH = HB$,

$3BF = 2FC$, and

$AF = 24$.

What is EF?

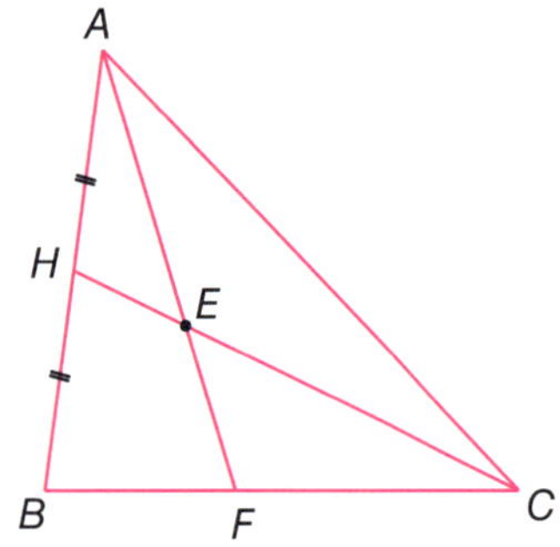

Solution 7.10

Apply Menelaus' Theorem from vertex C.

$$\frac{3k}{5k} \cdot \frac{b}{b} \cdot \frac{24-x}{x} = 1 \Rightarrow \frac{3}{5} \cdot \frac{24-x}{x} = 1$$

$$x = 9.$$

Hence, $EF = 9$.

Example 7.11

In $\triangle ABC$,

$AE = FE$,

$HB = 5$ cm,

$BF = 4$ cm, and

$FC = 6$ cm.

Find the length of AB.

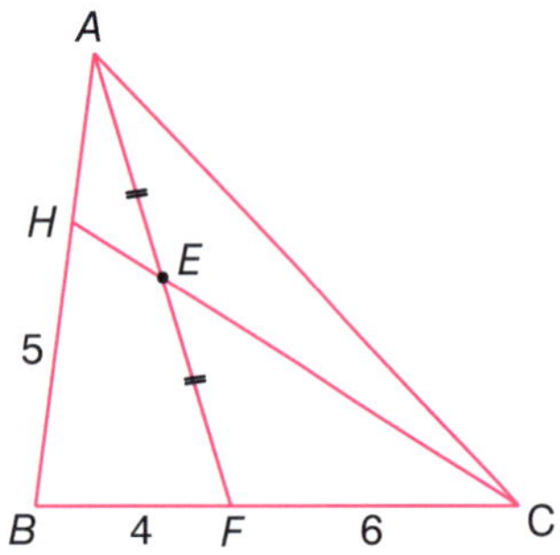

Solution 7.11

By the Menelaus Theorem

$$\frac{6}{10} \cdot \frac{5}{AH} \cdot \frac{a}{a} = 1 \Rightarrow AH = 3.$$

Therefore, $AB = 8$.

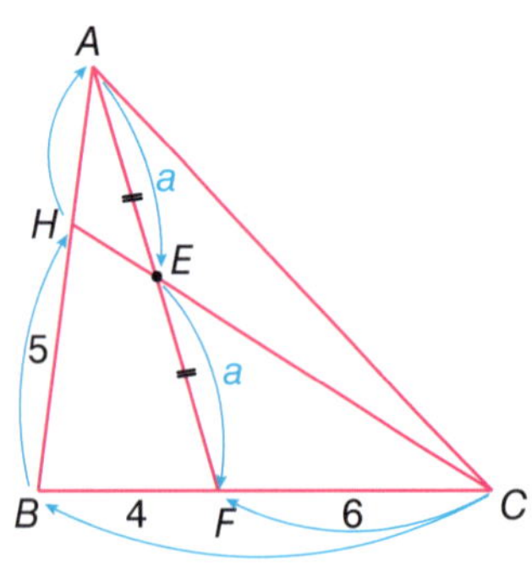

Try This

In the figure,

$DC = a$, $BD = 2a$

$BF = FE$,

$AF = 12$, and $FD = x$.

What is FD?

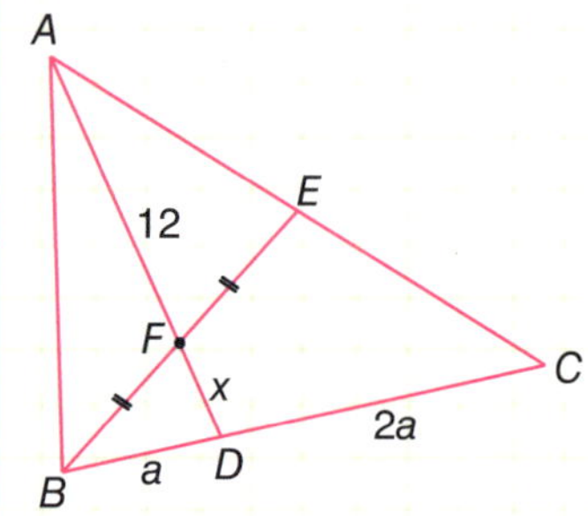

My Notes

My Notes

Example 7.12

In ΔABC,

$AF = FD$,

$BD = 8$,

$DC = 6$, and

$AC = 22$.

Find the length of AE.

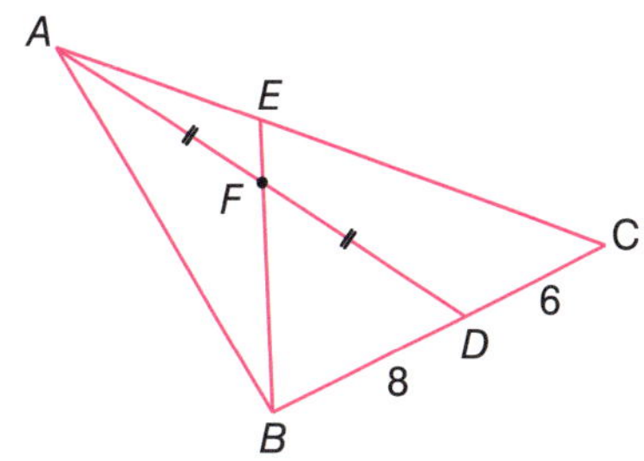

Solution 7.12

By the Menelaus Theorem from point B

$$\frac{8}{14} \cdot \frac{y}{x} \cdot \frac{a}{a} = 1 \Rightarrow \frac{x}{y} = \frac{4}{7}$$

$x = 4k$, and $y = 7k$.

$11k = 22 \Rightarrow k = 2$

Therefore, $AE = 4k = 4 \cdot 2 = 8$.

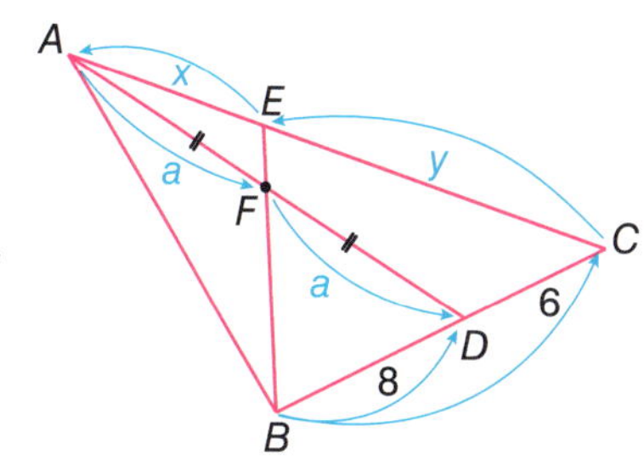

Definition 7.1

Mass Points Method

This method is based on the balance principle (or fulcrum principle).

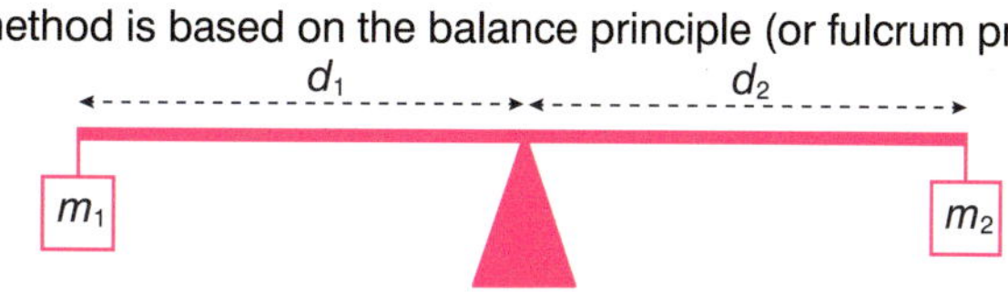

In the diagram of a lever above, m_1 and m_2 are weights, and d_1 and d_2 are their respective distances from the fulcrum. For the lever to be in balance, the product m_1d_1 must be equal to the product m_2d_2.

For example, consider a segment divided in the ratio of 1:2.

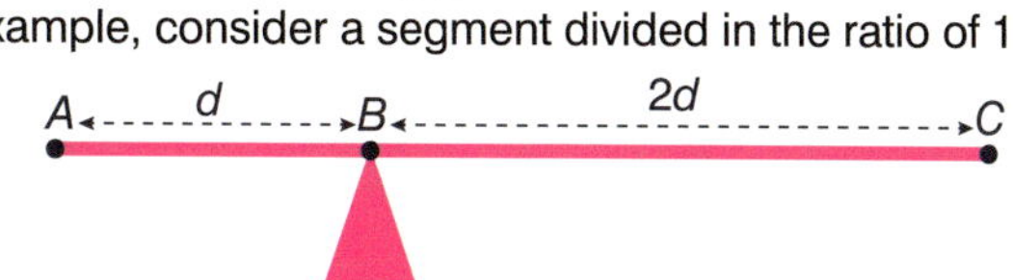

We can assign "masses" of 2 and 1 to points A and C respectively to "balance" the segment ($2 \cdot d = 1 \cdot 2d$).

Total Mass:

We can then assign a mass of 3 -the total of the masses of the end points- to point B, the "fulcrum". The completed mass-point diagram of the segment will look like this:

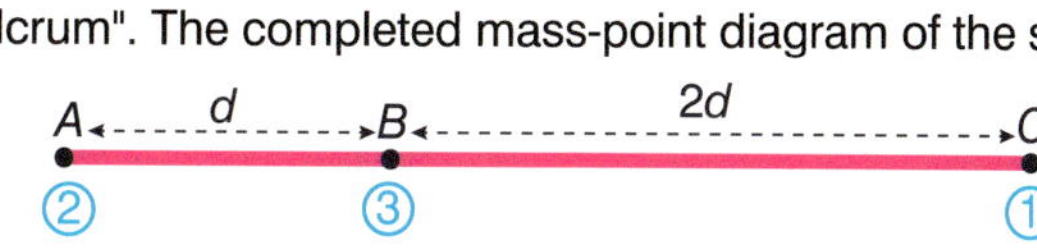

My Notes

Exercise 7.1

A segment is divided in the ratio 2:3. We can assign "masses" of 3 and 2 to points A and C. By the total mass idea, we can also assign a mass of 5 point to B.

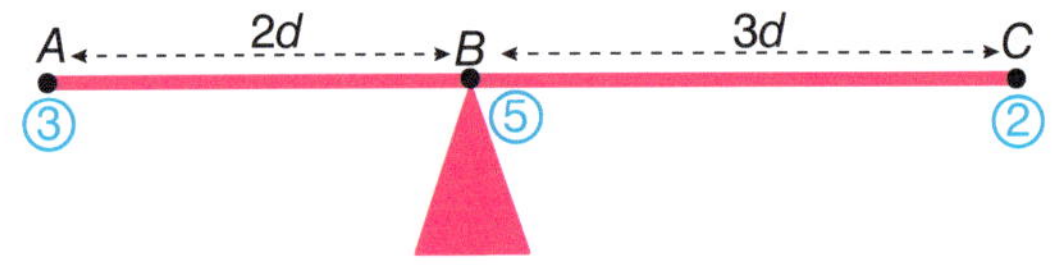

Exercise 7.2

In a blended system, if we put a mass of 2 at point A then a mass of 1 must be place point B. From the total mass rule, we can assign a mass of 3 to point C. Now let's consider a lever DC. Since the ratio of DE to EC is 3:4, then we put a mass of 4 to point D ($4m \cdot 3k = 3m \cdot 4k$). Finally, from the total mass rule, we can assign a mass of 7 to point E.

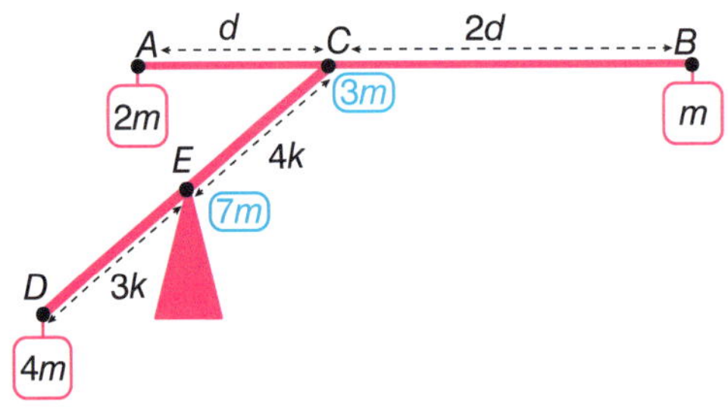

Exercise 7.3

Centroid-Mass points

Since point D, E, and F are the midpoints of three sides, then we can assign a mass of 1 to all vertices. By the total mass rule, points D, E, and F get a mass of 2.

So, $\dfrac{BP}{PE} = \dfrac{2}{1} = 2.$ $[1 \cdot PB = 2 \cdot PE]$

Similarly, $\dfrac{CP}{PF} = \dfrac{AP}{PD} = 2.$

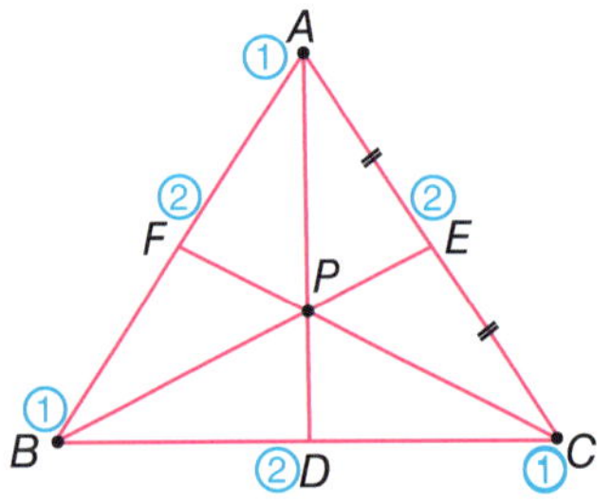

Example 7.13

In the figure, five segment lengths are given.

Find EC by using the mass points method.

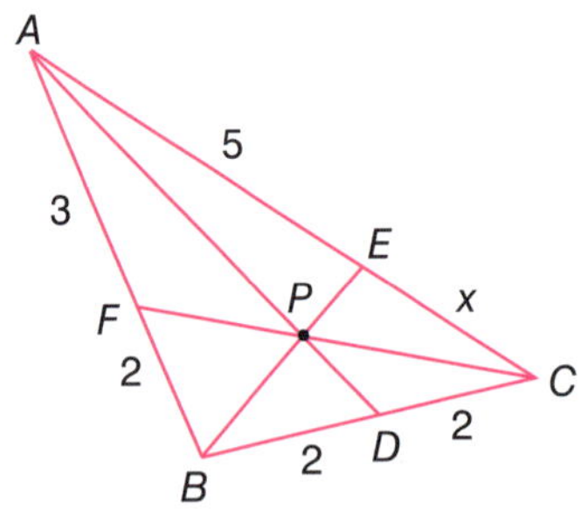

Solution 7.13

In order to avoid having fraction masses, we put a mass of 3 at B and a mass of 2 at A. So, point C must get a mass of 3 since B and C have the equal distance from each other. In order to balance A and C, we can set up an equation $2 \cdot 5 = 3 \cdot x$, therefore $x = \dfrac{10}{3}$.

Catriona Agg @Cshearer41
Geometry puzzle

$a + b$ is a right angle.

What is the area of the parallelogram?

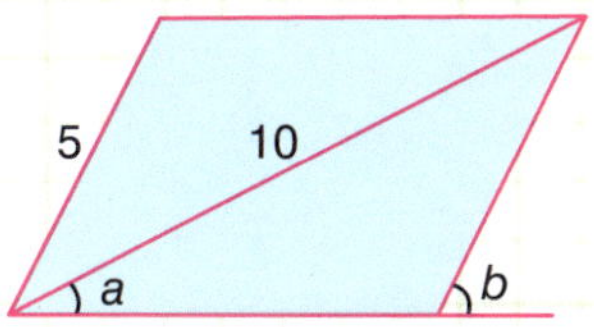

My Notes

Example 7.14

In $\triangle ABC$, point D, E, and F are on BC, CA, and AB, respectively. Suppose segments AD, BE, and CF intersect at P.

If $\frac{PD}{AD} = \frac{1}{3}$ and $\frac{PE}{BE} = \frac{1}{4}$, find $\frac{FP}{FC}$.

Solution 7.14

We assign masses of 9 and 3 to points E and B respectively, to balance BE. By the total mass rule, point P get a mass of 12.

Masses of A and D must add up to 12 and must be from 1:2 ratio. So, point A must get a mass of 4 and point D must get a mass of 8. By the total mass rule applying AC, C gets mass of 5 and similarly F gets mass of 7.

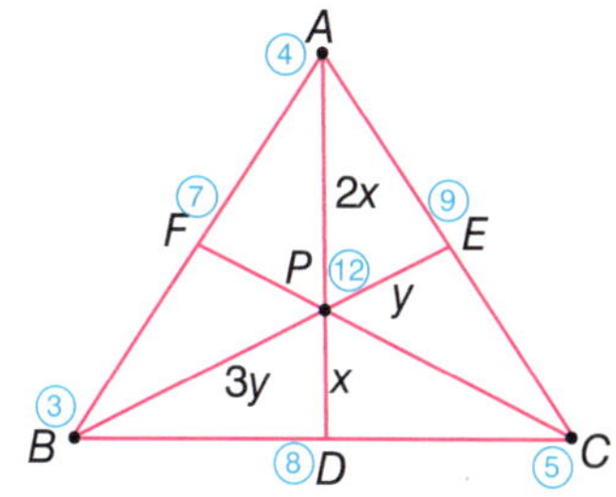

Hence, $7FP = 5PC \Rightarrow \frac{FP}{PC} = \frac{5}{7}$ $\cdot$Therefore, $\frac{FP}{FC} = \frac{5}{12}$.

Example 7.15

In the figure,

$AD = AF = 4$,

$AG = 5$, $DB = 2$, and

$FC = 6$.

What is GE?

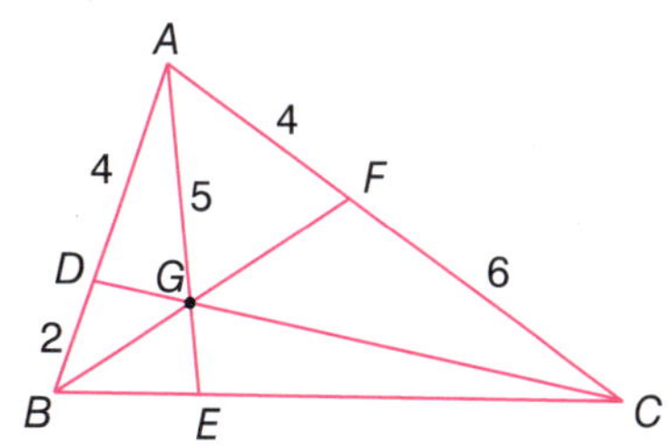

Solution 7.15

We assign a mass of 6 to A and a mass of 12 to B. So, point D gets the total mass of 18. By the total mass rule applying to AC, point C weights 4 and point F weights 10. Thinking of points B and C, point E gets the total mass of 16. Finally, point G gets total mass 22 by the total mass rule.

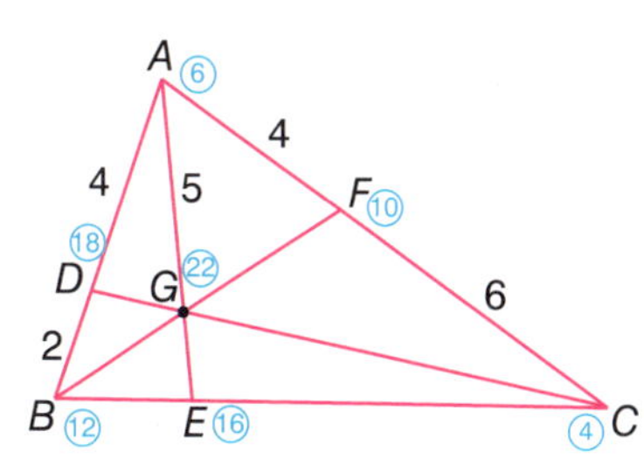

From AE, $6\,AG = 16\,GE$

$$6 \cdot 5 = 16\,GE \Rightarrow GE = \frac{30}{16} = \frac{15}{8}.$$

Example 7.16

In ΔABC, $AB = 16$, $BC = 10$, and $AC = 12$. D is on AB such that $AD = 6$. E is the midpoint of BC. Let F be the intersection of lines ED and AC (Out side of ΔABC).

What is FA?

Solution 7.16

We can create this figure.

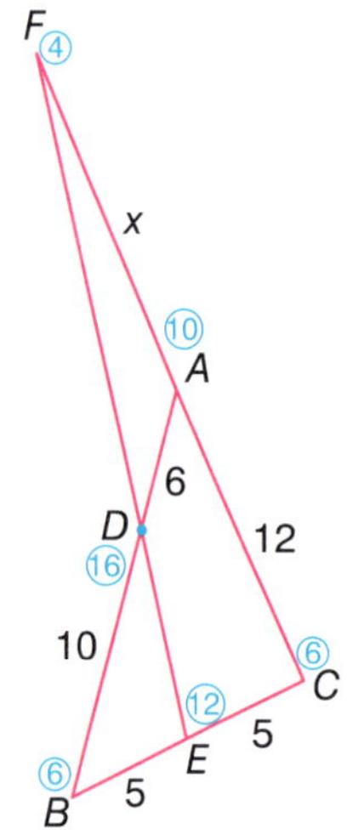

Method 1: By Menelaus' Theorem (From point F)

$$\frac{x}{x+12} \cdot \frac{5}{5} \cdot \frac{10}{6} = 1 \Rightarrow x = 18.$$

Method 2: From the Mass Points, we place a mass of 10 at point A, a mass of 6 at point B, and a mass of 6 at point C. E will have the total mass of 12 and D will at the total mass of 16, therefore, F will have a mass of 4. By looking at FC, we can write a the balanced equation

$$4FA = 6CA$$

$$4 \cdot x = 6 \cdot 12 \Rightarrow x = \frac{72}{4} = 18 = AF.$$

Example 7.17 (2009 AIME I #5)

Triangle ABC has $AC = 450$ and $BC = 300$. Points K and L are located on $\overline{AC}$ and $\overline{AB}$ respectively so that $AK = CK$, and CL is the angle bisector of angle C. Let P be the point of intersection of $\overline{BK}$ and $\overline{CL}$, and let M be the point on line BK for which K is the midpoint $\overline{PM}$.

If $AM = 180$, find LP.

Solution 7.17

By Angle Bisector Theorem

$$\frac{AL}{LB} = \frac{450}{300} = \frac{3}{2}.$$

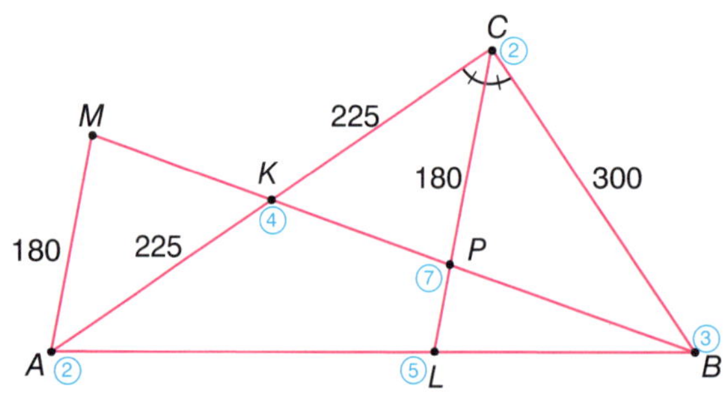

So, we can assign weight A as 2 and B as 3 and L as 5. Since K is the midpoint of A and C, we can assign mass of 2 to C and mass of 4 to K

[ΔAMK and ΔCPK congruent $AM = PC = 180$]. By the total mass, we can assign mass of 7 to P. Finally, looking at CL, we can write $5PL = 2PC$

$$PL = PC\frac{2}{5} \Rightarrow PL = 180\frac{2}{5} = 72.$$

My Notes

1. G is the centroid of ΔABC.

$AC = 15$ cm,

$BG = 5$ cm, and

$CG = 6$ cm.

Find the area of ABC.

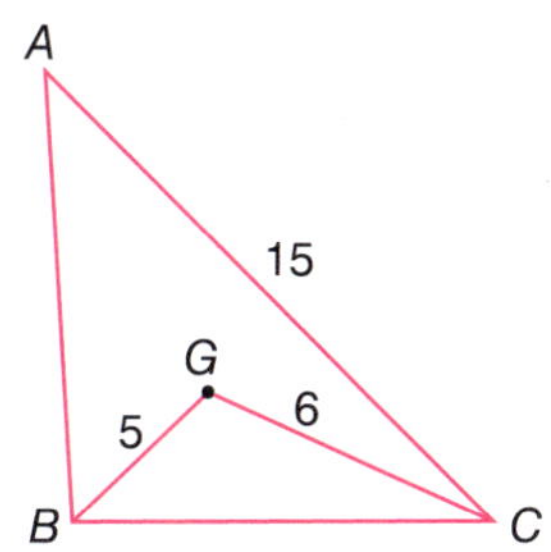

2. In the figure, ΔRST is an isosceles triangle with

$RS = RT = 10$ and

$ST = 16$.

If medians RZ, TX, and SY meet at Q, then find $RQ + QT$.

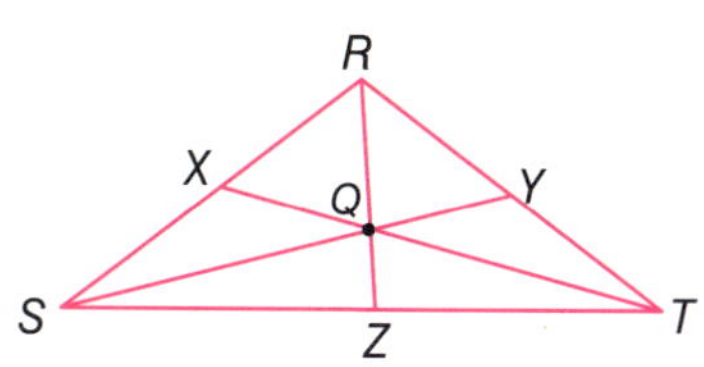

3. G is the centroid of ABC and A, G, and D are collinear (are on the same line).

$3\,GD = DC$,

$AB = 2\sqrt{17}$, and

$AC = 16$.

What is AG?

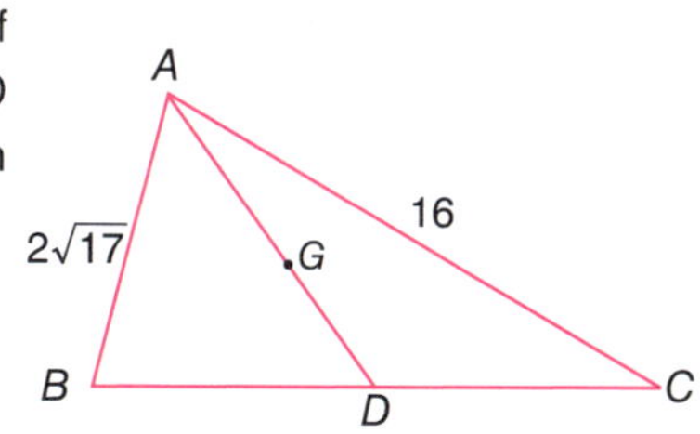

4.

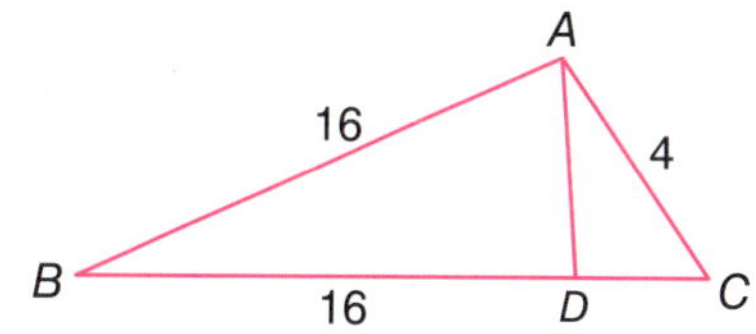

In the figure $m\angle ACB = 2m\angle CAD$, $AB = BC = 16$, and $AC = 4$.

What is DC?

5.

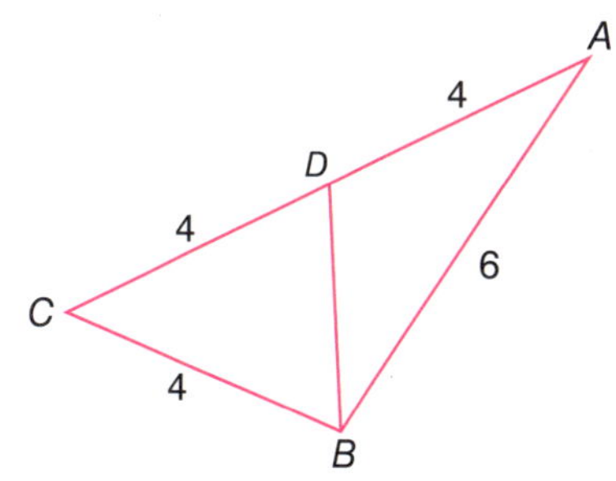

In the figure, $AD = DC = BC = 4$ and $AB = 6$.

What is the value of DB?

6. In the figure, BF and CD intersect at E.

$AD = DB$, $BE = 2\,EF$, and

$DE = 2$ cm.

What is CD?

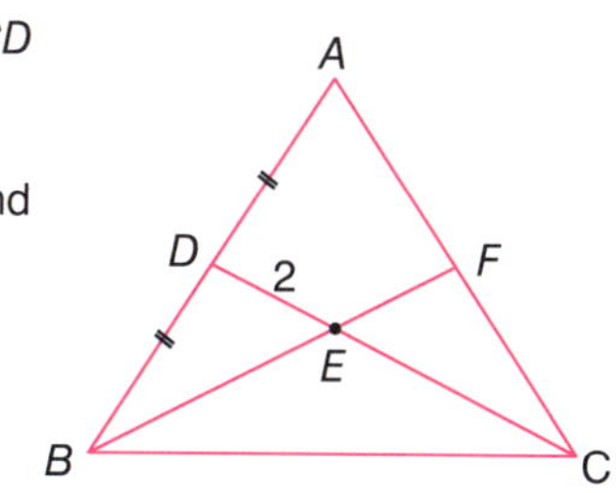

7. In the figure,

$AB = 6\sqrt{2}$ cm,

$AC = 8\sqrt{2}$ cm, and

$BD = DC = 5\sqrt{2}$ cm.

What is AD?

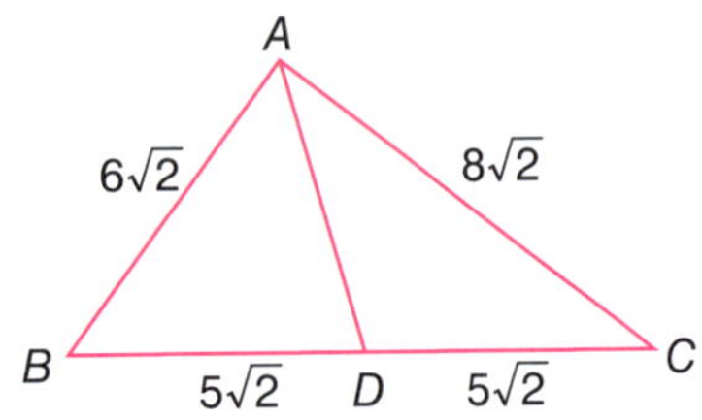

8. In $\triangle MNP$, medians PM, NA, and MC intersect at centroid Q.

a) If $MQ = 8$, find QC.

b) If $QB = 3$, find PQ.

c) If $AQ = 3.5$, find QN.

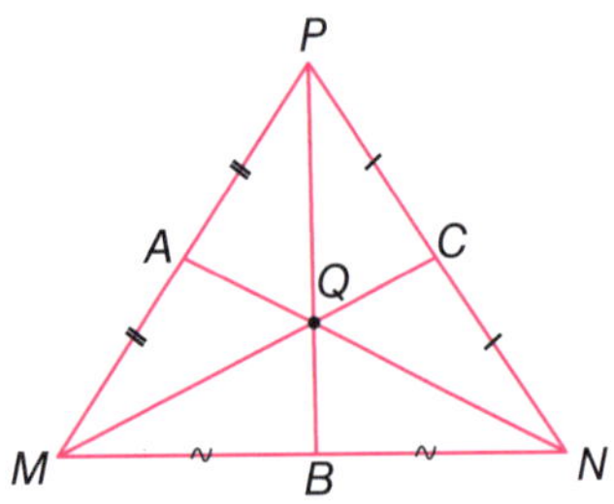

9. In the figure, G is the centroid of $\triangle ABC$.

$AB \perp GD$

$AD = BD$,

$GD = 4$, and

$AC = 13$.

Find the area of ABC.

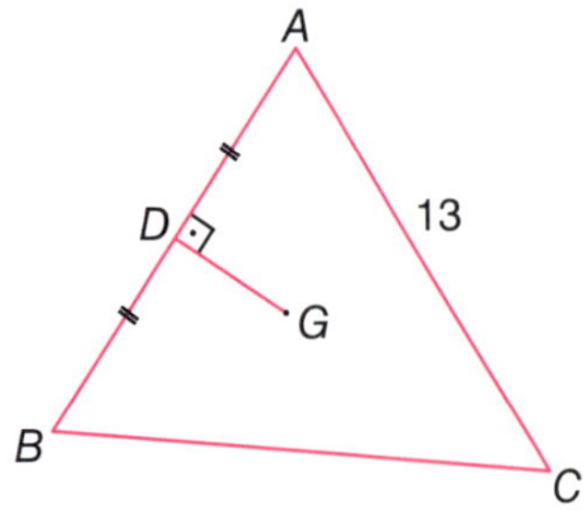

10. G is the centroid of $\triangle ABC$.

$BG \perp GC$,

$AG = 3x+4$, and

$BC = 6x-8$.

What is AG?

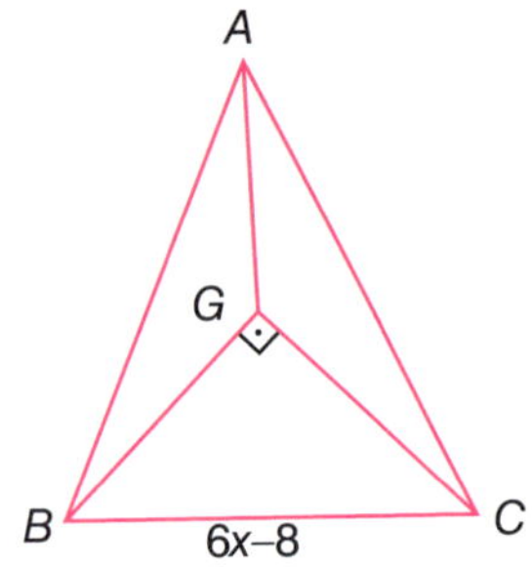

11.

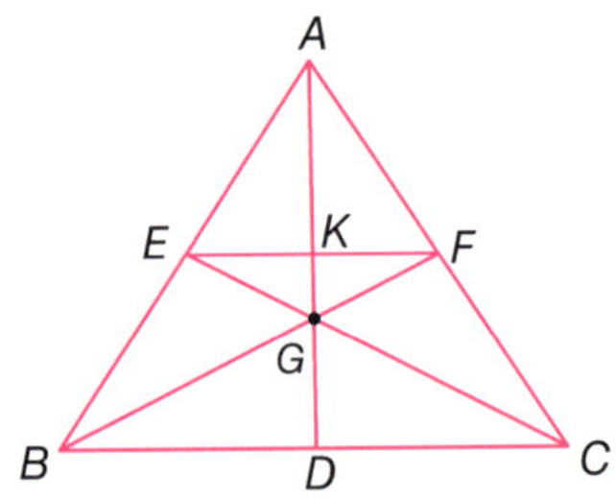

In the figure, AD, BF, and CE are medians of $\triangle ABC$.

Given that $KG = 2$ and $BC = 8$, find $EG + GD$.

12.

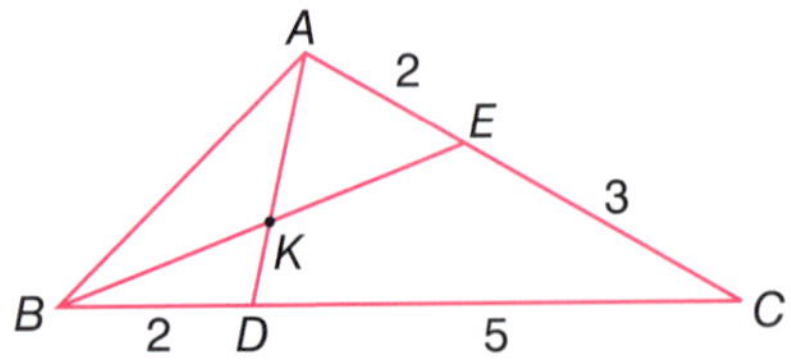

In the figure,

$BD = AE = 2$, $CB = 5$, and $EC = 3$.

What is the ratio of BK to KE?

1. Points M, N, and K belong to side AB, BC, and CA respectively.

$\frac{AM}{MB} = \frac{1}{4}$,
$\frac{BN}{NC} = \frac{2}{3}$.

It is known that the cevians AN, BK, and CM are concurrent.

What is the ratio $\frac{CK}{KA}$?

2. In triangle ABC, the medians BE and CD intersect at F and are perpendicular to each other.

If $BE = 18$ and $CD = 24$, what is AF?

3.

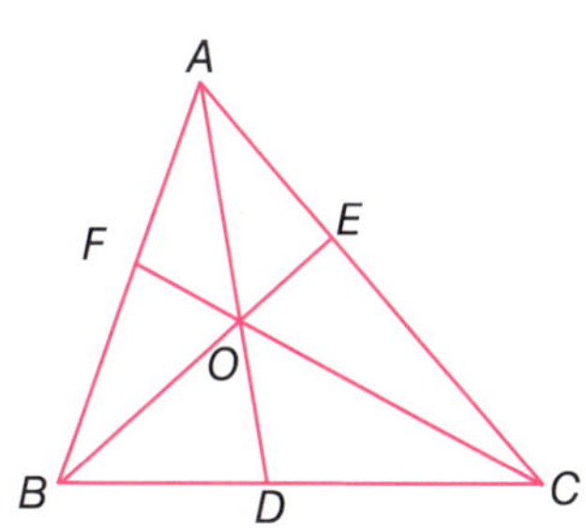

In the figure, $\frac{AF}{AB} = \frac{4}{9}$ and $\frac{AE}{EC} = \frac{2}{3}$.

Find $\frac{AO}{AD}$.

4. Point D is in the interior of triangle ABC. Lines AD,BD, and CD intersect the sides AC, AB, and BC at points E, F, and G, respectively.

Find the ratio $\frac{BF}{FA}$, if it is known that BC = p and $\frac{EC}{AE} = q$.

5. G is the centroid of $\triangle ABC$ with $\angle BGC = 90°$.

If $AG = 6$, then find BC.

6. In the figure, G is the centroid of $\triangle ABC$

$AD = DC = 8$,

$GD = 4$, and

$GC = 10$.

What is AB?

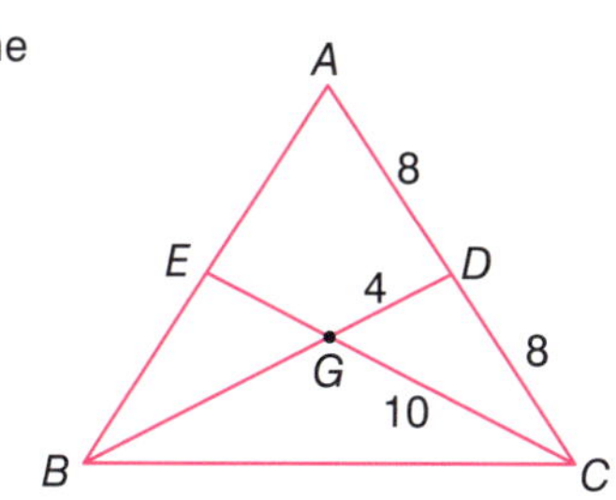

7. ABC is a right triangle with $m\angle A = 90°$.

G is the centroid of $\triangle ABC$ with $m\angle AGC = 90°$

If $BC = 12$, then find GC.

10. In the figure,

$EC = 2\,AE$,

$BK = KE$, and

$AF = 6$.

What is FB?

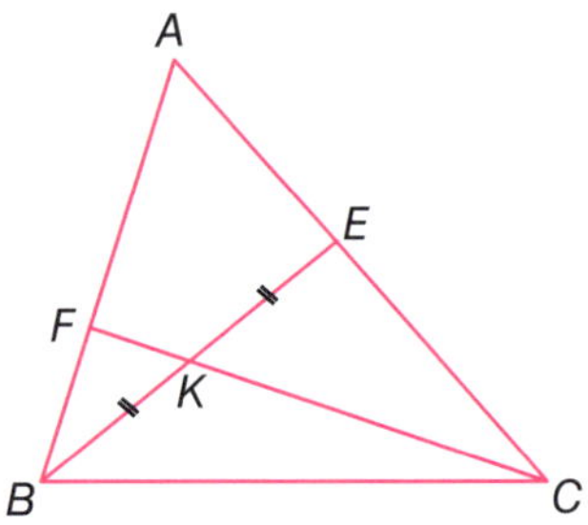

8.

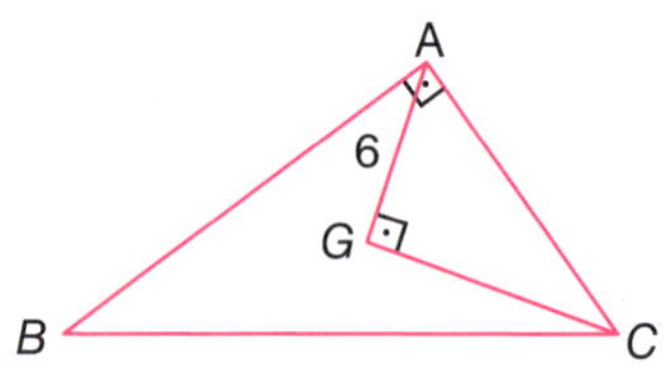

In the figure, point G is the centroid of $\triangle ABC$, with $m\angle BA\ \ C = 90°$ and $AC \perp GC$.

If $AG = 6$ cm, then find AC.

11. In $\triangle ABC$, points D, E, and F on BC, AC, and AB respectively. Suppose segments AD, BE, and CF intersect at P.

Given that $\frac{PD}{AD} = \frac{2}{3}$ and $\frac{PE}{BE} = \frac{1}{4}$, find $\frac{PF}{CF}$.

9. V_a, V_b, and V_c are medians of sides a, b, and c of a $\triangle ABC$.

Given that $a = 12$, $V_b = 12$, and $V_c = 15$, find the value of V_a.

12. In the figure,

$AD = EC = 3$,

$DB = AF = 2$, and

$AE = 4$.

What is GE?

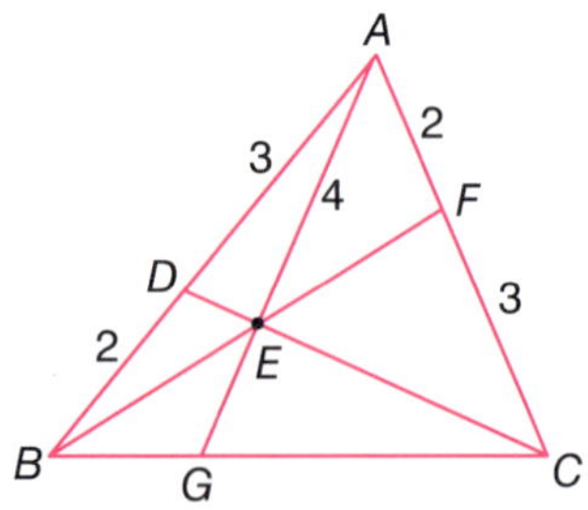

1. **(AlphaStar PiMC 2018 #16)**

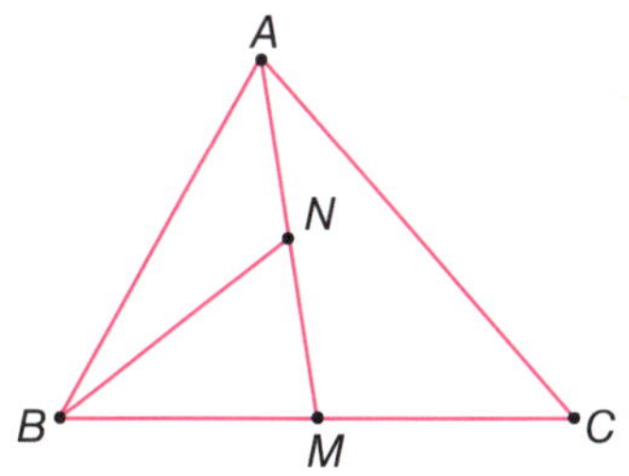

Let M be the midpoint of side BC on triangle ABC. Let N be the midpoint of AM.

What is the ratio of the area of ΔABC to the area of ΔBNM?

2. **(2015 AMC 10A #11)** The ratio of the length to the width of a rectangle is 4:3. If the rectangle has diagonal of length d, then the area may be expressed as kd^2 for some constant k.

What is k?

A) $\frac{2}{7}$ B) $\frac{3}{7}$ C) $\frac{12}{25}$ D) $\frac{16}{25}$ E) $\frac{3}{4}$

3. **(2019 AMC 8 #24)**

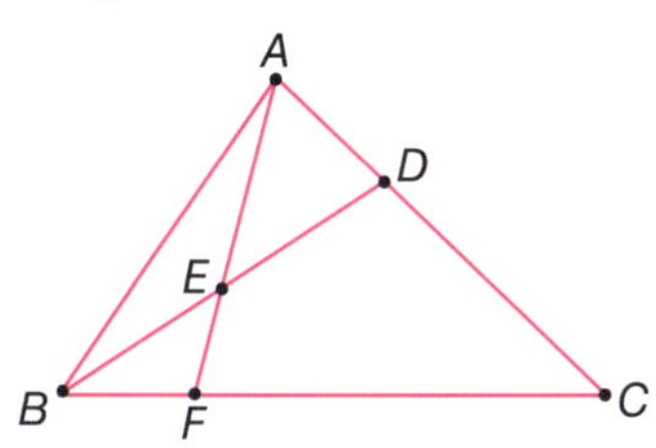

In triangle ABC, point D divides side $\overline{AC}$ so that $\frac{AD}{DC}=\frac{1}{2}$. Let E be the midpoint of $\overline{BD}$ and let F be the point of intersection of line BC and line AE.

Given that the area of ΔABC is 360, what is the area of ΔEBF?

A) 24 B) 30 C) 32 D) 36 E) 40

4. **(2004 AMC 10B #20)**

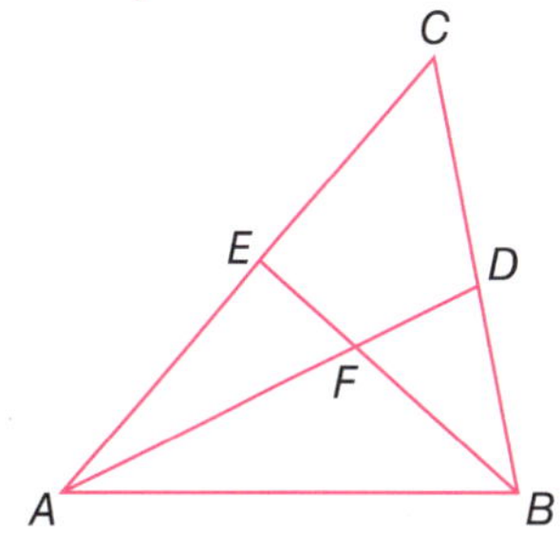

In ΔABC, $AB=6$, $BC=7$, and $CA=8$. Point D lies on $\overline{BC}$, $\overline{AD}$ bisects $\angle BAC$. Point E lies on $\overline{AC}$, and $\overline{BE}$ bisects $\angle ABC$. The bisectors intersect at F. What is the ratio $AF : FD$?

A) 3:2 B) 5:3 C) 2:1 D) 7:3 E) 5:2

5. **(2019 AMC 10B #16)** In ΔABC with a right angle at C, point D lies in the interior of $\overline{AB}$ and point E lies in the interior of $\overline{BC}$ so that $AC=CD$, $DE=EB$, and the ratio $\frac{AC}{DE}=\frac{4}{3}$.

What is the ratio $\frac{AD}{DB}$?

A) $\frac{2}{3}$ B) $\frac{2}{\sqrt{5}}$ C) $\frac{1}{1}$ D) $\frac{3}{\sqrt{5}}$ E) $\frac{3}{2}$

6. **(2016 AMC 10A #19)** In rectangle $ABCD$, $AB=6$ and $BC=3$. Point E between B and C, and point F between E and C are such that $BE=EF=FC$. Segments $\overline{AE}$ and $\overline{AF}$ intersect $\overline{BD}$ at P and Q, respectively. The ratio $BP : PQ : QD$ can be written as $r : s : t$ where the greatest common factor of r, s and t is 1.

What is $r+s+t$?

A) 7 B) 9 C) 12 D) 15 E) 20

7. (2012 AMC 10B #19) In rectangle $ABCD$, $AB = 6$, $AD = 30$ and G is the midpoint of $\overline{AD}$. Segment AB is extended 2 units beyond B to point E, and F is the intersection of $\overline{ED}$ and $\overline{BC}$.

What is the area of $BFDG$?

A) $\frac{133}{2}$ B) 67 C) $\frac{135}{2}$ D) 68 E) $\frac{137}{2}$

8. (2005 AMC 10B #23) In trapezoid $ABCD$ we have $\overline{AB}$ parallel to $\overline{DC}$, E as the midpoint of $\overline{BC}$, and F as the midpoint of $\overline{DA}$. The area of $ABEF$ is twice the area of $FECD$.

What is $\frac{AB}{DC}$?

A) 2 B) 3 C) 5 D) 6 E) 8

9. (2004 AMC 10B #20)

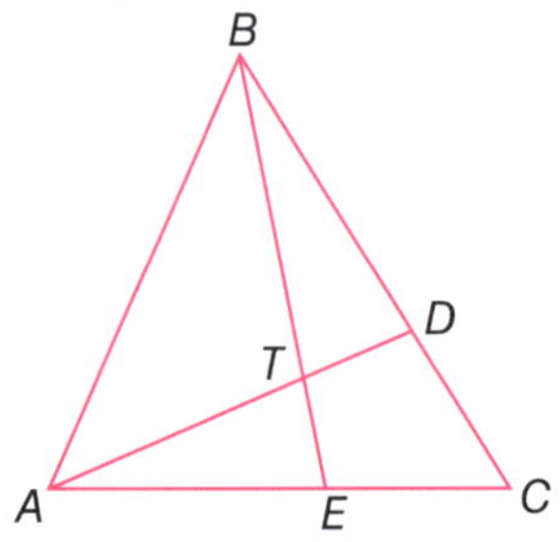

In ΔABC points D and E lie on BC and AC, respectively.

If AD and BE intersect at T so that $\frac{AT}{DT} = 3$ and $\frac{BT}{ET} = 4$, what is $\frac{CD}{BD}$?

A) $\frac{1}{8}$ B) $\frac{2}{9}$ C) $\frac{3}{10}$ D) $\frac{4}{11}$ E) $\frac{5}{12}$

10. (2002 AMC 10A #23) Points A, B, C and D lie on a line, in that order, with $AB = CD$ and $BC = 12$. Point E is not on the line, and $BE = CE = 10$. The perimeter of ΔAED is twice the perimeter of ΔBEC.

Find AB.

A) $\frac{15}{2}$ B) 8 C) $\frac{17}{2}$ D) 9 E) $\frac{19}{2}$

11. (MathCON 2019 Finals Gr 10 #32) In triangle ABC, the medians $\overline{BE}$ and $\overline{CD}$ intersect at F and are perpendicular to each other.

If $BE = 18$ and $CD = 24$, find the length AF.

A) 15 B) 20 C) 25 D) 30 E) None of the preceding

12. (2009 AIME I #4) In parallelogram $ABCD$, point M is on $\overline{AB}$ so that $\frac{AM}{AB} = \frac{17}{1000}$ and point N is on $\overline{AD}$ so that $\frac{AN}{AD} = \frac{17}{2009}$.

Let P be the point of intersection of $\overline{AC}$ and $\overline{MN}$. Find $\frac{AC}{AP}$.

CHAPTER 8

Notes

$45^\circ - 45^\circ - 90^\circ$

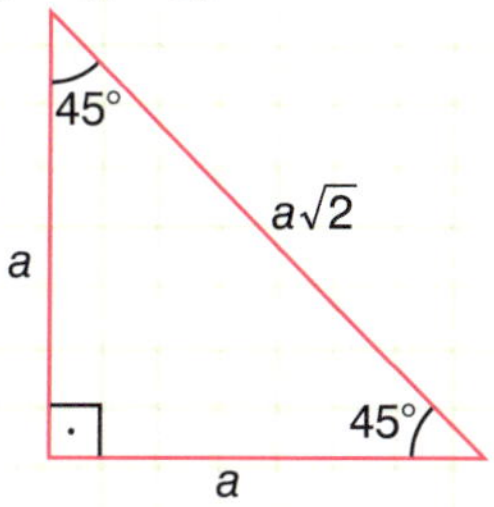

$30^\circ - 60^\circ - 90^\circ$

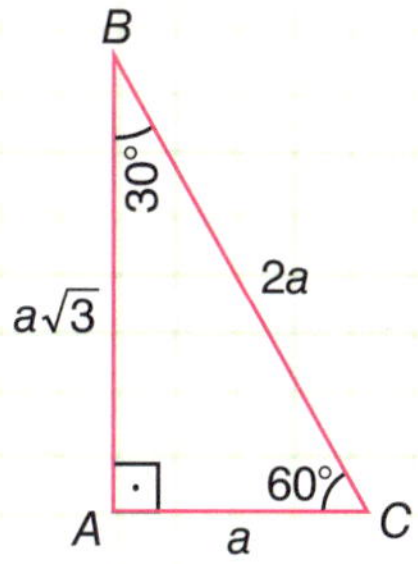

Exercise 8.1

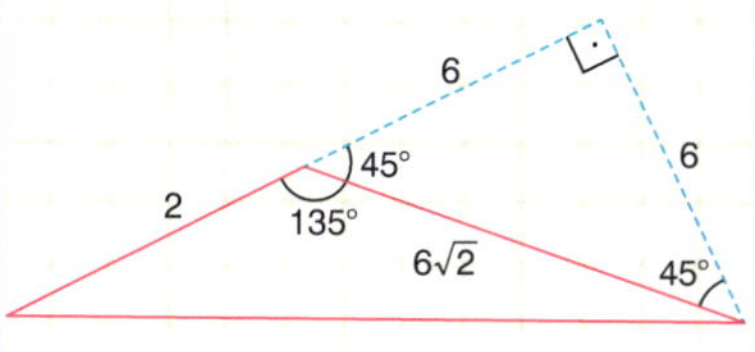

Exercise 8.2

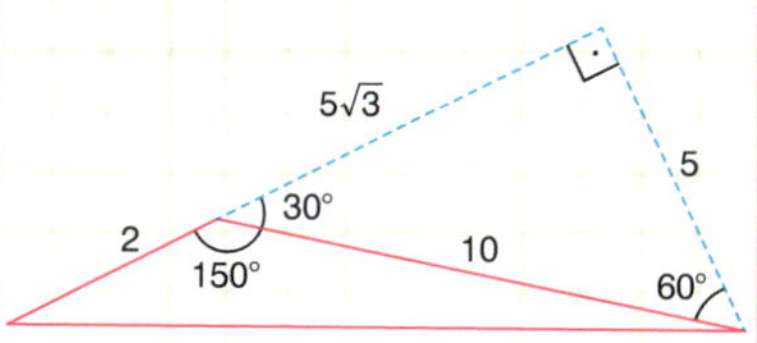

Area of Triangle

Theorem 8.1

Acute Triangle's Area

Suppose a,b, and c are the length of sides and h_a, h_b, and h_c are length of altitudes of a,b, and c.

$$\text{Area} = \frac{a \cdot h_a}{2} = \frac{b \cdot h_b}{2} = \frac{c \cdot h_c}{2}.$$

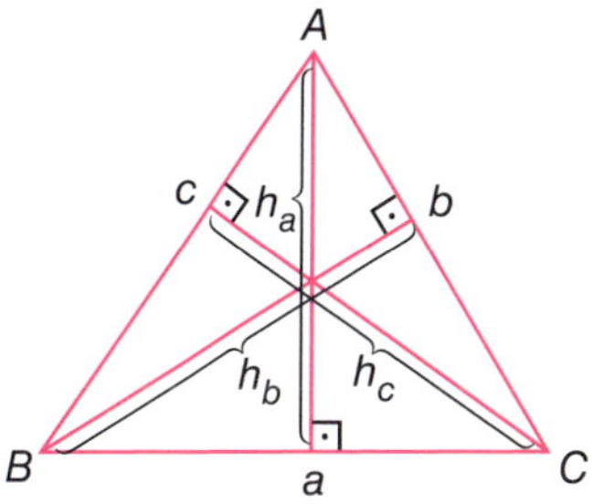

Theorem 8.2

Obtuse Triangle's Area

$$\text{Area} = \frac{a \cdot h_a}{2} = \frac{b \cdot h_b}{2} = \frac{c \cdot h_c}{2}.$$

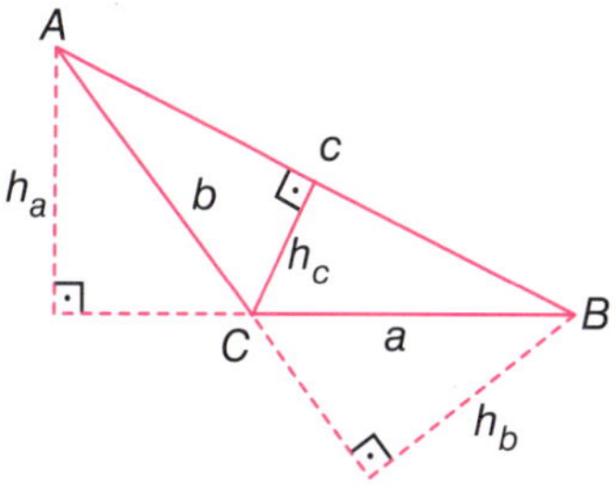

Example 8.1

Suppose $AB = 4$ units, $BC = 8$ unit ,and $m\angle B = 30^\circ$.

What is the area of ΔABC in square units?

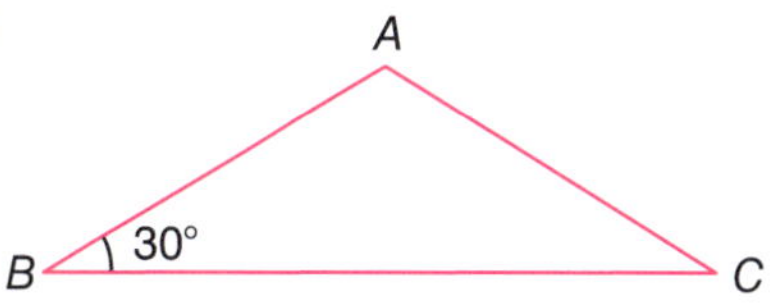

Solution 8.1

From $30^\circ - 60^\circ - 90^\circ$ triangle, the height of ΔABC is 2 units.

$$\text{Area } ABC = \frac{8 \cdot 2}{2} = 8 \text{ square units.}$$

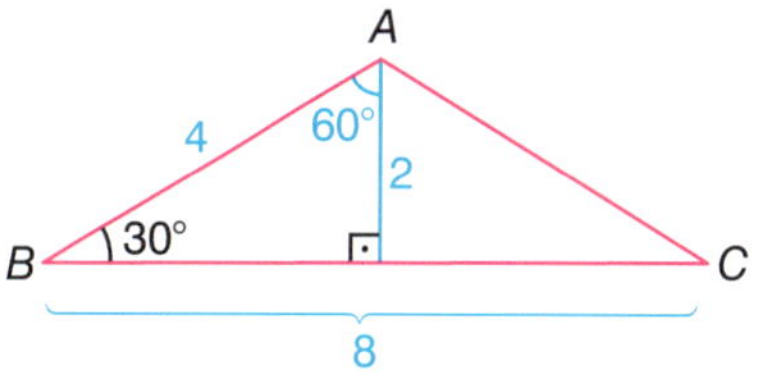

Example 8.2

Suppose $AB = 4$, $BC = 8$, and $m\angle A = 120^\circ$, then find the area of ΔABC.

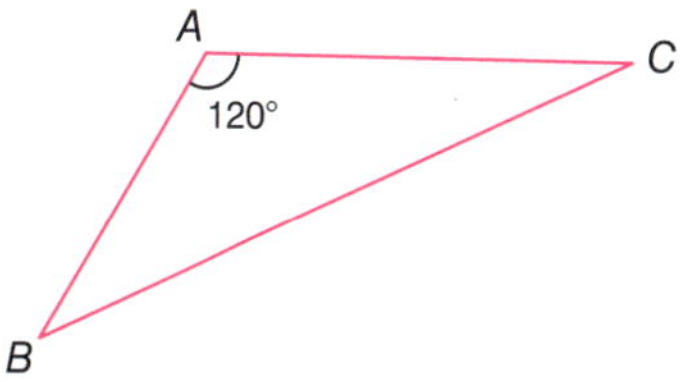

Solution 8.2

Since ΔABC is an obtuse triangle, we extend CA to create a right triangle. From $30^\circ - 60^\circ - 90^\circ$ triangle, the height of ΔABC is $4\sqrt{3}$. Therefore, the area of

$\Delta ABC = \dfrac{8 \cdot 4\sqrt{3}}{2} = 16\sqrt{3}$.

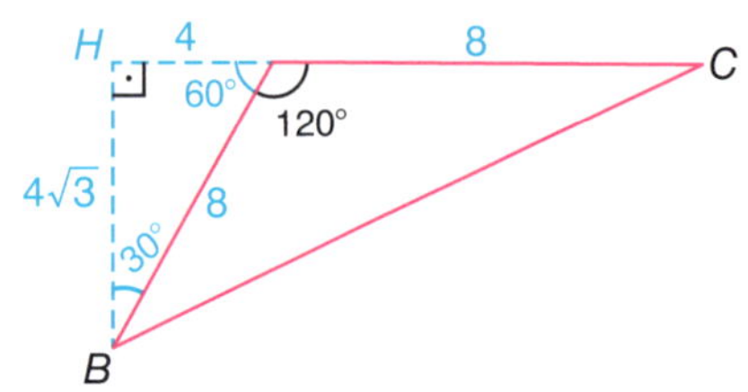

Try This

In the figure,

$AD \perp BA$, $AB = AC$,

$AD = 6$, and $DC = 1$.

Find the area of ADC.

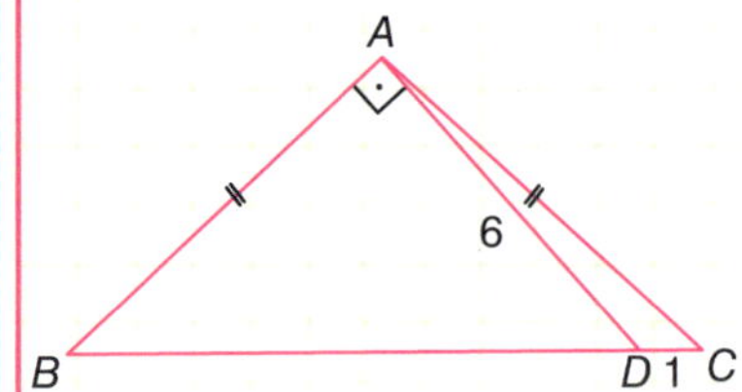

Example 8.3

Suppose

$AB = 8\sqrt{2}$,

$BC = 8$, and $m\angle B = 135^\circ$,

then find the area of ΔABC.

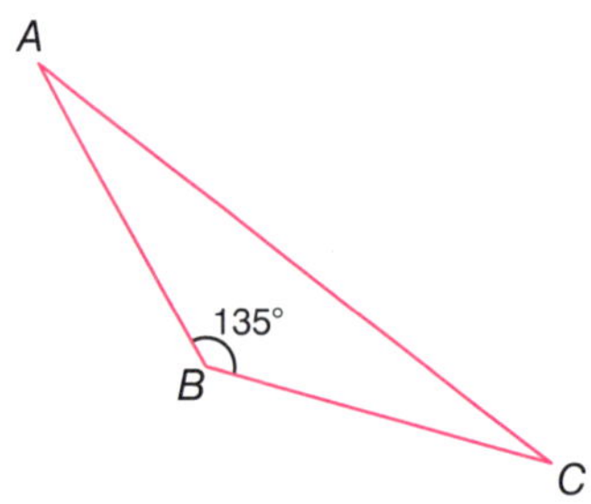

Solution 8.3

First method

We extend CB to point H to make a right triangle. By using ΔAHB $(30^\circ - 60^\circ - 90^\circ)$. We get $AH = HB = 8$.

Therefore, Area $\Delta ABC = \dfrac{8 \cdot 8}{2} = 32$.

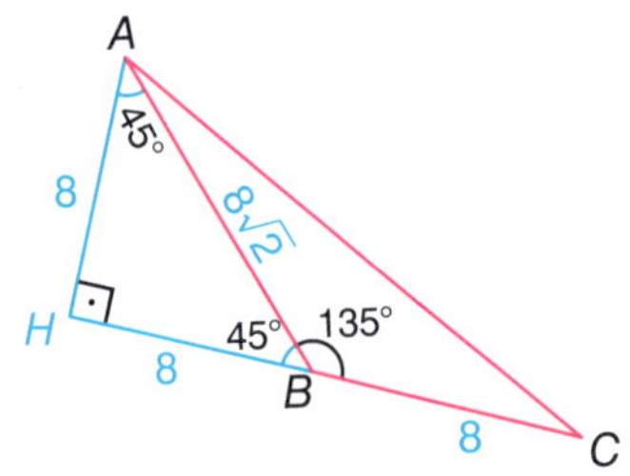

Second method

We extend AB to make a right triangle and we get height of $4\sqrt{2}$ by using ΔBHC.

Therefore, Area $\Delta ABC = \dfrac{8\sqrt{2} \cdot 4\sqrt{2}}{2} = 32$.

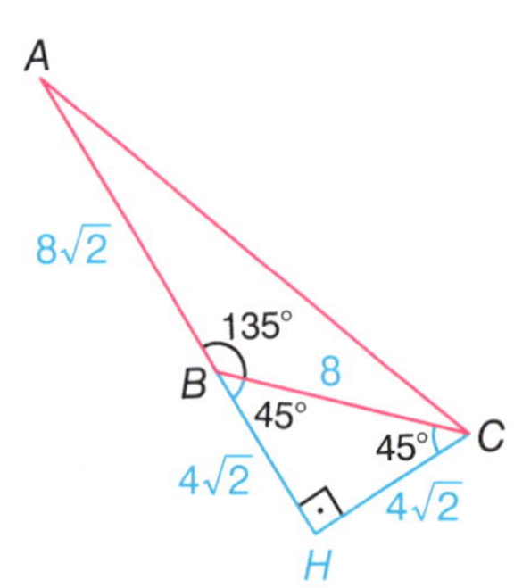

Example 8.4

In the figure,

$CH = 4$,

$AD = 6$, and

$BC = 10$.

What is AB?

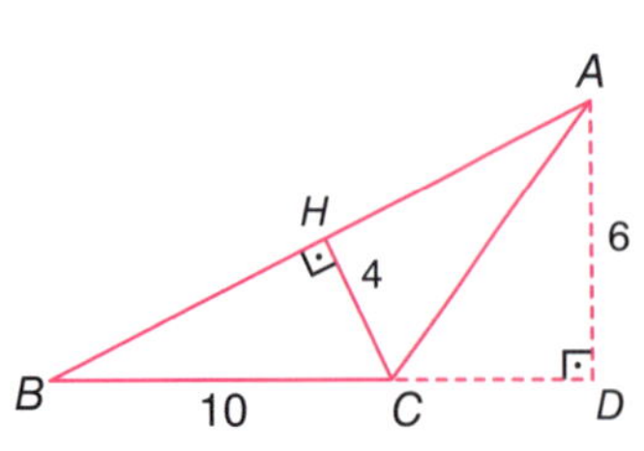

My Notes

Try This

$ABCD$ is a right triangle.

Given that

$a+b+c = 28$ cm and

$a^2+b^2+c^2 = 288$.

Find the area of the triangle.

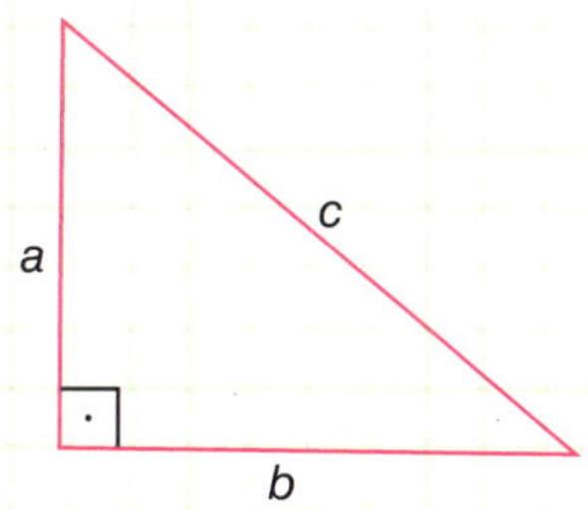

My Notes

Solution 8.4

Area of ABC can be written as two different expressions.

If the base is 10, then $\dfrac{10 \cdot 6}{2} = 30$ (1)

If the base is AB, then $\dfrac{AB \cdot 4}{2}$ (2)

(1) = (2), $30 = \dfrac{AB \cdot 4}{2} \Rightarrow AB = 15$.

Example 8.5

Prove that the area of the trapezoid is

$$\left(\frac{AB+DC}{2}\right) \cdot AH.$$

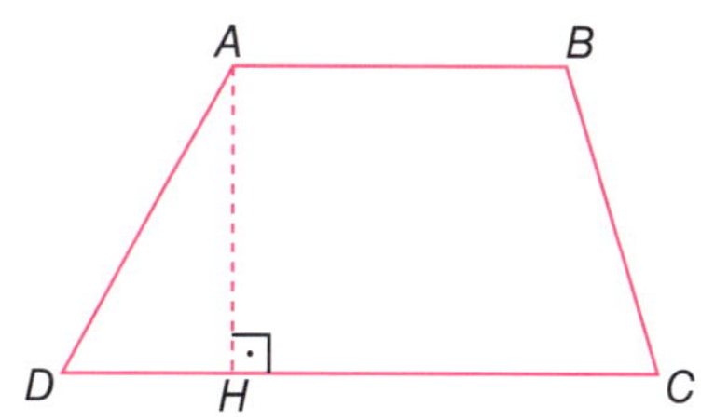

Solution 8.5

Draw a diagonal DB. Let 1 represent ΔABD and 2 represent ΔDBC.

Area of 1 triangle is $\left(\dfrac{AB \cdot AH}{2}\right)$.

Area of 2 triangle is $\left(\dfrac{DC \cdot BK}{2}\right)$.

Combine 1 and 2

$$\left(\frac{AB \cdot AH}{2}\right) + \left(\frac{DC \cdot AH}{2}\right) = \left(\frac{AB+DC}{2}\right) \cdot AH.$$

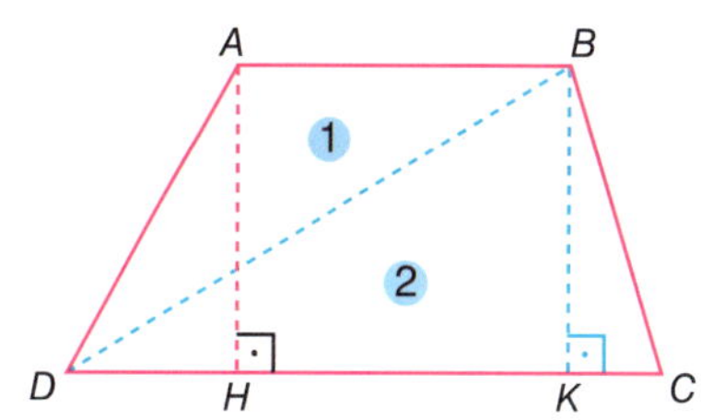

Example 8.6

E can be any point on DC. Prove that the area of the parallelogram is

Area$(ABCD) = 2$ Area(ΔAED).

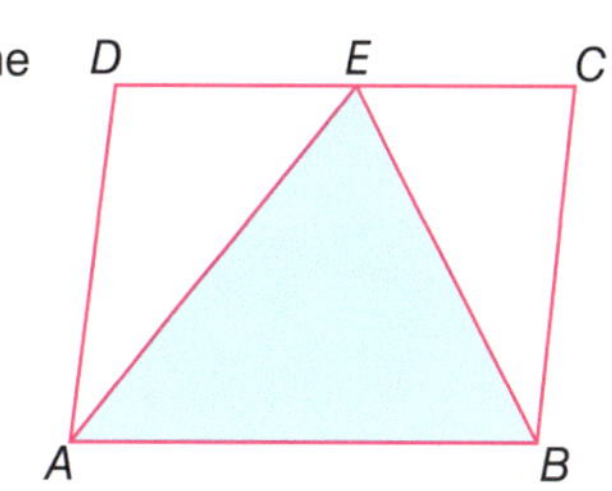

Solution 8.6

We draw a dashed line and the lines is parallel to AD or CB. The shaded area is $A+S$.

The area of parallelogram is $2A+2S$

Thus, area$(ABCD) = 2$ area(ΔAED).

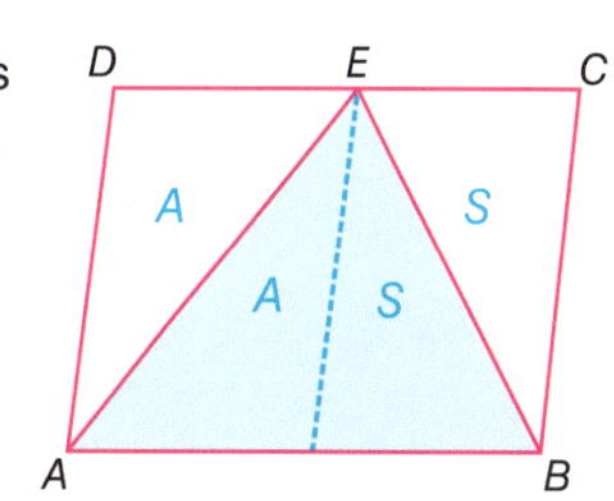

Example 8.7

In the figure, $AB = AC$ and point P is inside the triangle.

Prove that $PH_1 + PH_2 = BH$.

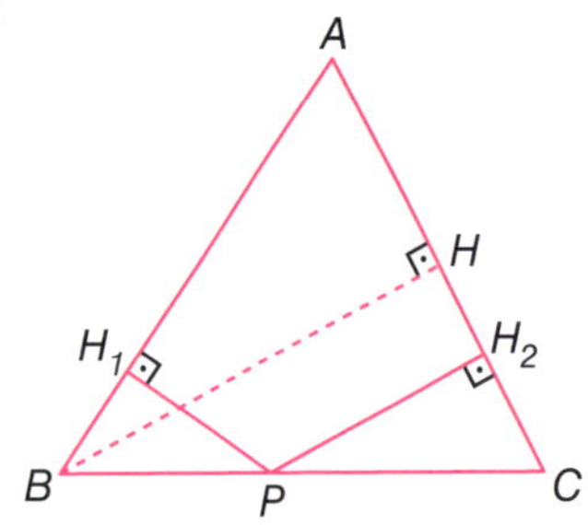

Solution 8.7

$\text{Area}(\Delta ABC) = \text{Area}(\Delta ABP) + \text{Area}(\Delta APC)$

$$\left(\frac{AC \cdot BH}{2}\right) = \left(\frac{AB \cdot PH_1}{2}\right) + \left(\frac{AC \cdot PH_2}{2}\right) \qquad (AB = AC)$$

Thus, $BH = PH_1 + PH_2$.

Example 8.8

ABC is an equilateral triangle and point O is inside the triangle.

Prove that $h_1 + h_2 + h_3 = h$ where h is the height of the triangle.

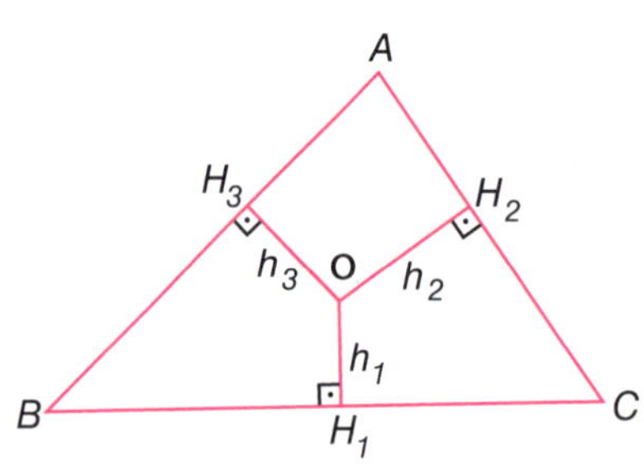

Solution 8.8

$\text{Area}(\Delta ABC) = \text{Area}(\Delta AOB) + \text{Area}(\Delta AOC) + \text{Area}(\Delta BOC)$.

Let $AB = AC = BC = a$. Then,

$$\frac{a \cdot h}{2} = \frac{a \cdot h_1}{2} + \frac{a \cdot h_2}{2} + \frac{a \cdot h_3}{2}$$

$$\frac{a}{2} \cdot h = \frac{a}{2} \cdot (h_1 + h_2 + h_3)$$

Therefore, $h = h_1 + h_2 + h_3$.

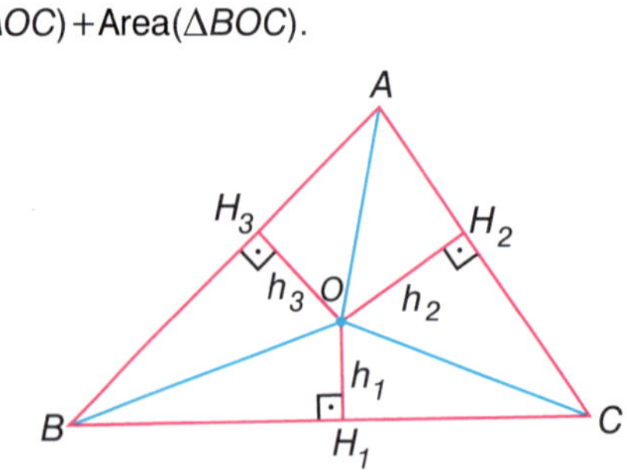

Prove This

XYZ is a triangle and $KYML$ is a parallelogram. Given that A, B, and C represent areas of indicated regions, prove that

$$B^2 - 4AC = 0.$$

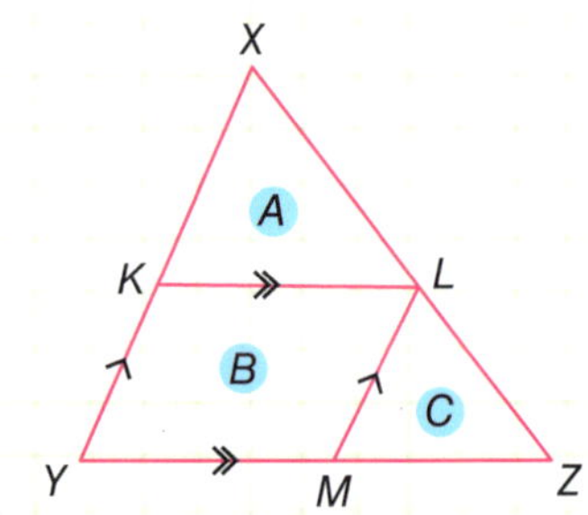

My Notes

Try This

In the figure, $BK = KD$ and the areas of the three triangles are

$Area\Delta BDK = 6$,

$Area\Delta KEC = 14$, and

$Area\Delta BKC = 14$.

What is the area of quadrilateral $ADKE$?

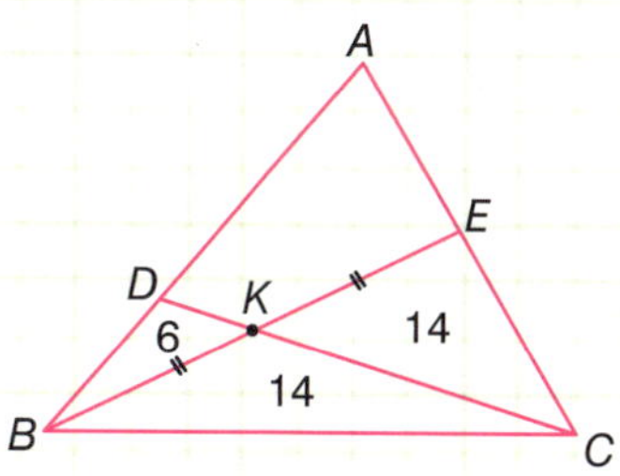

My Notes

Remark 8.1

The ratio of the areas of two similar triangles is equal to a square of the scale factor of similarity or similarity ratio. Let the similarity ratio be k, then $\frac{\text{Area}(\Delta ABC)}{\text{Area}(\Delta DEF)} = k^2$.

Proof 8.1

We know that

$$\frac{a}{e} = \frac{h_1}{h_2} = k,$$

$$\frac{\text{Area}(\Delta ABC)}{\text{Area}(\Delta DEF)} = \frac{\frac{1}{2} a \cdot h_1}{\frac{1}{2} e \cdot h_2}$$

$$= \frac{a}{e} \cdot \frac{h_1}{h_2} = k\,k = k^2.$$

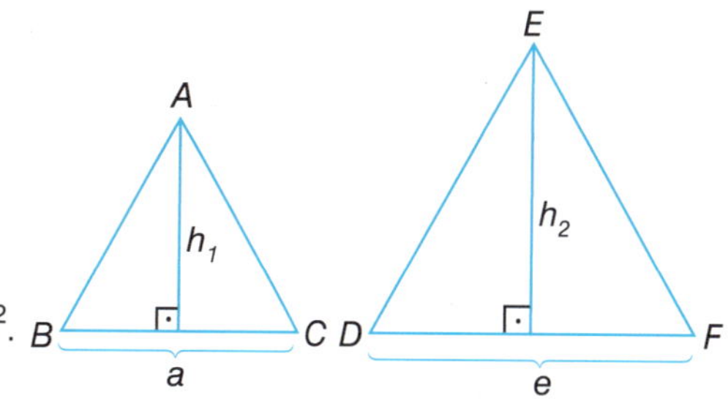

Example 8.9

ΔABC and ΔMNP are two similar triangles with a similarity ratio of $\frac{BC}{NP} = \frac{3}{2}$.

If Area(ΔABC) = 45 cm^2, then what is the area of ΔMNP?

Solution 8.9

From Remark 8.1

$$\frac{\text{Area}\Delta ABC}{\text{Area}\Delta MNP} = \left(\frac{3}{2}\right)^2 = \frac{9}{4},$$

$$\frac{45}{\text{Area}\Delta MNP} = \frac{9}{4} \Rightarrow \text{Area}\Delta MNP = 20 \text{ cm}^2.$$

Example 8.10

In the figure, $DE//FG//BC$, $AD = DF = FB$, and the area of $BCFG$ is 15 cm^2.

Find the area of quadrilateral $FGED$.

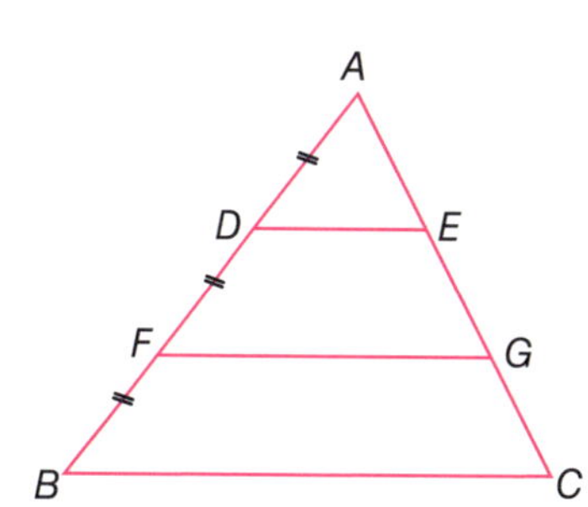

Solution 8.10

$$\frac{\text{Area}(\Delta ADE)}{\text{Area}(\Delta AFG)} = \left(\frac{1}{2}\right)^2 = \frac{1}{4} = \frac{S}{4S}$$

$$\frac{\text{Area}(\Delta ADE)}{\text{Area}(\Delta ABC)} = \left(\frac{1}{3}\right)^2 = \frac{1}{9} = \frac{S}{9S}$$

If area$(BCGF) = 5 \cdot S = 15 \Rightarrow S = 3$.

So, area$(FGED) = 3 \cdot S = 3 \cdot 3 = 9$ cm^2.

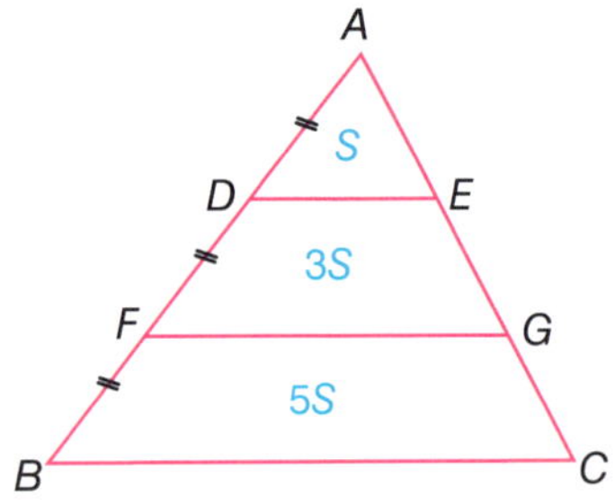

Try This

In the figure,

$AB = 5$, $DC = 10$,

$AC = 3$, and $BD = 4$.

Find the area of $ABDC$.

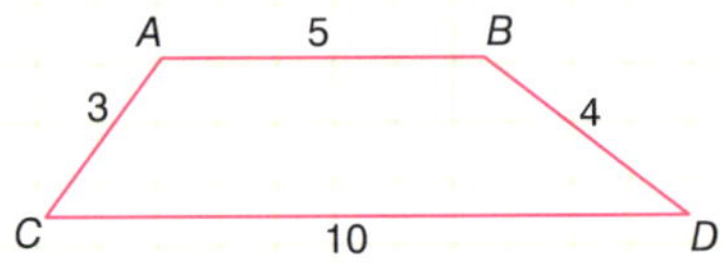

My Notes

Example 8.11

In the figure,

$BC = 10$, $DC = 4$.

What is value of $\frac{\text{Area } \Delta ABC}{\text{Area } \Delta DEC}$?

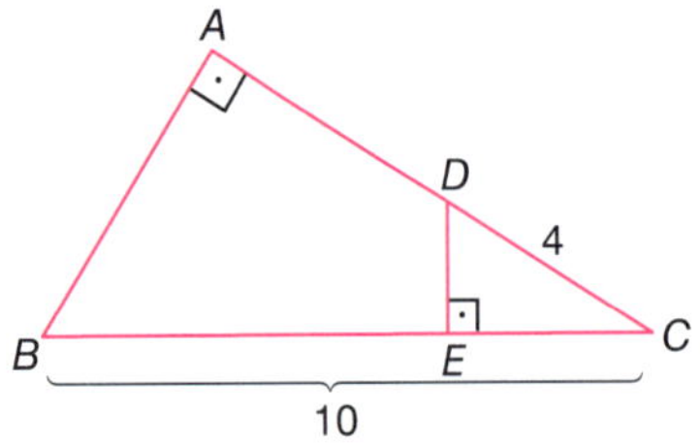

Solution 8.11

The similarity ratio of ΔDEC to ΔABC is $k = \frac{4}{10} = \frac{2}{5}$.

Therefore, $\frac{\text{Area}(\Delta ABC)}{\text{Area}(\Delta DEC)} = k^2 = \left(\frac{5}{2}\right)^2 = \frac{25}{4}$.

Theorem 8.3

Heron's Formula

Given that a, b, and c are sides of a triangle and s is the semi perimeter of the triangle, then the area of the triangle is $\sqrt{s \cdot (s-a) \cdot (s-b) \cdot (s-c)}$.

Example 8.12

Find the area of a triangle with sides of 7, 8, and 9.

Try This

In the figure, *ABCD* is a trapezoid and *DE* is an angle bisector

$AB = 8$, $AD = 6$, and $DC = 10$.

Find the area of *ABED*.

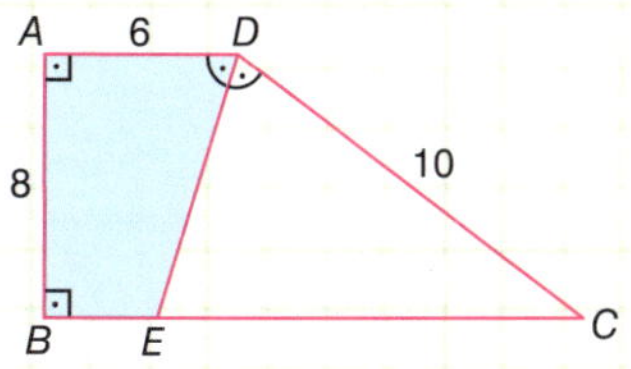

My Notes

Solution 8.12

From Heron's formula.

The semi perimeter $= \dfrac{7+8+9}{2} = 12.$

Then replace all into the formula

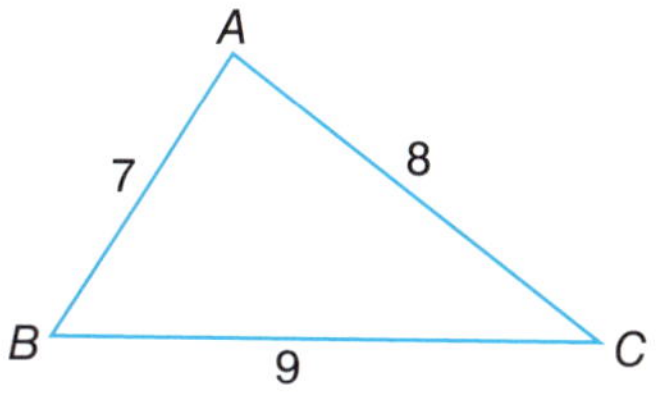

$$\sqrt{12\cdot(12-7)\cdot(12-8)\cdot(12-9)} = \sqrt{12\cdot5\cdot4\cdot3}$$
$$= \sqrt{36\cdot4\cdot5}$$
$$= 6\cdot2\sqrt{5}$$
$$= 12\sqrt{5}.$$

Example 8.13

In the figure,

$AB = 4$, $DC = 24$,

$AD = 12$, and $BC = 16$.

What is BH?

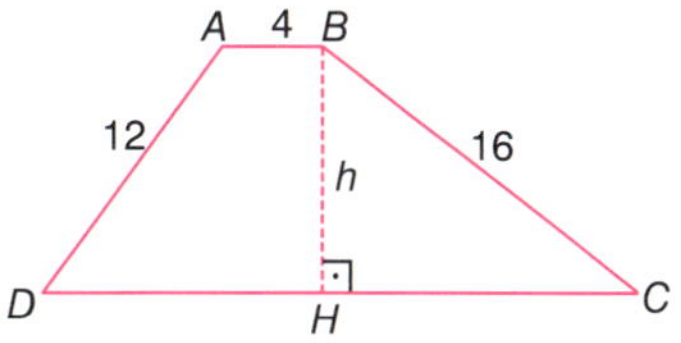

Solution 8.13

We create a parallelogram so that *DE* and *EC* become 4 and 20. By Heron's formula,

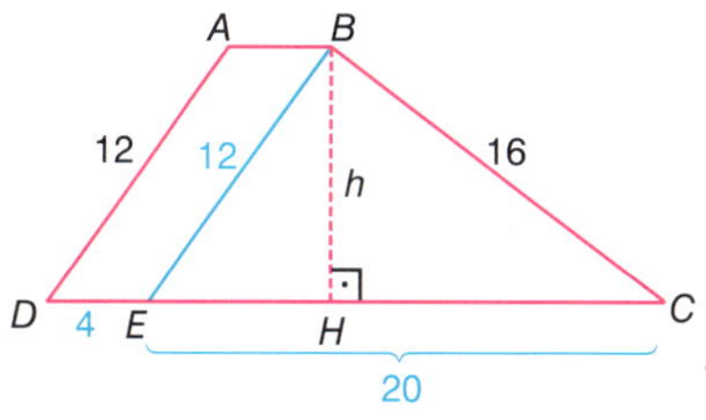

$$\text{Area}(\Delta BEC) = \sqrt{24\cdot12\cdot8\cdot4}$$
$$= 96.$$

This area is also equal to $\dfrac{20\cdot h}{2}$.

So, $\dfrac{20\cdot h}{2} = 96$. Therefore, $h = \dfrac{48}{5}$.

Remark 8.2

a. If two triangles have equal height, then ratio of their areas is the ratio of their bases.

b. If two triangles have equal bases, then ratio of their areas is the ratio of their height.

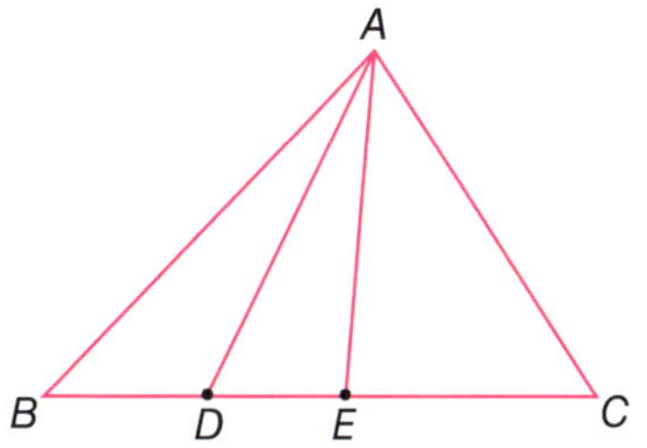

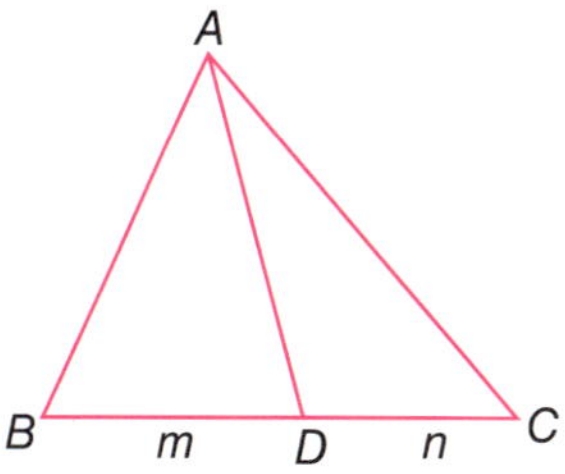

$\text{Area}(ABD):\text{Area}(ADE):\text{Area}(AEC) = BD:DE:EC$

$$\frac{\text{Area}(ABD)}{\text{Area}(ADC)} = \frac{m}{n}$$

Example 8.14

In the figure,

$2\,AD = DC$,

$2\,EF = 2\,FC = BE$.

Find the ratio of $\dfrac{\text{Area}(\Delta DFC)}{\text{Area}(\Delta ABD)}$.

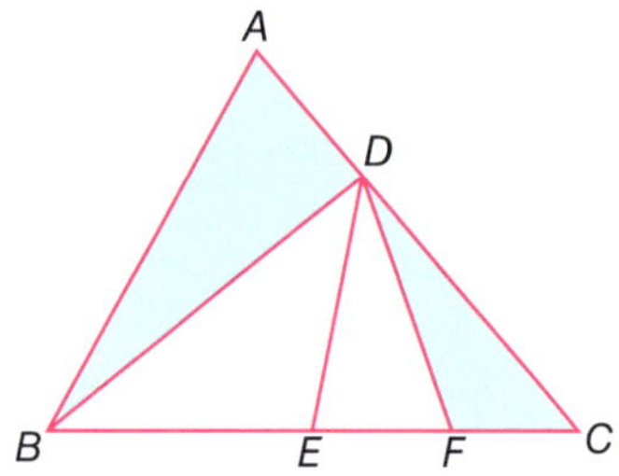

Solution 8.14

We can replace the related area sizes based on their base length. We start by replacing Area(ΔDFC) = S, then Area(ΔDEF) = S, and Area(ΔBED) = $2S$.

Since, Area(ΔDCB) = 2Area(ΔADB), then Area(ΔABD) = $2S$.

So, $\dfrac{\text{Area}(\Delta DFC)}{\text{Area}(\Delta ABD)} = \dfrac{S}{2S} = \dfrac{1}{2}$.

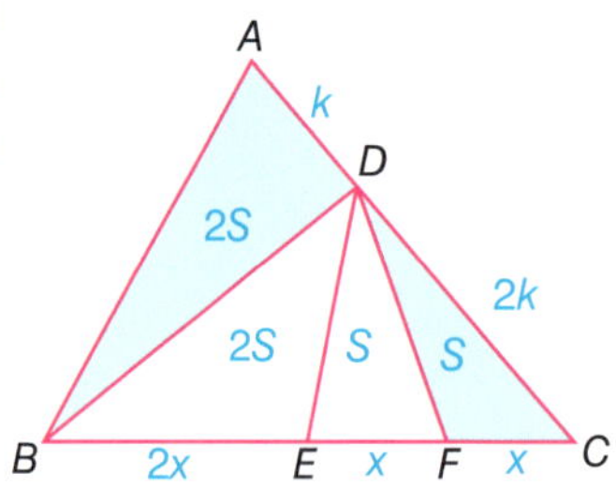

Example 8.15

In the figure,

$AB = 8$, $DC = 6$,

$AC = 7$, and $BD = 3$.

Find the area of ΔABD.

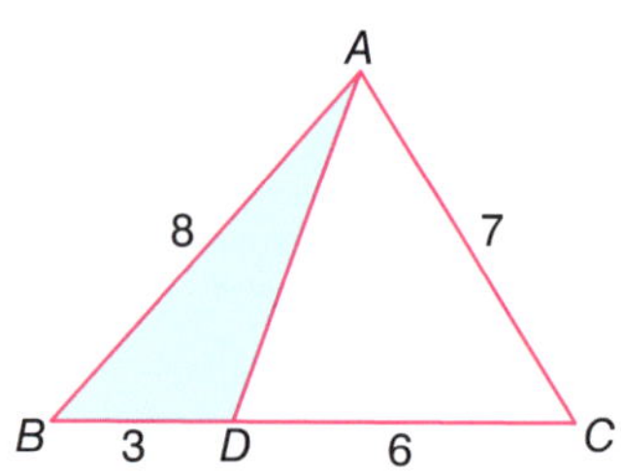

Solution 8.15

From Heron's formula

Area of $\Delta ABC = \sqrt{12 \cdot 4 \cdot 5 \cdot 3}$

$= 12\sqrt{5}$.

From the base area connections $\dfrac{BD}{BC} = \dfrac{3}{9} = \dfrac{1}{3}$.

Therefore, area of ABD is $4\sqrt{5}$.

Try This

In the figure, $ABCD$ is a square, M is the midpoint of AD, Q is the area of the square, and P is the shaded area.

If $P = \dfrac{Q}{7}$, then find AN.

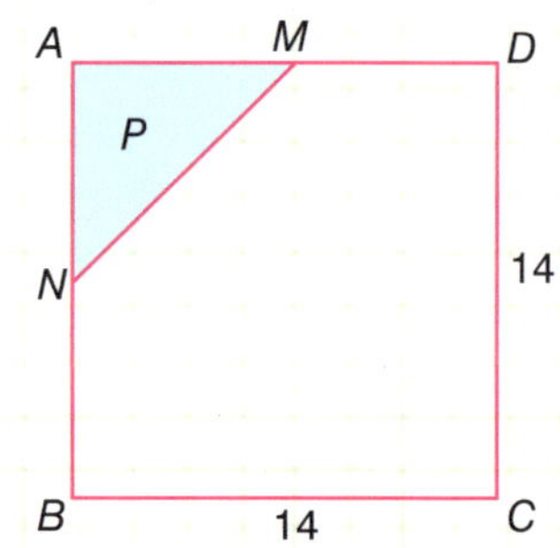

My Notes

Catriona Agg @Cshearer41

Geometry puzzle

Three squares inside a triangle.

What is the sum of the two shaded areas?

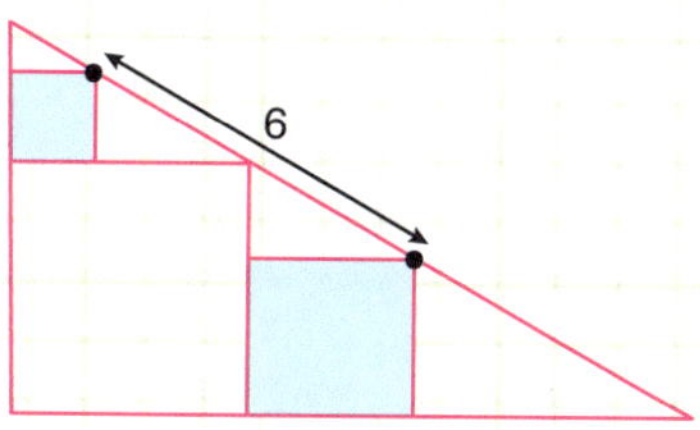

My Notes

Theorem 8.4

Trapezoid Area Formula

i) $Area\Delta AED = Area\Delta BEC$

ii) $A^2 = X{\cdot}Y$, or equivalently

$\text{Area}(\Delta ADE) = \sqrt{\text{Area}(\Delta AEB){\cdot}\text{Area}(\Delta DEC)}$.

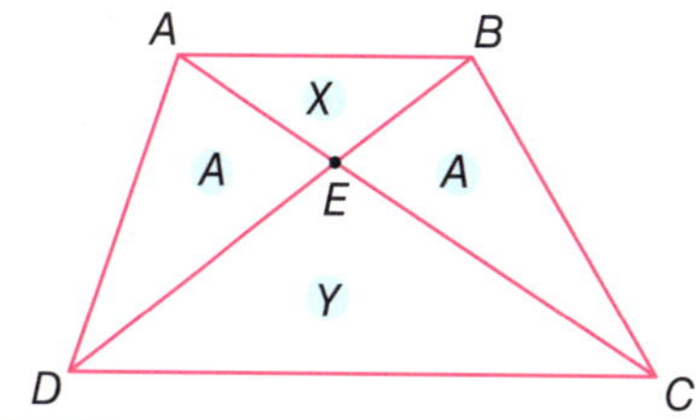

Proof 8.1

Using remark 8.2,

$$\frac{CE}{AE} = \frac{\text{Area}(CEB)}{\text{Area}(AEB)} = \frac{\text{Area}(CDE)}{\text{Area}(ADE)}$$

$$\Rightarrow \quad \frac{A}{X} = \frac{Y}{A}$$

$$\Rightarrow \quad A^2 = X{\cdot}Y.$$

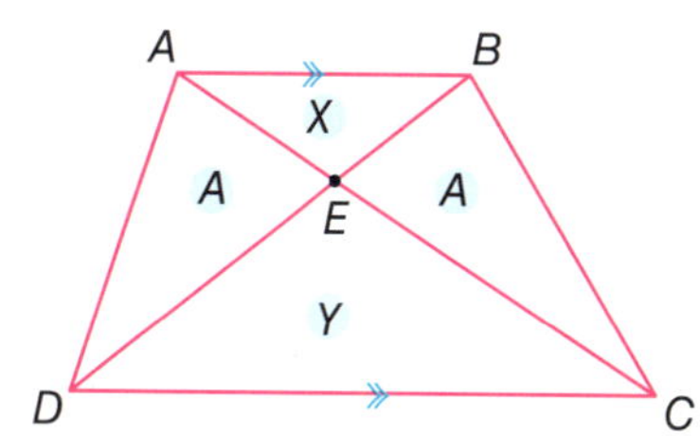

Example 8.16

In the figure,

$\text{Area}(ABE) = 4$ and

$\text{Area}(DEC) = 9$.

Find the area of the trapezoid.

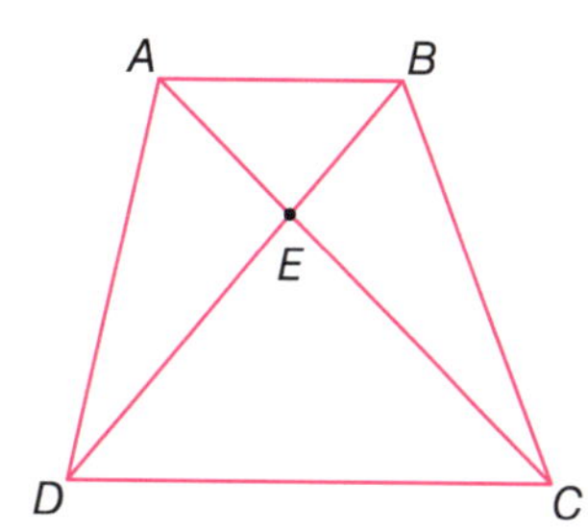

Solution 8.16

$\text{Area}(ADE) = \text{Area}(BEC) = S,$

$S^2 = 9{\cdot}4$ (Theorem 8.1)

$S = 3{\cdot}2$

$S = 6$

So, the Area of $ABCD = 4+9+6+6 = 25$.

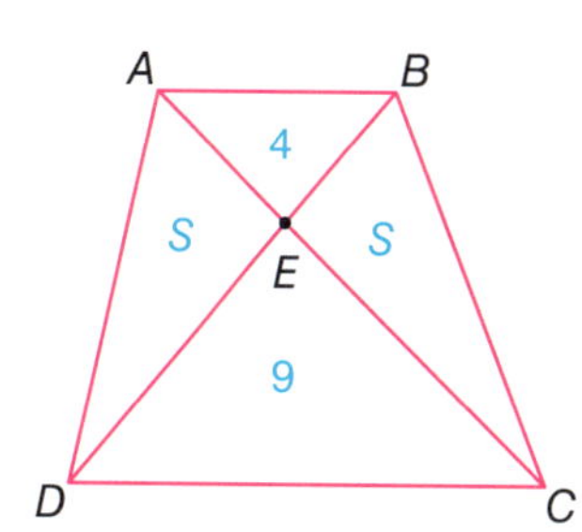

Example 8.17

In the figure,

$AE = DC = 5$ cm,

$EB = 3$ cm, and $BD = 6$ cm.

If the area of ΔAED *is* 15 cm^2, then what is the area of ΔABC in cm^2?

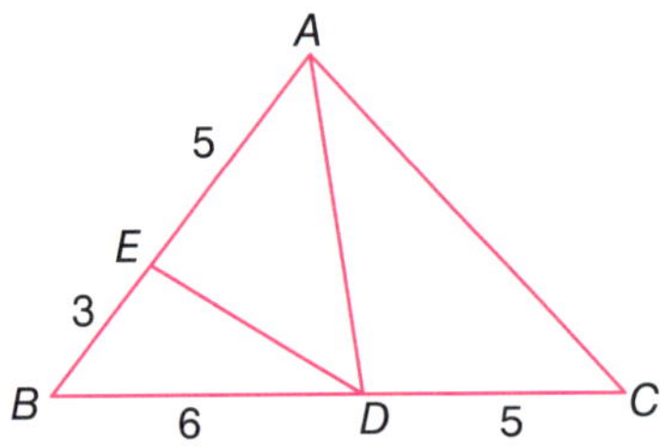

Solution 8.17

$$\frac{5}{3} = \frac{15}{\text{Area}(BED)} \Rightarrow \text{Area}(BED) = 9 \text{ cm}^2$$

$$\frac{\text{Area}(ABD)}{\text{Area}(ADC)} = \frac{6}{5} = \frac{24}{\text{Area}(ADC)} \Rightarrow$$

$\text{Area}(ADC) = 20 \text{ cm}^2$

The area of $A(ABC) = 9+15+20 = 44$ cm^2.

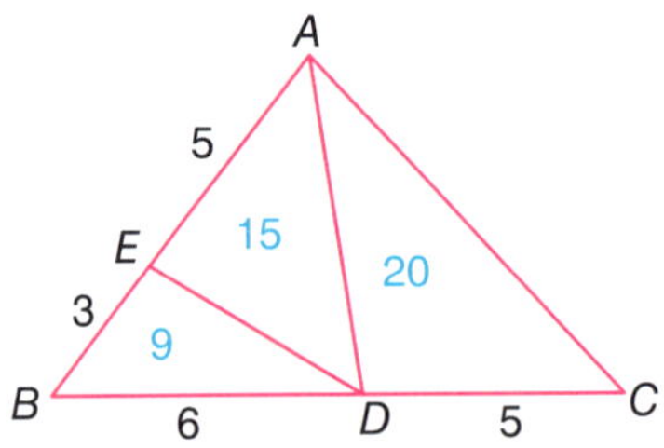

Example 8.18

In the figure, $AE \perp BC$,

$BC = 12$ cm and

$DE = 4$ cm.

Given that $\Delta ABC = 30$ cm^2, find AD.

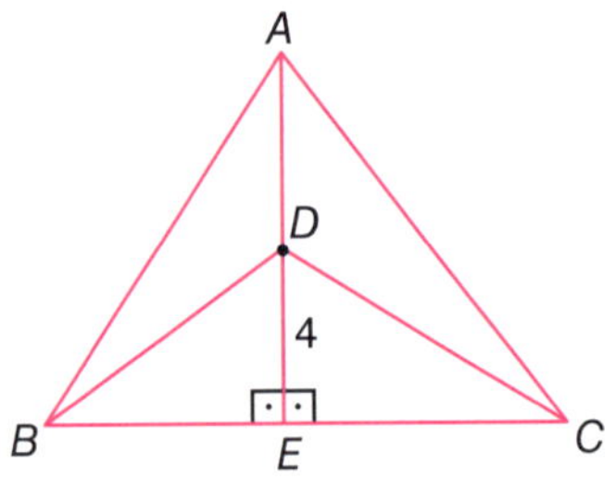

Solution 8.18

From the area formula

$$\text{Area}(\Delta ABC) = \frac{(x+4)\cdot 12}{2} = 30.$$

So, $(x+4) = 5 \Rightarrow x = 1$.

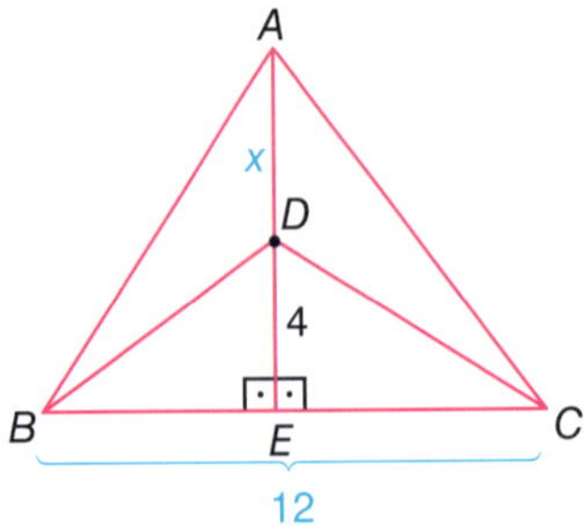

Catriona Agg @Cshearer41

Geometry puzzle

Three of the triangles in this stack are equilateral.

Find the sum of the pink and green areas.

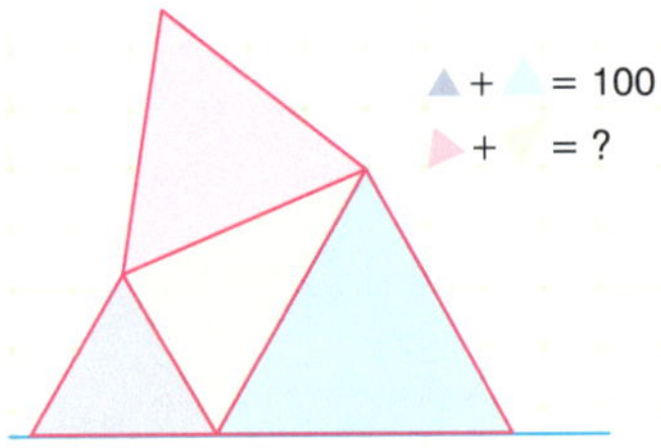

My Notes

1. In a triangle ABC,

$BC = 14$ cm,

$AC = 8$ cm, and

$m\angle ACB = 60°$.

What is the area of ΔABC?

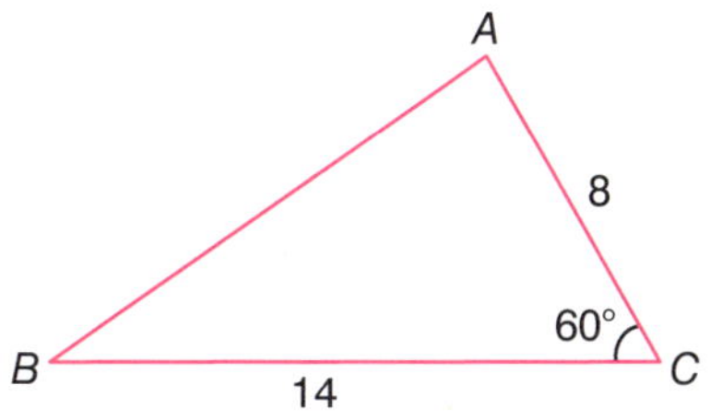

2. In the figure,

$AB = AD = 6$ cm,

$BD = 8$ cm, and

$DC = 4$ cm.

Find the area of ΔADC.

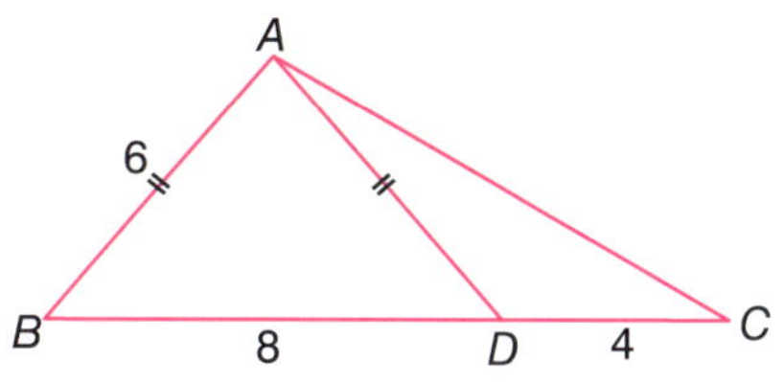

3. ABC is an isosceles triangle with

$AB = AC$,

$m\angle A = 120°$ and

$BC = 18$.

Find the area of triangle ΔABC.

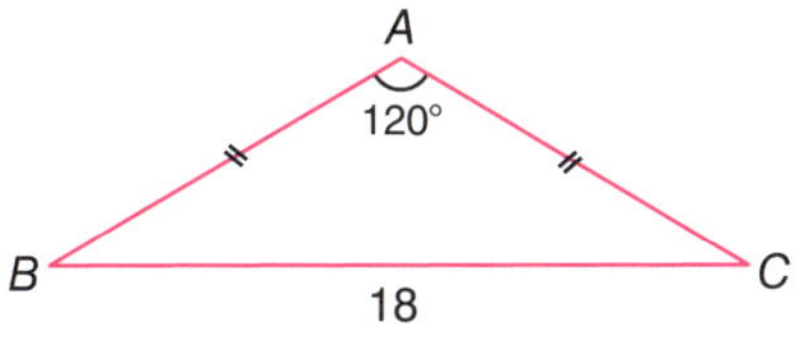

4.

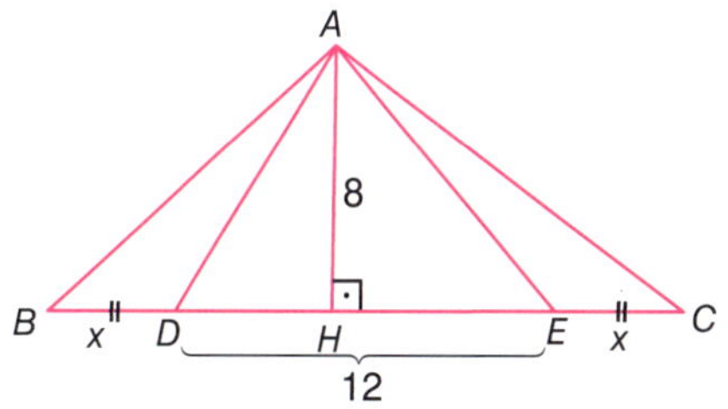

In the figure, Area(ΔADC)= 56 cm^2, $DE = 12$ cm, and $AH = 8$ cm.

What is BC?

5. In the figure,

Area$\Delta ABC = 70$ cm^2,

BD= 4 cm, and

$DC = 10$ cm.

Find the area of ΔABD.

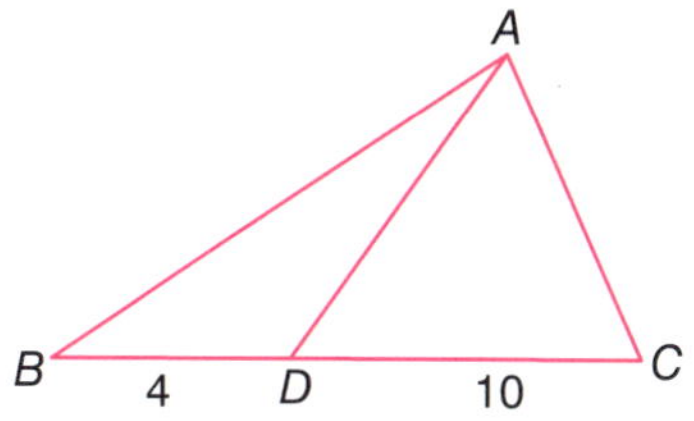

6. In the figure,

$AE = 5\ EC$,

$BC = 5\ DC$,

$m\angle B = 30°$, and

$AB = 18$ cm.

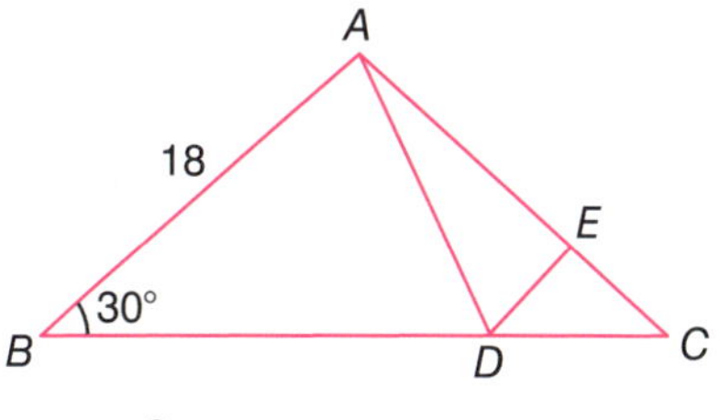

If the area of ΔADE is 15 cm^2,find the length of BD.

7. The base of an isosceles triangle is 10 units and the other sides are 13 units.

Find the area of the triangle.

8. In ΔABC,

$BC = 8$,

$AC = 6$, and

$m\angle C = 45°$.

Find the area of ΔABC.

9. In the figure, G is the centroid of ΔABC, $AB = 4AD$, and Area$\Delta ABC = 84$.

Find the area of ΔADG.

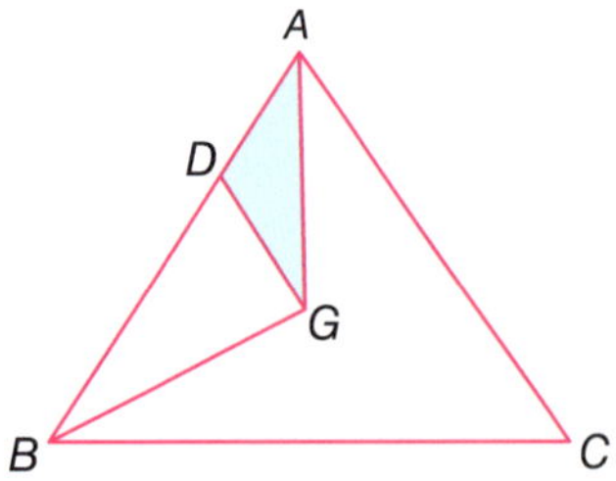

10.

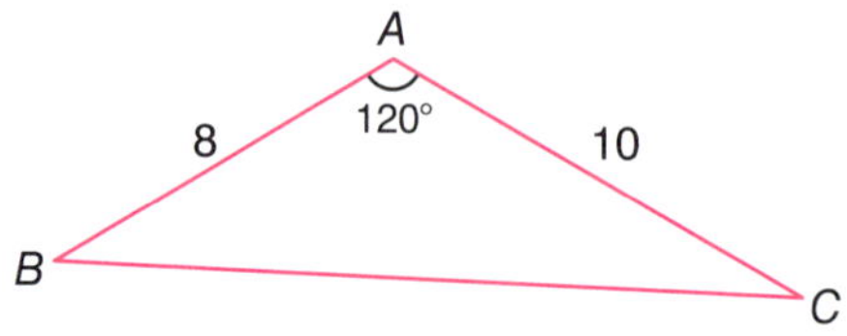

In the figure, $m\angle A = 120°$.

Given that $AB = 8$ and $AC = 10$, find the area of ΔABC.

11. One of the acute angles in a right triangle measures 22.5° and the length of the hypotenuse is 12 units.

Find the area of the triangle.

12. In the figure, $DE \parallel BC$.

Given that $AD = 5$, $DB = 3$, and Area $BCED = 39$, find the area of ABC.

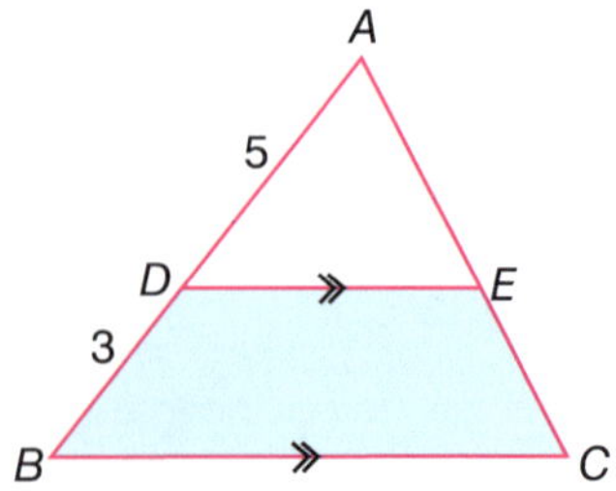

1. In the figure,

$AF = FB = BD = 4$,

$DC = 5$, $EC = 7$, and

$AE = 3$.

What is the ratio of the area of ΔABC to ΔDEF?

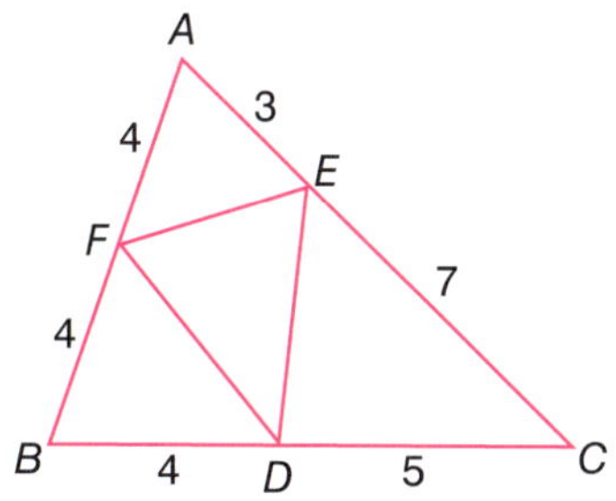

2. An equilateral triangle ABC is given. Point P lies on the base BC such that $m\angle APB = 75°$.

If one side of the triangle is 12 units, then find the area of ΔAPC.

3. In the figure,

$AB = 5$ cm, $BC = 12$ cm,

$CD = 15$ cm, and $AD = 14$ cm.

Find the area of the quadrilateral $ABCD$.

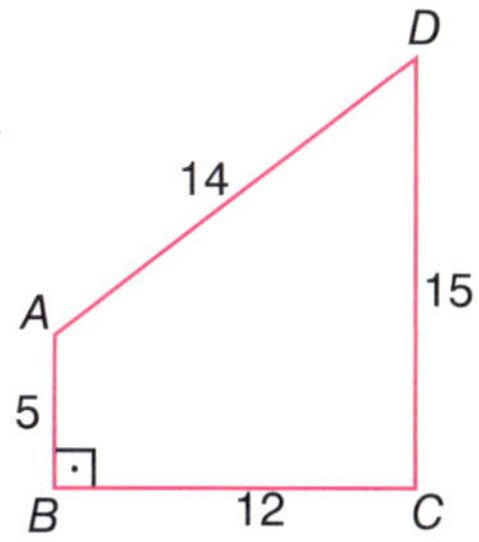

4. In the figure, point D lies on the altitude AH.

Given that

$BC = 10$, and

$AD = 6$,

find the area of $ABDC$.

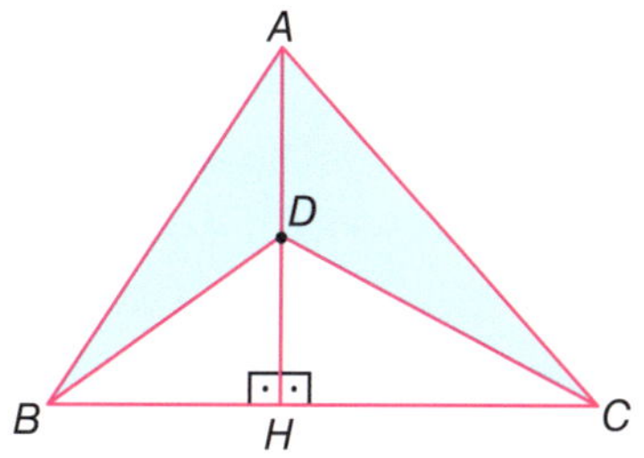

5. In the figure,

$AE = 6$, $EC = 3$, and

$BD = DC = 5$.

What is the ratio of the area of ΔABC to ΔEDC?

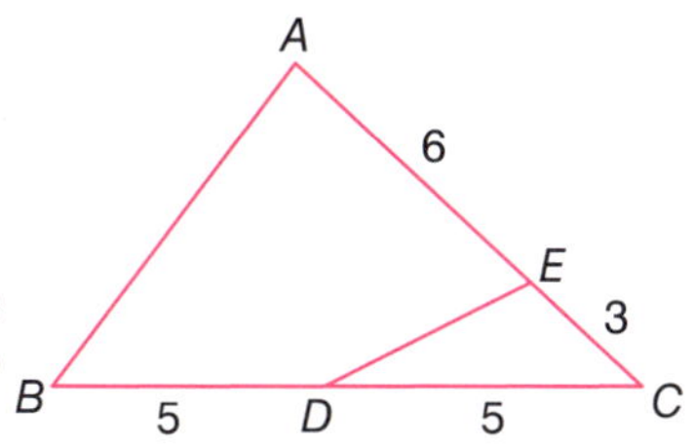

6. In the figure, $ABCD$ is a kite.

Given that $AB = AD$,

$BC = DC = 4\sqrt{3}$,

$m\angle BAD = 60°$, and

$m\angle BCD = 120°$.

What is the area of kite $ABCD$?

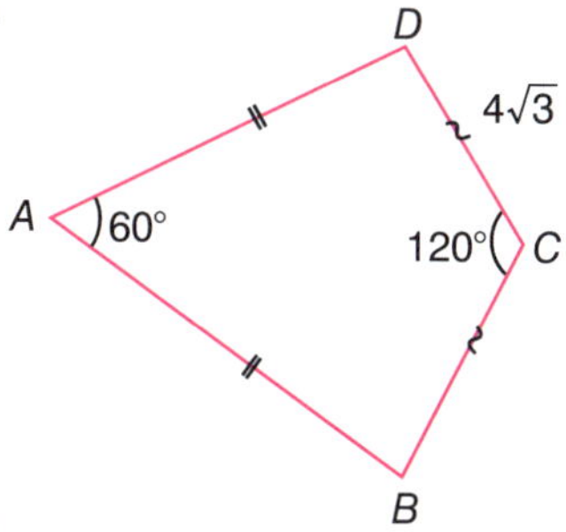

7. In $\triangle ABC$, BP and CP are angle bisectors.

$AB = 5$, $AC = 7$, and the area of $\triangle ABP$ is 20 cm^2.

Find the area of $\triangle APC$.

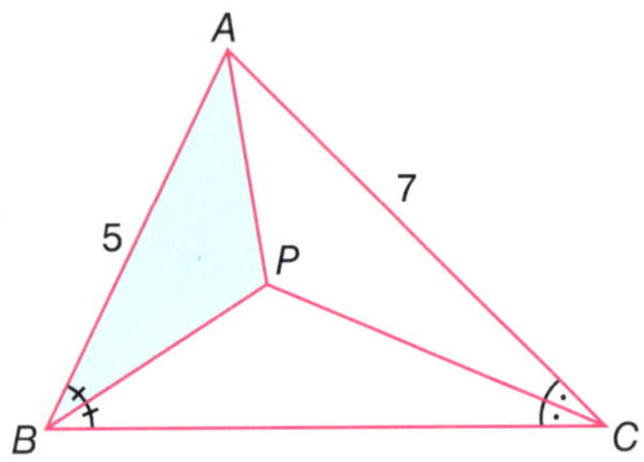

8.

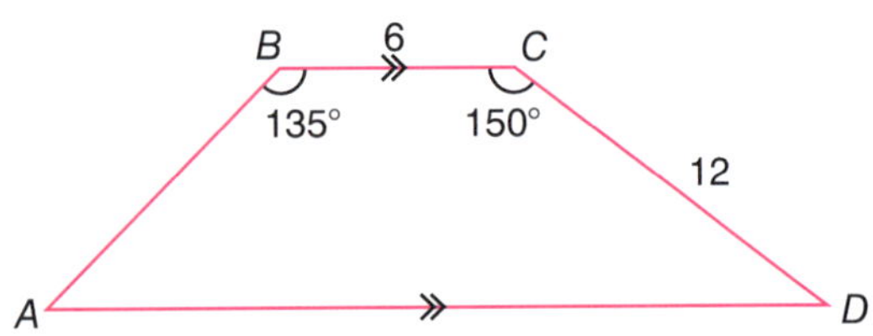

In the figure, $ABCD$ is a trapezoid with

$CD = 12$, $BC = 6$, $m\angle ABC = 135°$ and $m\angle BCD = 150°$.

What is the area of the trapezoid $ABCD$?

9.

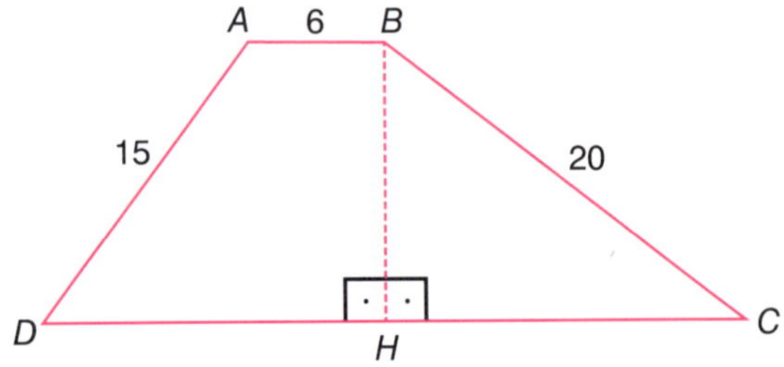

In the figure, $AB = 6$, $DC = 31$, $AD = 15$, and $BC = 20$.

Find the length of BH.

10.

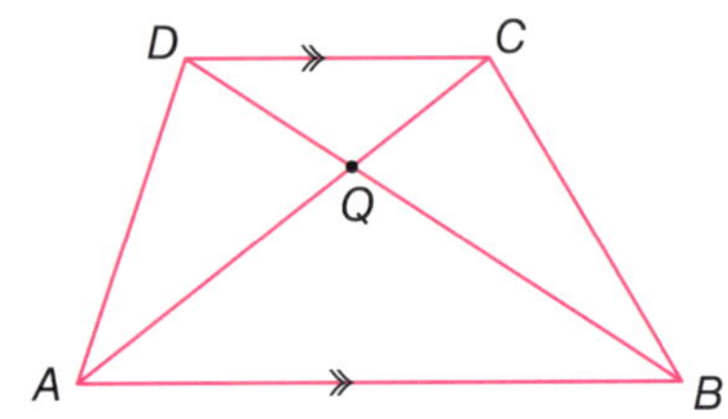

In the figure, $ABCD$ is a trapezoid and O is the intersection point of its diagonals. The area of $\triangle DQC$ is 9 cm^2 and the area of$\triangle AOB$ is 25 cm^2.

What is the area of trapezoid $ABCD$?

11.

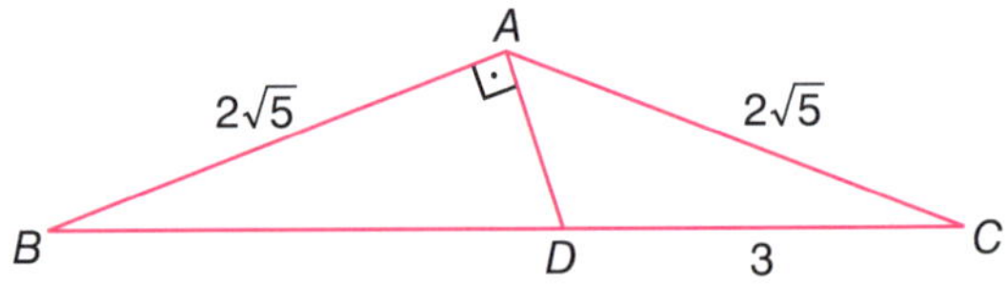

In the figure, $AD \perp BA$, $AB = AC = 2\sqrt{5}$, and $DC = 3$.

What is the area of ADC?

12.

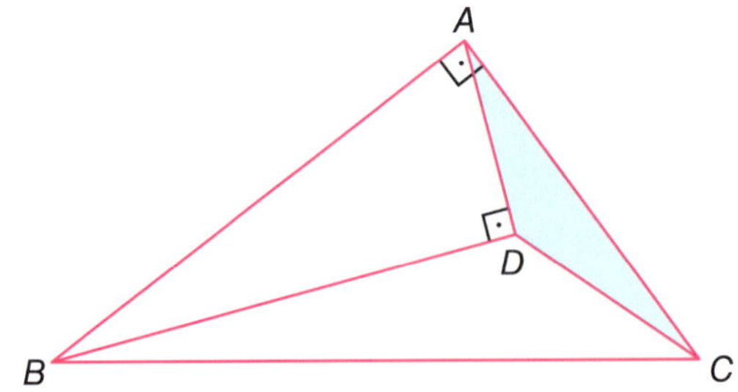

Suppose ABC is a right triangle, $AB \perp AC$ and $AD \perp BD$.

Given that $AB = AC = 20$ and $BD = 16$, find the area of $\triangle ADC$.

1. (CWML 2017-18 Meet 1 #2)

Triangle ABC is $30^\circ-60^\circ-90^\circ$ as shown.

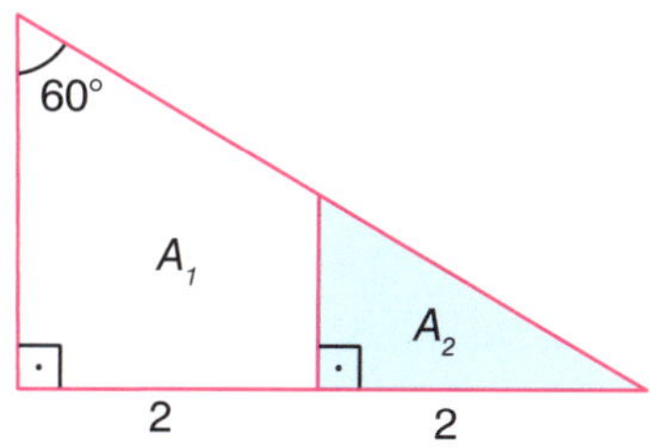

What is the ratio of areas $\frac{A_1}{A_2}$?

A) 1 B) $\sqrt{3}$ C) 2 D) 3 E) 4

2. (2005 AMC 10B # 14)

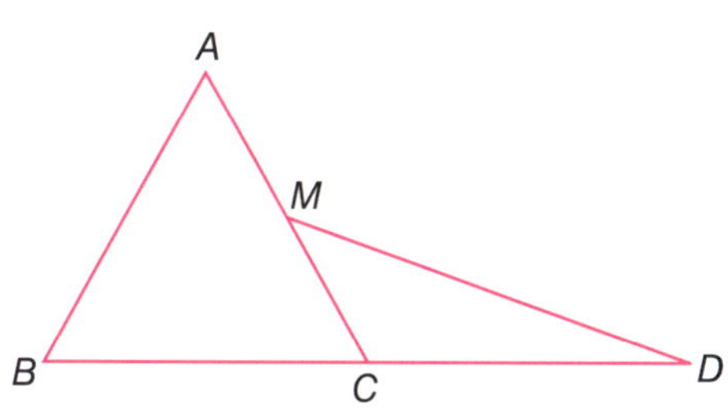

Equilateral ΔABC has side length 2, M is the midpoint of $\overline{AC}$, and C is the midpoint of $\overline{BD}$.

What is the area of ΔCDM?

A) $\frac{\sqrt{2}}{2}$ B) $\frac{3}{4}$ C) $\frac{\sqrt{3}}{2}$ D) 1 E) $\sqrt{2}$

3. (CWML 2015-16 Meet 1 #14) Triangle ABC is similar to DEF. The measure of side DE is 1.5 times the measure of side AB.

If the area of ΔABC is a 8 square inches find the area of ΔDEF.

4. (2009 AMC 10B # 18) Rectangle $ABCD$ has $AB = 8$ and $BC = 6$. Point M is the midpoint of diagonal $\overline{AC}$, and E is on AB with $ME \perp AC$.

What is the area of ΔAME?

A) $\frac{65}{8}$ B) $\frac{25}{3}$ C) 9 D) $\frac{75}{8}$ E) $\frac{85}{8}$

5. (2018 AMC 12B # 13)

Square $ABCD$ has side length 30, Point P lies inside the square so that AP $=12$ and $BP = 26$. The centroids of ΔABP, ΔBCP, ΔCDP, and ΔDAP are the vertices of a convex quadrilateral.

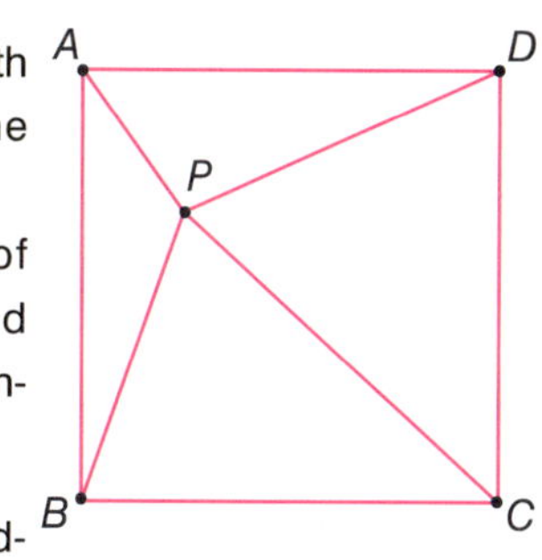

What is the area of that quadrilateral?

A) $100\sqrt{2}$ B) $100\sqrt{3}$ C) 200 D) $200\sqrt{2}$ E) $200\sqrt{3}$

6. (2018 AMC 10A # 9)

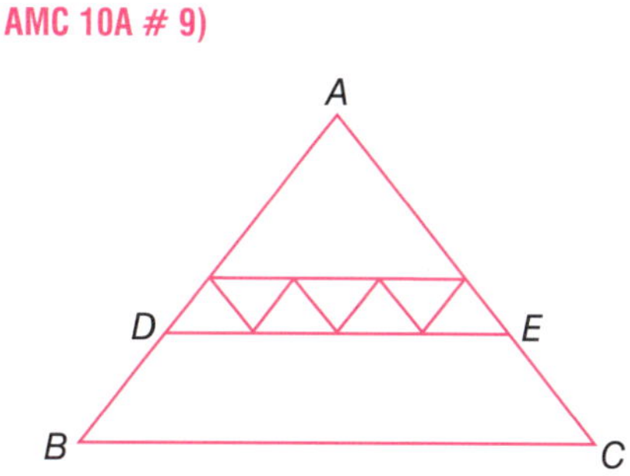

All of the triangles in the diagram below are similar to isosceles triangle ABC, in which $AB = AC$. Each of the 7 smallest triangles has area 1, and ΔABC has area 40.

What is the area of trapezoid $DBCE$?

A) 16 B) 18 C) 20 D) 22 E) 24

7. **(AMC–Fermat 2018 #14)** Two similar triangles have areas 5 and 720.

What is the ratio of the perimeter of the larger triangle to the perimeter of the smaller triangle?

8. **(MathCON 2019 Final GR 10 #36)**

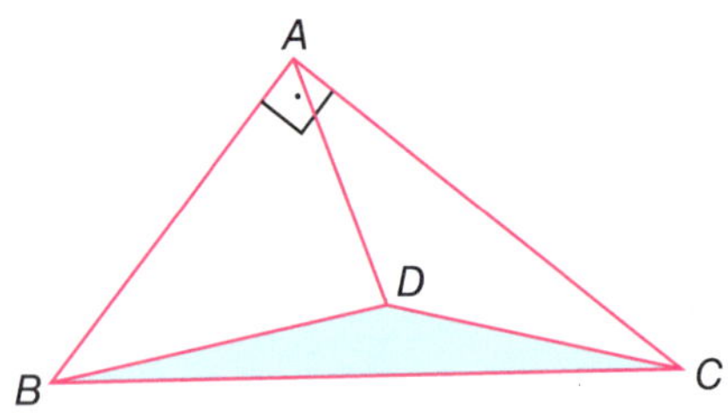

Suppose ABC is a triangle, $\overline{AB} \perp \overline{AC}$ and $\overline{AD} \perp \overline{BD}$.

If $AB = AC = 20$ and $BD = 16$, find the area of triangle BCD.

A) 24 B) 28 C) 30 D) 32 E) 36

9. **(2017 AMC 10B # 19)** Let ABC be an equilateral triangle. Extend side $\overline{AB}$ beyond B to a point B' so that $BB' = 3AB$. Similarly, extend side $\overline{BC}$ beyond C to a point C' so that $CC' = 3BC$, and extend side $\overline{CA}$ beyond A to a point A' so that $AA' = 3CA$.

What is the ratio of the area of $\Delta A'B'C'$ to the area of ΔABC?

A) 9:1 B) 16:1 C) 25:1 D) 36:1 E) 37:1

10. **(2013 AMC 10B # 16)**

In triangle ABC, medians AD and CE intersect at P, $PE = 1.5$, $PD = 2$, and $DE = 2.5$.

What is the area of $AEDC$?

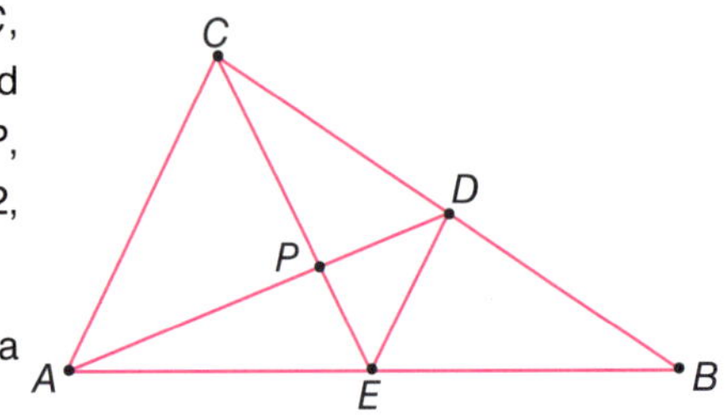

A) 13 B) 13.5 C) 14 D) 14.5 E) 15

11. **(2006 AMC 10B # 23)**

A triangle is partitioned into three triangles and a quadrilateral by drawing two lines from vertices to their opposite sides. The areas of the three triangles are 3, 7, and 7 as shown.

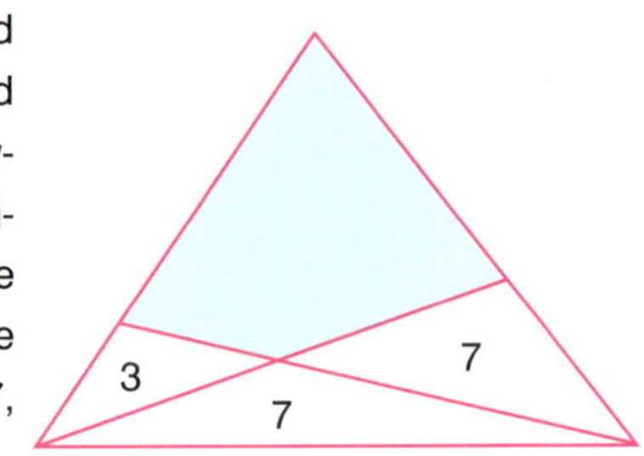

What is the area of the shaded quadrilateral?

A) 15 B) 17 C) $\frac{35}{2}$ D) 18 E) $\frac{55}{3}$

12. **(2005 AMC 10A # 25)** In ABC we have $AB = 25$, $BC = 39$, and $AC = 42$. Points D and E are on AB and AC respectively, with $AD = 19$ and $AE = 14$.

What is the ratio of the area of triangle ADE to the area of the quadrilateral $BCED$?

A) $\frac{266}{1521}$ B) $\frac{19}{75}$ C) $\frac{1}{3}$ D) $\frac{19}{56}$ E) 1

CHAPTER 9

Polygons

A polygon with n vertices (and n sides) is known as an n-gon.

If all sides and are equivalent, the polygon is called regular. Polygons can be convex, concave, or star. The word "polygon" derives from the Greek *poly*, "many," and *gonia*, meaning "angle."

The most familiar type of polygon is the regular polygon, which is a convex polygon with equal sides lengths and angles. The generalization of a polygon into three dimensions is called a polyhedron, into four dimensions is called a polychoron, and into n dimensions is called a polytope.

My Notes

Definition 9.1

A polygon is a closed figure with at least 3 straight sides.

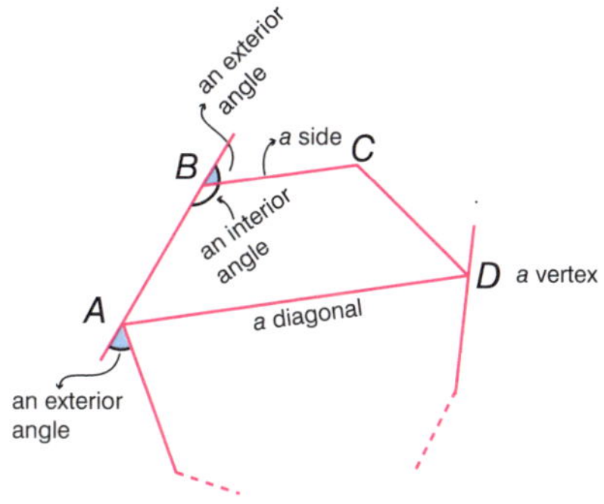

Convex and Concave Polygons

A polygon is called convex if the line segment AB also contains all the points of the segment with the interior of the polygon.

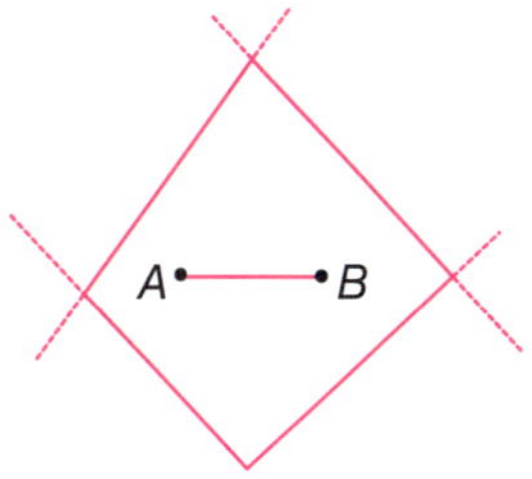

A polygon is called concave if the line segment AB has points that outside of the polygon.

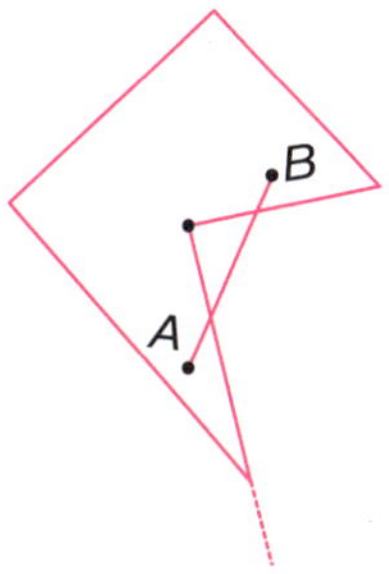

Names of Polygons with n-sides

Number of sides	Name
3	Triangle
4	Quadrilateral
5	Pentagon
6	Hexagon
7	Heptagon
8	Octagon
9	Nonagon
10	Decagon

Definition 9.2

A diagonal of a polygon is a line segment that connects two nonconsecutive vertices. A triangle has no diagonal.

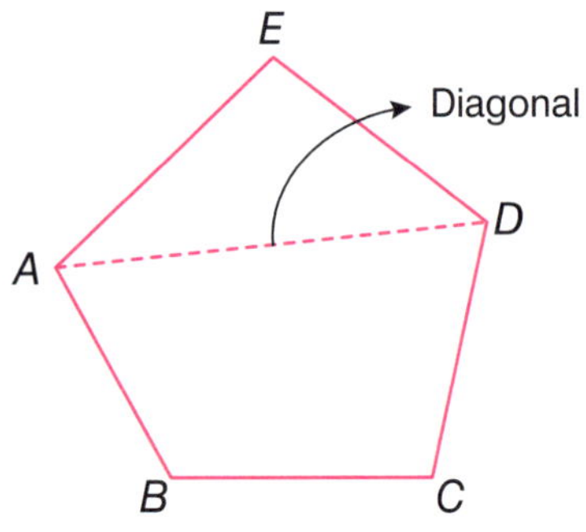

Theorem 9.1

A convex polygon with n sides has $\frac{n(n-3)}{2}$ diagonals.

My Notes

Proof 9.1

Proof 1:

An n-sided polygon has n vertices, and from each vertex, we can draw $(n-3)$ diagonals. At each vertex, there are $(n-3)$ diagonals an endpoint. Since there are n vertices we count $n(n-3)$ diagonals, but we counted each endpoint of the diagonal twice. So, we simply divide by 2 to get our formula $\frac{n(n-3)}{2}$.

Proof 2:

An n-sided polygon has n vertices. Consider all with n points. All n vertices can be connected with $\binom{n}{2} = \frac{n(n-1)}{2}$ lines $C(n,2) = \frac{n(n-1)}{2}$. Since n of these lines are sides of a polygon, then we have to exclude n. So the number of diagonals is

$$C(n,2) - n = \frac{n(n-1)}{2} - n = \frac{n^2-3n}{2} = \frac{n(n-3)}{2}.$$

Proof 3:

n vertices can be connected with.

$1+2+3+\ldots+(n-1)$ lines segments.

$$1+2+3+\ldots+(n-1) = \frac{(n-1)\,n}{2} \qquad \overset{\textbf{Gauss formula}}{\left[1+2+3+\ldots+n = \frac{n(n+1)}{2}\right]}$$

those line segments, we exclude n since sides of the polygon. So, the number of diagonals is

$$\frac{n(n-1)}{2} - n = \frac{n(n-3)}{2}.$$

My Notes

Exercise 9.1

The number of diagonals for an 11-sided polygon is

$$\frac{11\cdot(11-3)}{2} = \frac{11\cdot 8}{2} = 44.$$

Example 9.1

Find the number of sides of a polygon which has 9 diagonals.

Solution 9.1

From the formula

$$\frac{n(n-3)}{2} = 9 \Rightarrow n(n-3) = 18.$$

So, $n = 6$ satisfies the equation.

Example 9.2

In a convex polygon, the number of diagonals is equal to 5 less than twice as much as the number of sides.

What is the number of sides of this polygon?

Solution 9.2

Let n be the number of sides. Then, we setup the following equation.

$$2n-5 = \frac{n(n-3)}{2}$$

$$4n-10 = n(n-3)$$

$$4n-10 = n^2-3n$$

$$n^2-7n+10 = 0 \Rightarrow (n-5)(n-2) = 0$$

$n = 2$ (impossible)

$n = 5$

The answer is a 5-sided polygon.

Theorem 9.2

The sum of the measures of the interior angles of any n-sided convex polygon is $(n-2)\cdot 180°$.

Note

For any n-gon, there are exactly $(n-2)$ non-overlapping triangles, each of which have angles that sum to 180°. Therefore, the sum of all interior angles in a convex polygon is $(n-2)\cdot 180°$.

My Notes

Proof 9.2

Proof 9.2.1

We can create $(n-2)$ triangles drawing from one vertex of a n-sided polygon.

4-sided 2Δ

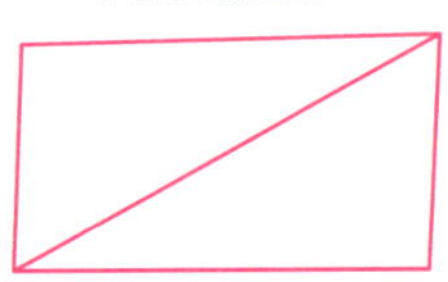

5-sided 3Δ

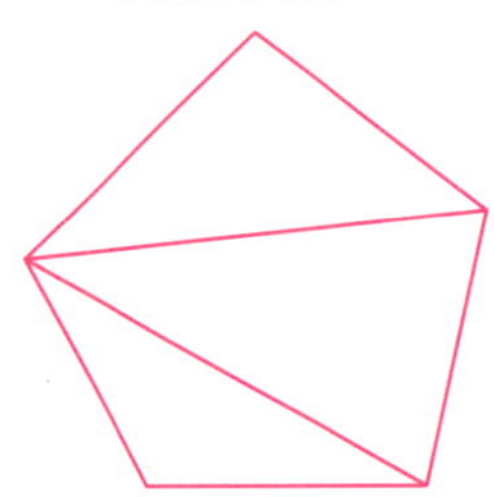

So, the sum of all measures of the interior angles of the polygon is $(n-2)\cdot 180°$.

Proof 9.2.2

In the figure, we have an n-sided polygon and n triangles. So the sum of all is $180°\cdot n$. But the inner parts' angles do not belong to interior angles of the polygon. So, $180\cdot n - 360 = (n-2)\cdot 180$.

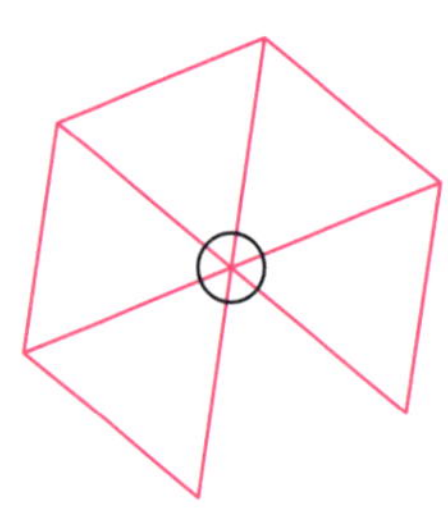

Example 9.3

a) Find the sum of the measures of the interior angles of a polygon with 12 sides.

b) The sum of the interior angles of a polygon is 1620°. How many sides does this polygon have?

c) If the sum of a regular polygon's interior angles is 1080°, then find one interior angle's measure.

Solution 9.3

a) $(12-2)180 = 10\cdot 180 = 1800$

b) $1620 = (n-2)180 \Rightarrow (n-2) = \dfrac{1620}{180} \Rightarrow n-2 = 9 \Rightarrow n = 11.$

c) $1080 = (n-2)\cdot 180 \Rightarrow n-2 = 6 \Rightarrow n = 8.$

Since this is a regular polygon, one exterior angle is $\dfrac{1080}{8} = 135°$. Therefore, one interior angles measure is 45°.

Explanation

Exterior angles an exterior angle for a polygon is formed by extending one side of the polygon from one of its end points.

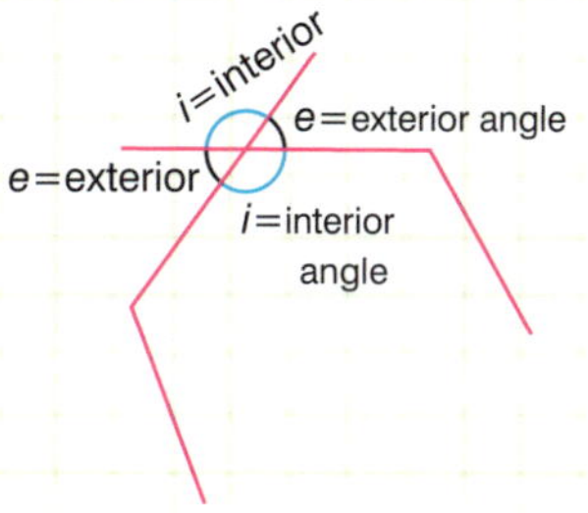

$e+i = 180°$

My Notes

Theorem 9.3

Exterior Angles Polygon Sum Theorem.

The sum of the measures of the exterior angles of any convex polygon.

Proof 9.3

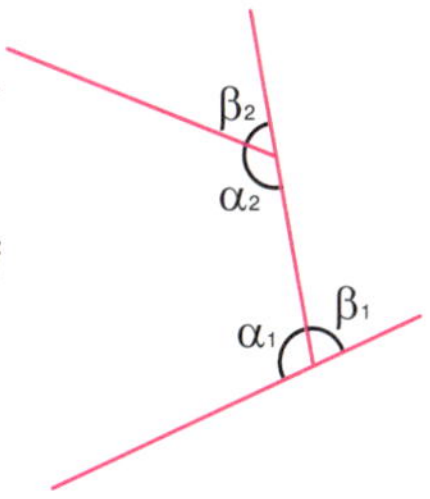

Let $\alpha_1, \alpha_2, \alpha_3, \ldots, \alpha_n$ be the interior angles of an n-sided polygon.

Let $\beta_1, \beta_2, \beta_3, \ldots, \beta_n$ be the corresponding exterior angles of the polygon.

$$\left.\begin{aligned} \alpha_1+\beta_1 &= 180° \\ \alpha_2+\beta_2 &= 180° \\ &\ldots\ldots\ldots \\ \alpha_n+\beta_n &= 180° \end{aligned}\right\} \quad \underbrace{\alpha_1+\alpha_2+\alpha_3+\ldots+\alpha_n}_{180°\cdot(n-2)} + \underbrace{\beta_1+\beta_2+\beta_3+\ldots+\beta_n}_{x} = 180°\cdot n$$

$$180°\cdot(n-2)+x = 180°\cdot n$$

$$180°\cdot n-360°+x = 180°\cdot n$$

$$x = 360°$$

Example 9.4

Suppose the sum of all measures of interior angles of a polygon is 1980°.

a) How many sides does this polygon have?

b) How many diagonals does this polygon have?

Solution 9.4

a) $(n-2)\ 180 = 1980$

$$n-2 = \frac{1980}{180} = 11 \Rightarrow n = 13$$

b) $\frac{n(n-3)}{2} = \frac{13\cdot 10}{2} = 65.$

Example 9.5

The measures of the interior angle of a heptagon are in the ratio 4:5:6:7:7:8:8.

Find the sum of the measures of the smallest interior angle and the smallest exterior angle of the heptagon.

Solution 9.5

$4k+5k+6k+7k+7k+8k+8k = (7-2)180$

$45k = 5 \cdot 180$

$k = 20$

So, the smallest interior angle is $4k$ which equals $4 \cdot 20 = 80$. The smallest exterior angle is $180-8k = 180-8 \cdot 20$ which equals 20. Therefore, the answer is 100.

Example 9.6

Determine the number of sides of a regular polygon if

a) one interior angle measures 170°.

b) one exterior angle measures 36°.

Solution 9.6

a) $170° \cdot n = 180° \cdot (n-2) \Rightarrow 170° \cdot n = 180° \cdot n - 360°$

$10° \cdot n = 360° \Rightarrow n = 36°.$

So, 36-sides polygon.

Alternate solution: One exterior angle's measure is $180°-170° = 10°$. So, $\frac{360°}{10°} = 36°$.

b) By using previous alternate solution, $\frac{360°}{n} = 36° \Rightarrow n = \frac{360°}{36°} = 10°$.

So, 10-sided polygon.

Try This

Boomerang Shape Angle

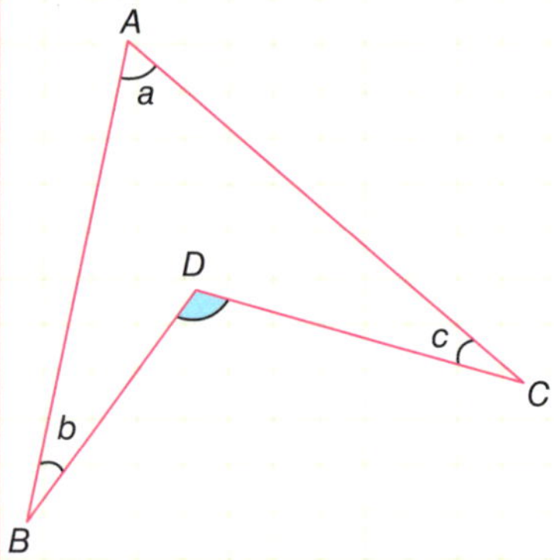

Show hat $m\angle BDC = a+b+c$.

My Notes

My Notes

Example 9.7

ABCDE is a regular pentagon and ΔAFB is an equilateral triangle.

What is $m\angle AFE$?

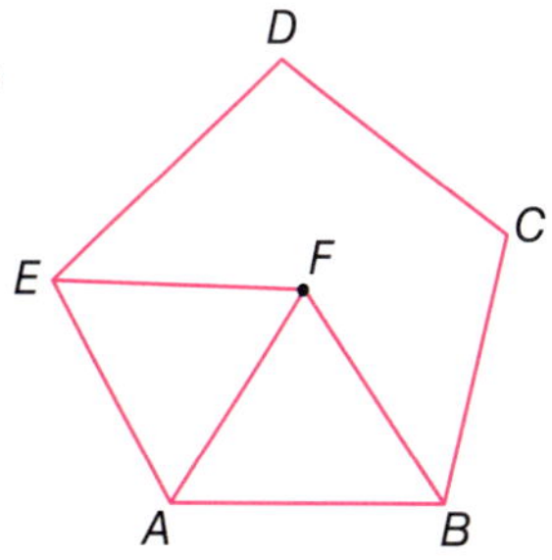

Solution 9.7

$AE = AF$

$m\angle EAF = 48°$

$m\angle AFE = \frac{180° - 48°}{2} = 66°.$

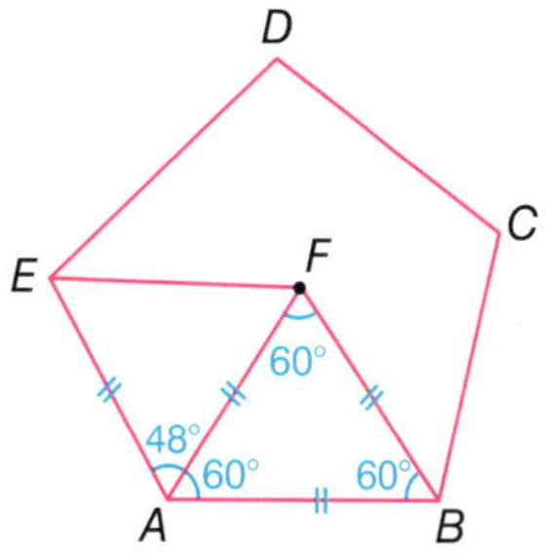

Example 9.8

In the figure, *ABCDE* is a regular polygon with $m\angle BDE = 120°$.

How many sides does this polygon have?

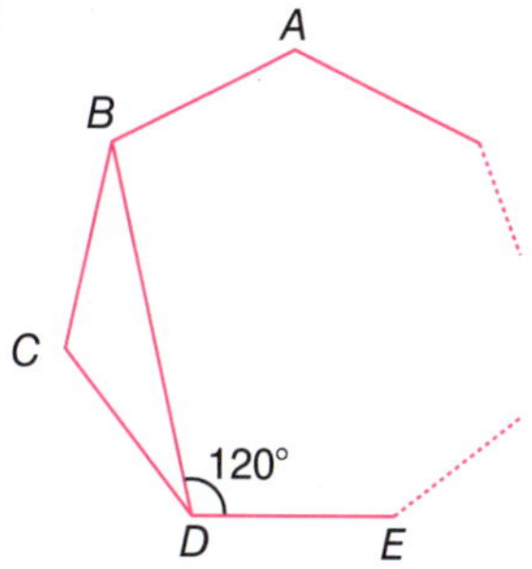

Solution 9.8

$120° + x + x + x = 180°$

$3x = 60° \Rightarrow x = 20°.$

If one interior angle's measure is 140°, then one exterior angle's measure is 40°.

So, $n = \frac{360°}{40°} \Rightarrow n = 9.$

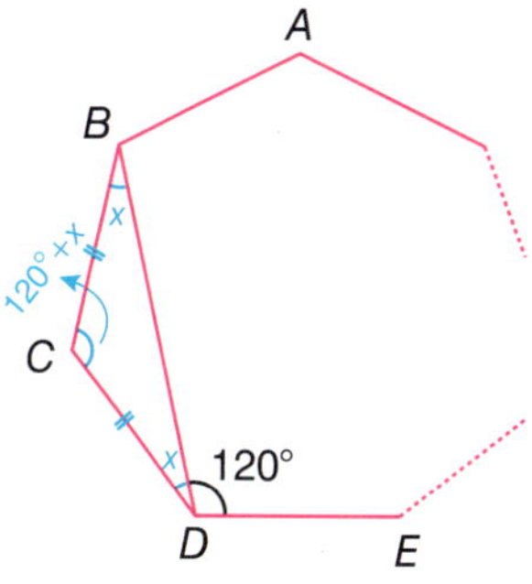

Example 9.9

In the figure, *ABCDEF* is a regular polygon. *ABKL* is a square and points *F, L,* and *M* are collinear.

What is $m\angle MLK$?

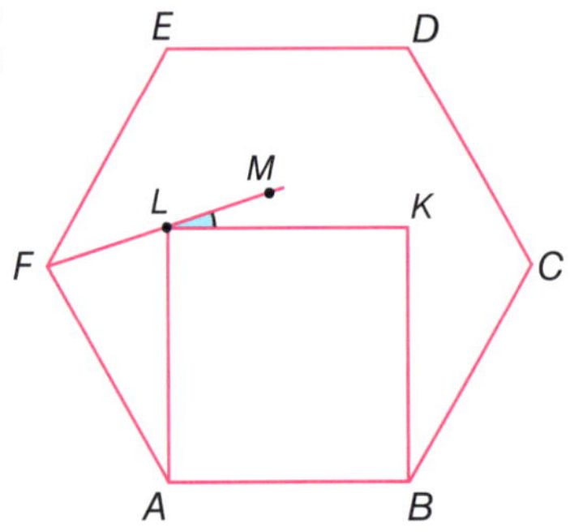

Solution 9.9

Note that ΔAFL is an isosceles triangle with

$AF = AL$,

$m\angle ALF = 75°$, and

$m\angle FAL = 30°$.

Hence, $75°+90°+\alpha = 180°$

Therefore, $\alpha = 15°$

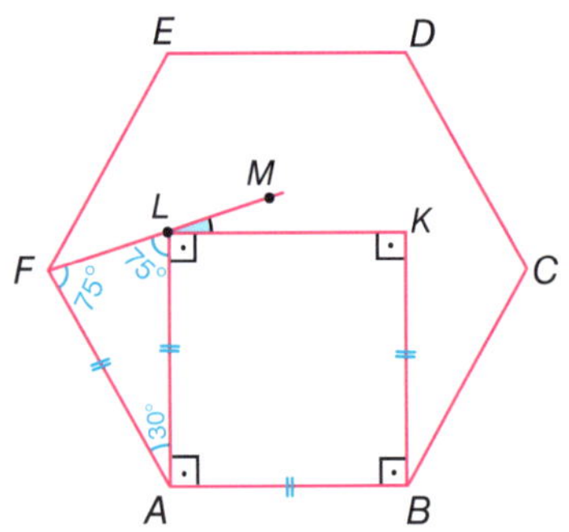

Example 9.10

In the figure, *ABCDEF* is a regular hexagon with

$DH = 2$ and

$HC = 4$.

What is *AH*?

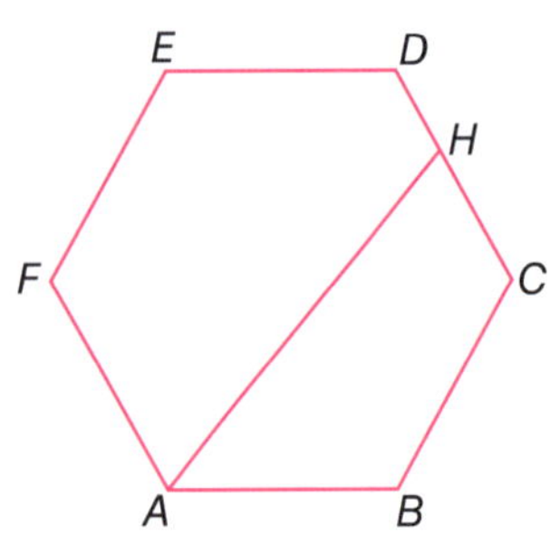

Solution 9.10

We connect points *A* and *C*. Since $AB = AC$ and $m\angle ABC = 120°$, then $m\angle ACB = m\angle CAB = 30°$. So, ΔHCA is a right triangle.($m\angle BCD = 120°$)

From the Pythagorean Theorem.

$x^2 = 4^2+(6\sqrt{3})^2$

$x^2 = 16+108$

$x^2 = 124$

$x = \sqrt{124} \Rightarrow x = 2\sqrt{31}$.

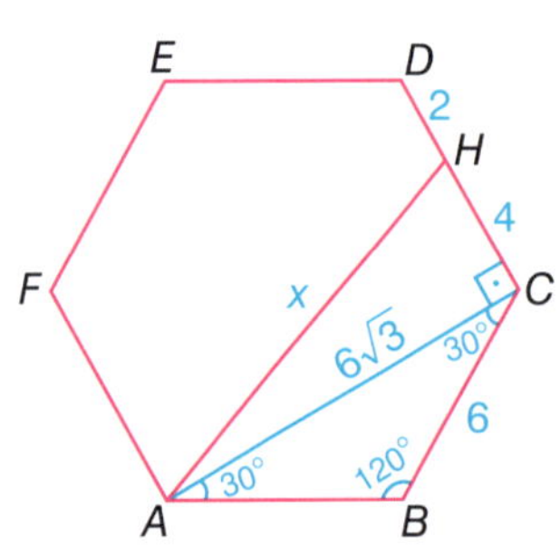

Challenge

The three hexagons are regular with side a length of 6.
What is the length of *XY*?

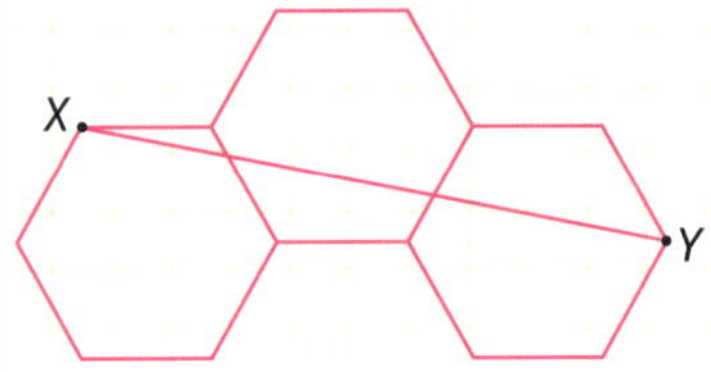

My Notes

Explanation

Inscribed Quadrilateral

An inscribed quadrilateral is any four sided figure whose vertices all lie on a circle. (The sides are therefore chords in the circle!) This conjecture give a relation between the opposite angles of such a quadrilateral. It says that these opposite angles are in fact supplements for each other. In other words, the sum of their measures is 180 degrees.

For the quadrilateral shown below, this conjecture says that:

$\angle A+\angle C = 180°$

$\angle B+\angle D = 180°$

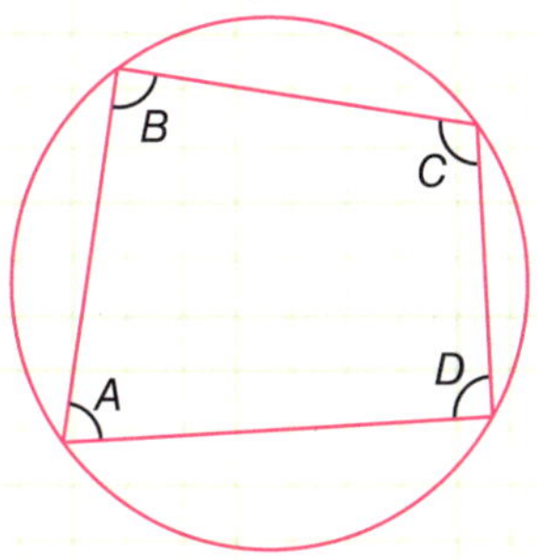

My Notes

Quadrilaterals

Definition 9.3

A quadrilateral is a four sided polygon with four angles.

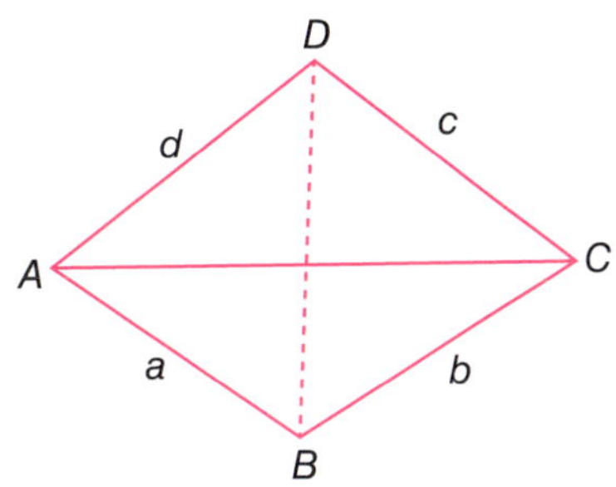

Convex quadrilateral

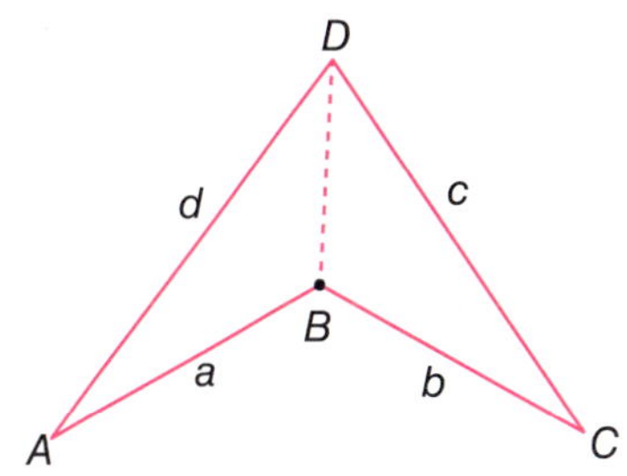

Concave quadrilateral

Theorem 9.4

In a convex quadrilateral below,

$m\angle A+m\angle B+m\angle C+m\angle D = 360°$ and

$m\angle A'+m\angle B'+m\angle C'+m\angle D' = 360°$.

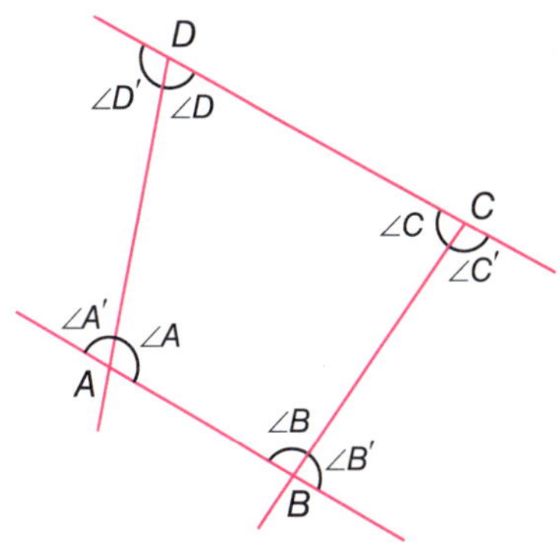

Example 9.11

Suppose $x+y = 105°$, then find the value of $m+n$.

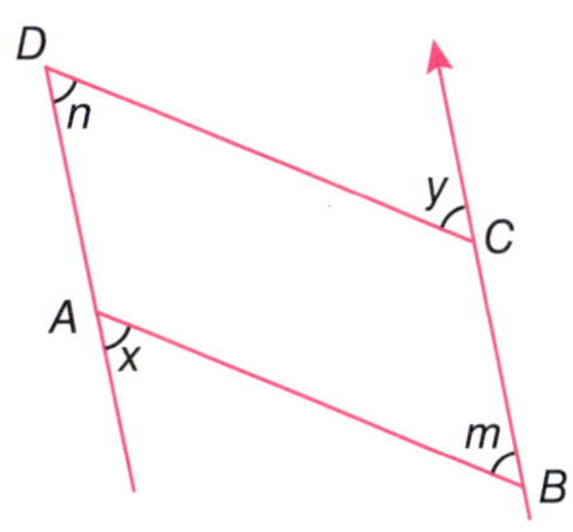

Solution 9.11

$180°-x+180°-y+m+n = 360°$

$360°-(x+y)+m+n = 360°$

$m+n = 105°$

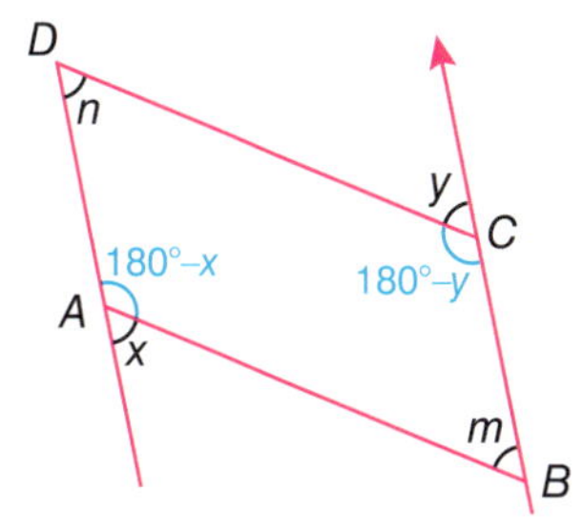

Theorem 9.5

ABCD is a convex quadrilateral. *AK* and *KB* are bisectors of $\angle A$ and $\angle B$.

$m\angle AKB = \dfrac{m\angle C + m\angle D}{2}$.

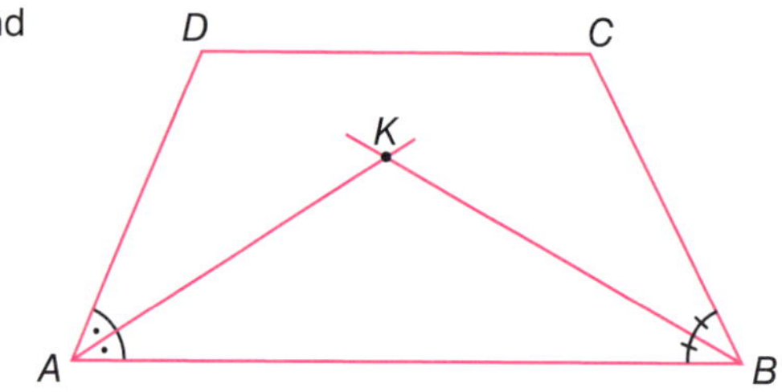

Proof 9.5

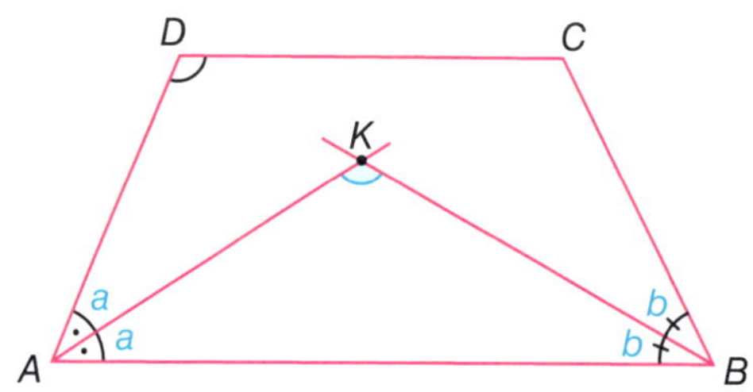

$2a+2b+m\angle D+m\angle C = 360°$ *ABCD*

$m\angle D+m\angle C = 360°-(2a+2b)$

$m\angle K+a+b = 180°$ *AKB*

$m\angle K = 180°-(a+b)$

$2m\angle K = 360°-2(a+b)$

$2m\angle K = m\angle D+m\angle C$

$m\angle K = \dfrac{m\angle C+m\angle D}{2}$

Theorem 9.6

ABCD is a convex quadrilateral.

AK and *KC* are bisectors of $\angle A$ and $\angle C$.

$m\angle AKL = \dfrac{|m\angle D - m\angle B|}{2}$.

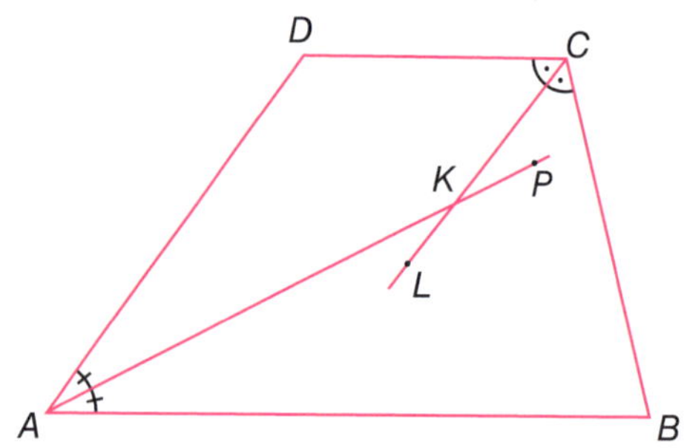

Proof 9.6

If $d>b$, then using *ADCK*, we get

(1) $a+d+c+180°-x = 360°$.

Similarly, using *ADCB*, we get

(2) $2a+d+2c+b = 360°$

Subtracting (2) from 2 times (1), we get,

$d-b+360°-2x = 360°$.

Therefore, $x = \dfrac{d-b}{2}$.

Similarly, if $d<b$, then $x = \dfrac{b-d}{2}$.

Hence, $m\angle AKL = \dfrac{|m\angle D-m\angle B|}{2}$.

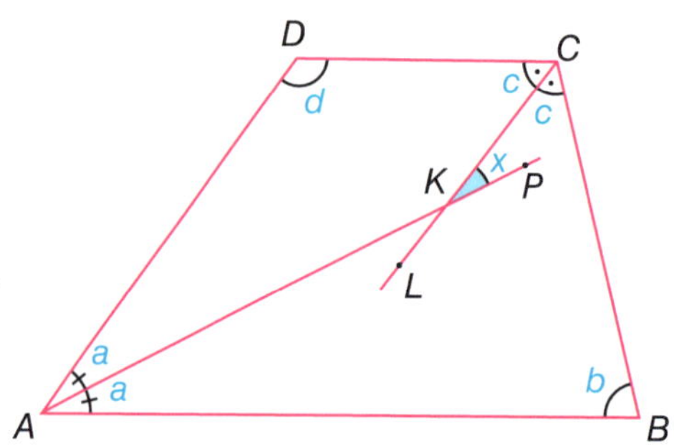

Prove This

In the figure, *AF, BF, CE* and *DE* are bisectors of $\angle A$, $\angle B$, $\angle C$ and $\angle D$, respectively.

Prove that the quadrilateral *EKFM* is an inscribed quadrilateral.

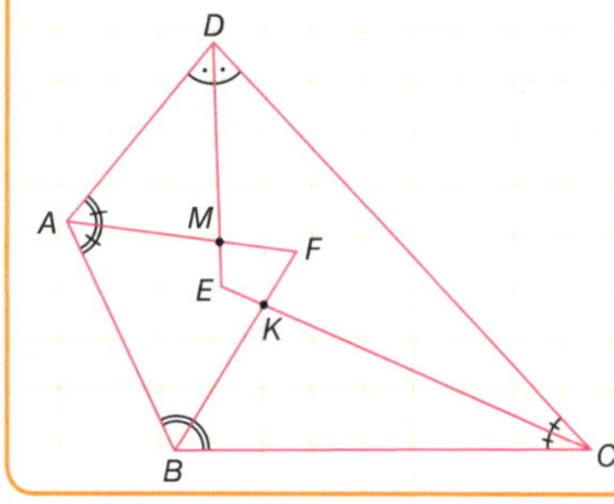

My Notes

Catriona Agg @Cshearer41
Geometry puzzle

Both hexagons are regular.

How long is the blue line?

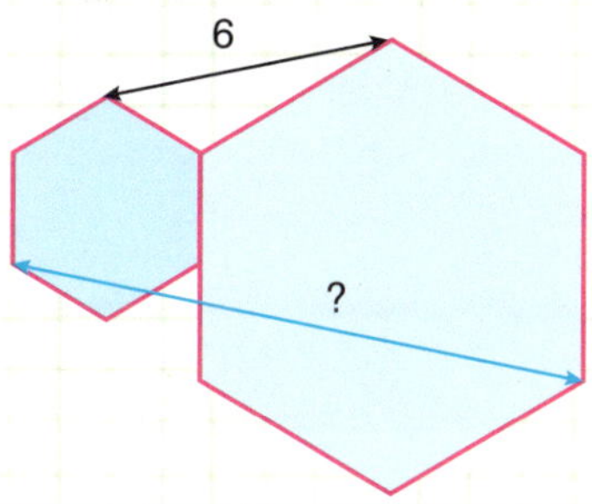

My Notes

Theorem 9.7.

ABCK is a concave quadrilateral,

$m\angle BKC = a+b+c$.

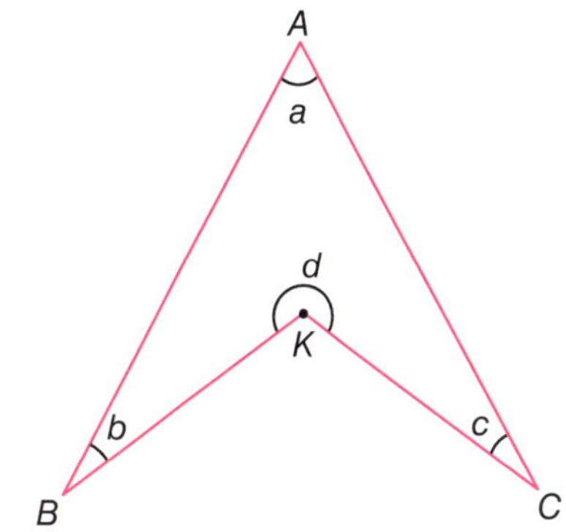

Theorem 9.8

$a^2+c^2 = b^2+d^2$.

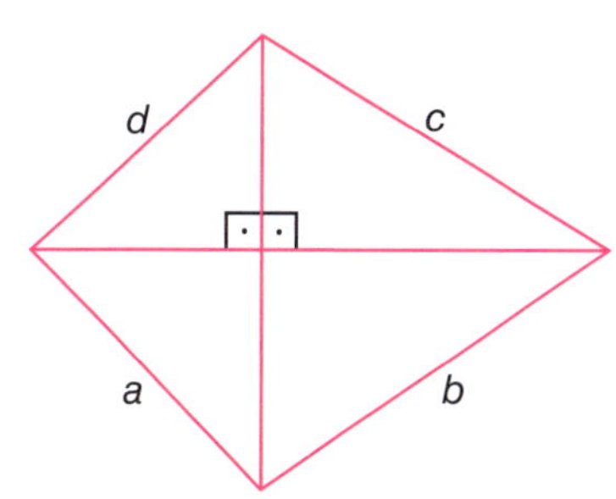

Proof 9.8

$x^2+y^2 = d^2$ (1)

$x^2+z^2 = c^2$ (2)

$y^2+t^2 = a^2$ (3)

$t^2+z^2 = b^2$ (4)

(2)+(3) $x^2+z^2+y^2+t^2 = a^2+c^2$

(1)+(4) $x^2+y^2+t^2+z^2 = d^2+b^2$

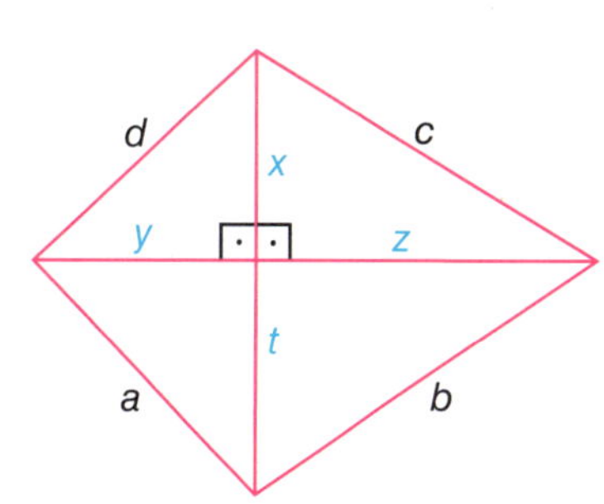

Example 9.12

In the figure,

$BK = KC$,

$AD = 6$,

$DC = 10$, and

$AB = 4$.

Find EK.

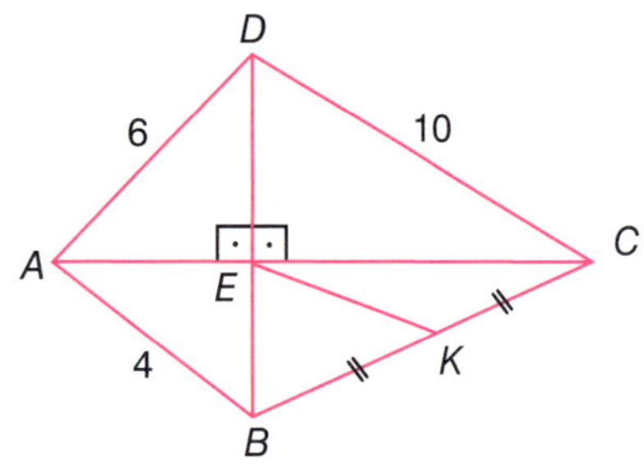

Solution 9.12

$$4^2+10^2 = 6^2+BC^2$$

$$16+100 = 36+BC^2$$

$$BC^2 = 80$$

$$BC = \sqrt{80} = 4\sqrt{5} \Rightarrow BK = KC = FC = 2\sqrt{5}.$$

Theorem 9.9

Varignon's Theorem

M, N, K, and L are midpoints of their sides.

The perimeter of $MNKL$ is equal to the sum of DB and AC.

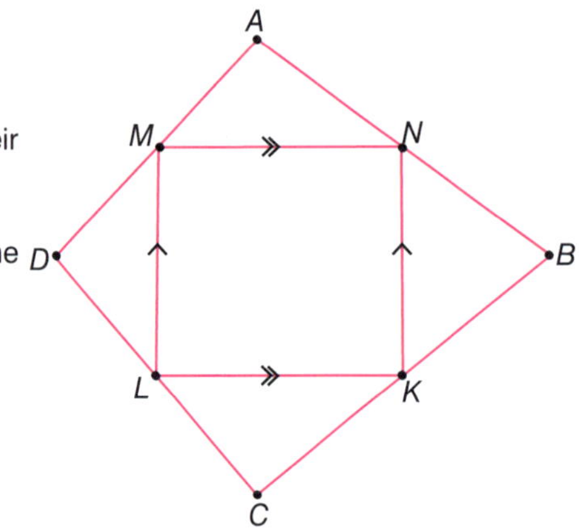

Proof 9.9

We first draw two diagonals.

Let $DB = 2a$ and $AC = 2b$. It can be seen that $ML // AC // NK$ and $MN // DB // LK$.

Therefore, $MN = LK = a$ and $ML = NL = b$.

The perimeter of $MNKL = a+b+b+a$
$= 2a+2b$

$DB+AC = 2a+2b$

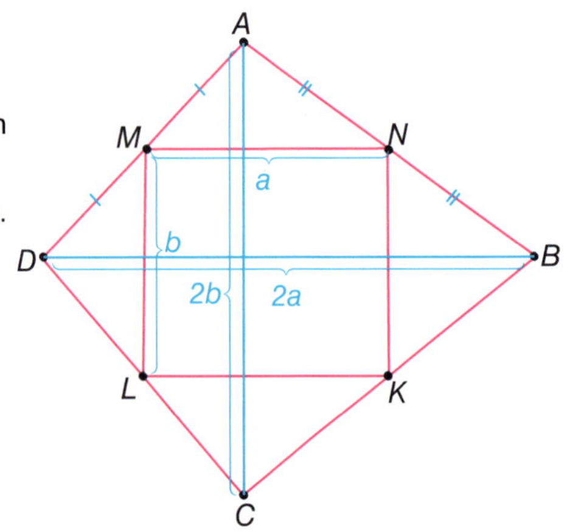

Theorem

Varignon's Theorem

The midpoint of the sides of an arbitrary quadrilateral form a parallelogram. If the quadrilateral is convex, then the area of the parallelogram is half as large as the area of the quadrilateral.

My Notes

Try This

ABCDEFGH a regular 8-sided polygon.

FD and *AE* are diagonals.

If $AE = 16$, then find *FK*.

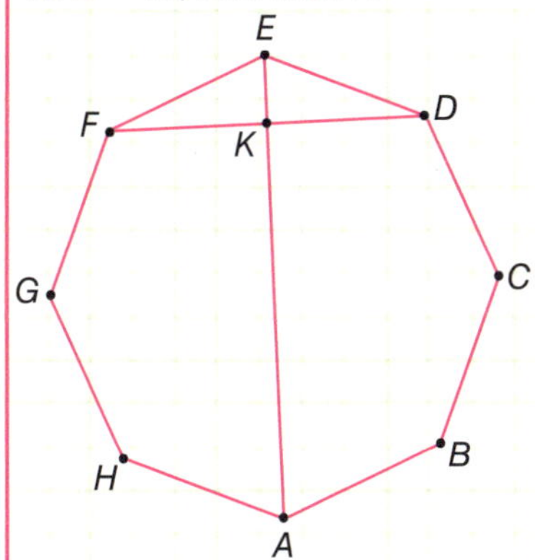

My Notes

Example 9.13

ABCDEF is a regular hexagon.

$BD = 3DK$, and

$AB = 2\sqrt{3}$ cm.

Find *EK*.

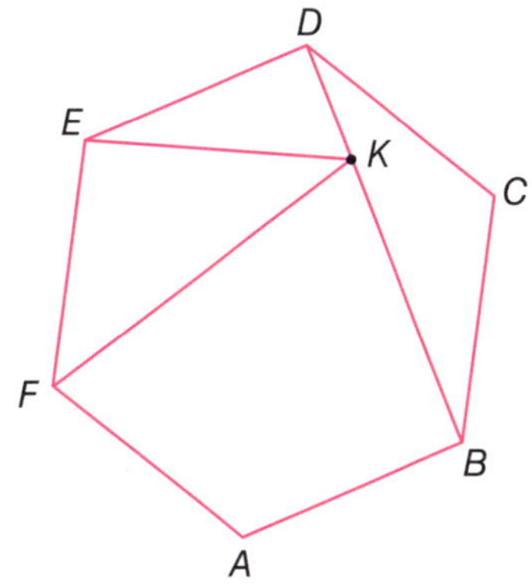

Solution 9.13

We connect points *F* and *C*. *FC* is a line of symmetry of the hexagon. Thus $DB \perp CF$. From right triangle ΔFOK.

$$x^2 = 1^2 + (3\sqrt{3})^2$$

$$x^2 = 1 + 27$$

$$x = \sqrt{28}$$

$$x = 2\sqrt{7}.$$

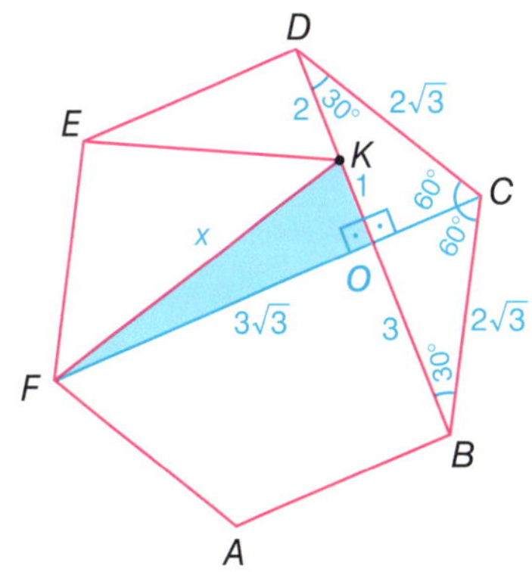

Example 9.14

ABCD is a quadrilateral. Points *E* and *F* are the midpoint of *AB* and *DC*.

If $DB = 24$ and $AC = 18$, then find *EF*.

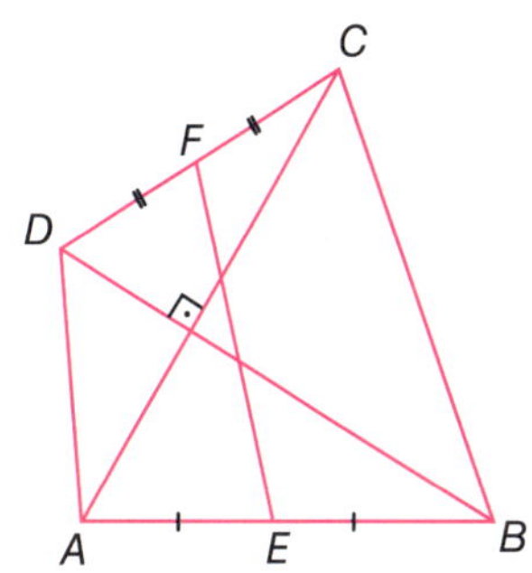

Solution 9.14

Let point *G* be the midpoint of *AD*. We create two parallel lines, *GF*//*AC* and *GE*// *DB*. Since $AC \perp DB$ then $m\angle FGE = 90°$. From the triangle similarities.

$$GF = \frac{AC}{2} = \frac{18}{2} = 9 \text{ and}$$

$$GE = \frac{DB}{2} = \frac{24}{2} = 12.$$

From the right triangle ΔFGE.

$$EF = \sqrt{9^2 + 12^2}$$

$$EF = 15.$$

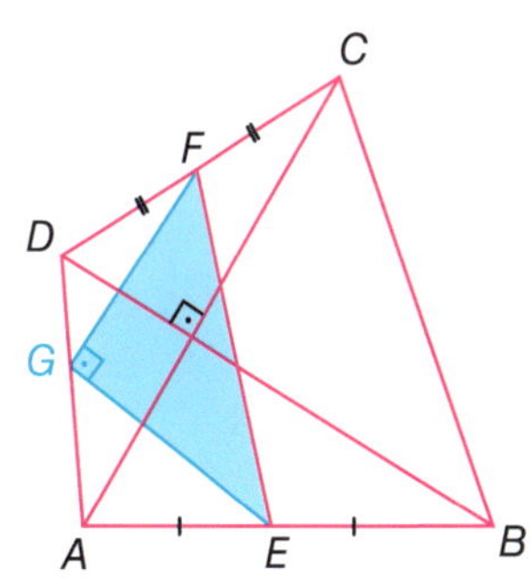

Example 9.15

In the figure, *ABCDEFG* is a regular heptagon.

Prove that $\frac{1}{a}=\frac{1}{b}+\frac{1}{c}$.

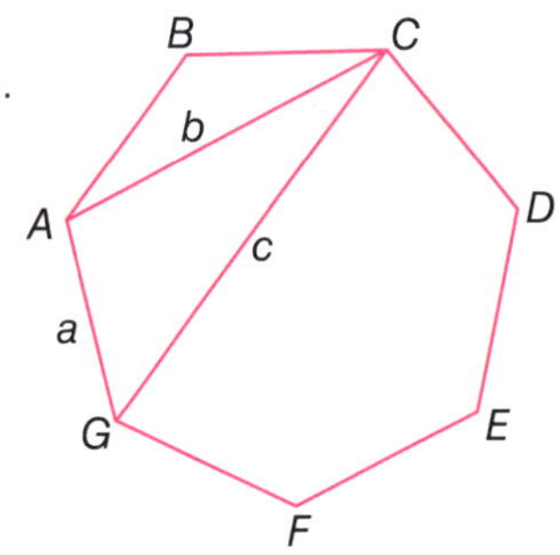

Solution 9.15

Method 1 (proof without words)

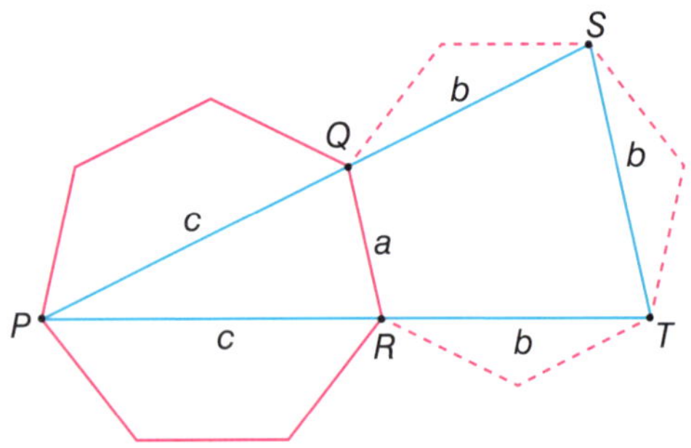

$\Delta PQR \sim \Delta PST$

$$\frac{c}{c+b}=\frac{a}{b} \Rightarrow cb=ac+ab \quad (\div abc)$$

$$\frac{cb}{abc}=\frac{ac}{abc}+\frac{ab}{abc} \Rightarrow \frac{1}{a}=\frac{1}{b}+\frac{1}{c}.$$

Catriona Agg @Cshearer41

Geometry puzzle

Two regular pentagons.

What is the angle?

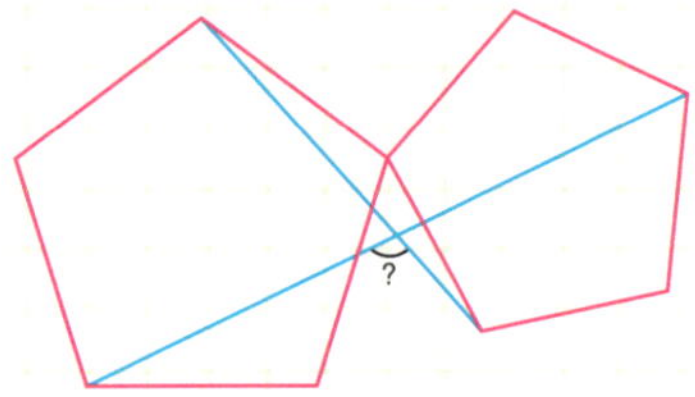

My Notes

1.

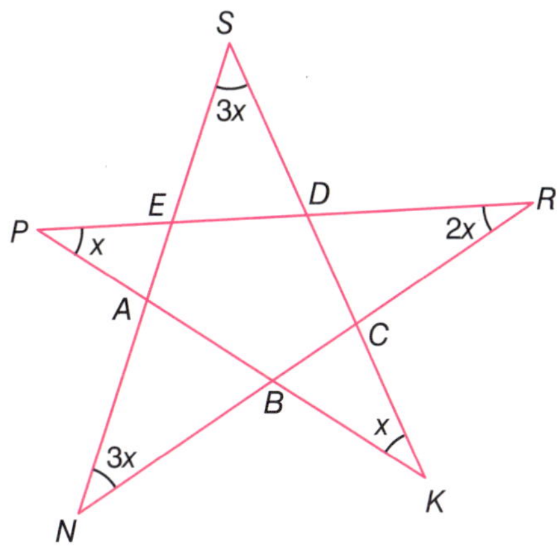

In the figure,

m∠S = m∠N = 3x, m∠R = 2x, and m∠P = m∠K = x.

Find the value of x.

2.

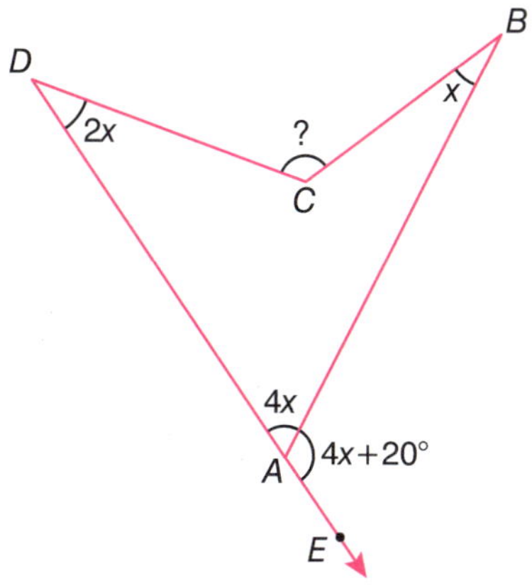

In the figure, ABCD is concave quadrilateral with

m∠DAB = 4x, m∠D = 2x, and m∠BAE = 4x+20°.

Find m∠BCD.

3. The measures of the interior angle of a pentagon are in the ratio 3:4:5:7:8.

Find the sum of the measures of the smallest interior angle and the smallest exterior angle of the pentagon.

4. In the figure, ABCDEF is a regular hexagon and ΔEND is an equilateral triangle

Find m∠AEN.

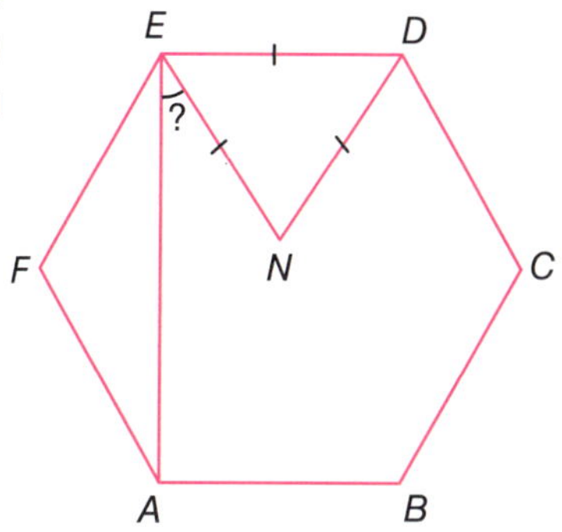

5. The interior angles of a polygon measures 80°, 115°, and 135°, and all the other interior angles measure 165°.

How many sides does this polygon have?

6. In the figure, ABCDEF is a hexagon with

m∠D = 105°,

m∠E = 100°,

m∠F = 110°,

m∠HAB = 40°, and

m∠BCG = 50°

Find m∠ABC.

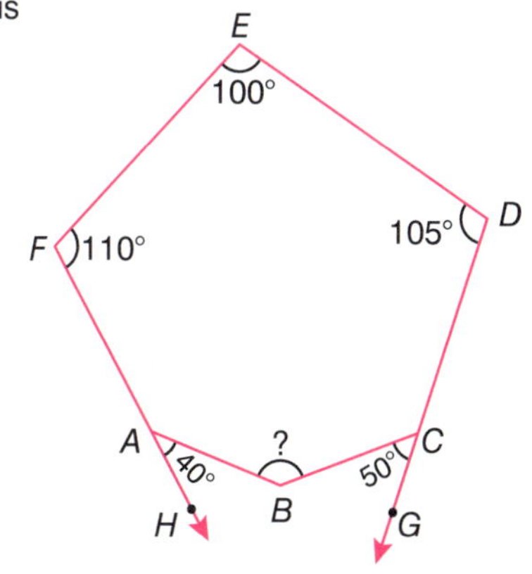

7.

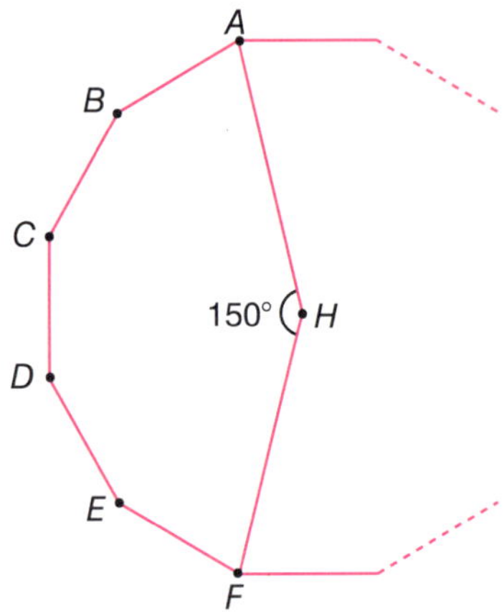

In the figure, *ABCDEF* is a regular polygon and line segments *AH* and *FH* bisect $\angle A$ and $\angle F$ respectively

If $m\angle AHF = 150°$, find the number of sides of the polygon.

8. A square *ABCD* and a regular pentagon *CEFGD* share a common side *CD* as shown.

What is the sum of the degree measures of $\angle CEB$ and $\angle CBE$?

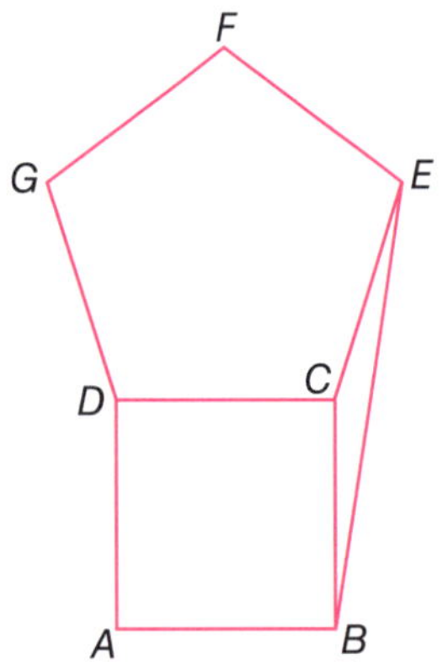

9. One measure of an interior angle of a regular polygon is equal to four times the measures of an exterior angle.

How many sides does this polygon have?

10. In the figure, *ABCDEF* is a regular hexagon with

$DH = 1$ cm,

$HC = 2$ cm.

Find the length of *AH*.

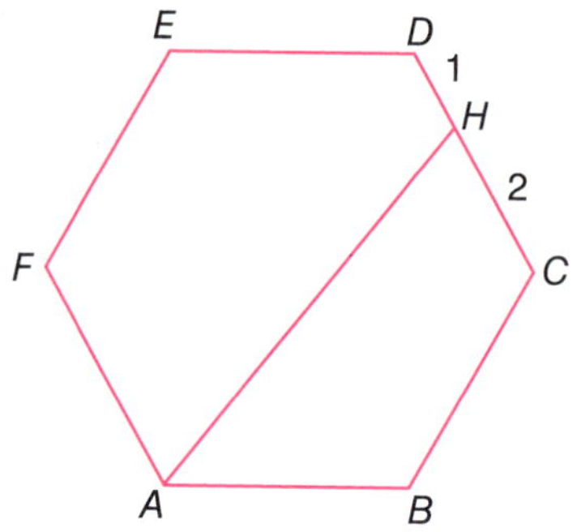

11. *ABCDE* is a regular pentagon with $FB \perp AB$

What is $\dfrac{DG}{DF}$?

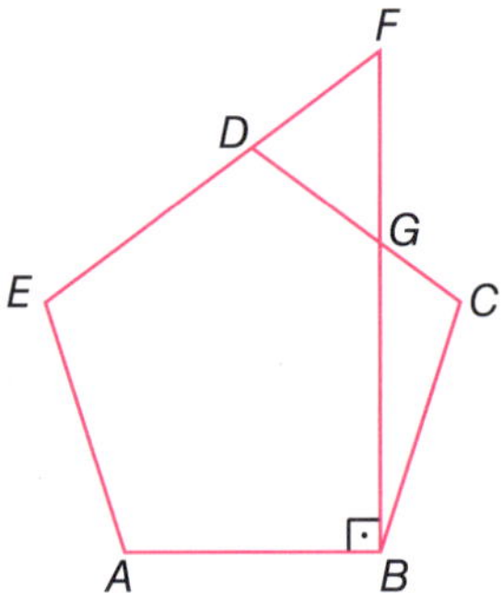

12. In the figure, *ABCD* is a regular polygon with $OH \perp AB$ and $OE \perp CD$. O is at the center of the polygon and $m\angle HOE = 80°$.

How many sides does the polygon have?

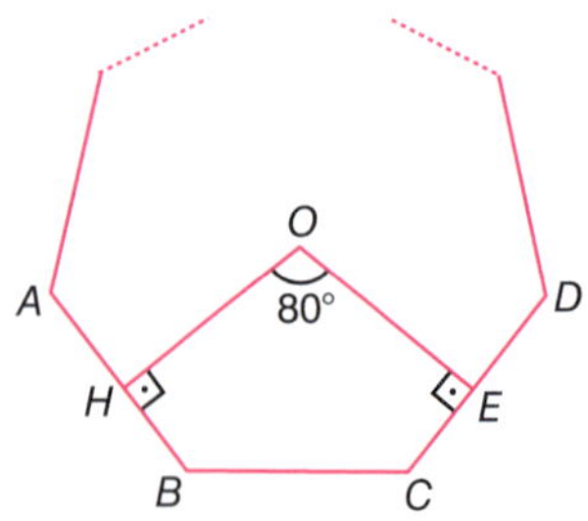

1. The sum of the measures of five interior angles of a regular polygon is 270° more than the sum of the measures of all its other interior angles.

   How many sides does this polygon have?

2. In the figure, $ABCDE$ is a regular pentagon, DB is diagonal and AF is the bisector of $\angle A$.

   What is $m\angle AFB$?

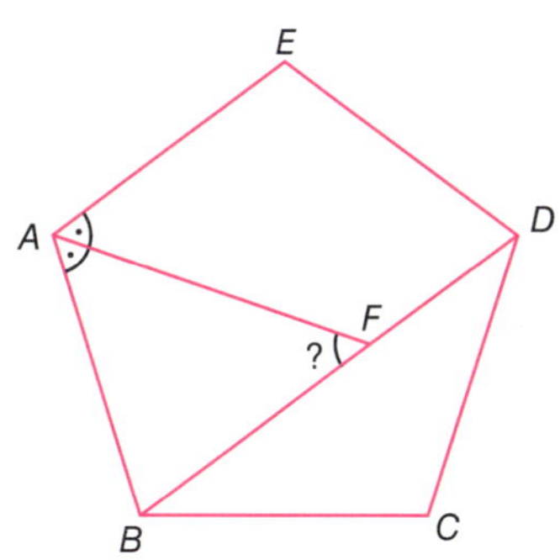

3. In the figure, $ABCDE$ is a regular pentagon. With $AF \perp BF$. Given that A, F and D are collinear.

   What is the measures $m\angle FBD$.

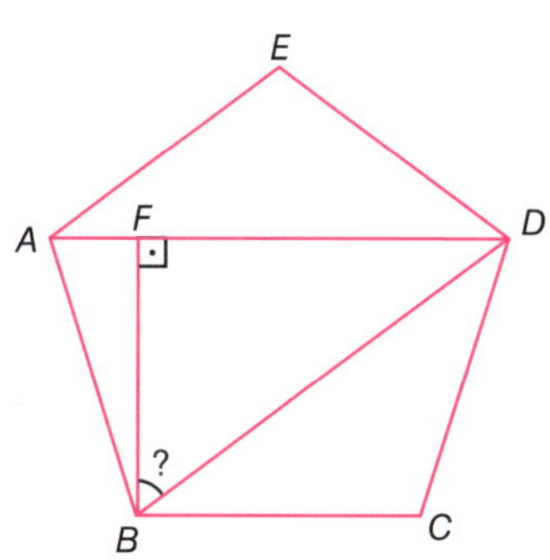

4. In the figure, diagonals AC and BD are perpendicular to each other,

   $AB = 7$ cm,

   $BC = 9$ cm,

   $AD = 6$ cm, and

   $DF = FC$.

   What is the length of EF?

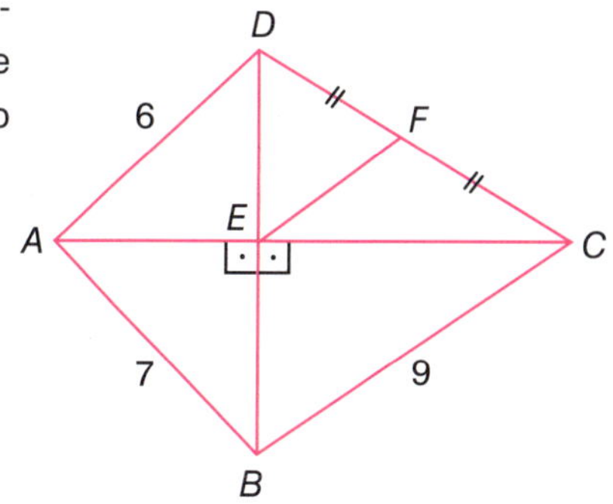

5. In the figure, $ABCDEF$ is a regular polygon.

   If $m\angle CEF = 135°$, what is the measure of one interior angle of the polygon?

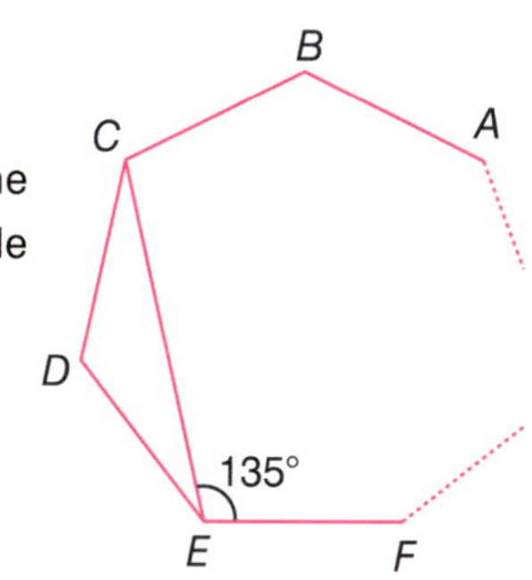

6. In the figure, $ABCDE$ is a regular pentagon and $BCGF$ is a square.

   Find $m\angle FAB$.

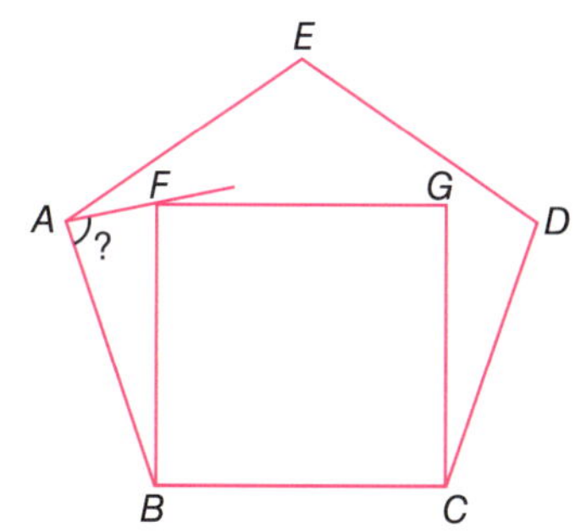

7. In the figure, $ABCD$ is a kite with

$BK = KC$,

$AD = 3$,

$DC = 5$, and

$AB = 2$.

Find EK.

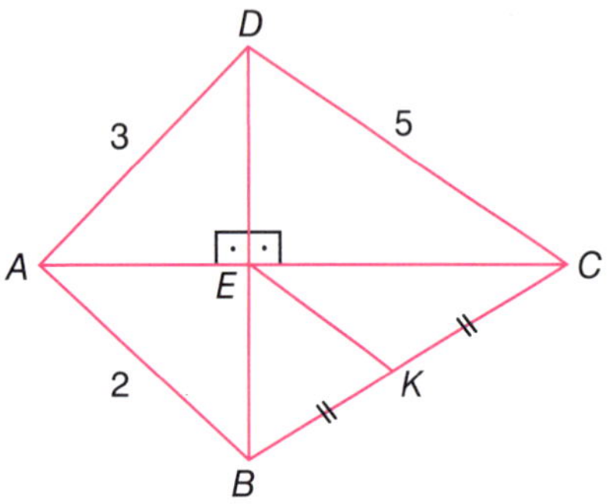

8. In the figure, $ABCDEF$ is a regular hexagon with

$DH = 3$ and

$HC = 6$.

Find AH.

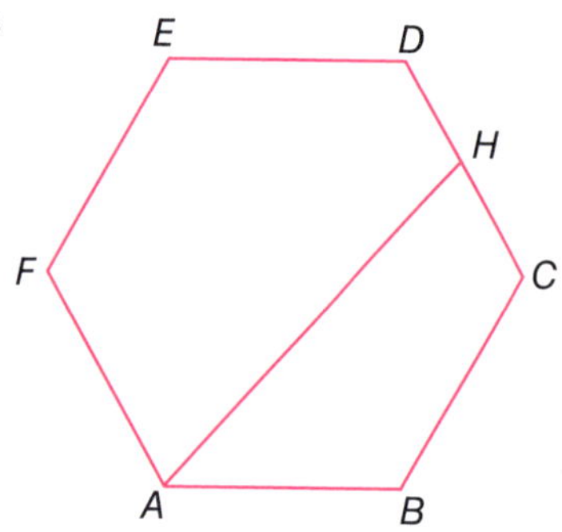

9. In the figure, $ABCDEF$ is a regular hexagon.

$BD = 3DK$,

$AB = 4\sqrt{3}$.

Find FK.

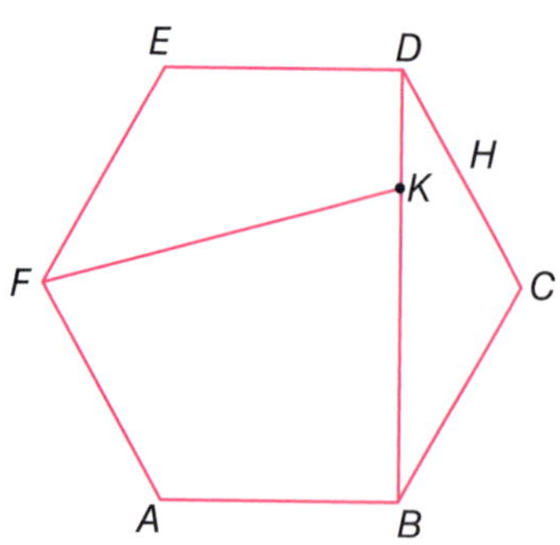

10. In the figure, $ABCDE$ is a regular pentagon with $AB \perp FH$.

What is the value of $\beta - \alpha$?

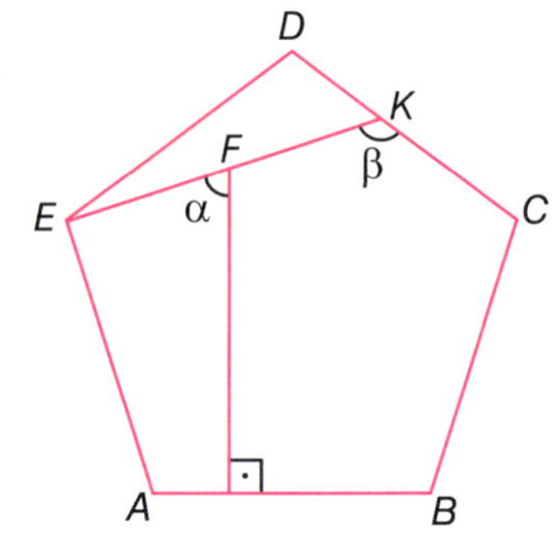

11. In the figure, $ABCDEFG$ is a regular heptagon, $AG = 3$, $AC = 4$.

Find CG.

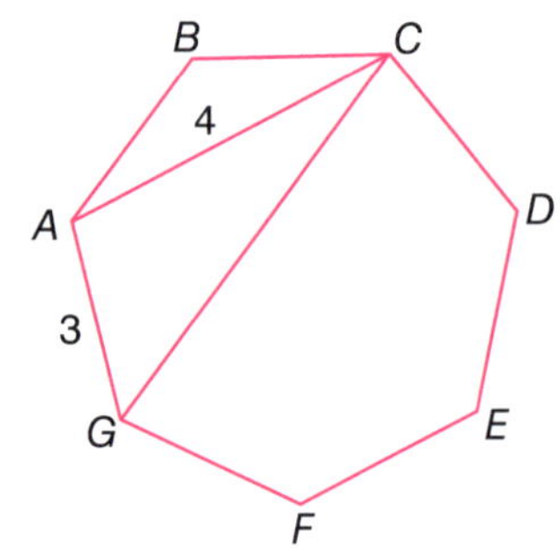

12. $ABCD$ is a quadrilateral. Points E and F are the midpoint of AB and DC.

If $DB = 12$, $AC = 16$, then find EF.

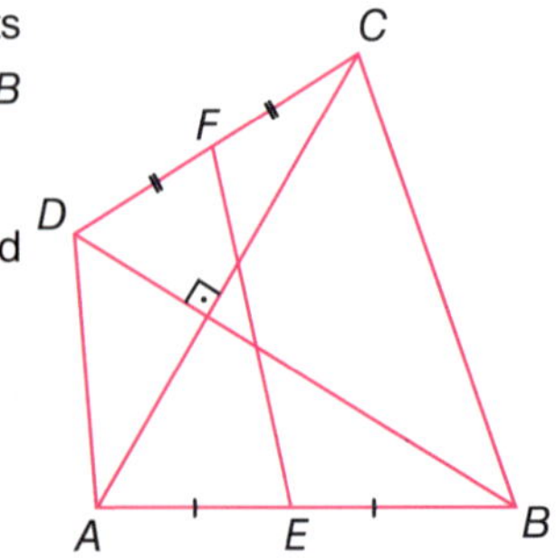

1. **(2009 AMC 8 #9)** Construct a square on one side of an equilateral triangle. One on non-adjacent side of the square, construct a regular pentagon, as shown. One a non-adjacent side of the pentagon, construct a hexagon. Continue to construct regular polygons in the same way, until you construct an octagon.

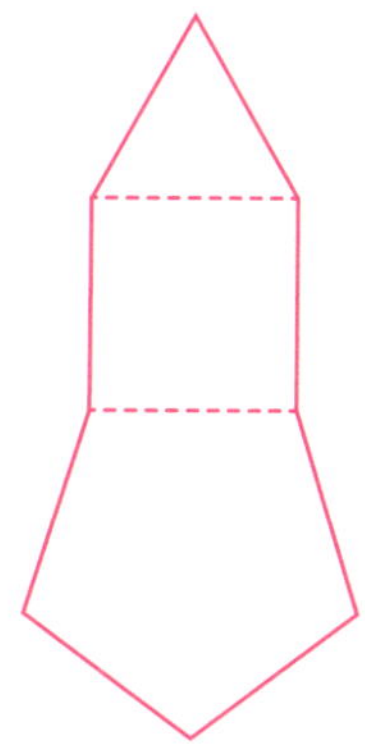

How many sides does the resulting polygon have?

A) 21 B) 23 C) 25 D) 27 E) 29

2. **(1999 AMC 8 # 21)**

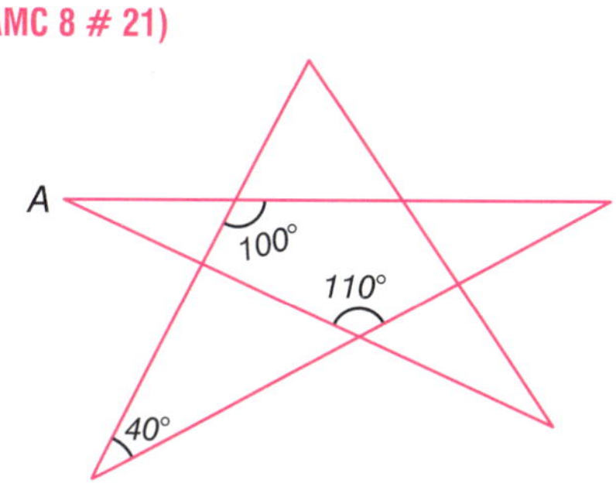

The degree measure of angle *A* is

A) 20 B) 30 C) 35 D) 40 E) 45

3. **(PIMC 2015-Indv. #22)**

In heptagon *SWAPNIL* with 7 size.

$\angle S = \angle W = \angle A = \angle P = \angle N = \angle I = 149°$.

How many degrees is *L*?

4. **(CWML 2017-18 Meet 2 #4)** *ABCDEFGHI* is a regular nonagon.

If sides *AB* and *HI* are extended to meet at point *J*, find the degree measure of angle *AJI*.

5. **(CWML 2014-15 Meet 3 #5)** The sides of a regular polygon of *n* sides, $n>4$, are extended to form a star.

The degree measure of the interior angle at each point of the star is:

A) $\frac{360}{n}$ B) $\frac{(n-4)180}{n}$ C) $\frac{(n-2)360}{n}$ D) $180-\frac{90}{n}$ E) $\frac{180}{n}$

6. **(CWML 2015-16 Meet 2 #6)**

The regular heptagon *ABCDEFG* has sides *EF* and *AG* extended to intersect at point *H* as shown.

What is the degree measure of angle *GHF*?

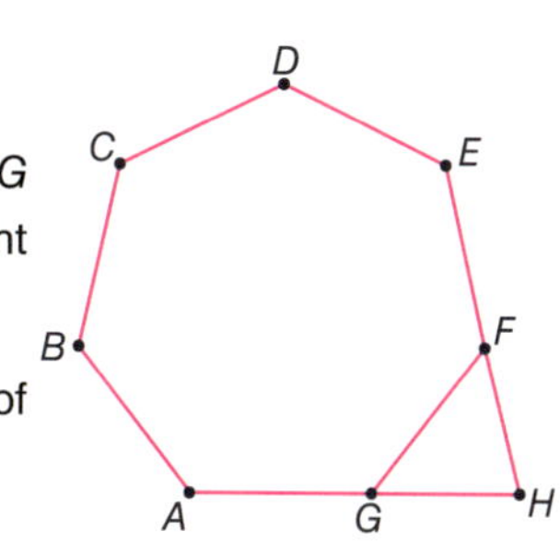

7. **(2020 AMC 12A #14)** Regular octagon $ABCDEFGH$ has area n. Let m be the area of quadrilateral $ACEG$.

What is $\frac{m}{n}$?

A) $\frac{\sqrt{2}}{4}$ B) $\frac{\sqrt{2}}{2}$ C) $\frac{3}{4}$ D) $\frac{3\sqrt{2}}{5}$ E) $\frac{2\sqrt{2}}{3}$

8. **(2008 AMC 10B #24)** Quadrilateral $ABCD$ has $AB = BC = CD$, angle $ABC = 70$ and angle $BCD = 170$.

What is the measure of angle BAD?

A) 75 B) 80 C) 85 D) 90 E) 95

9. **(2011 AMC 10B #16)**

A dart board is a regular octagon divided into regions as shown. Suppose that a dart thrown at the board is equally likely to land anywhere on the board.

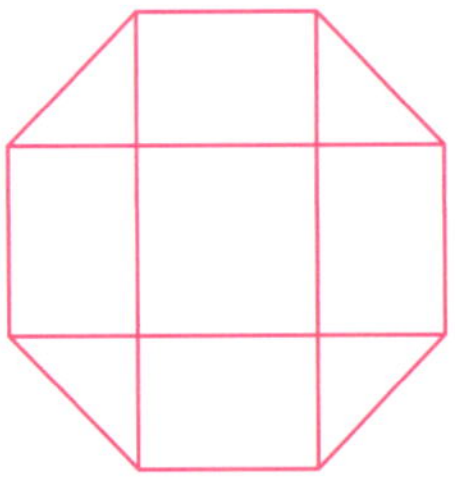

What is the probability that the dart lands within the center square?

A) $\frac{\sqrt{2}-1}{2}$ B) $\frac{1}{4}$ C) $\frac{2-\sqrt{2}}{2}$ D) $\frac{\sqrt{2}}{4}$ E) $2-\sqrt{2}$

10. **(2009 AMC 12A #20)** Convex quadrilateral $ABCD$ has $AB = 9$ and $CD = 12$. Diagonals AC and BD intersect at E, $AC = 14$, and ΔAED and ΔBEC have equal areas.

What is AE?

A) $\frac{9}{2}$ B) $\frac{50}{11}$ C) $\frac{21}{4}$ D) $\frac{17}{3}$ E) 6

11. **(2005 AMC 10A #20)** An equiangular octagon has four sides of length 1 and four sides of length $\frac{\sqrt{2}}{2}$, arranged so that no two consecutive sides have the same length.

What is the area of the octagon?

A) $\frac{7}{2}$ B) $\frac{7\sqrt{2}}{2}$ C) $\frac{5+4\sqrt{2}}{2}$ D) $\frac{4+5\sqrt{2}}{2}$ E) 7

12. **(2003 AMC 10B #23)** A regular octagon $ABCDEFGH$ has an area of one square unit. What is the area of the rectangle $ABEF$?

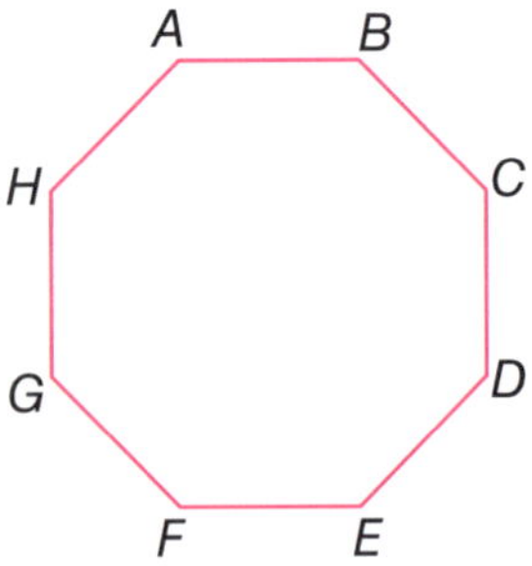

A) $1-\frac{\sqrt{2}}{2}$ B) $\frac{\sqrt{2}}{4}$ C) $\sqrt{2}-1$ D) $\frac{1}{2}$ E) $\frac{1+\sqrt{2}}{2}$

CHAPTER 10

A Parallelogram

My Notes

Definition 10.1

A parallelogram is a quadrilateral which two pairs of opposite parallel sides.

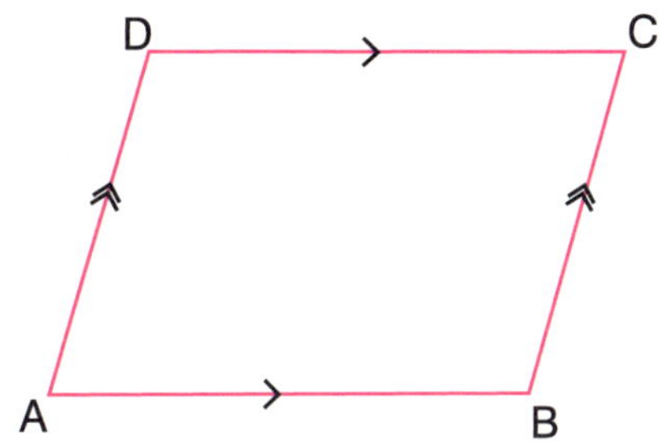

Theorem 10.1

Opposite sides of a parallelogram are congruent.

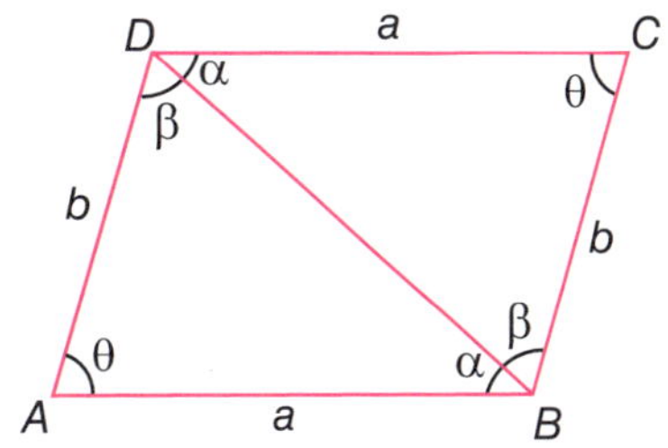

Example 10.1

In the figure, $ABCD$ is a parallelogram and BN is the angle bisector of $\angle B$.

If $ND = 4$ and $DC = 6$, then find BC.

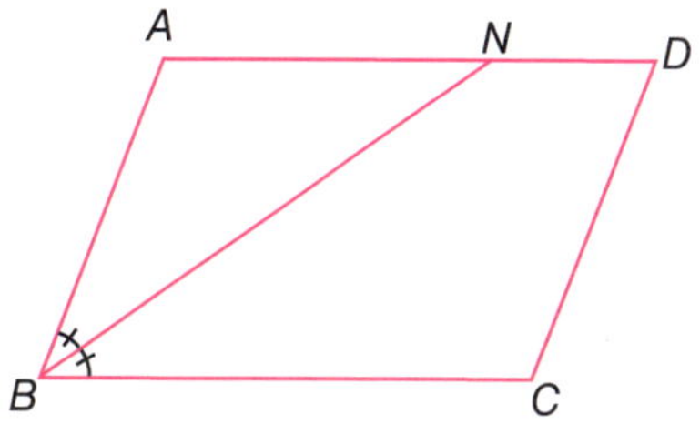

Solution 10.1

$m\angle ANB = m\angle ABN$ ($AN // BC$)
Therefore, ABN is an isosceles triangle and $AB = AN$.

$$AB = AN = DC = 6$$

$$BC = AN + ND$$

Therefore, $BC = 6+4 = 10$.

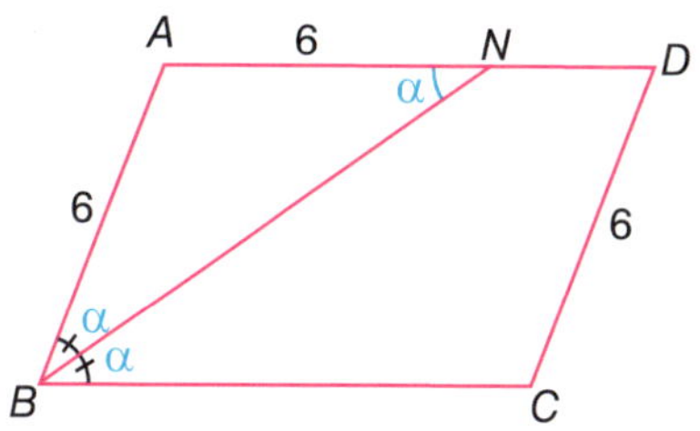

My Notes

Example 10.2

In the figure, $ABCD$ is a parallelogram and AF and DE are angle bisectors. If the perimeter of the parallelogram is 48 cm and $EF = 6$ cm, then find BC.

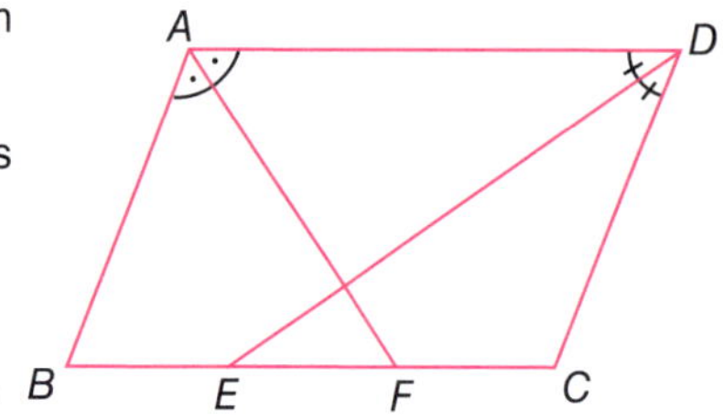

Solution 10.2

$$2(6+a)+2(2a+6) = 48$$
$$12+2a+4a+12 = 48$$
$$6a = 24$$
$$a = 4$$
$$BC = 14.$$

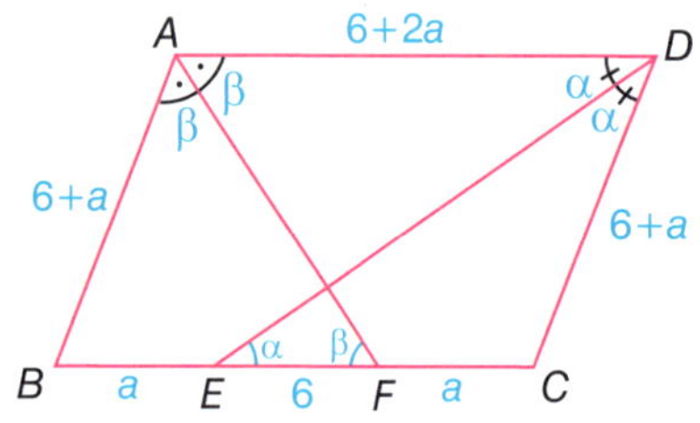

Theorem 10.2

In a parallelogram $ABCD$,

$m\angle A = m\angle C$ and $m\angle D + m\angle A = 180°$

Opposite vertex angles are equal to each other and adjacent angles are supplementary.

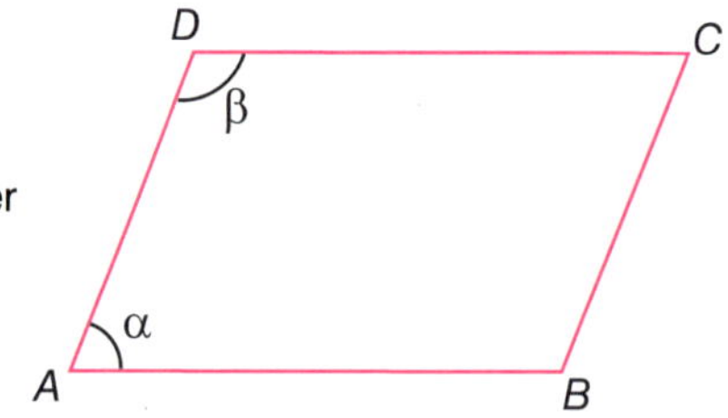

Exercise 10.1

$m\angle AEB = 180° - (\alpha+\beta)$

since $(\alpha+\beta) = 90°$ $(2\alpha+2\beta = 180°)$.

$m\angle AEB = 90°$.

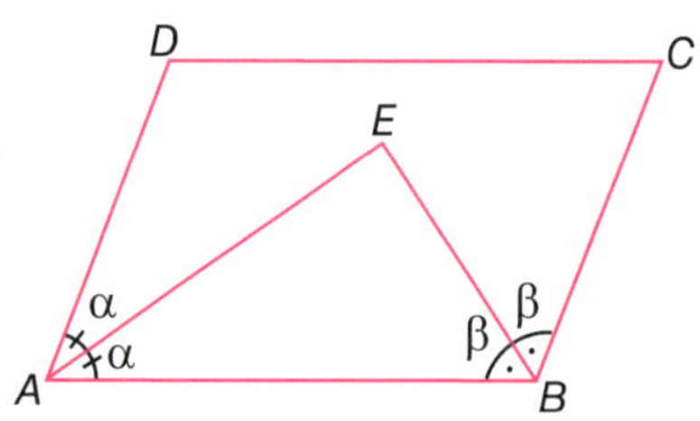

Theorem 10.3

The diagonals of a parallelogram bisect each other.

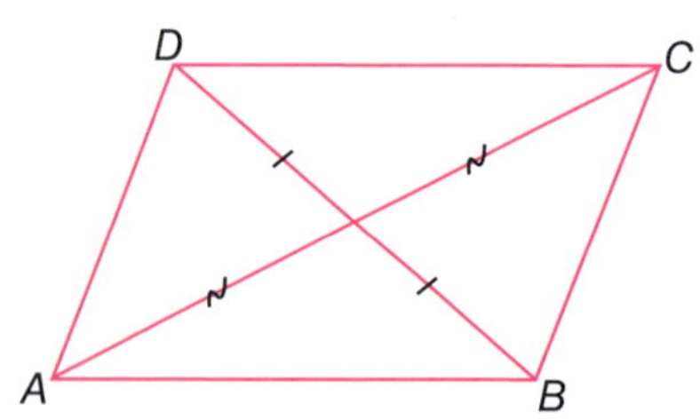

My Notes

Theorem 10.4

Let $ABCD$ be a parallelogram as shown, $e^2+f^2 = 2(a^2+b^2)$, where a and b are sides and e and f are the diagonals.

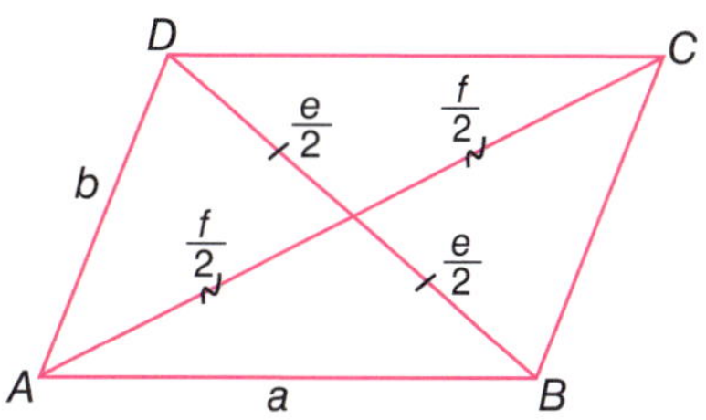

Proof 10.4

In ΔABD, by the Stewart's Theorem.

$$\left(\frac{f}{2}\right)^2 = \frac{a^2\frac{e}{2}+b^2\frac{e}{2}}{\frac{e}{2}+\frac{e}{2}} - \frac{e^2}{4} \Rightarrow \frac{f^2}{4} = \frac{2(a^2+b^2)}{4} - \frac{e^2}{4}$$

$$\frac{f^2}{\cancel{4}} + \frac{e^2}{\cancel{4}} = \frac{2(a^2+b^2)}{\cancel{4}}.$$

Therefore, $e^2+f^2 = 2(a^2+b^2)$.

Theorem 10.5

Let $ABCD$ be a parallelogram as shown. The sum of two opposite veracities' shortest distances to any line ($a+d$ and $b+c$) are equal to each other.

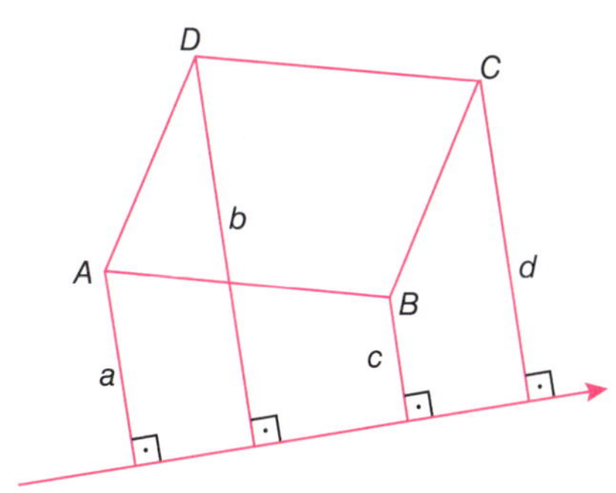

Proof 10.5

In $AA'C'C$, PK is middle line segment and equals to

$\frac{AA'+C'C}{2} = \frac{a+d}{2}$.

In $DD'B'B$, PK is also middle line segment and equals to

$\frac{DD'+B'B}{2} = \frac{b+c}{2}$.

Therefore, $a+d = b+c$.

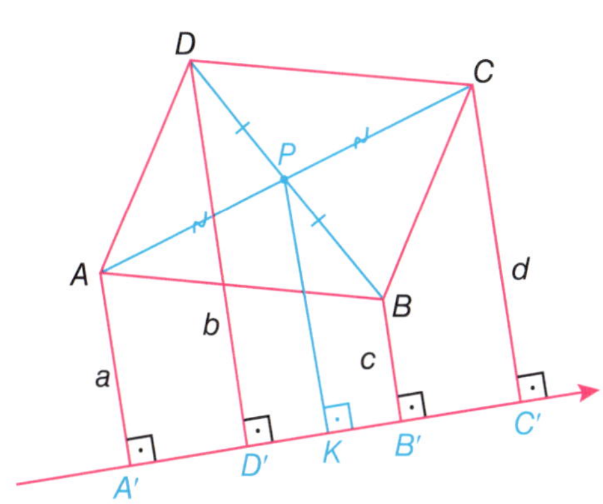

Example 10.3

In the figure, $ABCD$ is a parallelogram with $AE \perp BC$ and $AF \perp DC$. Given that

$BC = 10$ cm,

$AB = 6$ cm, and

$AE = 5$ cm.

What is AF?

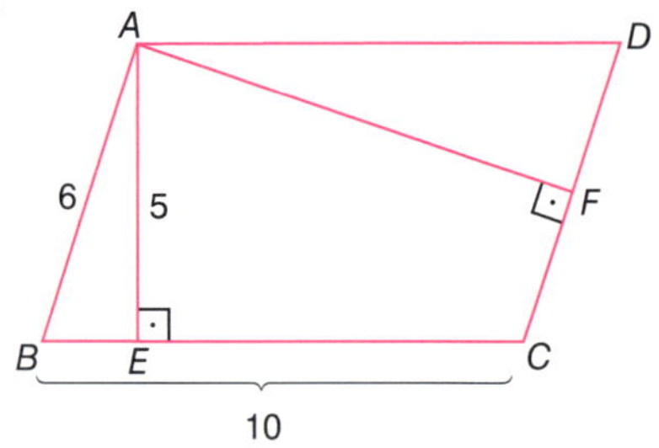

Solution 10.3

Area of $ABCD$ can be expressed in two ways;

$DC \cdot AF = 6 \cdot AF$ and $BC = AE = 10 \cdot 5 = 50$.

$$AF = \frac{50}{6} = \frac{25}{3}.$$

Example 10.4

In the figure, $ABCD$ is a parallelogram with

$AE = 1$,

$FB = 9$, and

$3DK = 7KF$.

Find CF.

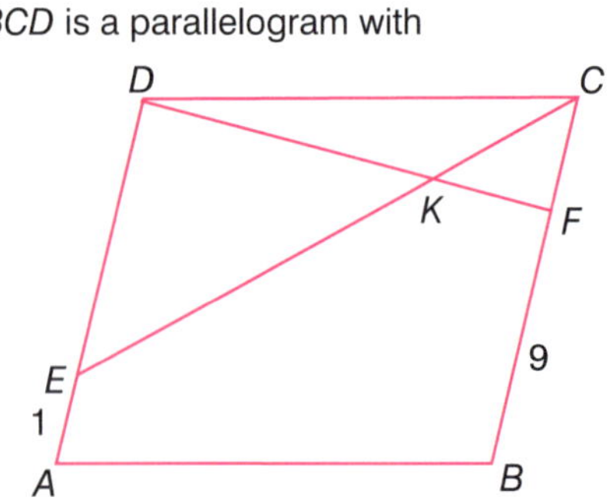

Solution 10.4

From the similarity (BOW-TIE), we can get

$CF = 3x$ and $DE = 7x$

$3x+9 = 7x+1 \Rightarrow 4x = 8 \Rightarrow x = 2$

$CF = 3x = 6$.

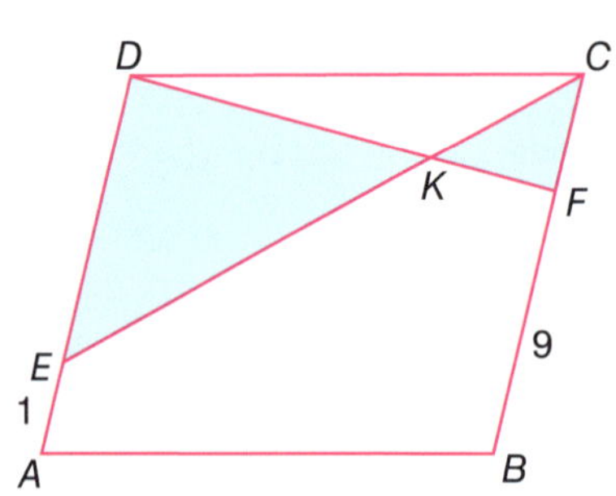

Prove This

$ABCD$ is a parallelogram. Prove that $AK^2 = EK \cdot KF$.

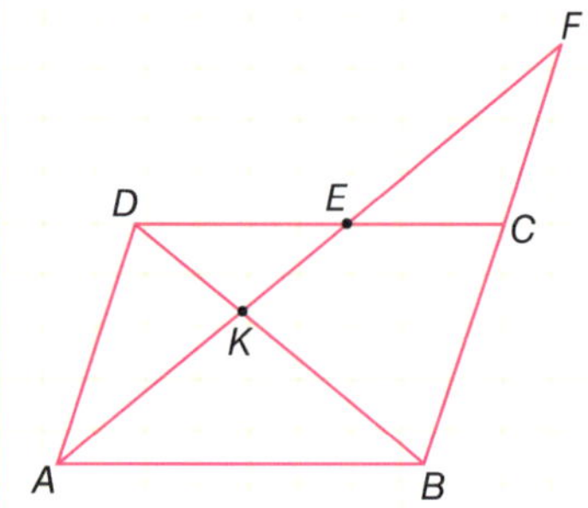

My Notes

Try This

In the figure, *ABCD* is a parallelogram with

$AK = 3\,KE$ and $DE = EC$.

If $KF = 6$, then find *DK*.

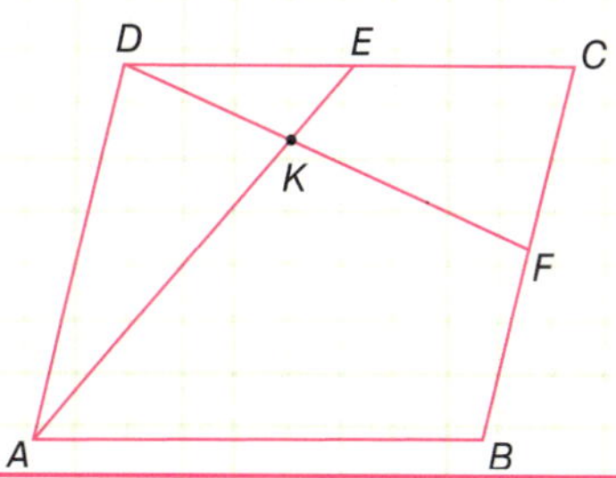

My Notes

Example 10.5

In the figure, *ABCD* is a parallelogram,

$CE = 2\,ED$, and

$AE = 24$ cm.

Find *KE*.

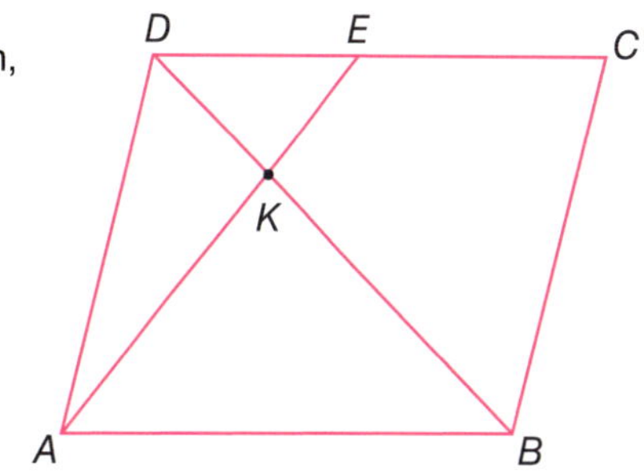

Solution 10.5

From the BOW-TIE similarity ($\Delta AKE \sim \Delta BKA$)

$3KE = AK$

$k + 3k = 24 \Rightarrow k = 6.$

So, $KE = 6$.

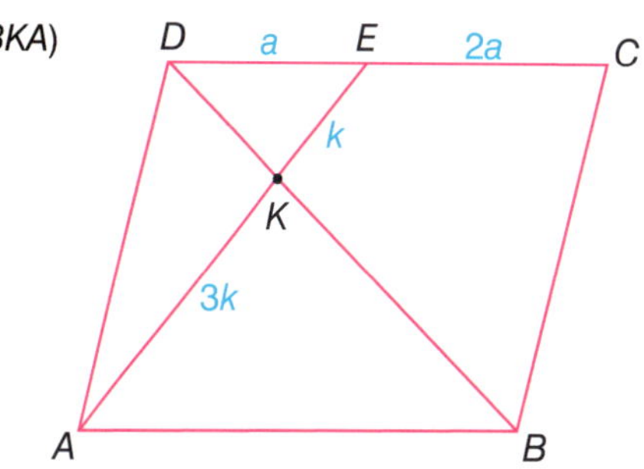

Example 10.6

In the figure, *ABCD* is a parallelogram with

$AH = HE$,

$GH = 4$,

$HF = 6$, and

$AF = 10$.

What is *DE*?

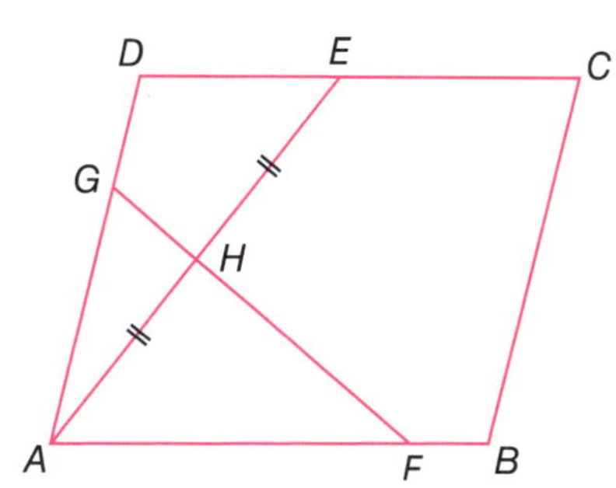

Solution 10.6

[BOW-TIE at point *H*] $GG' = 2$.

[BOW-TIE at point *G*] $G'D = 2$.

Since $\Delta HG'E$ and ΔAHF are congruent, then $G'E = AF = 10$. Therefore, $DE = 8$.

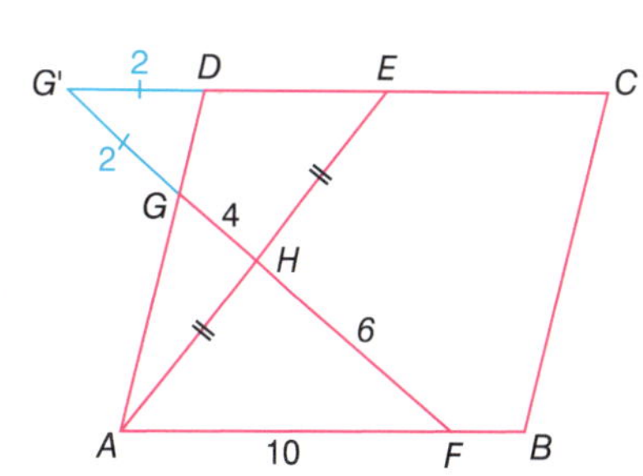

B Rectangles

My Notes

Definition 10.2

A rectangle is a parallelogram which has four right angles.

Theorem 10.6

The length of the diagonals of a rectangle are equal.

Theorem 10.7

British Flag Theorem

Let P be any point inside a rectangle is shown.

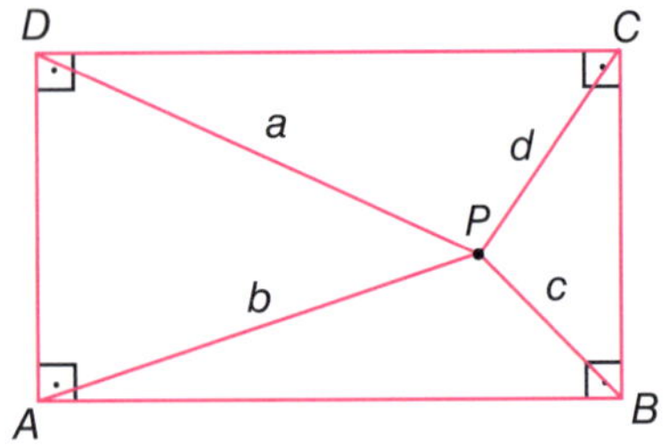

Let P be any point outside a rectangle is shown.

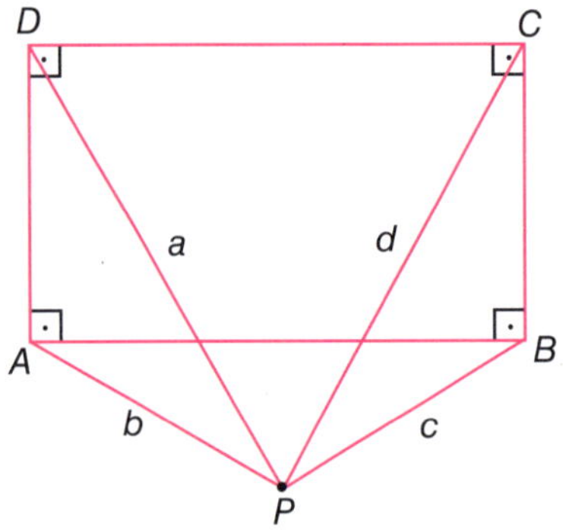

Then $a^2+c^2 = b^2+d^2$

Proof 10.7

In the figure, $ABCD$ is a rectangle,

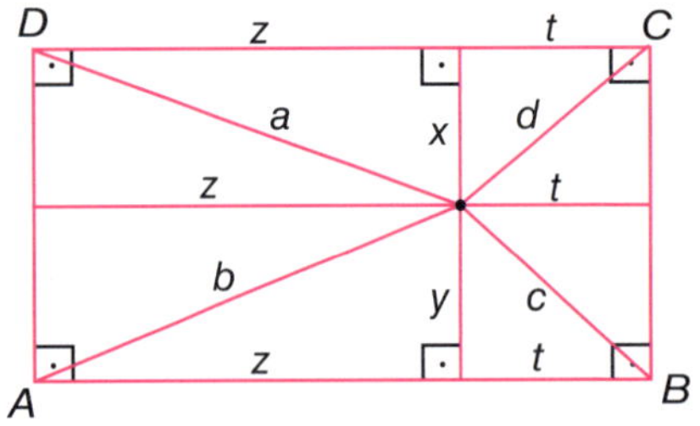

$a^2 = z^2+x^2$ and

$c^2 = y^2+t^2$.

So, $a^2+c^2 = z^2+x^2+ y^2+t^2$.

$d^2 = x^2+t^2$ and

$b^2 = z^2+y^2$.

So, $b^2+d^2 = z^2+x^2+ y^2+t^2$.

Therefore, $a^2+c^2 = b^2+d^2$.

Try This

In a rectangle $ABCD$,

$m\angle DAE = m\angle FAE$,

$AF = FE$, and

$AD = 8$ cm.

What is AF?

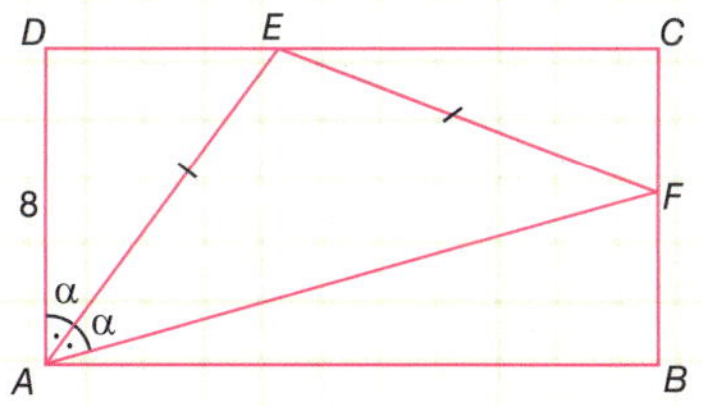

My Notes

Example 10.7

In the figure, $ABCD$ is a rectangle,

$AE = 9$,

$DE = 10$, and

$CE = 8$.

Find $BE = x$.

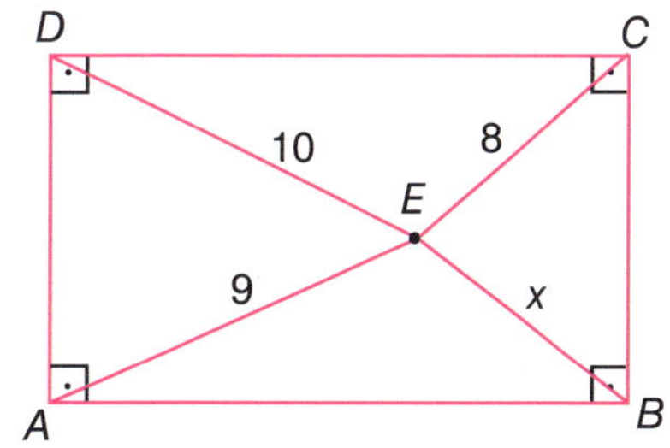

Solution 10.7

By the British flag Theorem,

$9^2+8^2= 10^2+x^2$, where EB is x.

Solving, we find $x^2 = 45$ and $x = 3\sqrt{5}$.

Example 10.8

$ABCD$ is a rectangle, $DH \perp AC$,

$DC = 20$, and $AD = 15$.

What is HB?

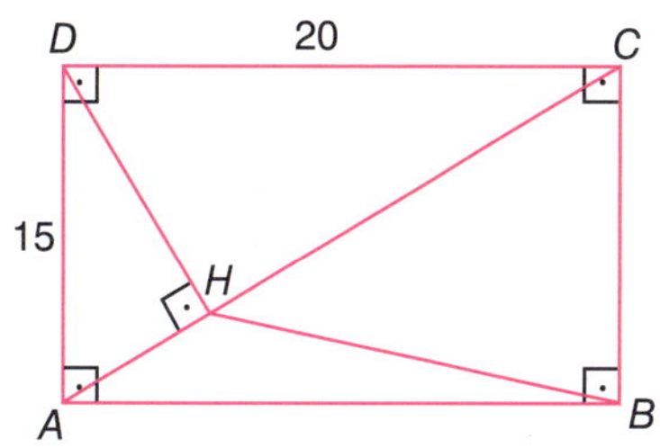

Solution 10.8

By using right triangles similarities

$15^2 = AH{\cdot}AC \qquad (k^2 = n(m+n))$

$225 = AH{\cdot}25 \Rightarrow AH = 9$, then $DH = 12$ and $HC = 16$.

By the British Flag Theorem

$$12^2+x^2 = 9^2+16^2 \Rightarrow x^2 = 193$$

Solving, we find $x = \sqrt{193}$.

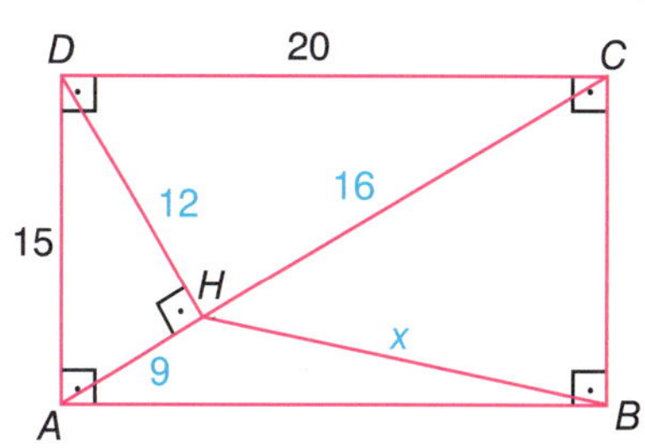

Example 10.9

ABCD is a rectangle,

$DH = 8$,

$HB = 4$, and

$AB = CH$.

Find *CB*.

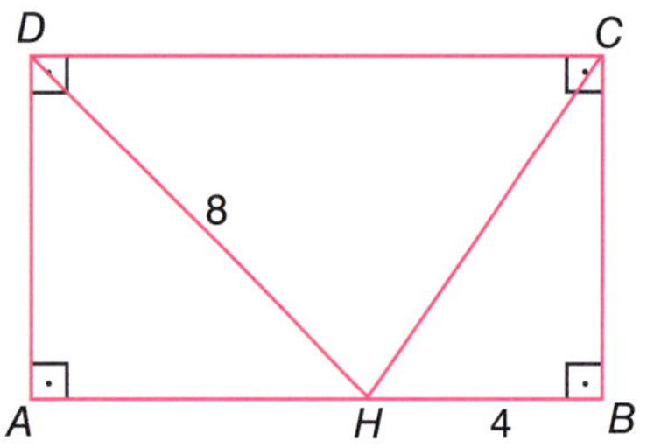

Solution 10.9

$x^2 = 8^2 - a^2$ (1)

$x^2 = (4+a)^2 - 4^2$ (2)

$64 - a^2 = 16 + 8a + a^2 - 16$

$2a^2 + 8a = 64$

$a^2 + 4a = 32 \Rightarrow a = 4$

so, $CB = 4\sqrt{3}$.

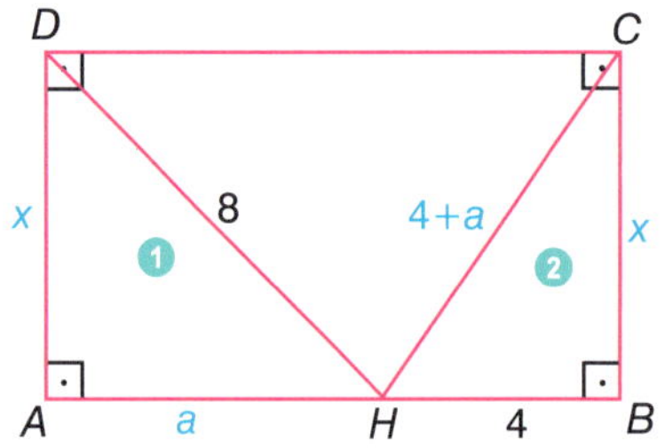

Example 10.10

ABCD is a rectangle,

$AE = EB$, $CE = 6\sqrt{2}$, and

EC is angle bisector.

Find *DC*.

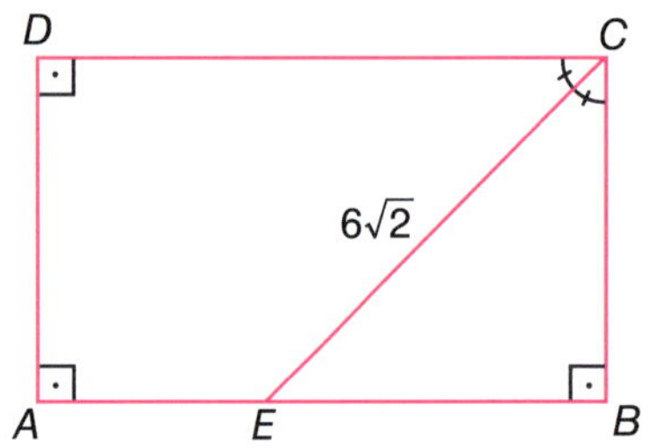

Solution 10.10

Notice that $m\angle CEB = 45°$.

Thus, $EB = CB = 6$.

Therefore, $DC = AE + EB$.

So, $DC = 12$.

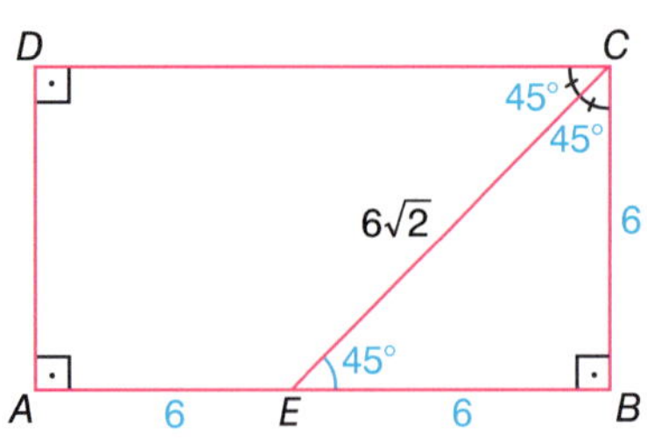

Try This

ABCD is a rectangle and point *P* is inside the rectangle.
Given that $PB = 5$, $PC = 10$, and $PD = 14$ what is *PA*?

My Notes

C Rhombus

My Notes

Definition 10.3

A rhombus is a parallelogram where four sides all have the same length.

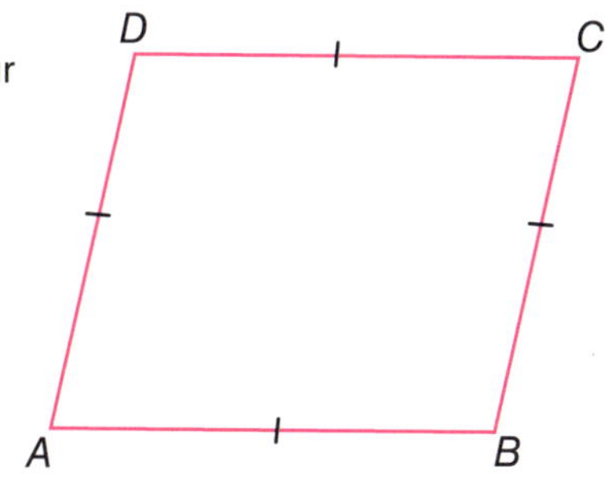

Theorem 10.8

Each diagonal of a rhombus bisects two angles of the rhombus. The diagonals of a rhombus are perpendicular to each other and bisect each other.

Theorem 10.9

In a rhombus the sum of the square of the lengths of the diagonals is equal to four times the square of the lengths of the sides.

Proof 10.9

$$\left(\frac{e}{2}\right)^2+\left(\frac{f}{2}\right)^2 = a^2$$

$$\frac{e^2}{4}+\frac{f^2}{4} = a^2$$

$$e^2+f^2 = 4a^2.$$

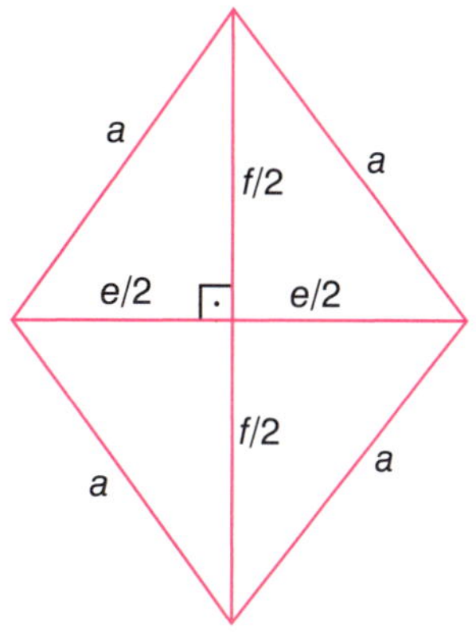

Example 10.11

Find the perimeter a rhombus where its diagonals measure 10 cm and 24 cm.

Solution 10.11

By drawing two diagonals of the rhombus.
One side is equal to 13 cm. (5–12–13).
Therefore, the perimeter of the rhombus is
$13 \cdot 4 = 52$ cm

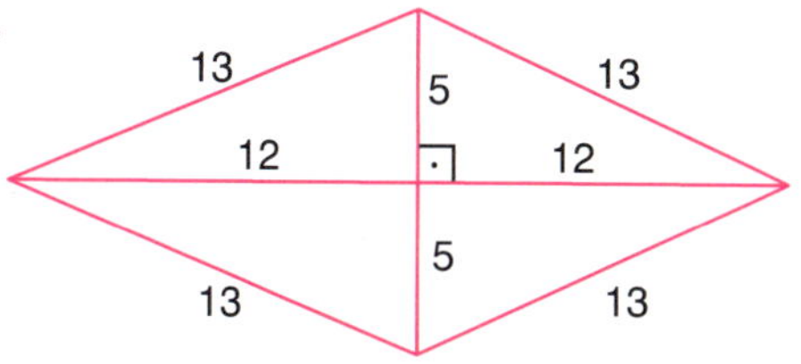

Example 10.12

In the figure, $ABCD$ is a rhombus with $DH \perp AB$,

$AH = 3$ cm, and

$HB = 2$ cm.

Find DB

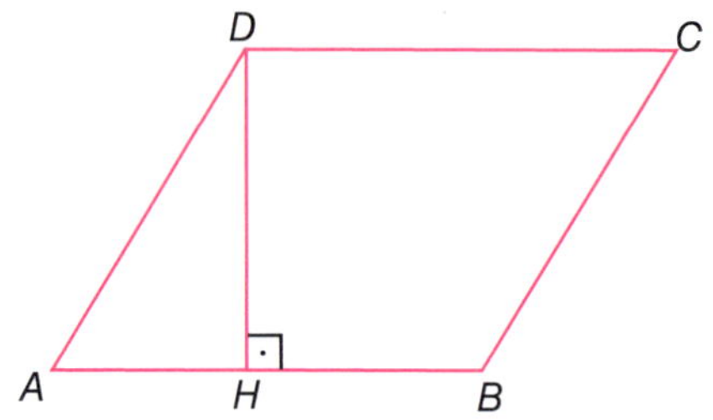

Solution 10.12

$x^2 = 2^2 + 4^2$

$x^2 = 20$

$x = \sqrt{20}$

$x = 2\sqrt{5}.$

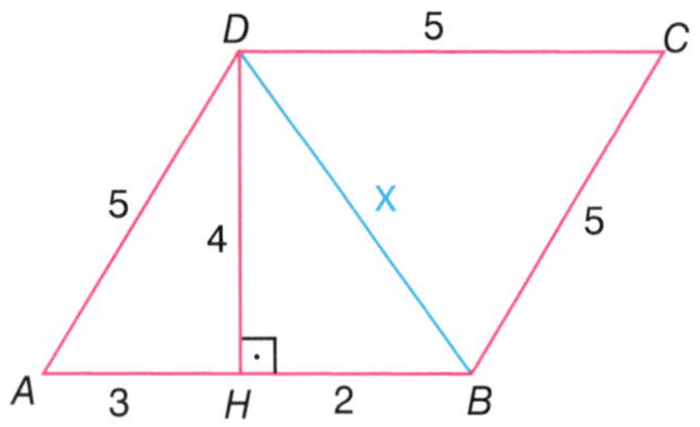

Therefore, $DB = 2\sqrt{5}$.

Example 10.13

In the figure, $ABCD$ is a rhombus with

$DE = 5$ cm,

$AE = 2$ cm, and

$AC = 10$ cm.

Find DB.

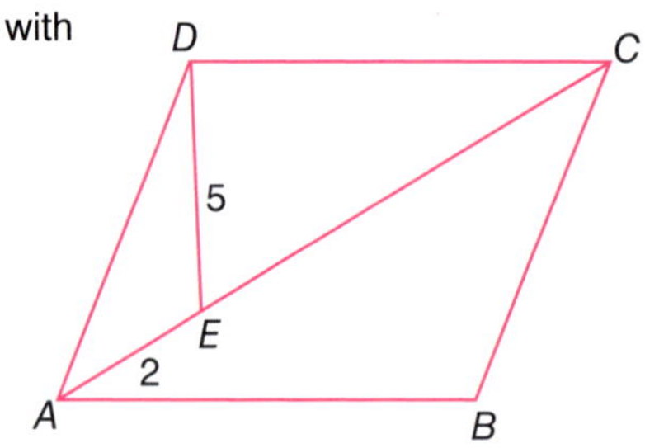

Solution 10.13

$2+a+2+a = 10$

$2a = 6$

$a = 3.$

Thus, $DK = 4$ and $DB = 8$.

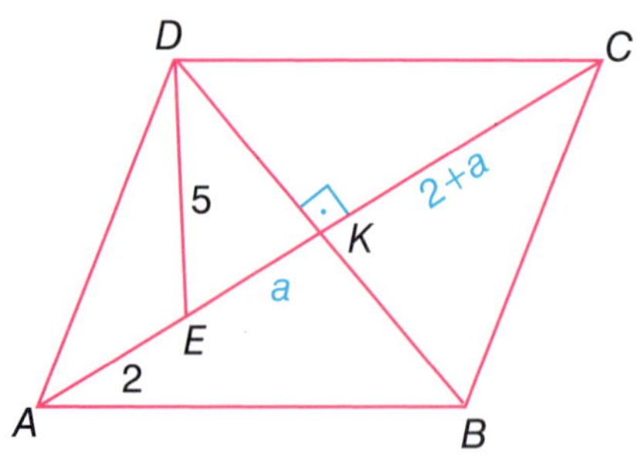

My Notes

D Square

My Notes

Definition 10.4

A square is a rectangle where sides are all equal.

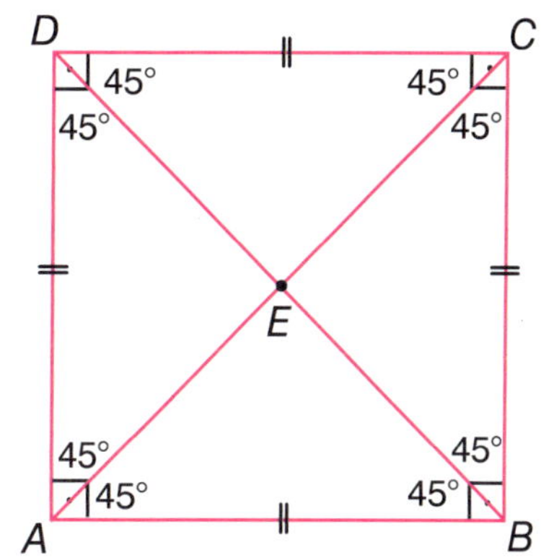

Theorem 10.10

Diagonals have the same length. Diagonals are perpendicular to each other. Diagonals bisect each other.

Example 10.14

ABCD is a square and ΔDEC is equilateral triangle.

Find $\angle BEC$.

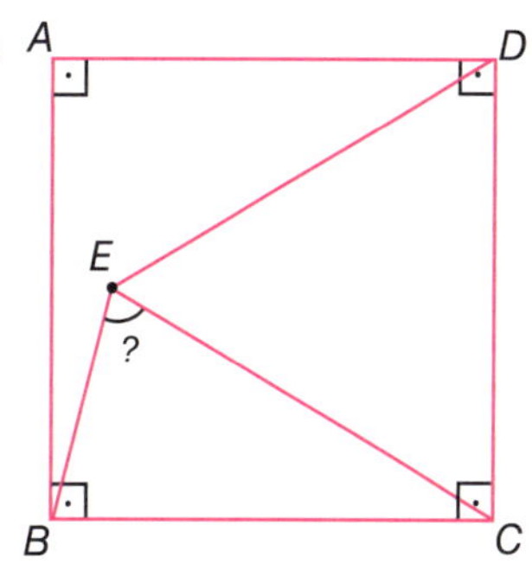

Solution 10.14

Observe that ΔBEC is an isosceles triangle with a top angle of 30°.

Therefore $m\angle BEC = 75°$.

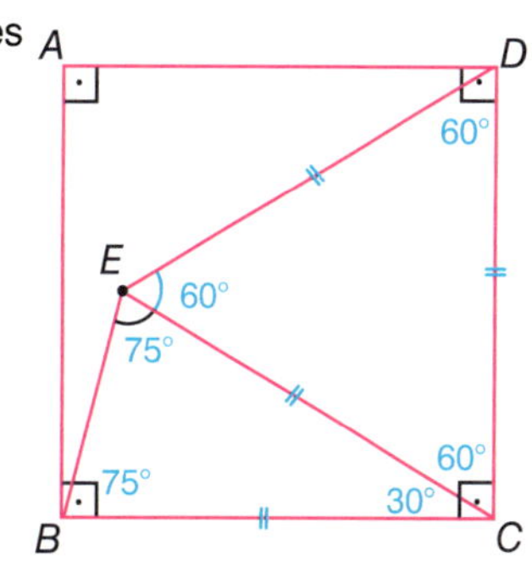

Example 10.15

In the figure, $ABCD$ is a square with

$m\angle DAE = 15°$, and

$AE = 2\sqrt{3}$.

What is AB?

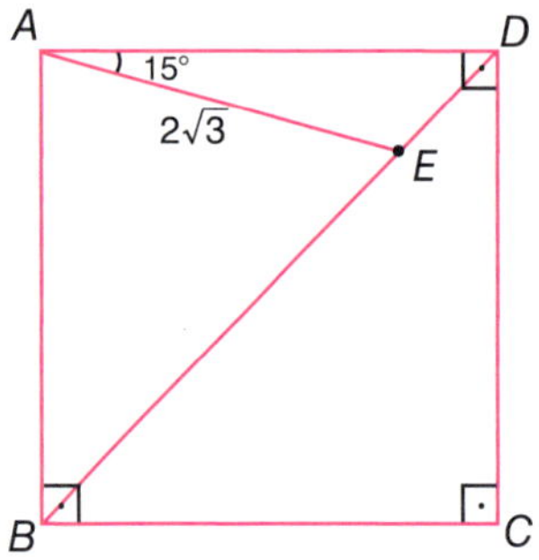

Solution 10.15

If we draw a diagonal AC we get ΔAKE at $30°-60°-90°$ degrees.

$a\sqrt{2} = 6 \Rightarrow a = 3\sqrt{2}$.

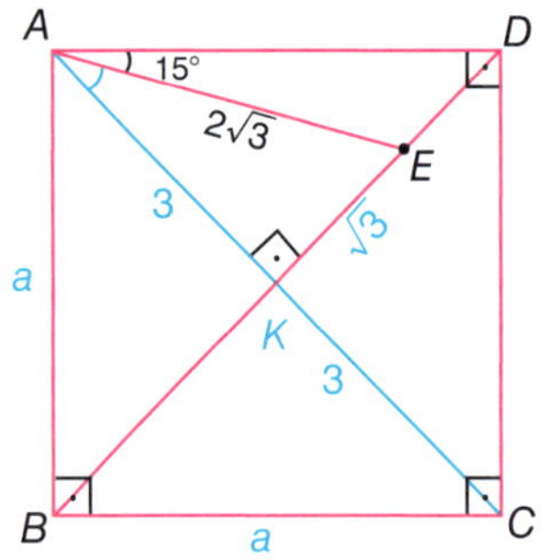

Example 10.16

In the figure, $ABCD$ is a square with

$DE = 3$, and

$BE = 21$.

What is AE?

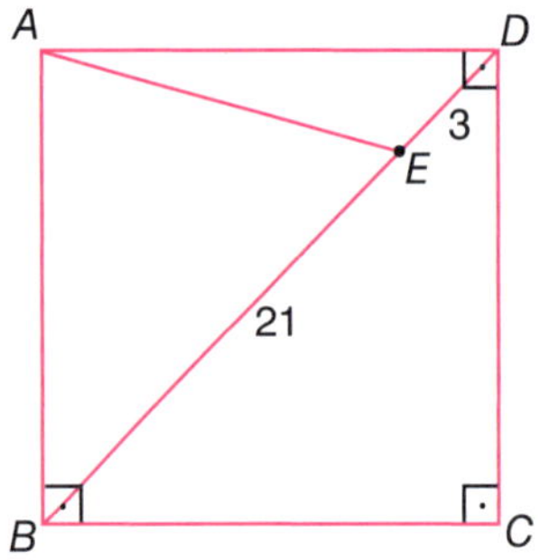

Solution 10.16

Similarly, we draw a diagonal AC then get a right triangle of ΔAKE.

Therefore, $AE = 15$. (9–12–15 triangle)

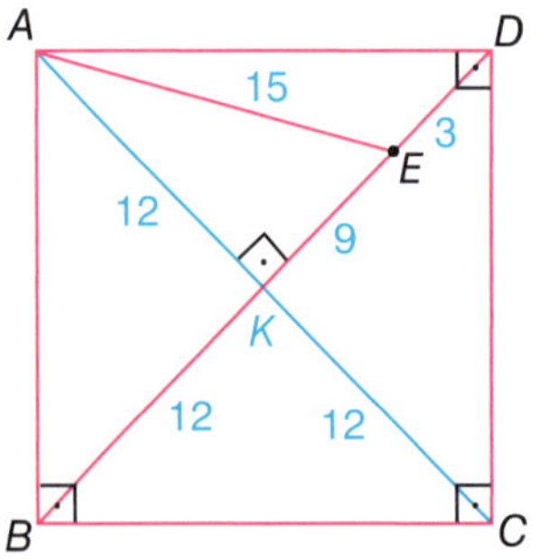

My Notes

Catriona Agg @Cshearer41
Geometry puzzle

This regular hexagon has been split into three polygons. Thirds imply equal pieces.

What is the angle?

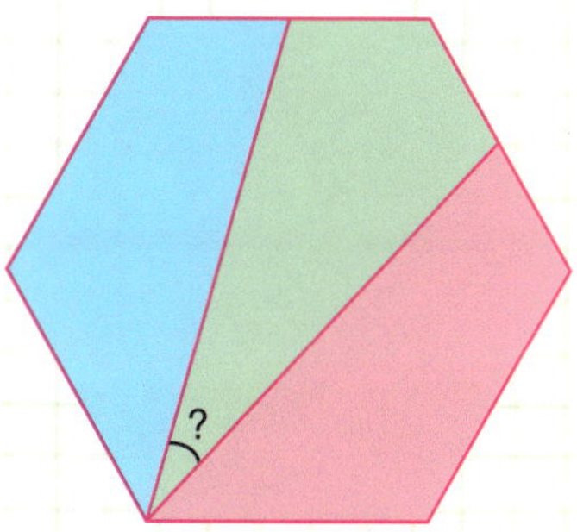

My Notes

Example 10.17

ABCD is a square, *A*, *C*, and *E* are collinear.

If $AC = BE$ find $m\angle CBE$.

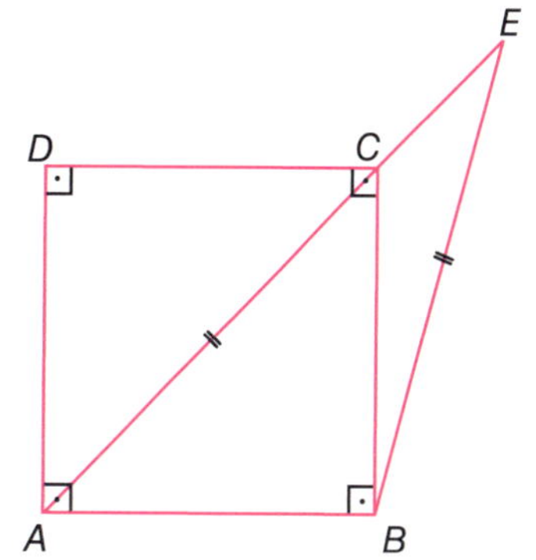

Solution 10.17

Since $FB = \frac{EB}{2}$ then $\angle E = 30°$.

Therefore, $m\angle CBE = 45° - 30° = 15°$.

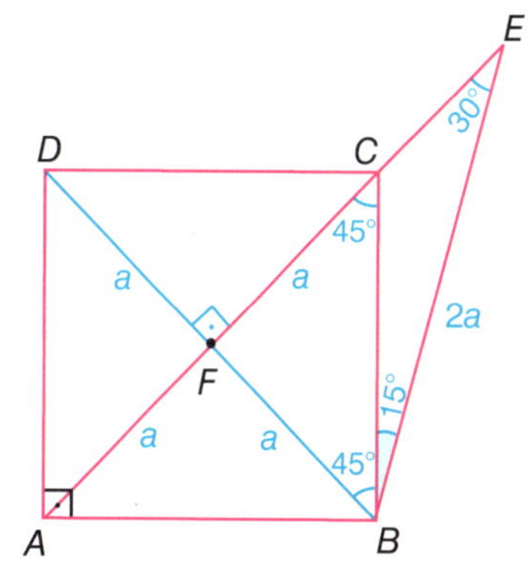

Example 10.17

ABCD is a square with

$BE = DE = 5$ cm,

$EC = 7$ cm.

Find *CB*.

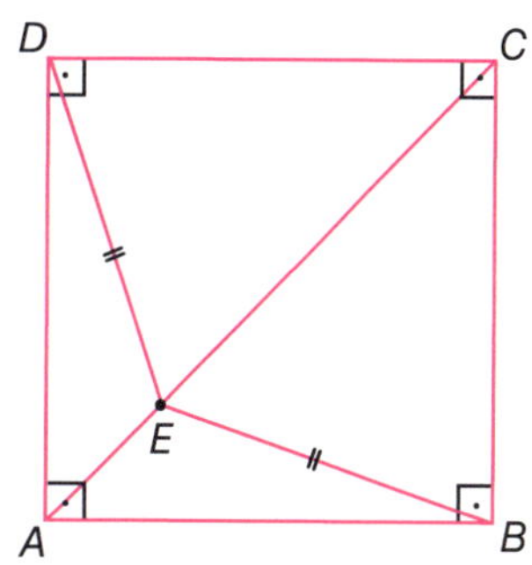

Solution 10.17

From ΔDEF and ΔEFB

$$5^2 = a^2 + (7-a)^2$$

$$25 = a^2 + 49 - 14a + a^2$$

$$2a^2 - 14a + 24 = 0$$

$$a^2 - 7a + 12 = 0$$

$$(a-4)\cdot(a-3) = 0$$

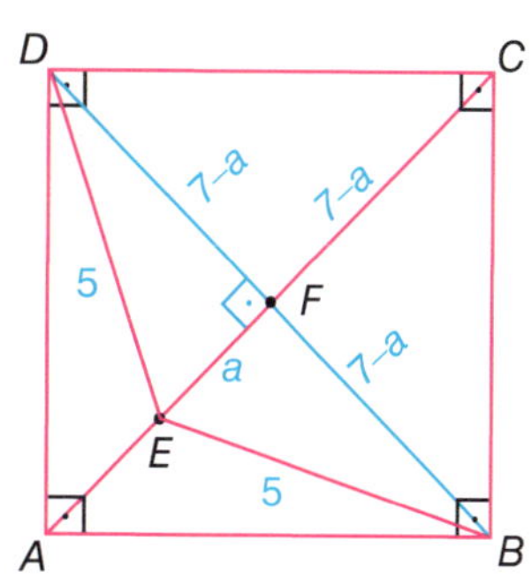

$a = 4$, $a = 3$. $a = 4$ does not work, Thus, $a = 3$ and $CB = 4\sqrt{2}$.

Example 10.18

ABCD is a square.

Given that

$DG = CH = BF = AE = 4$ and

$DE = AF = BH = CG = 2$.

What is the shaded area?

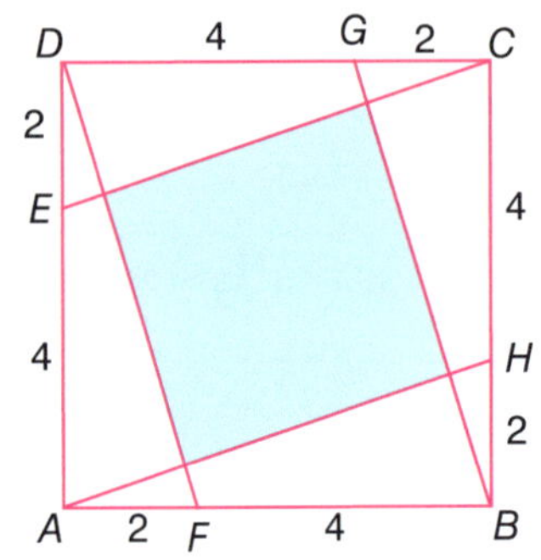

Solution 10.18

The one ratio is $\frac{1}{8}$.

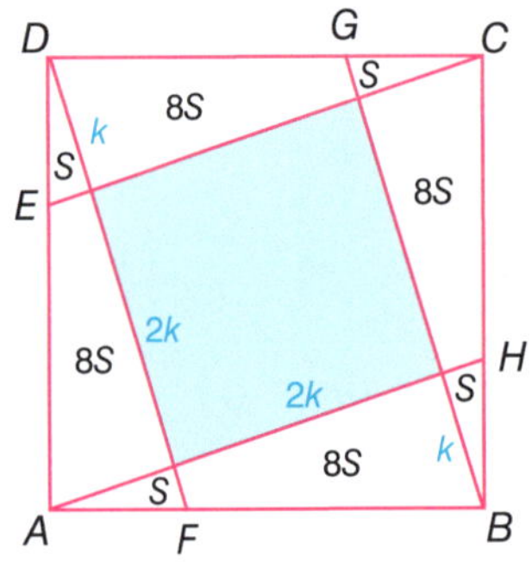

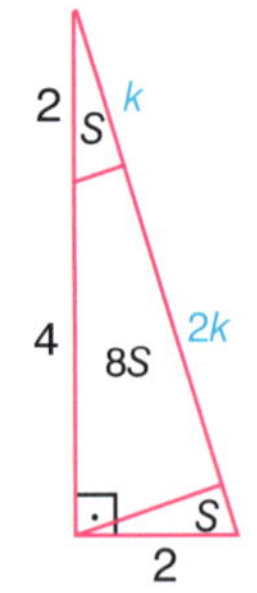

The area of the right triangle is $\frac{2 \cdot 6}{2} = 6$.

So, $10S = 6$, hence $S = \frac{3}{5}$.

Therefore, the shaded area is $36 - 36S = 36 - 36\frac{3}{5} = 14.4$.

Example 10.19 (2020 AMC 10B #21)

In square *ABCD*, points *E* and *H* lie on $\overline{AB}$ and $\overline{DA}$, respectively, so that $AE = AH$. Points *F* and *G* lie on $\overline{BC}$ and *CD*, respectively, and points *I* and *J* lie on $\overline{EH}$ so that $\overline{FI} \perp \overline{EH}$ and $\overline{GJ} \perp \overline{EH}$. See the figure below. Triangle *AEH*, quadrilateral *BFIE*, quadrilateral *DHJG*, and pentagon *FCGJI* each has area 1. What is FI^2?

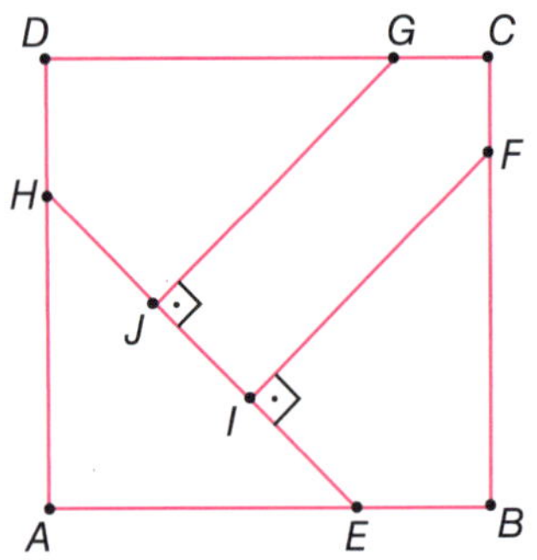

Solution 10.18

$$\text{Area}(FIP) - \text{Area}(EBP) = 1$$

$$\frac{x^2}{2} - \frac{(2-\sqrt{2})^2}{2} = 1$$

$$x^2 = 8 - 4\sqrt{2}.$$

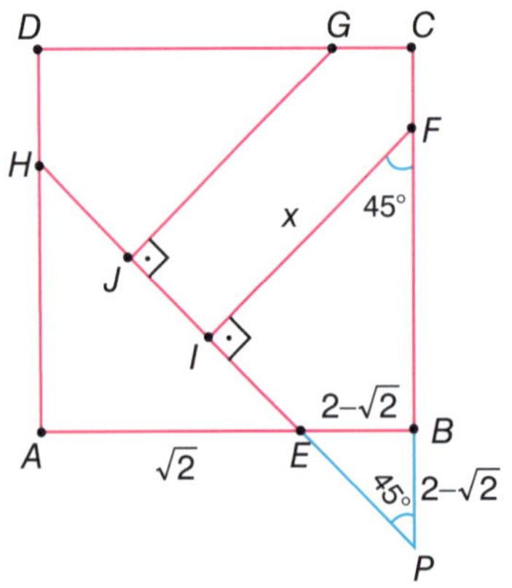

Try This

ABCD is a square $DH \perp EG$ and $FB \perp EG$.

Given that

$DH = 2$, $BF = 4$ and $HF = 4$.

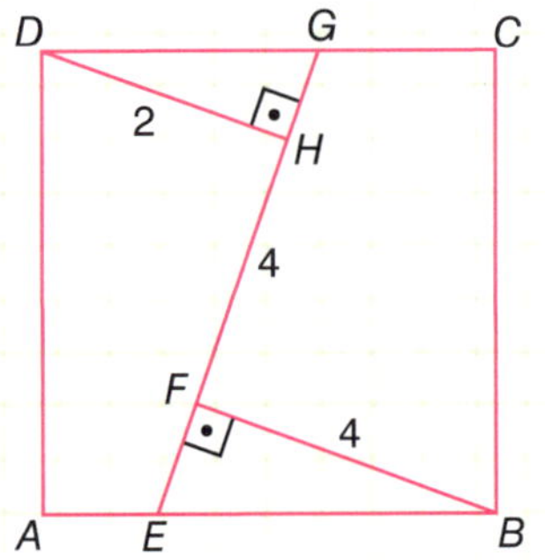

What is the area of the square?

My Notes

1. In the figure, $ABCD$ is a rectangle with $AE \perp EB$.

If $DE = 4$ and $CB = 6$ then what is EC?

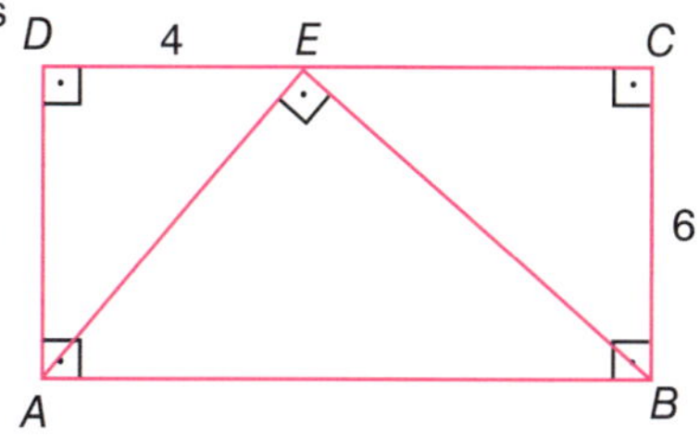

2. In the figure, $ABCD$ is a rectangle

Given that $AP = 4$,

$PC = 6$, and

$PB = 5$.

What is DP?

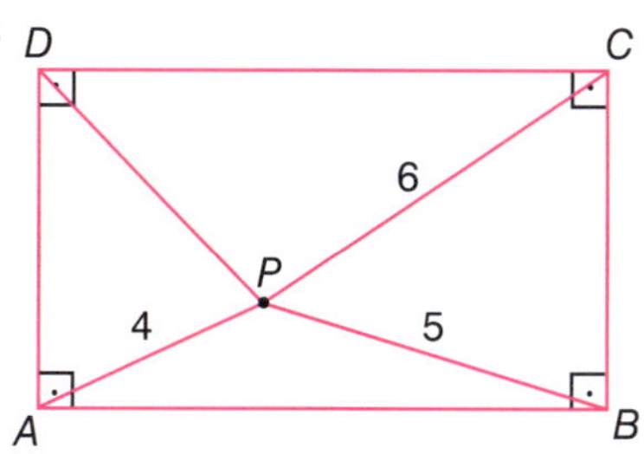

3. In the figure, $ABCD$ is a rectangle with AC diagonal angle is 11.

Given that $AE = EB$ and $CE = \sqrt{46}$.

What is DC?

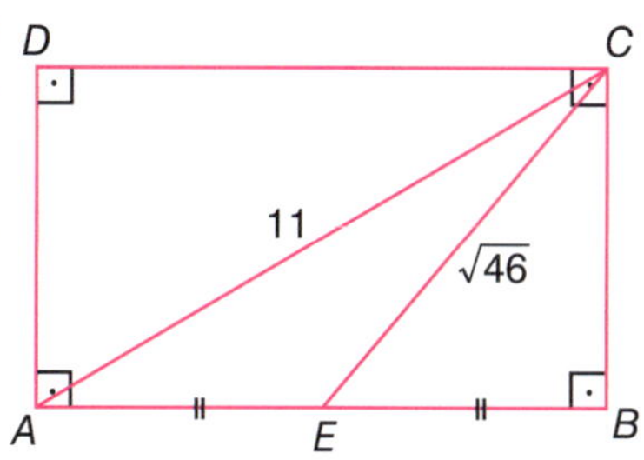

4. $ABCD$ is a square with

$DE = x+5$ and

$EB = 3x-13$.

What is the value of DE?

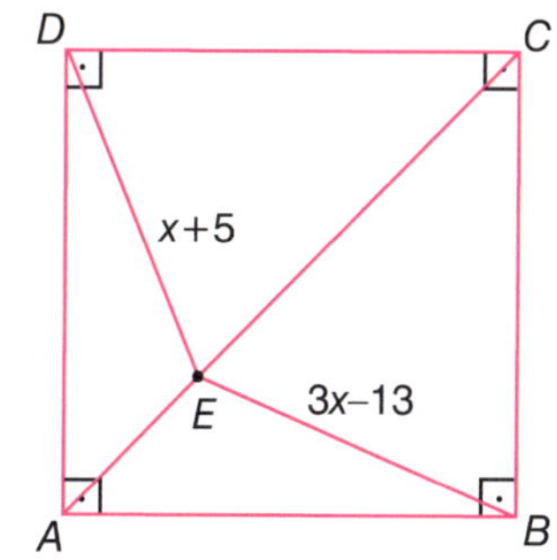

5. In the figure, $ABCD$ is a parallelogram and points E, F, G and B are collinear.

If $EF = 5$ and $GB = 6$, find the length of FG.

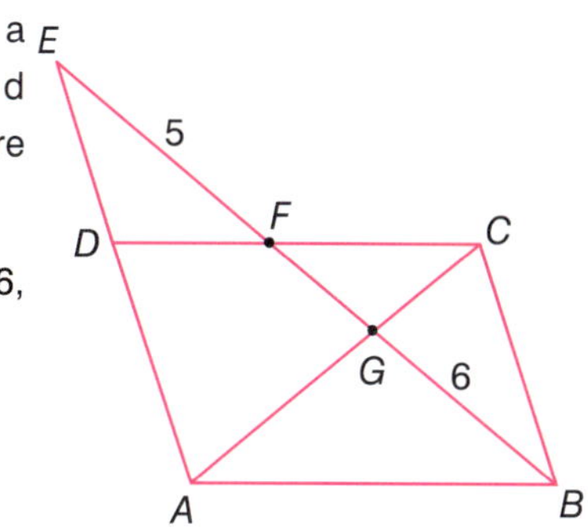

6. In the figure, $ABCD$ is a square and DAE is an equilateral triangle.

What is $m\angle AEB$?

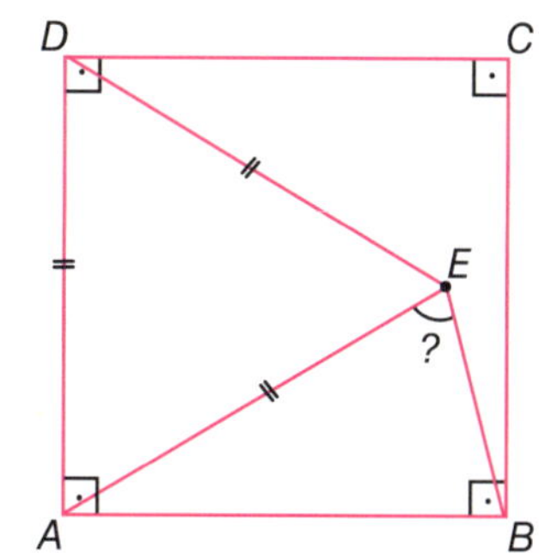

7. In the figure, $ABCD$ is a rhombus and point E *is* on the diagonal AC.

Given that $EC = 8$,

$AE = 28$, and

$AB = 30$.

What is DE?

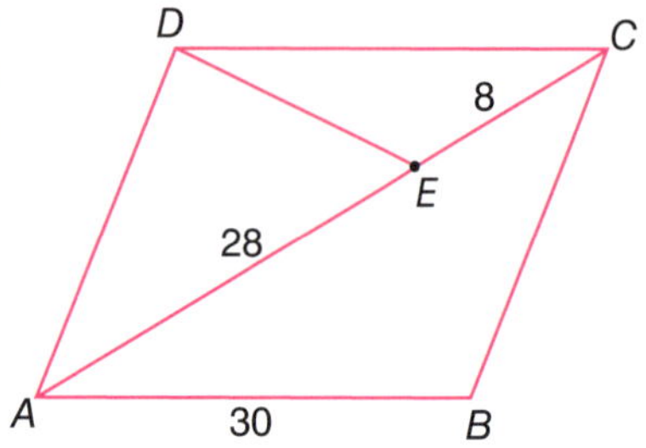

8. In the figure, $ABCD$ is a rhombus and points B, A, E and F, E,and C are collinear.

Given that $AE = 3$,
AB = 6, and

$EC = 27$.

What is FC?

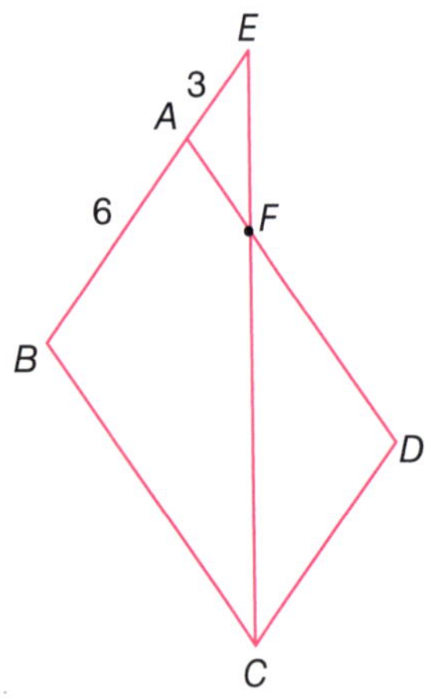

9. In the figure, $ABCD$ is a square with

$EC = 2$ and AE = 8.

What is DE?

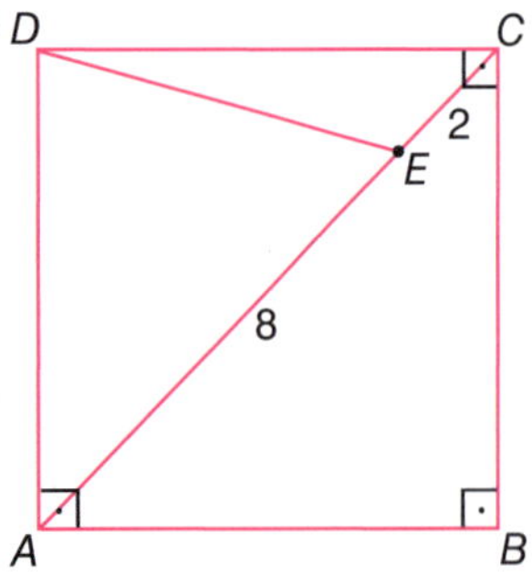

10. In the figure, $ABCD$ is a rhombus with

$DE = 10$,

$AE = 4$, and

$AB = 12$.

What is EC?

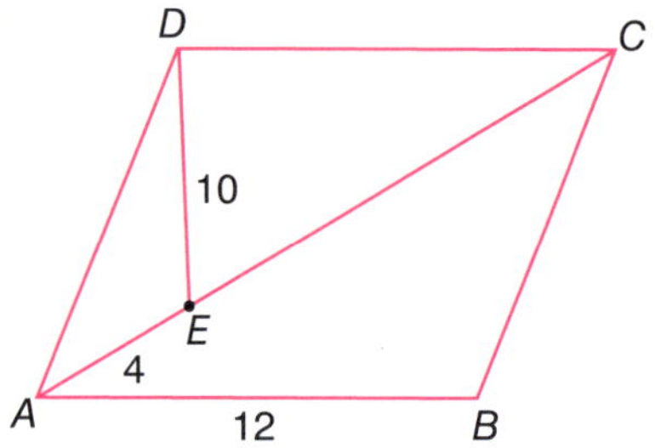

11. In the figure, $ABCD$ is a rectangle and EC angle bisector.

$AE = EB$ and

$EC = 6$,.

What is DC?

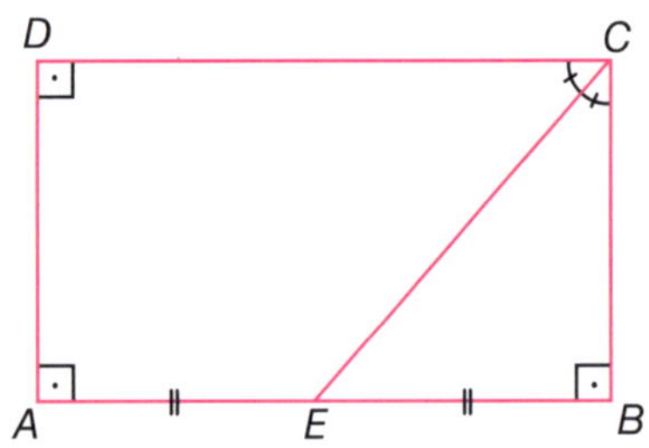

12. In the figure, $ABCD$ is a parallelogram

$CE = 2\,ED$ and

$AE = 18$.

Find KE.

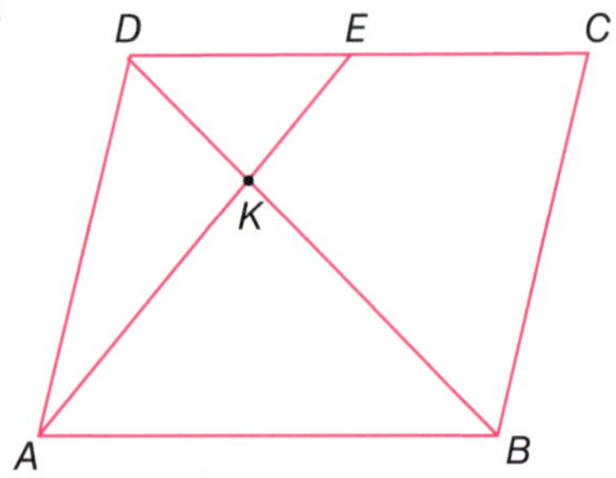

1. In the figure, *ABCD* is a parallelogram with

$AE = 2$,

$BF = 7$, and

$2DK = 3KF$.

What is *CF*?

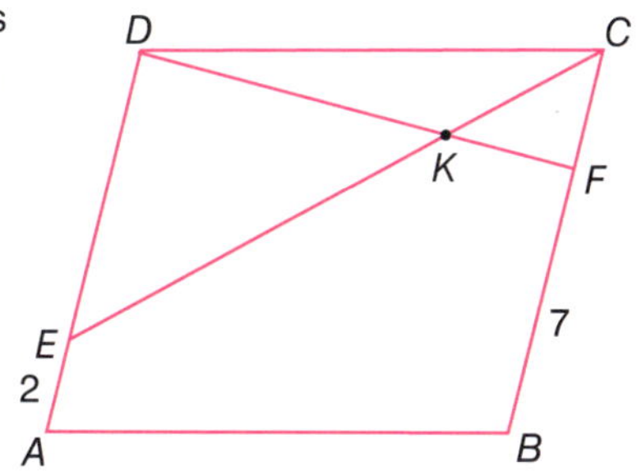

2. In the figure, *ABCD* is a square with

$DE = BE = 10$ cm,

$AE = 2$ cm.

What is *EC*?

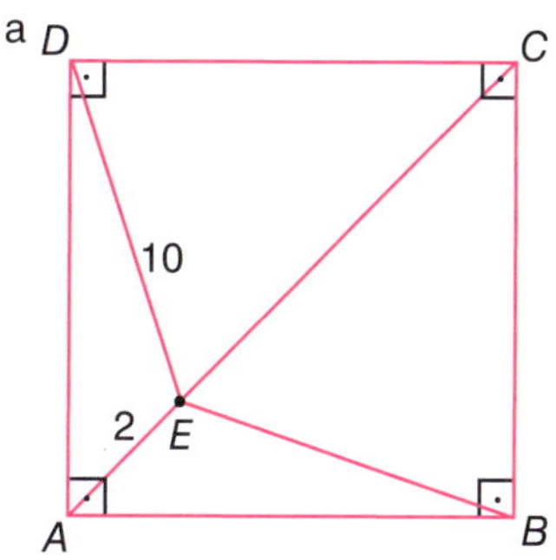

3. In the figure, *ABCD* is a rectangle $DH \perp AC$,

$AD = 2\sqrt{5}$, and

$DC = 4\sqrt{5}$.

What is *HB*?

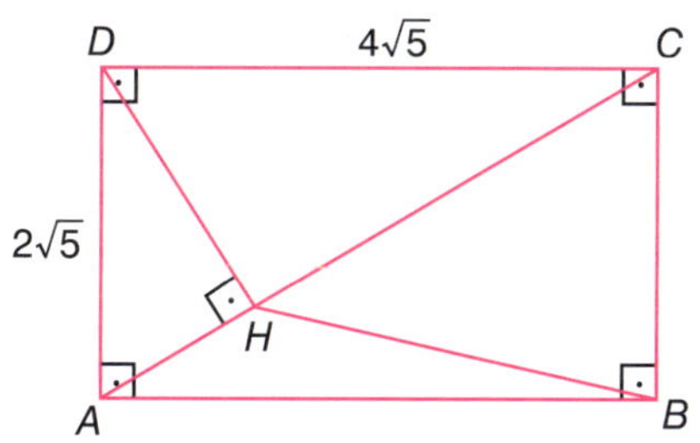

4. Suppose *ABCD* is a square with $AB = 30$.

If *K* is the intersection of its diagonals, and L is on *AB* such that $AL = 7$, what is the value of *KL*?

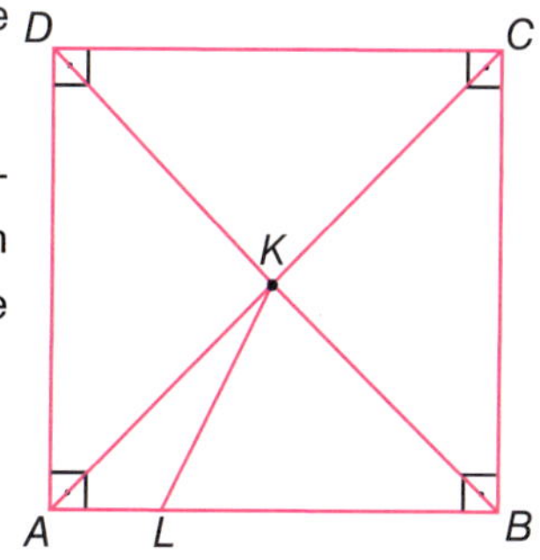

5. *ABCD* is a parallelogram.

Given that

$EC = BC = 3$ and

$DF = 2$ find the perimeter of the parallelogram.

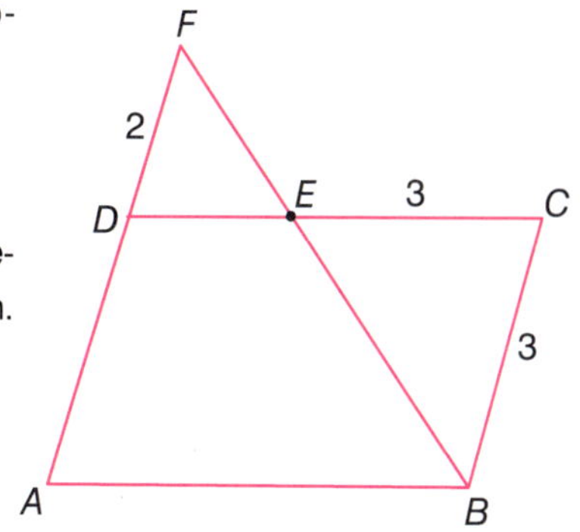

6.

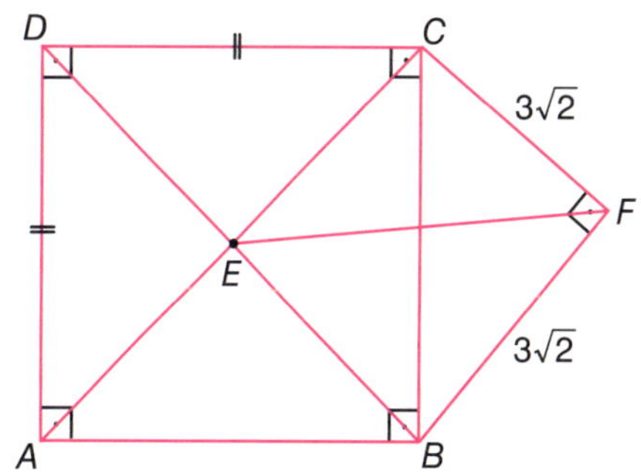

In the figure, *ABCD* is a square with point *E* intersection of its diagonals.

Given that $EB = FB = 3\sqrt{2}$ and *EFB* is a right triangle. What is AB?

7. $ABCD$ is a square with $EF \perp FC$.

If $EF = 51$ and

$FC = 68$.

What is DC?

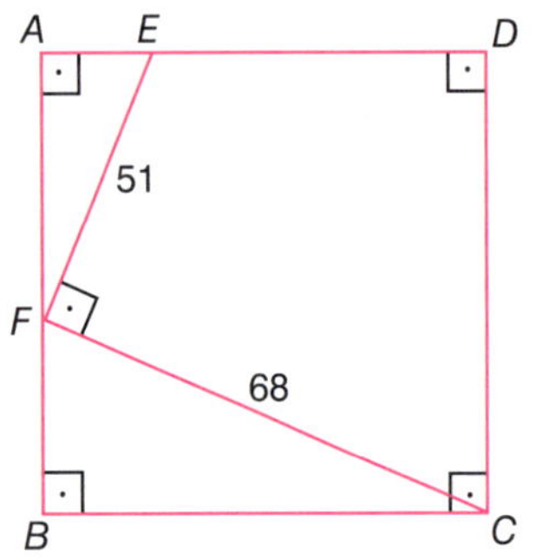

10. In the figure, $ABCD$ is a rectangle with $\angle DAF = \angle FAE$.

If $AF = FE$ and $AD = 8$.

find AE.

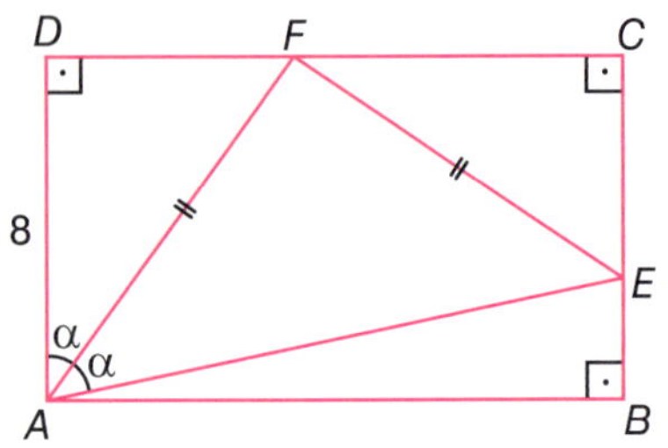

8.

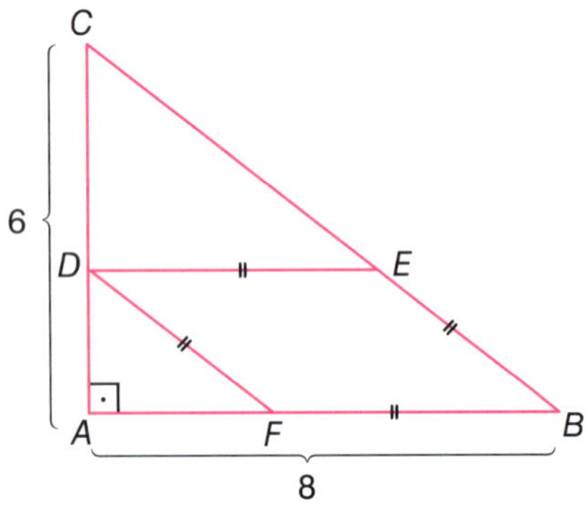

In the figure, ΔABC is a right triangle and $DEBF$ is a rhombus.

Given that $AB = 8$ cm and $AC = 6$ cm what is DE?

11. In the figure, $ABCD$ is a square, EDC is an equilateral triangle and DKE is a isosceles right triangle.

What is the measure of $\angle DAK$?

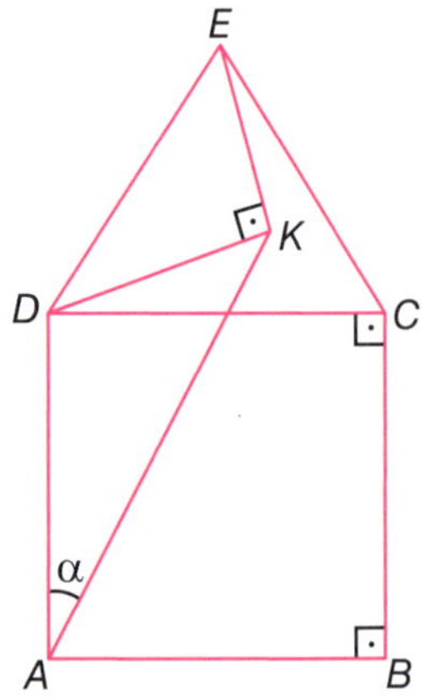

9. In the figure, $ABCD$ is a rectangle $DE \perp AC$ and $BF \perp AC$

If $AE = 2$, and $BF = 4$, find DF.

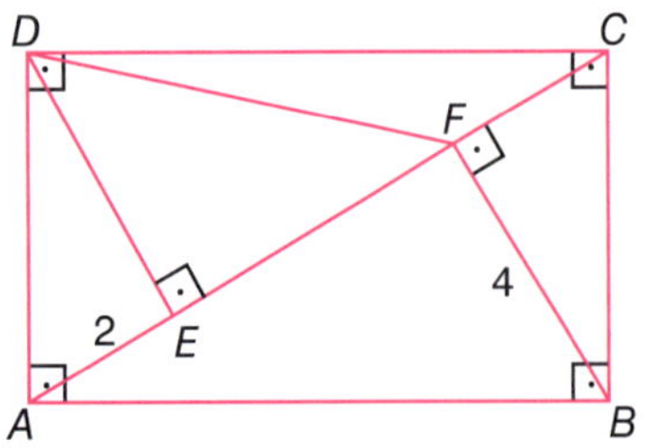

12.

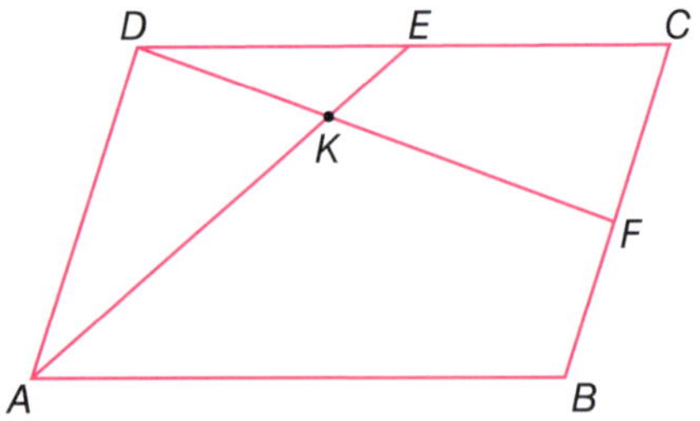

In the figure, $ABCD$ is a parallelogram. E and F are point BC and CD, respectively, with $AK = 4\,KE$, $DE = EC$, and $DK = 4$.

What is KF?

1. (2000 AMC 10 #7)

In rectangle $ABCD$, $AD = 1$, P is on $\overline{AB}$, and $\overline{DB}$ and $\overline{DP}$ trisect $\angle ADC$.

What is the perimeter of ΔBDP?

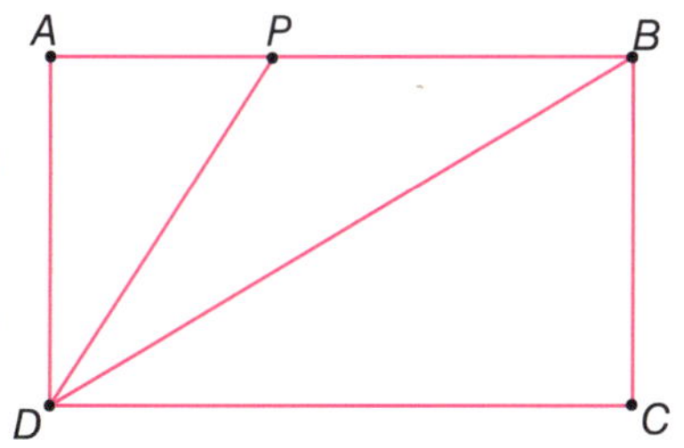

A) $3+\frac{\sqrt{3}}{3}$ B) $2+\frac{4\sqrt{3}}{3}$ C) $\frac{3+3\sqrt{5}}{2}$ D) $\frac{3+3\sqrt{5}}{2}$ E) $2+\frac{5\sqrt{3}}{3}$

2. (CWML-2014-15 Meet 2 #10)

The walls of a pole building are $x-6$ feet tall(see diagram). The wall height plus the roof rise $x+8$ feet.

If the length of the roof line from the peak of the roof to the top of each wall is 50 feet, what is the width (in feet) of the building?

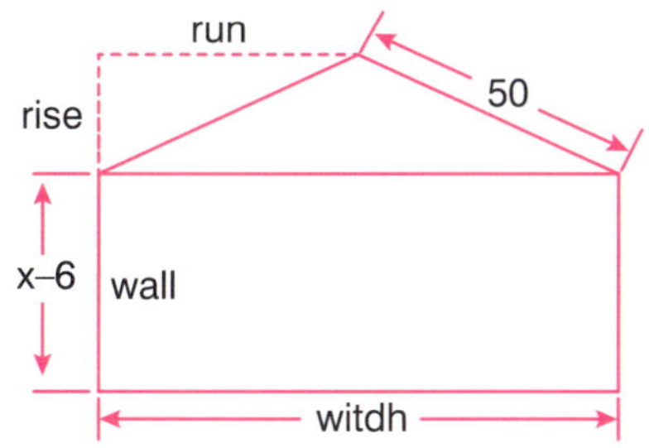

3. (2006 AMC 10B #15)

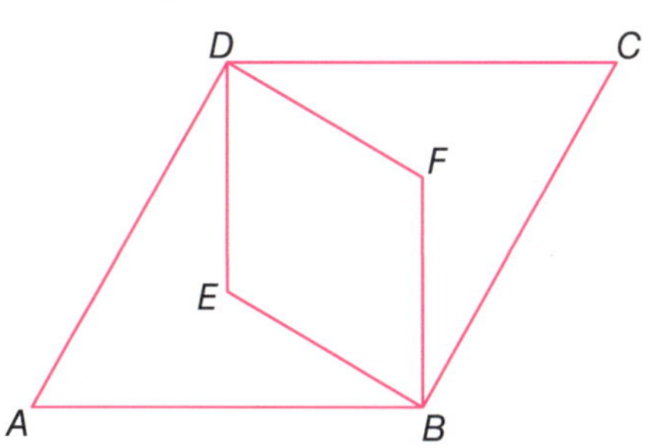

Rhombus $ABCD$ is similar to rhombus $BFDE$. The area of rhombus $ABCD$ is 24 and $\angle BAD = 60°$.

What is the area of rhombus $BFDE$?

A) 6 B) $4\sqrt{3}$ C) 8 D) 9 E) $6\sqrt{3}$

4. (2016 AMC 10B #19)

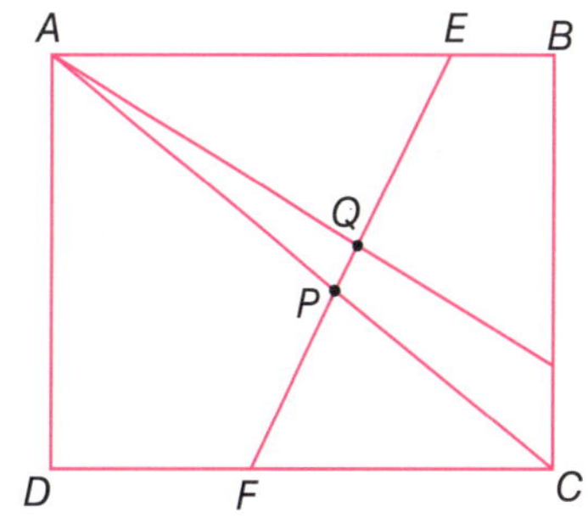

Rectangle $ABCD$ has $AB = 5$ and $BC = 4$. Point E lies on $\overline{AB}$ so that $EB = 1$, point G lies on $\overline{BC}$ so that $CG = 1$, and point F lies on $\overline{CD}$ so that $DF = 2$.

Segments AG and AC intersect EF at Q and P, respectively. What is the value of $\frac{PQ}{EF}$?

A) $\frac{\sqrt{13}}{16}$ B) $\frac{\sqrt{2}}{13}$ C) $\frac{9}{82}$ D) $\frac{10}{91}$ E) $\frac{1}{9}$

5. (2001 AMC 10 #20) A regular octagon is formed by cutting an isosceles right triangle from each of the corners of a square with sides of length 200.

What is the length of each side of the octagon?

A) $\frac{1}{3}(2000)$ B) $2000(\sqrt{2}-1)$ C) $2000(2-\sqrt{2})$ D) 1000 E) $1000\sqrt{2}$

6. (2009 AMC 10A #17) Rectangle $ABCD$ has $AB = 4$ and $BC = 3$. Segment EF is constructed through B so that EF is perpendicular to DB, and A and C lie on DE and EF, respectively.

What is EF?

A) 9 B) 10 C) $\frac{125}{12}$ D) $\frac{103}{9}$ E) 12

7. (AlphaStar 2019-Fermat # 8)

A rectangle with dimensions 32 and 72 is cut into three pieces and a square is formed by rearranging this pieces.

Find the side length of the square.

8. (2017 AMC 12B #8)

The ratio of the short side of a certain rectangle to the long side is equal to the ratio of the long side to the diagonal.

What is the square of the ratio of the short side to the long side of this rectangle?

A) $\frac{\sqrt{3}-1}{2}$ B) $\frac{1}{2}$ C) $\frac{\sqrt{5}-1}{2}$ D) $\frac{\sqrt{2}}{2}$ E) $\frac{\sqrt{6}-1}{2}$

9. (2006 AMC 10A #17)

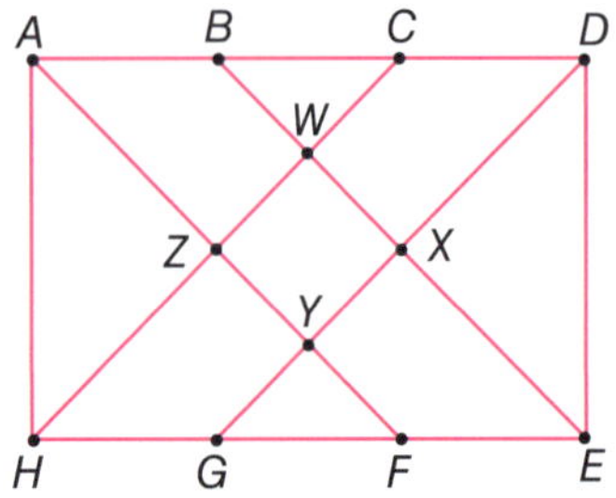

In rectangle $ADEH$, points B and C trisect $\overline{AD}$, and points G and F trisect $\overline{HE}$. In addition, $AH = AC = 2$, and $AD = 3$.

What is the area of quadrilateral $WXYZ$ shown in the figure?

A) $\frac{1}{2}$ B) $\frac{\sqrt{2}}{2}$ C) $\frac{\sqrt{3}}{2}$ D) $\frac{2\sqrt{2}}{2}$ E) $\frac{2\sqrt{3}}{2}$

10. (MathCON 2019-Final Gr 10 #28)

Suppose $ABCD$ is a square with $AB = 30$.

If K is the intersection of is the diagonals, and L is on $\overline{AB}$ such that $AL = 7$, what is the value KL?

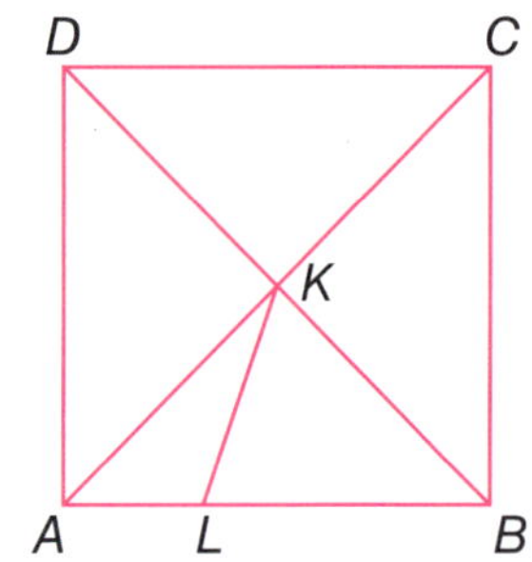

A) 13 B) 15 C) 17 D) $15\sqrt{2}$ E) None of the preceding

11. (2003 AMC 10A #22)

In rectangle $ABCD$, we have $AB = 8$, $BC = 9$, H is on BC with $BH = 6$, E is on AD with $DE = 4$, line EC intersects line AH at G, and F is on line AD with $GF \perp AF$.

Find the length of GF.

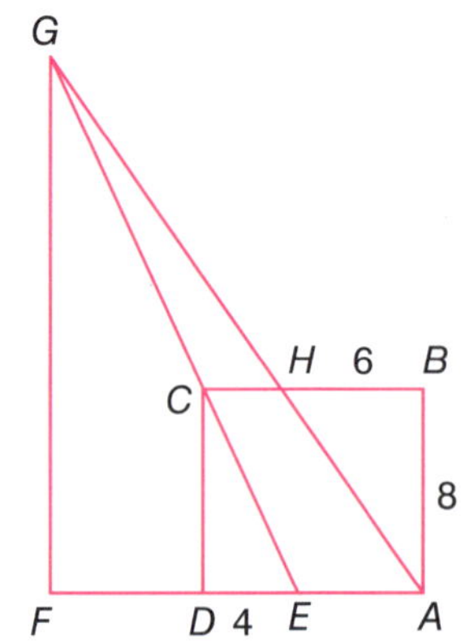

A) 16 B) 20 C) 24 D) 28 E) 30

12. (MathCON 2019-Final Gr 9 #40)

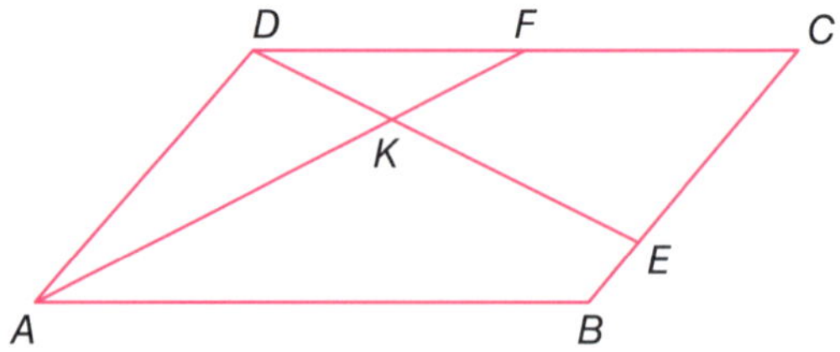

$ABCD$ is a parallelogram. E and F are points on BC and CD, respectively, with $CE = 3\,BE$, $CF = DF$, and $KF = 6$.

Find AK.

CHAPTER 11

E Trapezoid

My Notes

Definition 11.1

A trapezoid is a quadrilateral which has at least one pair of parallel sides.

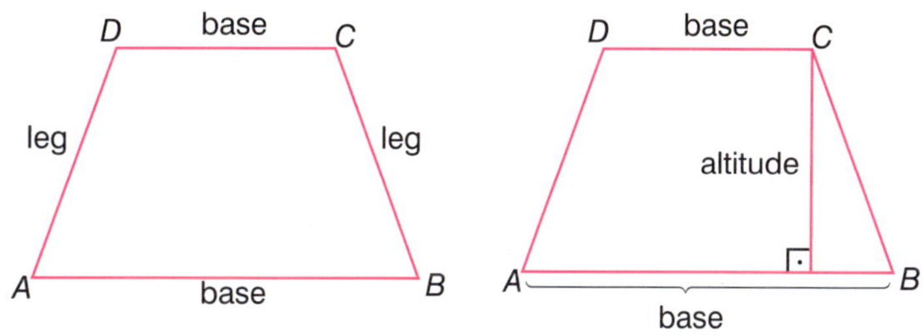

Theorem 11.1

In a trapezoid, two interior adjacent angles are supplementary each other.

$\alpha + \beta = 180°$

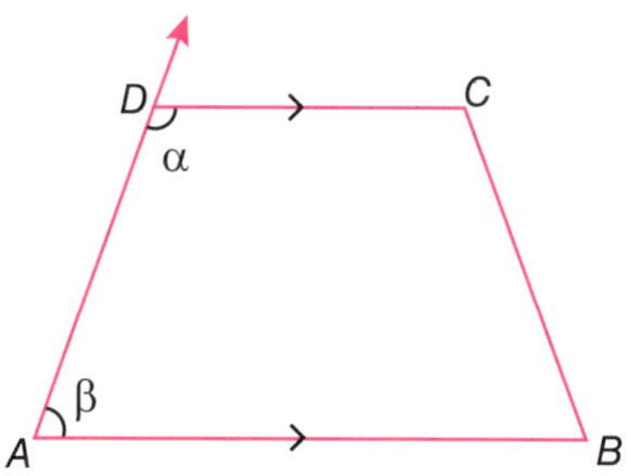

Theorem 11.2

The median of a trapezoid is the line segment that connects the midpoint of the legs.

$$\text{Median} = m = \frac{a+c}{2}$$

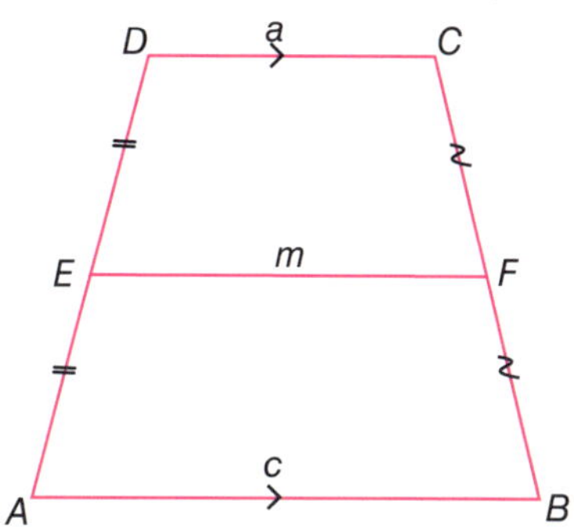

Proof 11.2

We create a parallelogram *DALC* then the mid segment $EK = a$. Since KF is the mid segment of the triangle *CLB*,

$KF = m - a = \dfrac{c-a}{2}$

From there, $EF = m = \dfrac{c-a}{2} + a$.

Thus, $m = \dfrac{c+a}{2}$.

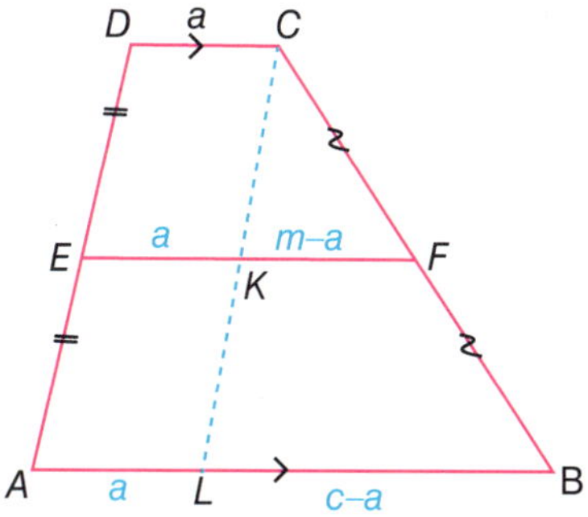

Example 11.1

Find the lengths of AB and DC.

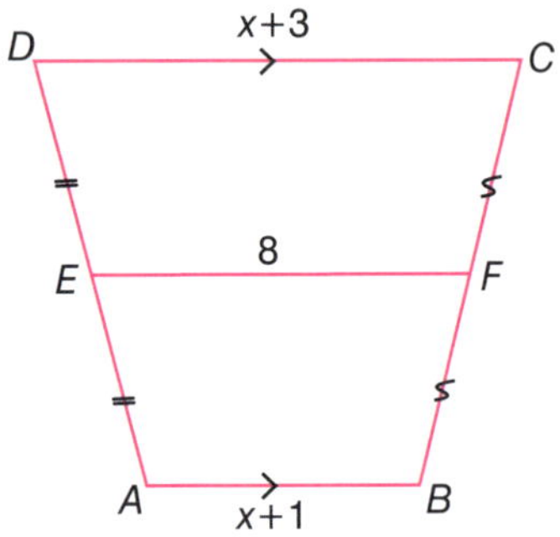

Solution 11.1

The median is $8 = \frac{x+1+x+3}{2} \Rightarrow 8 = \frac{2x+4}{2}$

$$8 = x+2$$

$$x = 6$$

$AB = x+1 = 6+1 = 7$ and $DC = x+3 = 6+3 = 9$.

Example 11.2

$ABCD$ is a trapezoid with $DE = EK = KA$ and $CF = FP = PB$.

Given that $EF = 6$ and $KP = 9$ then find the value of $AB \cdot DC$.

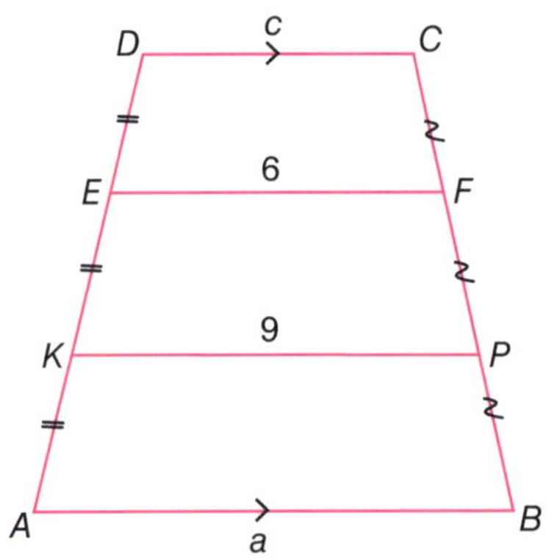

Solution 11.2

From the median formula on trapezoid $DKPC$.

$$6 = \frac{9+c}{2} \Rightarrow c = 3,$$

From the median formula on trapezoid $EABF$.

$$9 = \frac{a+6}{2} \Rightarrow a+6 = 18 \Rightarrow a = 18-6 = 12.$$

Therefore, $AB \cdot DC = 12 \cdot 3 = 36$.

Try This

In a trapezoid $ABCD$, $DC // AB$ and $AD = BC$, point E on DC and point H on AB.

If $EH \perp AB$, $AD = 17$, $EH = 8$, and $CE = 6$, what is HB?

My Notes

Try This

ABCD is a trapezoid, $AB // CD$,

$m\angle B = 48°$,

$m\angle D = 138°$,

$AB = 2DC = 4a$,

$AE = EB$, and

$DF = FC$.

Find the length of *EF* in terms of *a*.

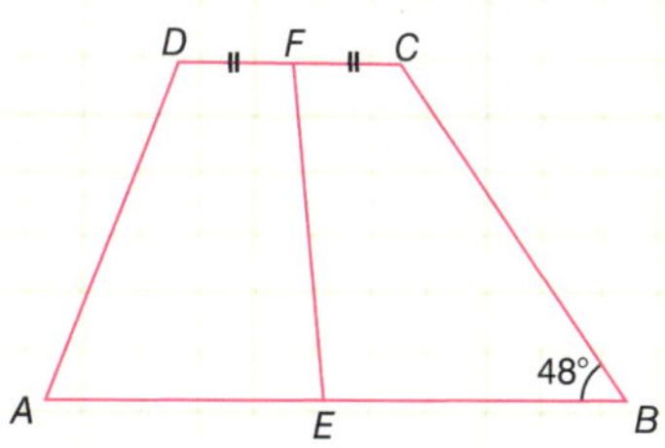

My Notes

Theorem 11.3

ABCD is a trapezoid with points *E* and *F* are the midpoint of two sides. Points *K* and *L* are on the diagonals on the mid segment on *EF*.

$$KL = \frac{a-c}{2}$$

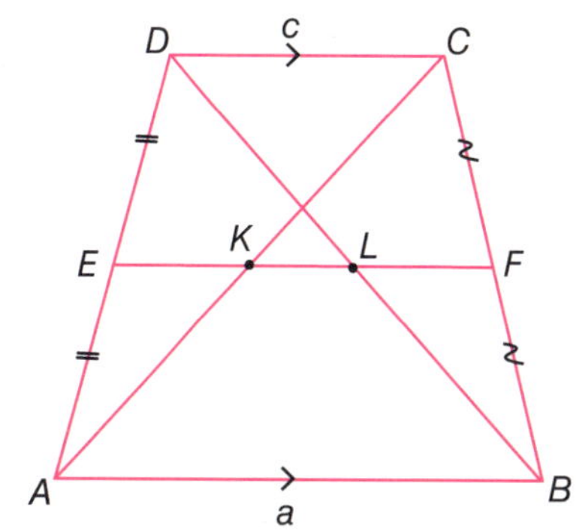

Proof 11.3

From $\triangle ADC$, $EK = \frac{c}{2}$.

From $\triangle BDC$, $LF = \frac{c}{2}$.

From the median formula, $EF = \frac{a+c}{2}$.

$KL = EF-(EK+LF) = \frac{a+c}{2} - \frac{c}{2} - \frac{c}{2} = \frac{a-c}{2}$.

Example 11.3

ABCD is a trapezoid and *EF* is the median of the trapezoid.

Given that $KL = 1$ cm and

$AB = 6$ cm.

What is *DC*?

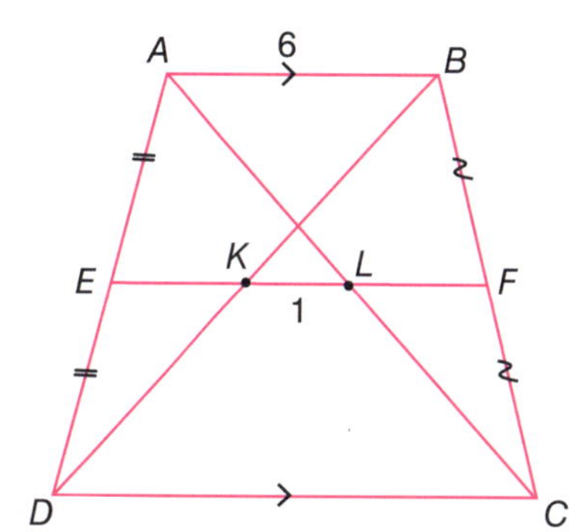

Solution 11.3

By applying Theorem 11.3

We get $1 = \frac{a-6}{2} \Rightarrow 2 = a-6$.

So, $a = 8$.

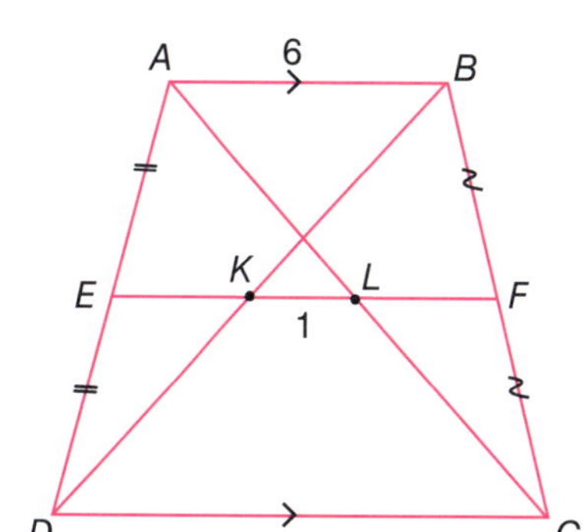

Example 11.4

ABCD is a trapezoid with $AB // CD$,

$m\angle D = 130°$, $m\angle B = 65°$, $AD = 12$, and $DC = 6$.

What is AB?

Solution 11.4

We create a parallelogram. $AC'CD$ and we notice that $\Delta C'CB$ is an isosceles triangle since $m\angle C'CB = m\angle C'BC$.

Thus, $C'C = C'B = 12$.

Therefore, $AB = 6+12 = 18$.

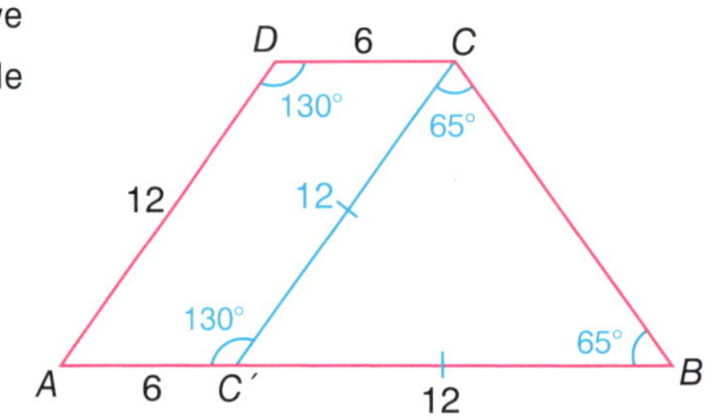

Example 11.5

ABCD is a trapezoid with

$BC = 24$ cm,

$DC = 16$ cm,

$AD = 4$ cm, and

$AB = 12$ cm.

What is the altitude of this trapezoid?

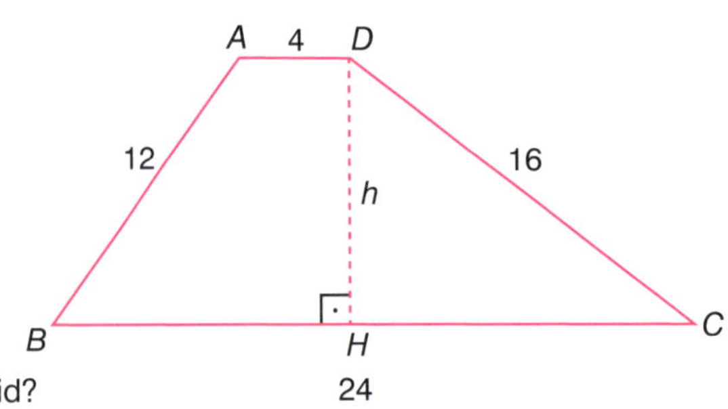

Solution 11.5

We draw a line segment DD' parallel to AB. We observe that $\Delta D'DC$ is a right triangle since its sides from the Pythagorean triple a pathogen triples (12,16,20).

From a triangle area formula

$$\frac{12 \cdot 16}{2} = \frac{h \cdot 20}{2}$$

$$h = \frac{12 \cdot 16}{20} = 9.6.$$

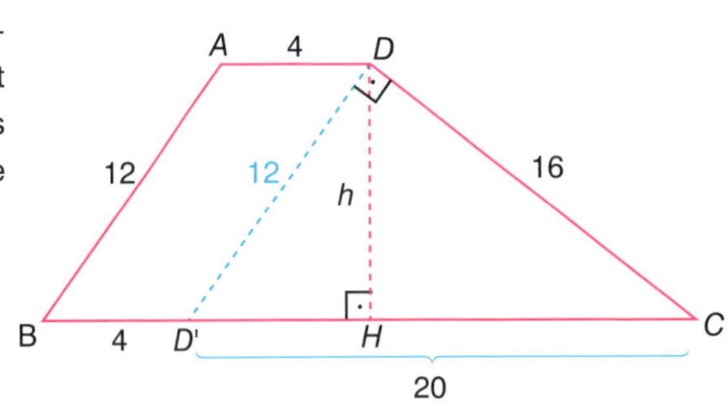

Try This

In trapezoid $ABCD$, $AB // DC$ and $m\angle DAB = 2m\angle ABC$.

Given that $AD = DC = 4$ and $AB = 9$, what is BC?

My Notes

Prove this

If $ABCD$ is a right trapezoid with perpendicular diagonals.

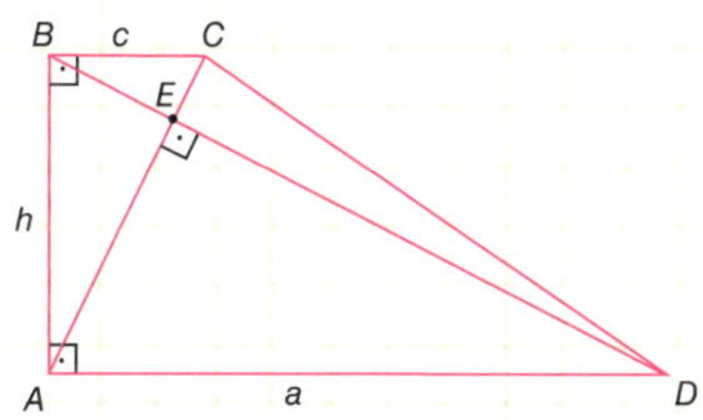

Prove that $h^2 = a \cdot c$.

My Notes

Example 11.6

$ABCD$ is a trapezoid with,

$BC = \sqrt{2}\,AD$,

$m\angle ADC = 135°$.

What is $m\angle CBA = \alpha$?

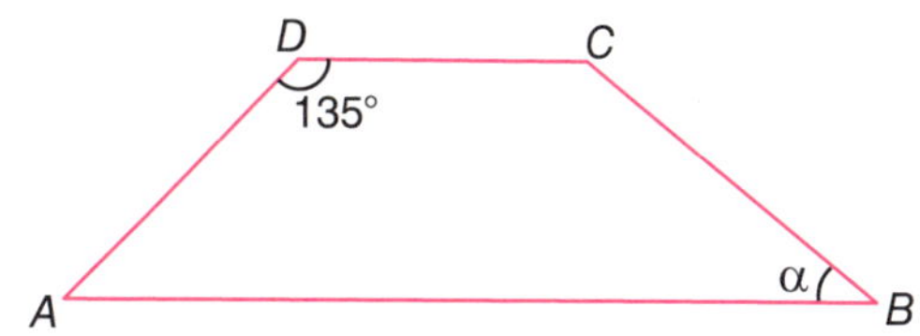

Solution 11.6

We draw a line segment CC' parallel to AD and then we draw a perpendicular food from C to AB. We can see that the length CH is half of the length CB. So, we conclude that $\angle B$ equals 30° (30°–60°–90° triangle).

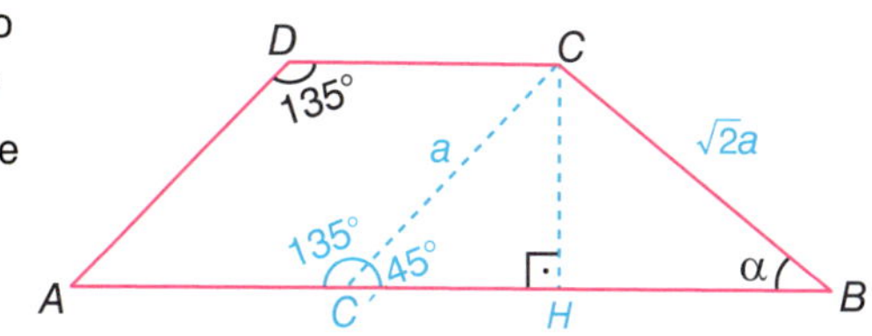

Let $C'C$ be a then $BC = a\sqrt{2}$.
From $\Delta C'HC$ (45°–45°–90°)

$C'H = CH = \dfrac{a\sqrt{2}}{2}$.

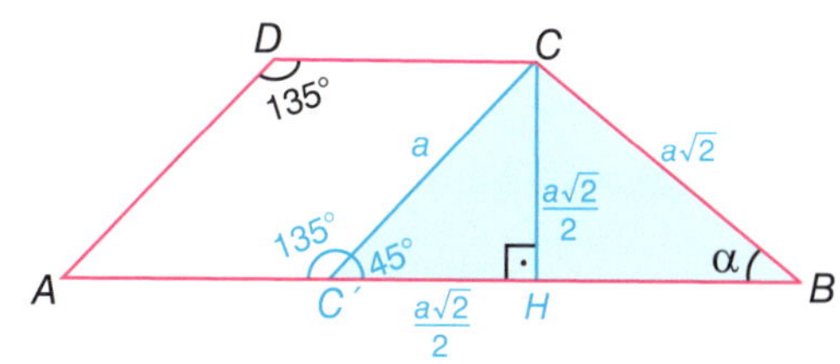

Example 11.7

$ABCD$ is a trapezoid with,

$DE = AE$,

$AB = 20$ cm,

$BE = 15$ cm, and

$BC = 18$ cm.

Find DC.

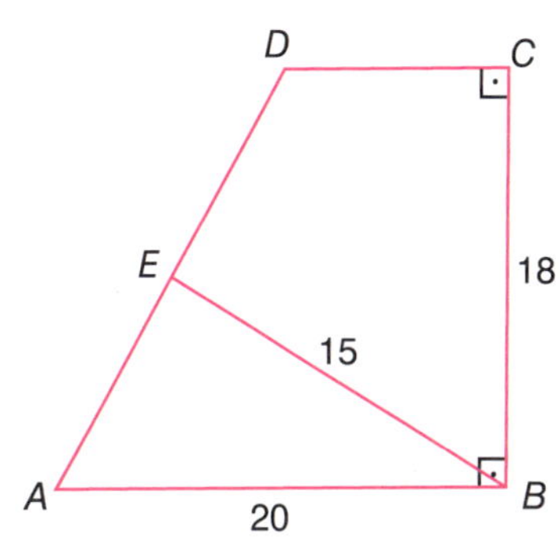

Solution 11.7

From ΔEFB, $EF = 12$, EF is also the median of the trapezoid.

$$12 = \frac{20+a}{2} \Rightarrow 20+a = 24$$
$$a = 4.$$

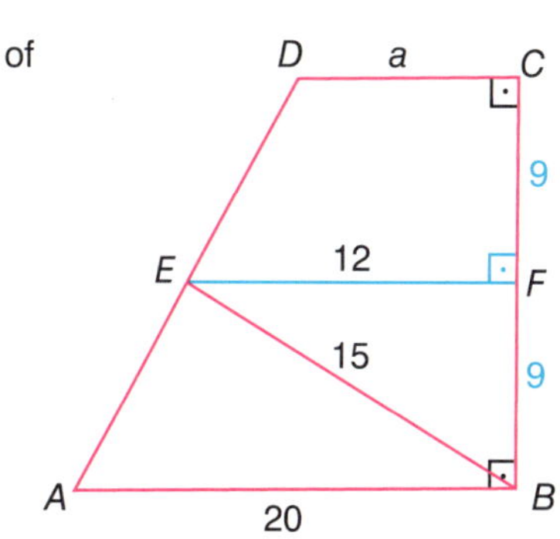

Theorem 11.4

If diagonals of an isosceles trapezoid are perpendicular to each other, then the height of trapezoid

$$CH = \frac{AB+DC}{2}.$$

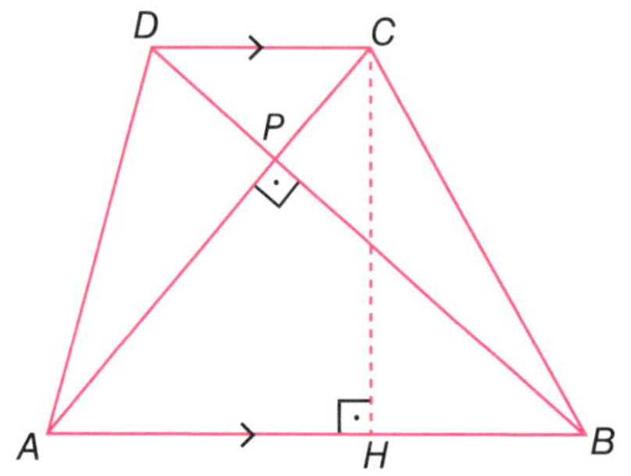

Proof 11.4

Let $DC = a$ and $AB = c$. We create a parallelogram $DBC'C$

$$m\angle ACC' = 90°$$

since our trapezoid is isosceles, so, the diagonals are equal each other. So, $AC' = a+c$.

Therefore, $CH = \frac{a+c}{2}$.

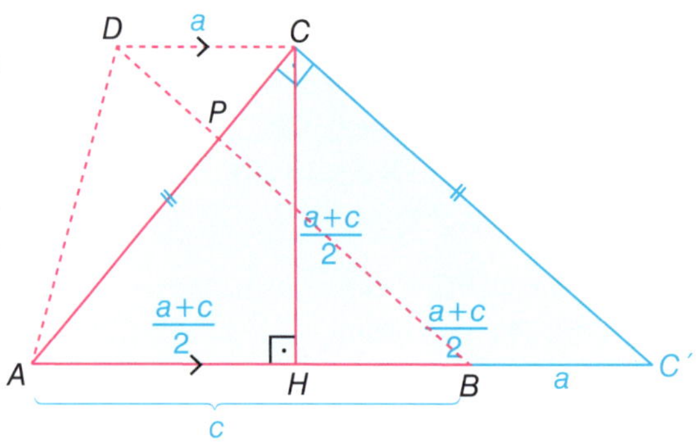

Theorem 11.5

If diagonals of a right trapezoid are perpendicular to each other then the height is equal to the square root of the product of the lengths of the bases.

$$d = \sqrt{a \cdot c}$$

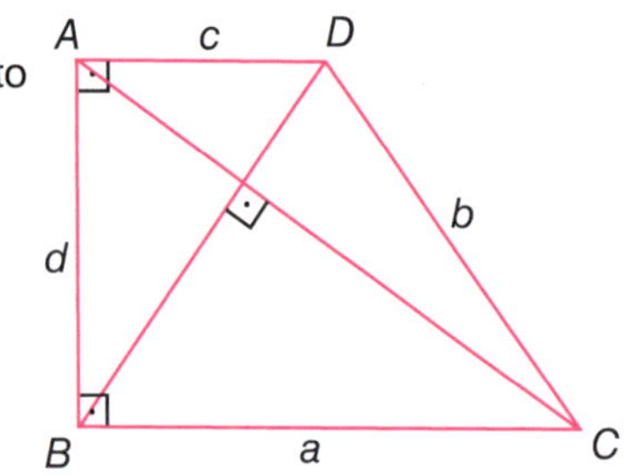

Proof 11.5

We used a similar idea used in the proof 11.4.

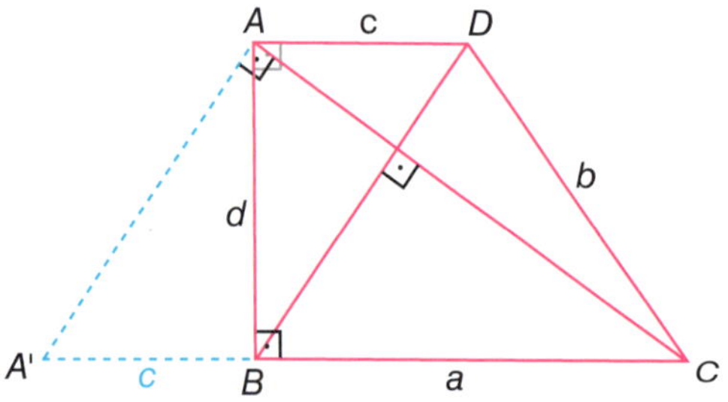

$\Delta A'BA$ and ΔABC are similar triangles.

$$\frac{d}{c} = \frac{d}{c} \Rightarrow d^2 = a \cdot c.$$

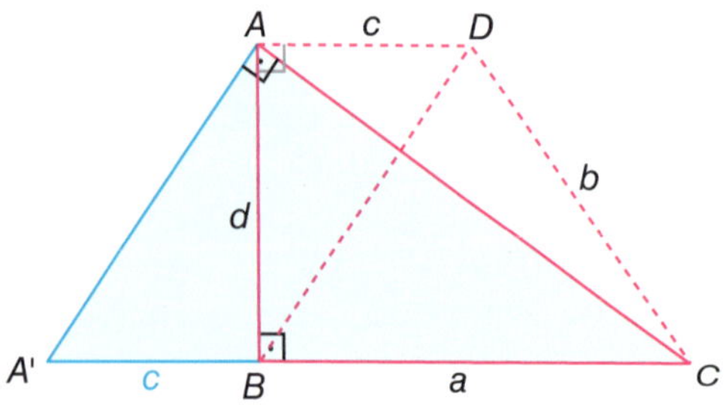

Try This

In the figure, $ABCD$ is a trapezoid with $KN // AB // DC$.
Given that $AB = a$, $DC = c$ than prove the following:

I) $KO = ON$.

II) $KN = \frac{2ac}{a+c}$.

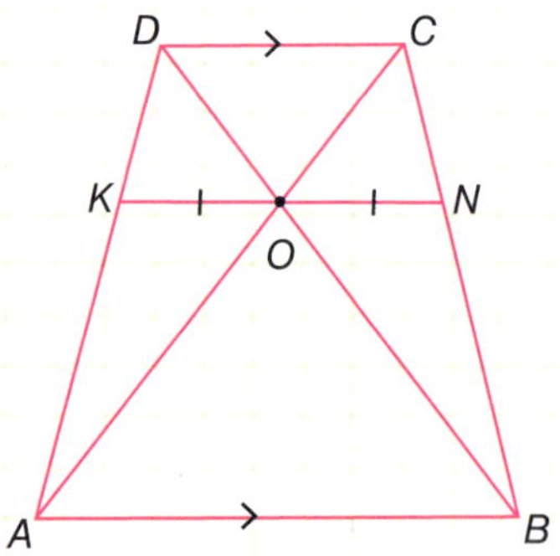

My Notes

Try This

ABCD is a right trapezoid with perpendicular diagonals.

Given that $AD = 2$ cm and $BC = 6$ cm. What is DC?

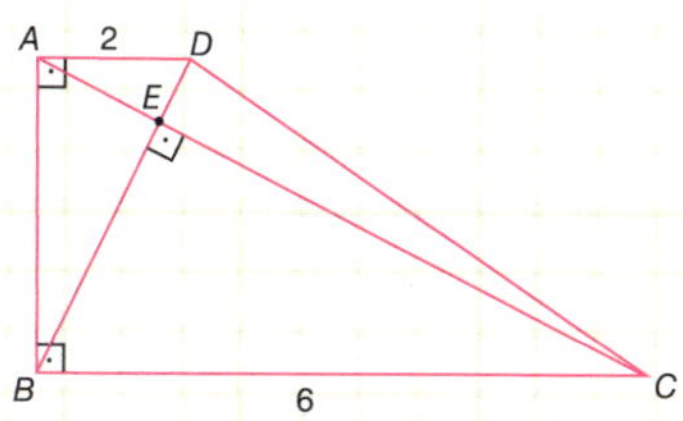

My Notes

Example 11.8

ABCD is a right trapezoid with perpendicular diagonals.

Given that $AD = 1$ cm and

$BC = 4$ cm.

What is DC?

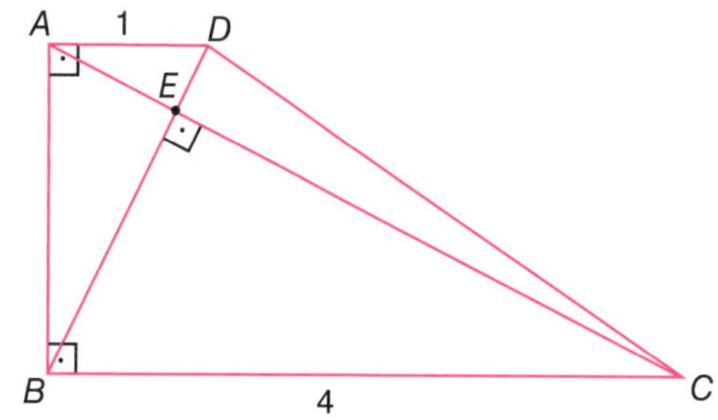

Solution 11.8

From Theorem 11. 5

$AB^2 = 1 \cdot 4 \Rightarrow AB = 2$ then $HD = 2$.

$x^2 = 2^2 + 3^2$ (Pythagorean theorem)

$x = \sqrt{13}$.

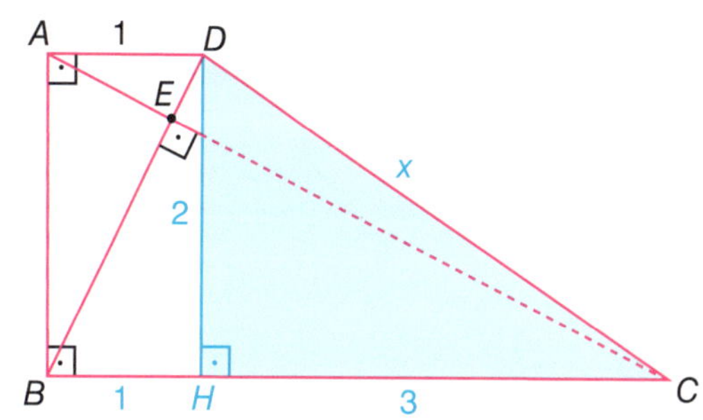

Example 11.9

In trapezoid *ABCD*, $\overline{AB} \parallel \overline{DC}$ and $m\angle DAB = 2m\angle ABC$.

Given that $AD = DC = 2$ and $AB = 6$.

What is BC?

Solution 11.9

Draw $\overline{AC}$ and let $m\angle DCA = \alpha$.

As $AD = DC$ and $\overline{AB} \parallel \overline{DC}$,

we have

$m\angle DCA = m\angle DAC = m\angle CAB = \alpha$.

As $m\angle DAB = 2m\angle ABC$ we also have

$m\angle DBA = \alpha$

Therefore, $\Delta ADC \sim \Delta ACB$, and so

$$\frac{x}{6} = \frac{2}{x}.$$

Hence, $x = 2\sqrt{3}$.

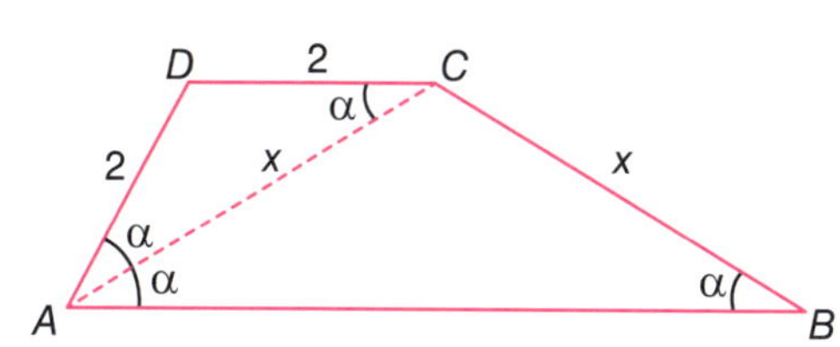

Definition 11.2

A Kite is a quadrilateral with two pairs of consecutive congruent sides and two non-congruent opposite sides.

Diagonals are perpendicular and one pair of opposite angles ($\angle BAD$ and $\angle BCD$) are congruent.

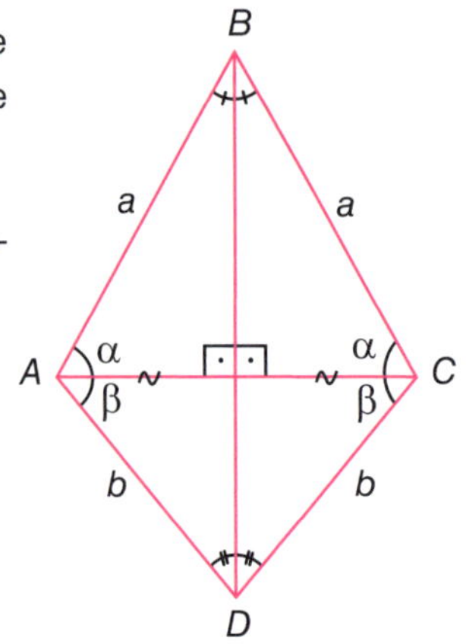

F Kite

Catriona Agg @Cshearer41
Geometry puzzle

The four sections in this square have the same area.

What is the missing length?

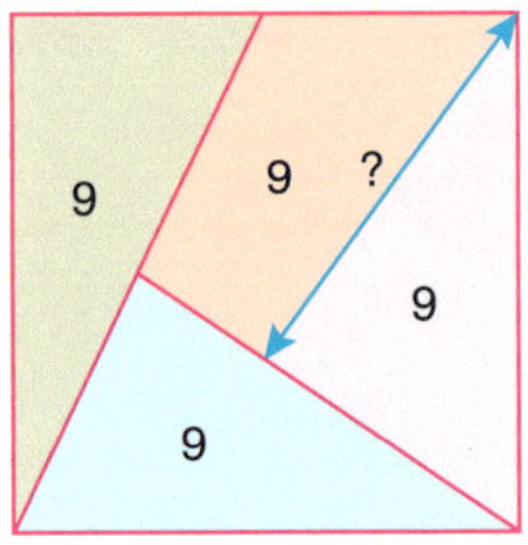

Example 11.9

In the figure, $ABCE$ is a kite with

$AE = EC$, and $AB = BC$.

Given that $AE = 6$ cm,

$ED = 9$ cm, and

$CD = 5$.

What is BC?.

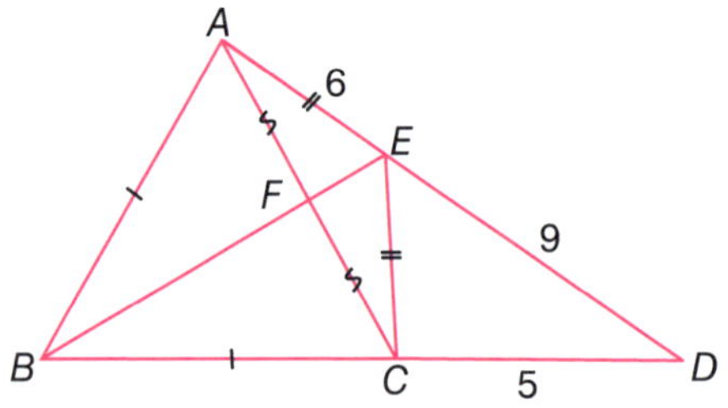

My Notes

Solution 11.9

BE bisects angle B since $ABCE$ is a kite.

From the interior angle bisector theorem

$$\frac{a}{6} = \frac{5+a}{9} \Rightarrow a = 5.$$

So, $BC = 5$.

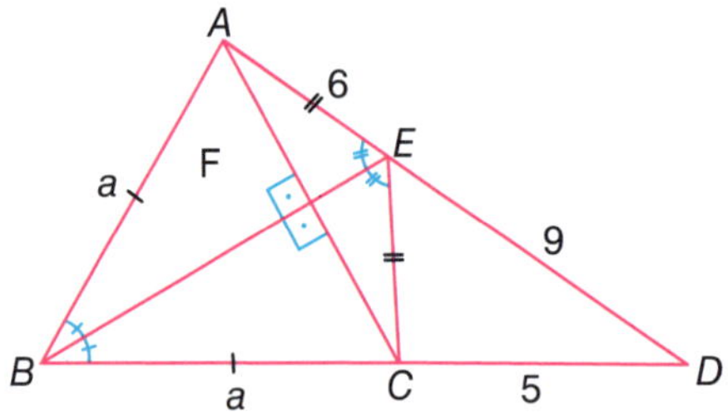

Catriona Agg @Cshearer41
Geometry puzzle

The area of the three square are given.

What the area of the shaded region?

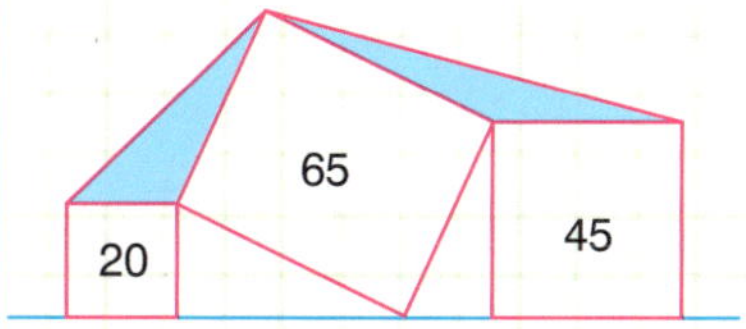

My Notes

Example 11.10

In the figure, *ABC* is a triangle *AFED* is a kite.

If $DB = 7$,

$BE = 8$,

$EC = 6$, and

$FC = 4$.

What is AC?

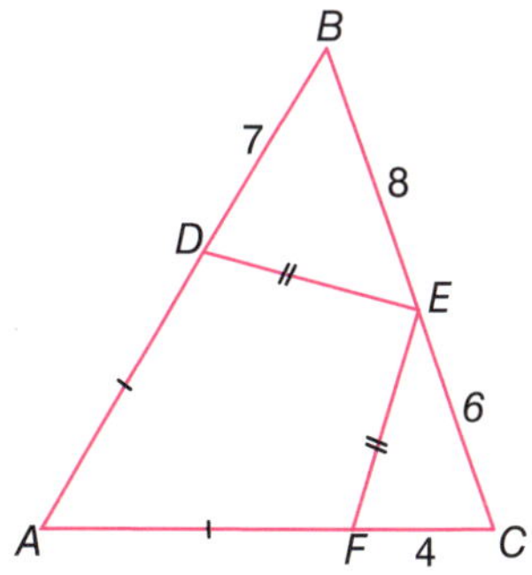

Solution 11.10

AE bisects $\angle A$ since *ADEF* is a kite.
By the angle bisector theorem

$$\frac{a+7}{8} = \frac{4+a}{6} \Rightarrow a = 5.$$

Therefore, $AC = 5+4 = 9$.

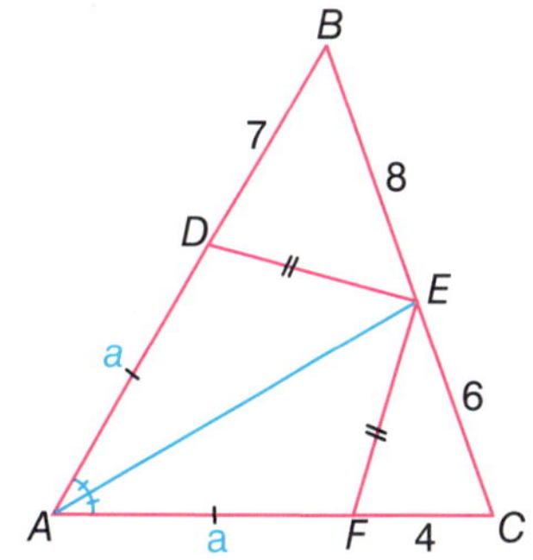

Example 11.11

In the figure, *ABCD* is a trapezoid, with $AB//CD$ and *ABCE* is a kite, with $EA = AB$.

If $CD = 17$,

$DE = 7$,

then find length of *AB*.

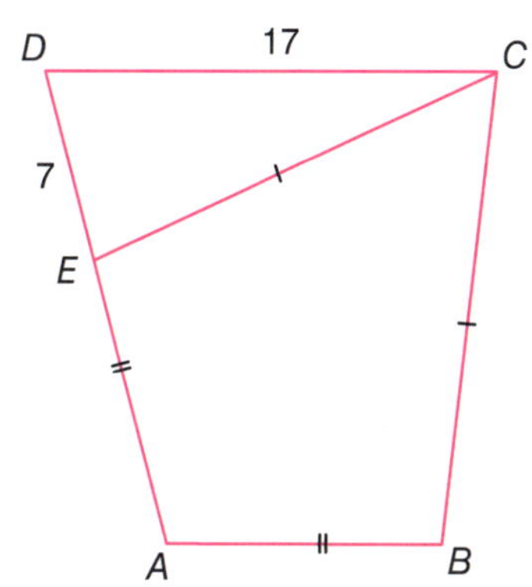

Solution 11.11

$m\angle DAC = m\angle DCA$ then $m\angle DCA$ is also α. $(DC//AB)$.

So, $CD = AD = 17$

Therefore, $7 + EA = 17$

$EA = AB = 10$.

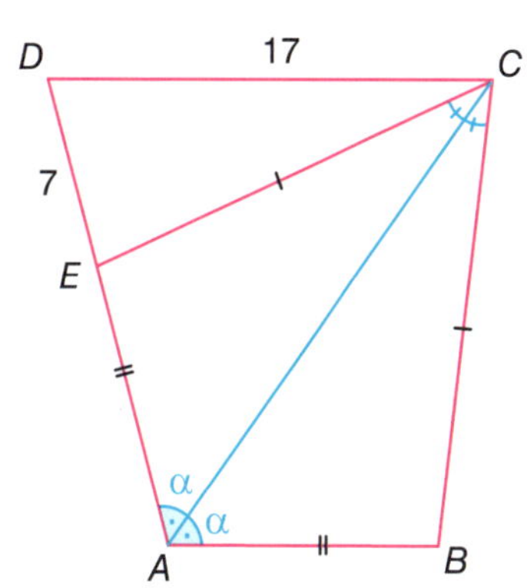

Example 11.12

In the figure, *ABCD* is a kite.

$m\angle BAD = 60°$,

$m\angle BCD = 120°$.

What is $\frac{x}{y}$?

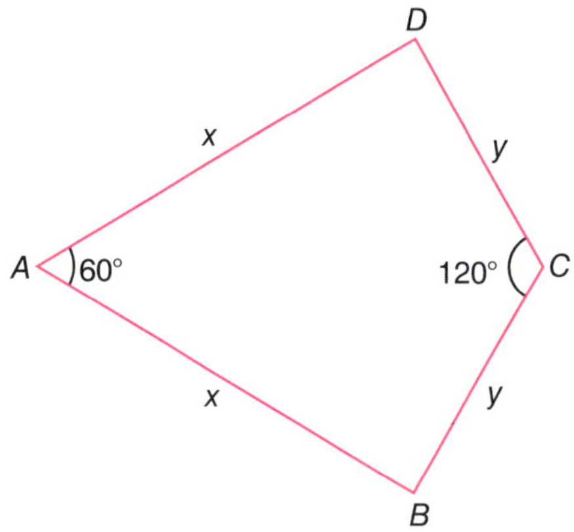

Solution 11.12

Diagonals AC and DB are perpendicular to each other.

$\frac{x}{2} = \frac{y}{2}\sqrt{3} \Rightarrow \frac{x}{y} = \sqrt{3}$.

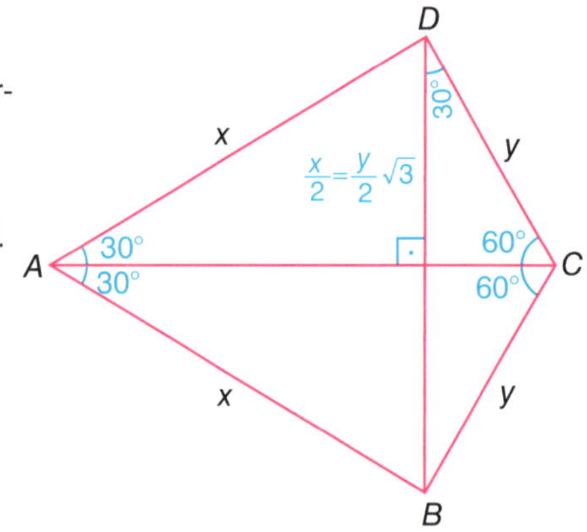

Example 11.13

In the figure, *ABCD* is a kite with

$AB = AD = 3$ cm,

$BD = DC = 9$ cm, and

$m\angle ADC = 120°$.

What is the area of the kite?

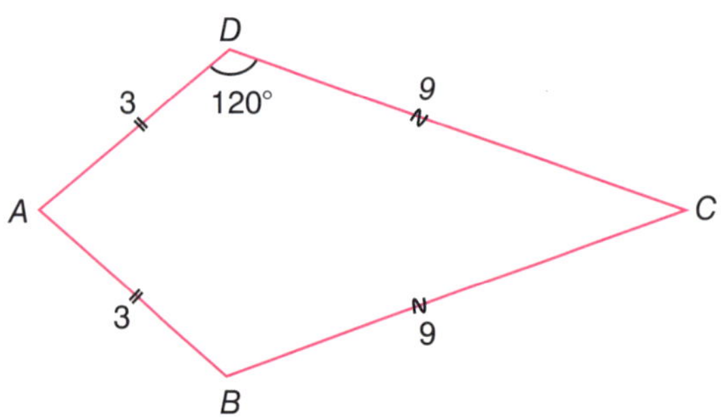

Solution 11.13

Diagonals *AC* makes two triangles.

Area $\Delta ADC = \frac{1}{2} 3 \cdot 9 \cdot \sin 120° = \frac{1}{2} 3 \cdot 9 \cdot \sin 60°$

Thus, $\frac{1}{2} \cdot 27 \cdot \frac{\sqrt{3}}{2} = \frac{27\sqrt{3}}{4}$

The area of the kite is $2 \cdot \frac{27\sqrt{3}}{4} = \frac{27\sqrt{3}}{2}$.

Catriona Agg @Cshearer41
Geometry puzzle

In this right-angled trapezium, the pink area is 6 more than the blue area.

What is the length of the sloping side?

My Notes

1. 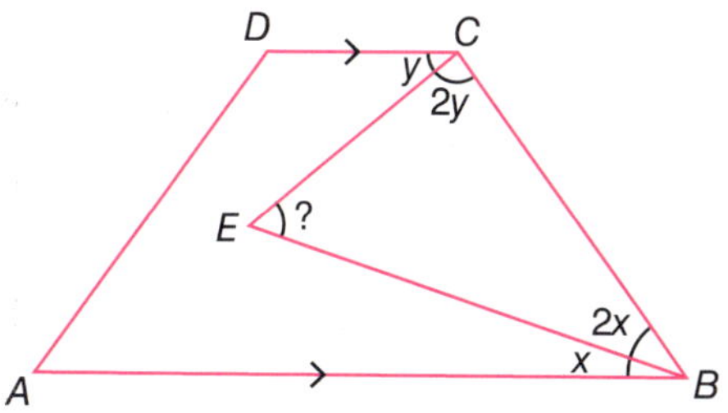

In the figure, $ABCD$ is a trapezoid. Point E lies inside trapezoid DC//AB.

$m\angle EBC = 2\,m\angle EBA = 2x$ and

$m\angle ECB = 2\,m\angle ECD = 2y$.

Find $m\angle CEB$.

2. $ABCD$ is a trapezoid with $AB//DC$. Points E and F on the midpoint of AD and BC, respectively.

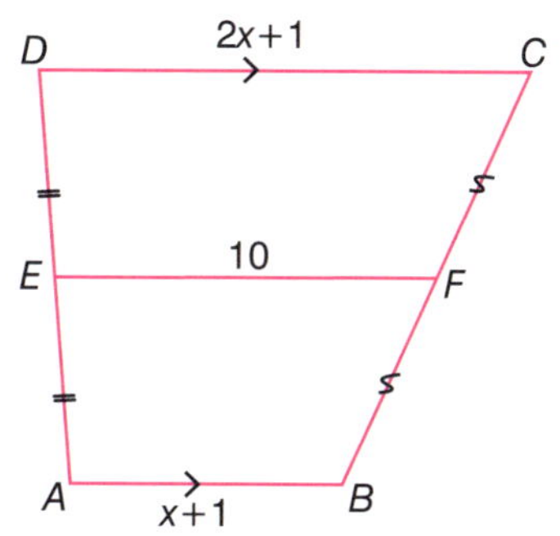

Given that

$AB = x+1$,

$DC = 2x+1$, and

$EF = 10$,

what is the length of AB?

3. $ABCD$ is a isosceles trapezoid with $AB//DC$.

Given $AB = 22$ cm, $DC = 12$ cm, and $BC = 13$ cm, find the height of the trapezoid.

4. In a kite $ABCD$, $AB = AD$, $CB = CD$ and E is the intersection point of the diagonals.

If $m\angle EAD = 30°$ and $BE = 3$ cm, find the length of AD.

5. The bases of an isosceles trapezoid measure 8 cm and 12 cm respectively.

If the diagonal of the trapezoid measures 14 cm, find the length of its sides.

6. $ABCD$ is a trapezoid with $DC//AB$. Point E midpoint of AD and $EC \perp BC$.

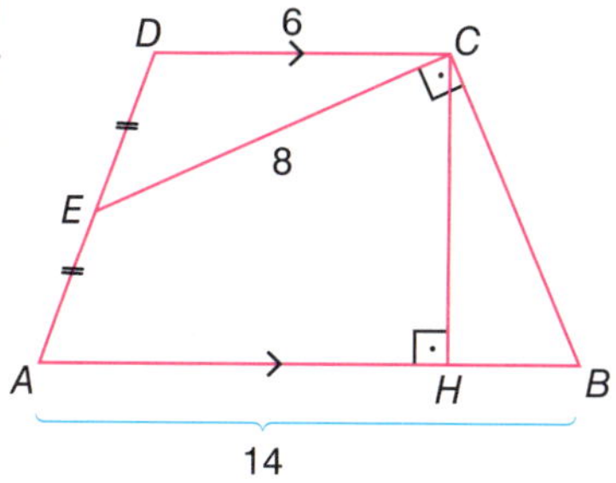

Given that

$DC = 6$,

$EC = 8$, and

$AB = 14$,

what is the value of HC?

7. In the figure, $ABCD$ is a right trapezoid with $AB//DC$ and $AC \perp BC$.

$AB = 13$ cm and

$DC = 4$ cm.

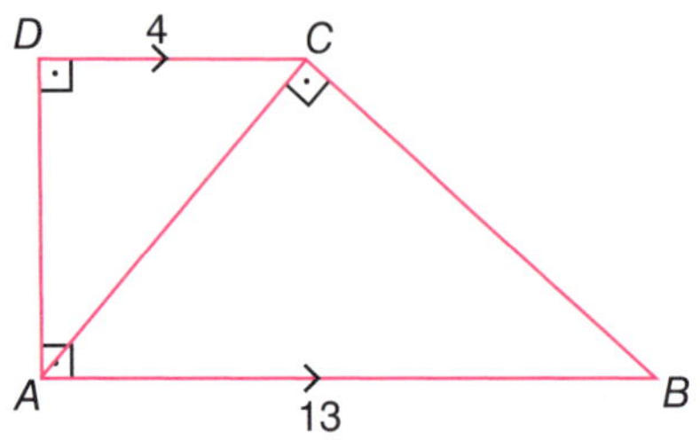

What is the height of trapezoid?

8. In the figure, $ABCD$ is a trapezoid. $AB//DC$, and AE and DE are the bisectors of A and D respectively.

If $CH \perp AB$, $AE = 4$ cm and $DE = 3$ cm, find the height CH.

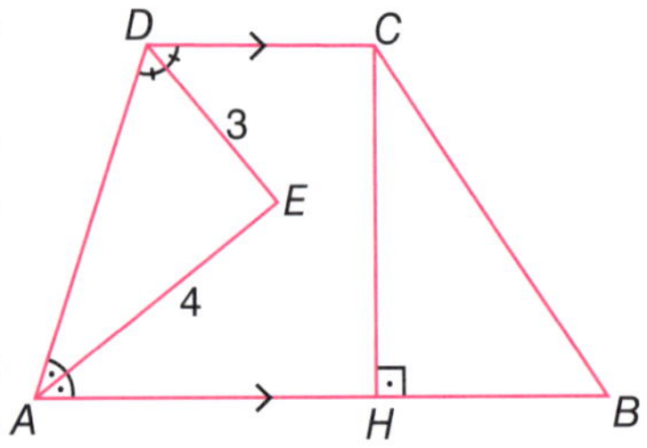

9. $ABCD$ is a kite and AE is the angle bisector of $\angle DAC$.

Given that
$AB = AD = 8$,
$AC = 12$, and
$m\angle BAC = m\angle AED$.
What is $\frac{BC}{EC}$?

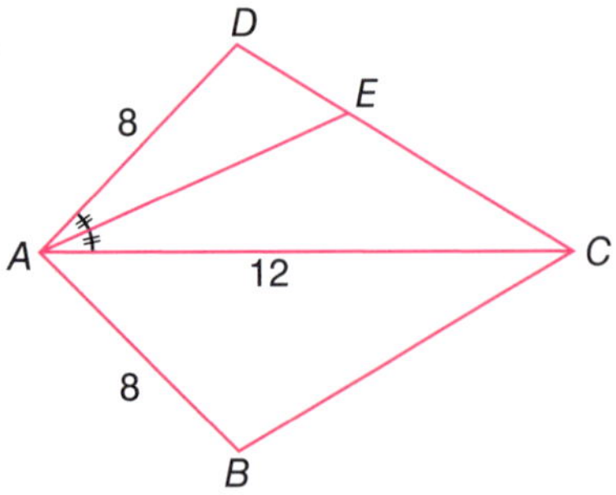

10. In the figure, $ABCE$ is a kite with $AE = EC$.

Given $AE = 8$ cm,

$ED = 10$ cm, and

$CD = 4$ cm.

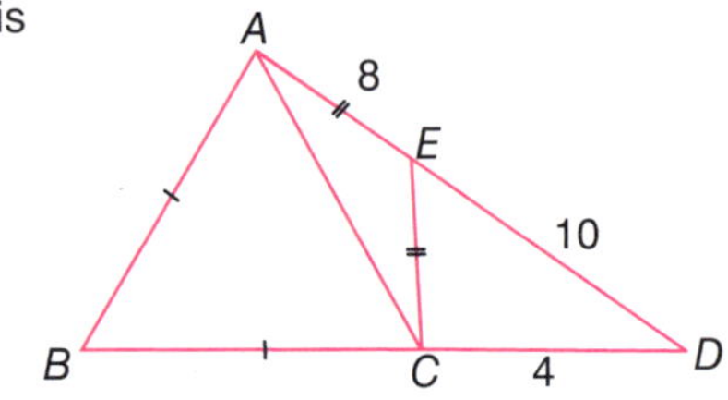

What is the value of AB?

11.

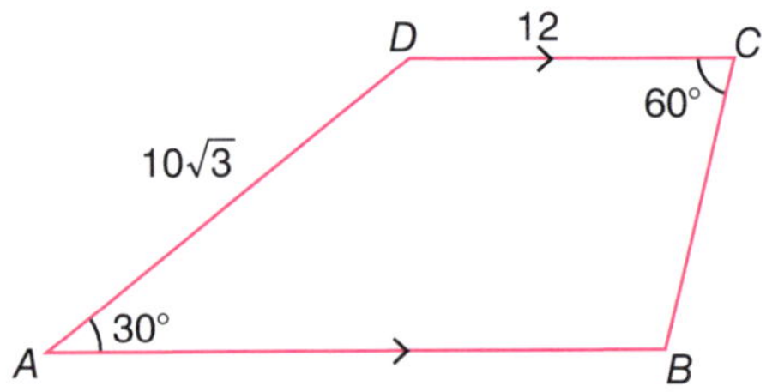

In the figure, $ABCD$ is a trapezoid with $AB//DC$.

$AD = 10\sqrt{3}$ cm, $DC = 12$ cm, $m\angle A = 30°$, and $m\angle C = 60°$.

What is the length of AB?

12. $ABCD$ is a trapezoid with $AB//DC$.

$AD = 10$ cm, $DC = 5$ cm, $m\angle D = 130°$, and $m\angle B = 65°$.

What is the length of AB?

1.

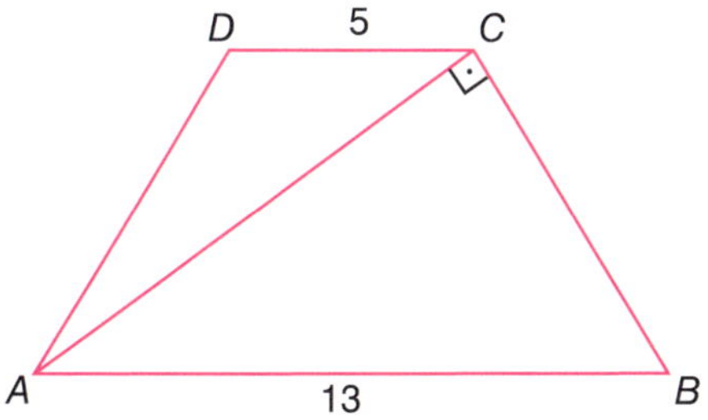

In the figure, $ABCD$ is a isosceles trapezoid with $AB//DC$ and $BC = DC$.

Given that AB = 13 cm and DC = 5, find height of the trapezoid.

2.

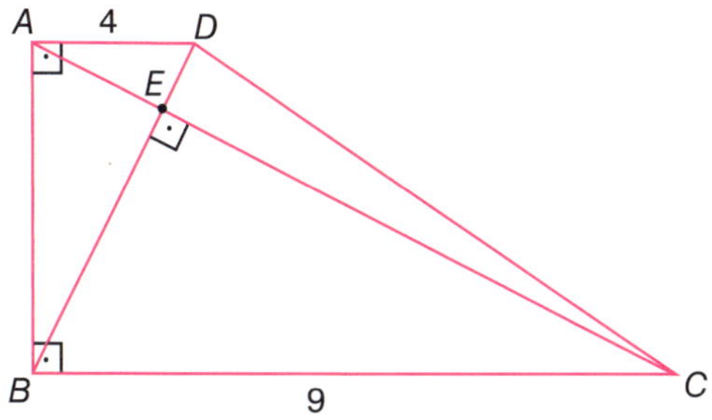

$ABCD$ is a right trapezoid with perpendicular diagonals.

Given that $AD = 4$ cm and, $BC = 9$ cm. What is DC?

3. $ABCD$ is a trapezoid with,

$DE = AE$,

$AB = 20$ cm,

$BE = 17$ cm, and

$BC = 16$ cm.

What is DC?

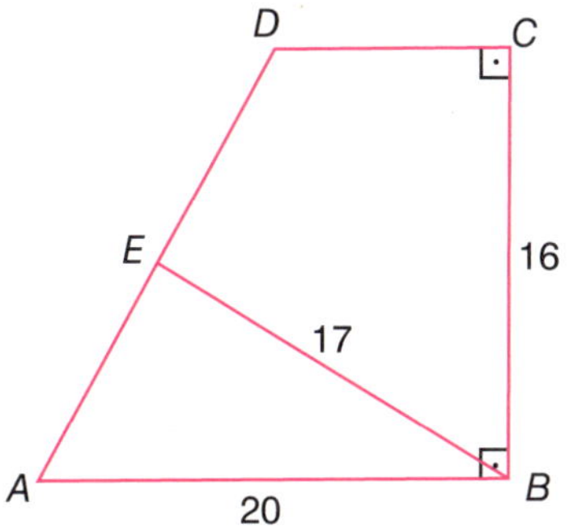

4. $ABCD$ is a trapezoid with $AB//CD$,

$m\angle C = 130°$, $m\angle A = 65°$, $CB = 12$, and $DC = 6$ cm.

Find the length of AB.

5. In the figure, $ABCD$ is a trapezoid, with $AB//CD$ and $ABCE$ is a kite, with $EA = AB$.

If $CD = 17$ and $DE = 7$, find the length of AB.

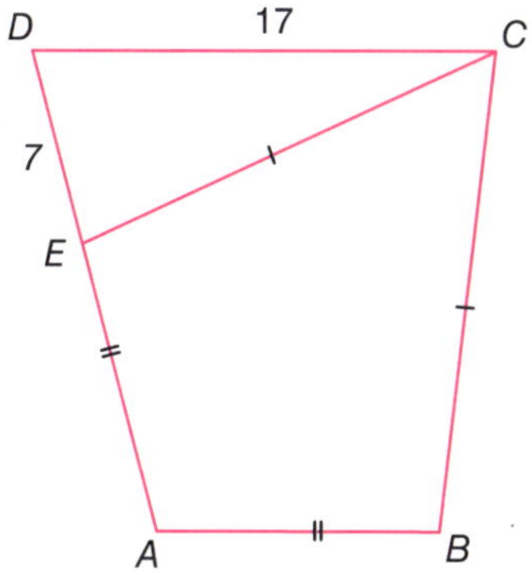

6.

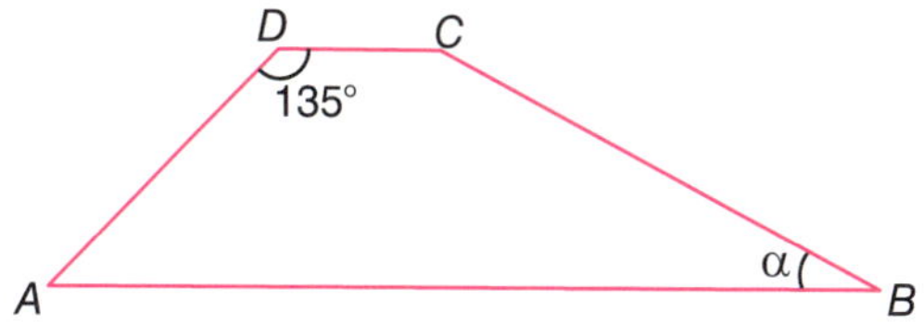

$ABCD$ is a trapezoid with, $BC = 8\sqrt{2}$, $AD = 8$, and $m\angle ADC = 135°$

What is $m\angle CBA = \alpha$?

7. In the figure, ABC is a triangle $AFED$ is a kite.

If $DB = 7$,

$BE = 8$,

$EC = 6$, and

$FC = 4$.

What is AC?

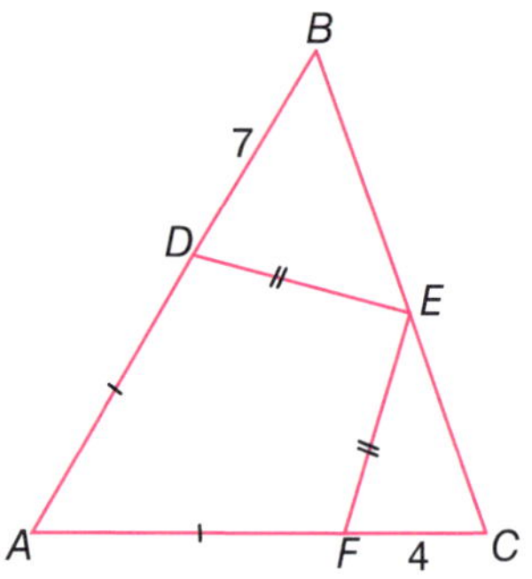

8. In the figure, $ABCD$ is a trapezoid and $EFCD$ is a kite.

Given that $DC = x$, and $AF = y$, find $\dfrac{DE}{EA}$.

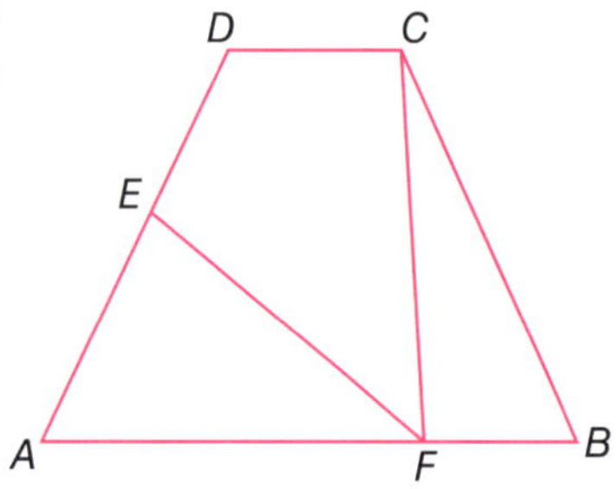

9. If x and y are lengths of bases of a trapezoid.

Find the length of the segment that connects the midpoints of its diagonals.

10. $ABCD$ is a trapezoid such that $AB//CD$.

$m\angle DAB = 2m\angle ABC$,

$AD = DC = 2$, and $AB = 3$.

Find BC.

11. In trapezoid $ABCD$ with bases $AB = 46$ and $CD = 29$.

Given that $AD = 15$ and $CB = 8$. What is the height of the trapezoid?

12. Let $ABCD$ is a trapezoid, P and S are the midpoints of AD and BC such that $AD//BC$.

If $AC = 12$, $BD = 16$, and $PS = 10$, find the area of $ABCD$.

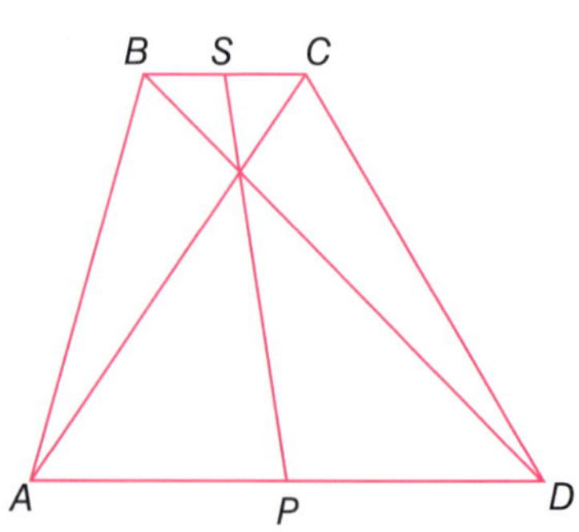

1. (2005 AMC 8 #19)

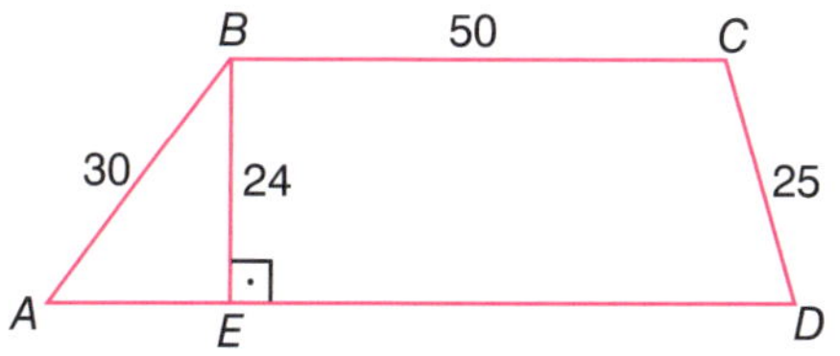

What is the perimeter of trapezoid $ABCD$?

A) 180 B) 188 C) 196 D) 200 E) 204

2. (1999 AMC 8 # 14)

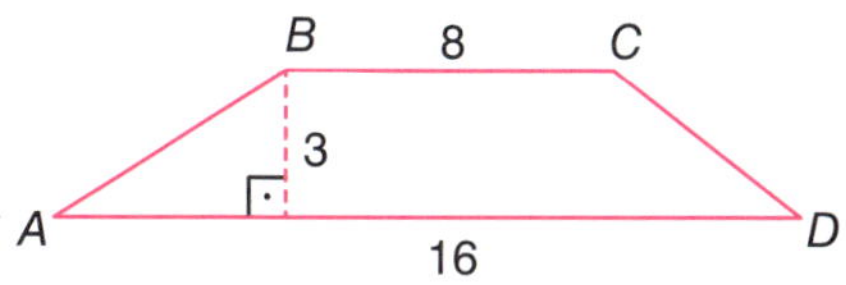

In trapezoid $ABCD$, the sides AB and CD are equal. The perimeter of $ABCD$ is

A) 27 B) 30 C) 32 D) 34 E) 48

3. (2011 AMC 8 #20)

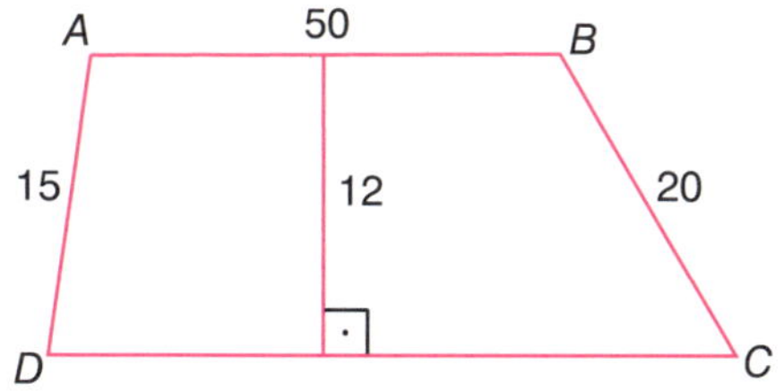

Quadrilateral $ABCD$ is a trapezoid, $AD = 15$, $AB = 50$, $BC = 20$, and the altitude is 12.

What is the area of the trapezoid?

A) 600 B) 650 C) 700 D) 750 E) 800

4. (2009 AMC 12B #16) Trapezoid $ABCD$ has $AD//BC$, $BD = 1$, $\angle DBA = 23°$, and $\angle BDC = 46°$.

The ratio $\dfrac{BC}{AD}$ is $\dfrac{9}{5}$.

What is CD?

A) $\dfrac{7}{9}$ B) $\dfrac{4}{5}$ C) $\dfrac{13}{15}$ D) $\dfrac{8}{9}$ E) $\dfrac{14}{15}$

5. (2014 AMC 10B #21) Trapezoid $ABCD$ has parallel sides $\overline{AB}$ of length 33 and $\overline{CD}$ of length 21. The other two sides are of lengths 10 and 14. The angles A and B are acute.

What is the length of the shorter diagonal of $ABCD$?

A) $10\sqrt{6}$ B) 25 C) $8\sqrt{10}$ D) $18\sqrt{2}$ E) 26

6. (MathCON 2019 GR 11 #36) Suppose $ABCD$ is a trapezoid such that $AB//CD$.

$m\angle C + m\angle D = 270°$,

$AD = 8$, $DC = 11$, and $BC = 15$.

Find the value of AB.

A) 26 B) 28 C) 30 D) 32 E) 34

7. **(2005 AMC 10B P#23)** In trapezoid $ABCD$ we have $\overline{AB}$ parallel to $\overline{CD}$, E as the midpoint of $\overline{BC}$, and F as the midpoint of DA. The area of $ABEF$ is twice the area of $FECD$.

What is $\frac{AB}{DC}$?

A) 2 B) 3 C) 5 D) 6 E) 8

8. **(2002 AMC 10A P #25)** In trapezoid $ABCD$ with bases AB and CD, we have

$AB = 52, BC = 12,$

$CD = 39$, and $DA = 5$.

The area of $ABCD$ is

A) 182 B) 195 C) 210 D) 234 E) 260

9. **(2013 AMC 12A #11)**

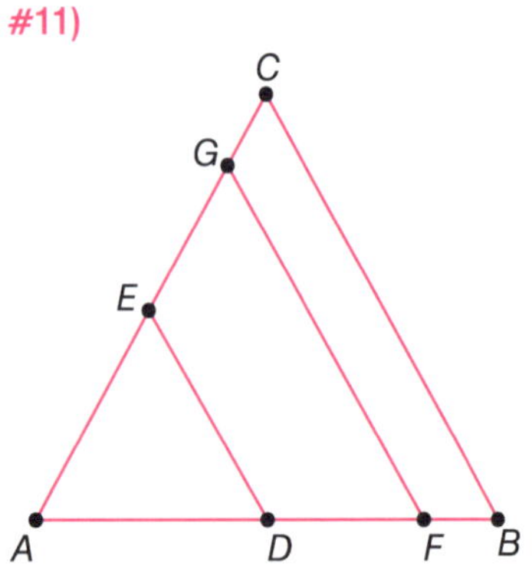

Triangle ABC is equilateral with $AB=1$. Points E and G are on $\overline{AC}$ and points D and F are on $\overline{AB}$ such that both $\overline{DE}$ and $\overline{FG}$ are parallel to $\overline{BC}$. Furthermore, triangle ADE and trapezoids $DFGE$ and $FBCG$ all have the same perimeter.

What is $DE+FG$?

A) 1 B) $\frac{3}{2}$ C) $\frac{21}{13}$ D) $\frac{13}{8}$ E) $\frac{5}{3}$

10. **(2008 AMC 12B #25)** Let $ABCD$ be a trapezoid with $AB//CD$, $AB = 11, BC = 5, CD = 19$, and $DA = 7$. Bisectors of $\angle A$ and $\angle D$ meet at P, and bisectors of $\angle B$ and $\angle C$ meet at Q.

What is the area of hexagon $ABQCDP$?

A) $28\sqrt{3}$ B) $30\sqrt{3}$ C) $32\sqrt{3}$ D) $35\sqrt{3}$ E) $36\sqrt{3}$

11. **(2012 AMC 12B #20)** A trapezoid has side lengths 3, 5, 7, and 11. The sums of all the possible areas of the trapezoid can be written in the form of $r_1\sqrt{n_1} + r_2\sqrt{n_2} + r_3$, where r_1, r_2, and r_3 are rational numbers and n_1 and n_2 are positive integers not divisible by the square of any prime.

What is the greatest integer less than or equal to $r_1+r_2+r_3+n_1+n_2$?

A) 57 B) 59 C) 61 D) 63 E) 65

12. **(2000 AIME II #8)** In trapezoid $ABCD$, leg $\overline{BC}$ is perpendicular to bases $\overline{AB}$ and $\overline{CD}$, and diagonals $\overline{AC}$ and $\overline{BD}$ are perpendicular.

Given that $AB = \sqrt{11}$ and $AD = \sqrt{1001}$, find BC^2.

CHAPTER 12

A Area of a Convex Polygon

Note

Commonly used trigonometric angles

$\sin\alpha = \sin(180° - \alpha)$

$\sin 150° = \sin 30° = \frac{1}{2}$

$\sin 120° = \sin 60° = \frac{\sqrt{3}}{2}$

$\sin 135° = \sin 45° = \frac{\sqrt{2}}{2}$

$\sin 90° = 1$

My Notes

Area of a Convex Polygon

Theorem 12.1

The area of a convex polygon is equal to half of the product of the length of the diagonals and the sine of the angle between diagonals.

In the figure,

Area $\Delta ABCD = \frac{1}{2} AC \cdot DB \sin\alpha$

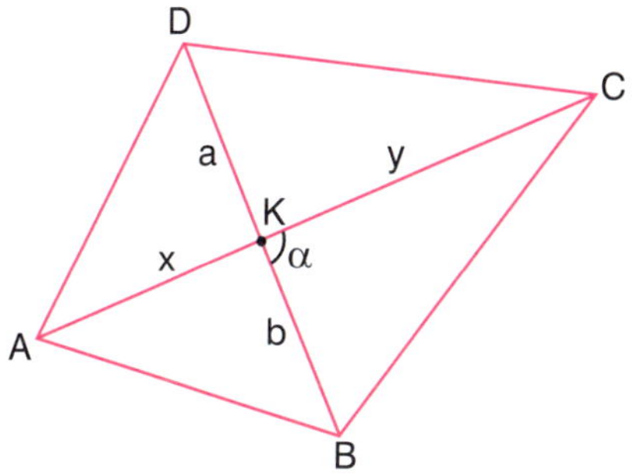

Proof 12.1

From the triangle area formula,

Area $\Delta ABC = \frac{1}{2} AB \cdot AC \sin\alpha$.

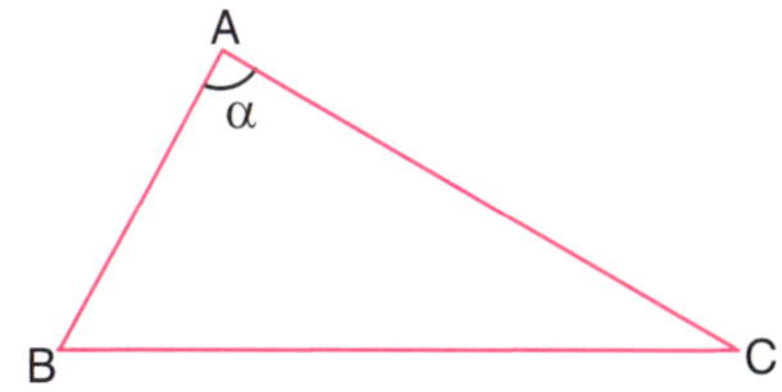

By expressing all four regions area,

$\frac{1}{2} ax \sin\alpha + \frac{1}{2} b \cdot y \sin\alpha + \frac{1}{2} bx \sin(180° - \alpha)$

$+ \frac{1}{2} a \cdot y \sin(180° - \alpha)$ $\quad [\sin(180° - \alpha) = \sin\alpha]$

$= \frac{1}{2} a(x+y)\sin\alpha + \frac{1}{2} b(x+y) \sin\alpha$

Thus the area of

$ABCD = \frac{1}{2} (a+b)(x+y) \sin\alpha$.

Therefore, Area $ABCD = \frac{1}{2} AC \cdot DB \sin\alpha$.

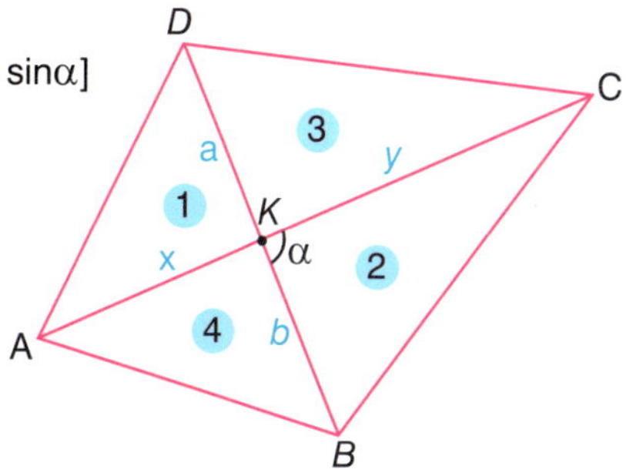

Remark 12.1

If $\alpha = 90°$, then $\sin\alpha = 1$.

Therefore, Area $ABCD = \frac{AC \cdot DB}{2}$.

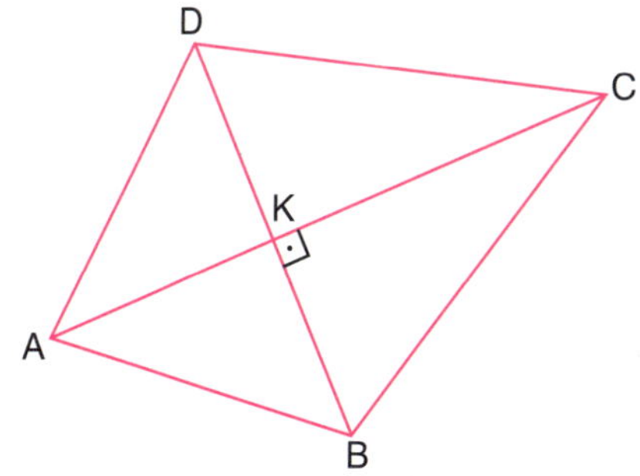

Example 12.1

In the figure,

$AC = 8\sqrt{3}$,

$BD = 12$, and

$m\angle AKB = 150°$.

What is the area of the quadrilateral *ABCD*?

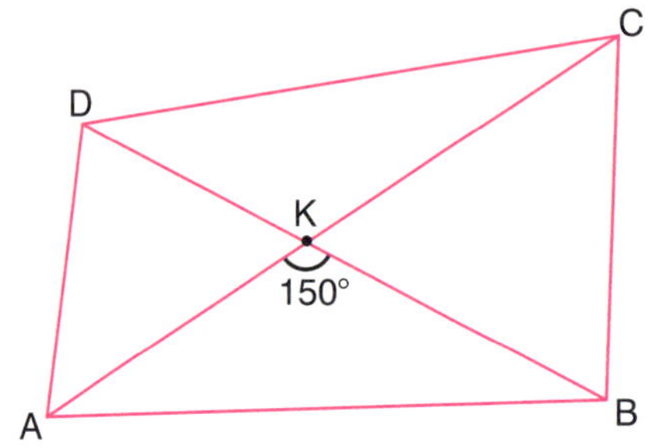

Solution 12.1

$$\text{Area } ABCD = \frac{1}{2} AC \cdot DB \sin 150° = \frac{1}{2} 8\sqrt{3} \cdot 12 \sin 30° \quad [\sin 150° = \sin 30° = \frac{1}{2}]$$

$$= \frac{1}{2} 8\sqrt{3} \cdot 12 \cdot \frac{1}{2} = 24\sqrt{3}.$$

Example 12.2

The diagonals of a quadrilateral *ABCD* are perpendicular to each other and $AC = 3\,BD$.

Given that the area of *ABCD* is 18, what is *BD*?

Solution 12.2

Let be $BD = x$ and $AC = 3x$

By using Area $ABCD = \dfrac{x \cdot 3x}{2} = 18$

$$3x^2 = 36$$

$$x^2 = 12 \Rightarrow x = 2\sqrt{3}.$$

So, $BD = 2\sqrt{3}$.

Try This

ABCD is a convex quadrilateral and $AD + BC = 4$.

Given that $AB = DC = 2$ and $m\angle DAB = m\angle DCB = 45°$ find the area of *ABCD*.

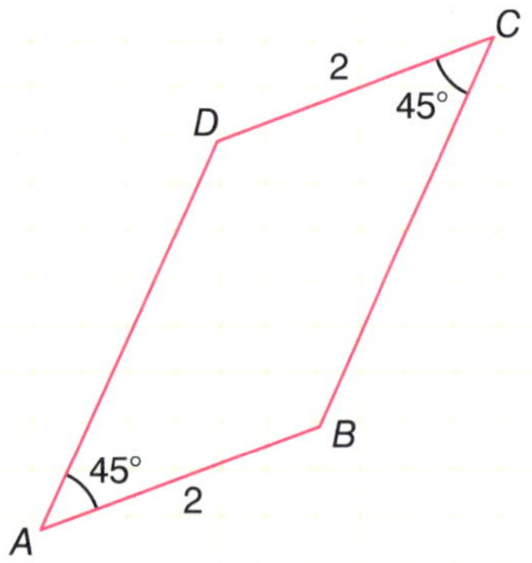

My Notes

My Notes

Theorem 12.2

If A_1, A_2, A_3, and A_4 are the areas of the regions as shown. Sections of the quadrilateral $ABCD$.

Then $A_1 \cdot A_3 = A_2 \cdot A_4$.

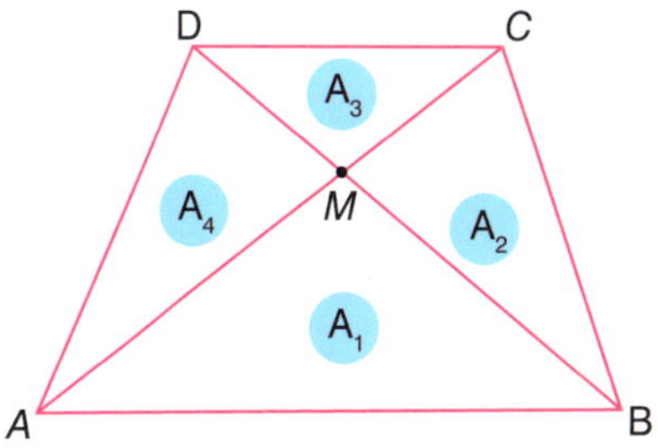

Proof 12.2

Let h_1 be as shown $A_3 = \frac{1}{2} a \cdot h_1$, $A_2 = \frac{1}{2} b \cdot h_1$.

Therefore, $\frac{A_3}{A_2} = \frac{a}{b}$. (1)

(From the right wing of the quadrilateral)

Let h_2 be as shown $A_4 = \frac{1}{2} a \cdot h_2$, $A_1 = \frac{1}{2} b \cdot h_2$

Therefore, $\frac{A_4}{A_1} = \frac{a}{b}$. (2)

(From the left wing of the quadrilateral)

(1) = (2) $\frac{A_3}{A_2} = \frac{A_4}{A_1} \Rightarrow A_1 \cdot A_3 = A_2 \cdot A_4$.

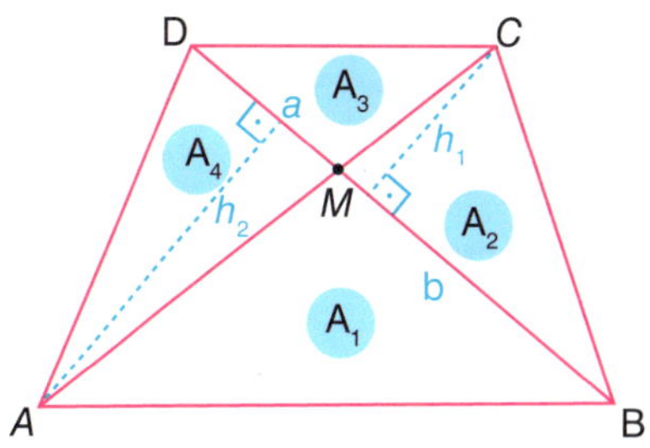

Example 12.3

In the figure,

$AC = 8$,

$DC = 6$, and

$BE = 2\,DE$.

What is the area of the quadrilateral $ABCD$?

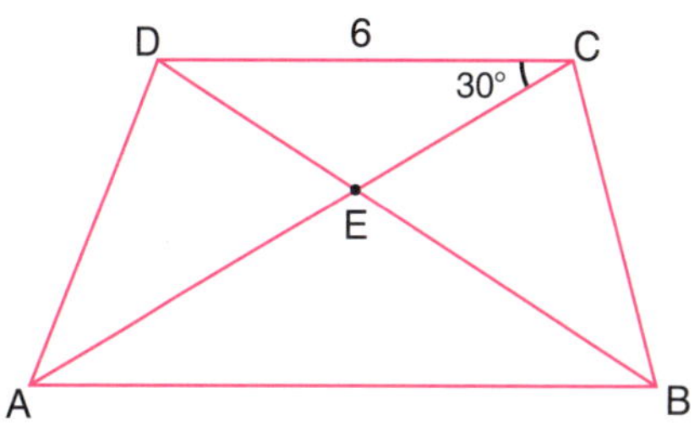

Solution 12.3

Area of $\Delta ACD = \frac{1}{2} 6 \cdot 8 \sin 30°$

$$= \frac{1}{2} 6 \cdot 8 \frac{1}{2} = 12.$$

Let A = Area(ΔDEC) = A and
let S = Area(ΔDAE).

Then Area(ΔCEB) = $2A$ and
Area(ΔAEB) = $2S$.

Hence, $12 = A+S$ and the $3A+3S = 36$.

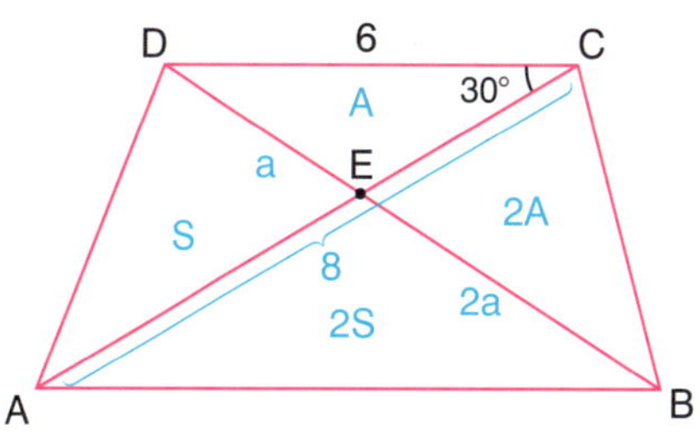

Example 12.4

ABCD is a convex quadrilateral. The diagonals *AC* and *BD* intersect at point *E*.

Given that Area $\Delta ADE = 12 \text{ cm}^2$,

Area $\Delta BCE = 16 \text{ cm}^2$, and

Area $\Delta ABE = 3$ AreaΔDEC.

Find the area of the quadrilateral *ABCD*.

Solution 12.4

From theorem 12.1.

$16 \cdot 12 = A \cdot 3A$

$16 \cdot 4 = A^2$

$8 = A.$

Therefore, the total area is 60 cm^2.

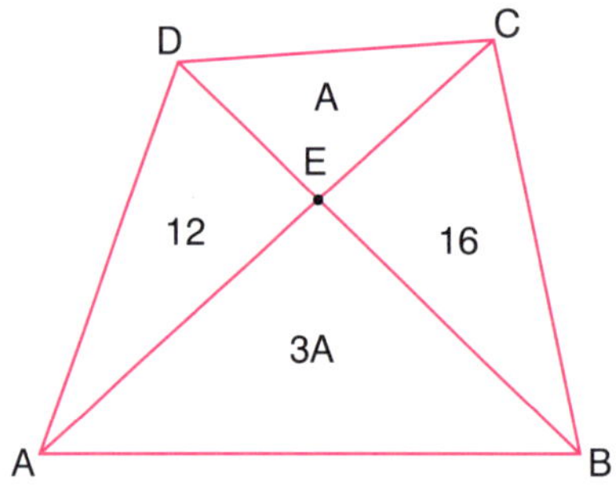

Area of Parallelogram

Theorem 12.3

The area of parallelogram *ABCD* is $a \cdot h_a = b \cdot h_b$.

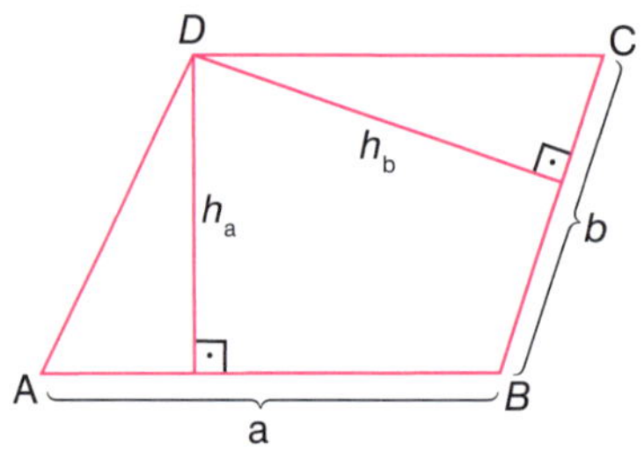

Proof 12.3

Area of $\Delta ADB = \frac{a \cdot h_a}{2}$

Area of $\Delta ABCD = 2 \cdot$Area of ΔADB

$= 2 \cdot \frac{a \cdot h_a}{2} = a \cdot h_a.$

Similarly, Area of $\Delta ABCD = 2 \cdot$Area of ΔDBC

$= 2 \cdot \frac{b \cdot h_b}{2} = b \cdot h_b.$

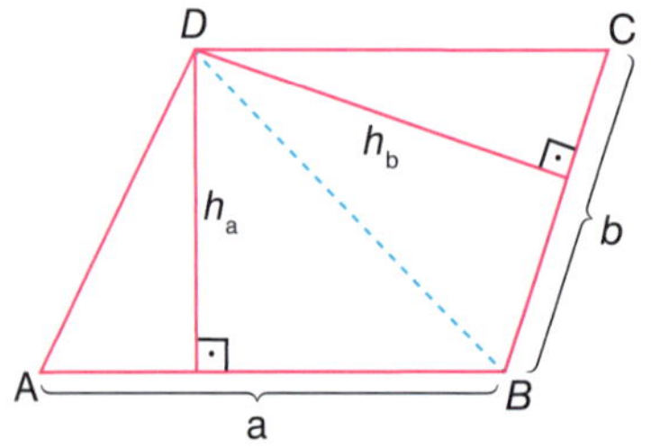

B Area of Parallelogram

My Notes

Prove This

ABCD is a parallelogram. Points *E* and *F* are the midpoints of *DC* and *BC*.

$$\text{Area } \triangle AEF = \frac{3}{8} \text{ Area } ABCD.$$

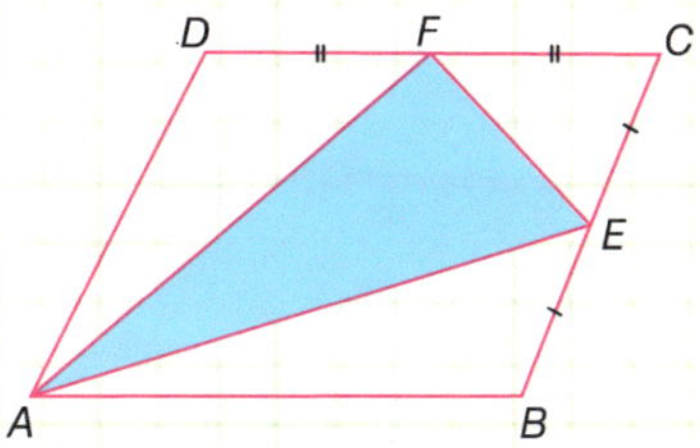

Theorem 12.4

If *ABCD* is parallelogram with sides *a* and *b* separated an angle *A*, then

Area $ABCD = a \cdot b \sin A$.

Proof 12.4

The sum of two congruent triangles' area

$$① + ② = \frac{1}{2} a \cdot b \sin\alpha + \frac{1}{2} a \cdot b \sin\alpha$$
$$= a \cdot b \sin\alpha$$

Therefore, Area $ABCD = a \cdot b \sin\alpha$.

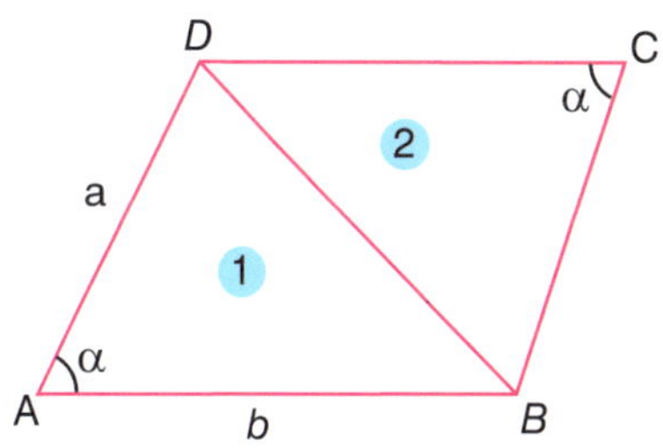

Remark 12.2

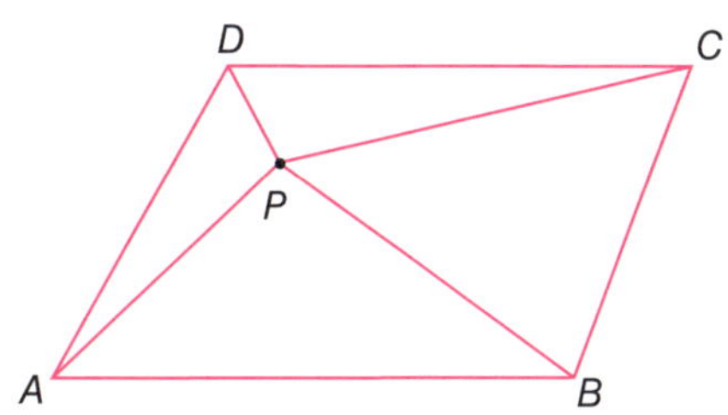

If P is any point inside a parallelogram *ABCD* then

Area(*APB*) + Area(*DPC*) = Area(*DAP*) + Area(*BPC*).

Remark 12.3

If the area of a parallelogram is *S*, then the following regions' area are;

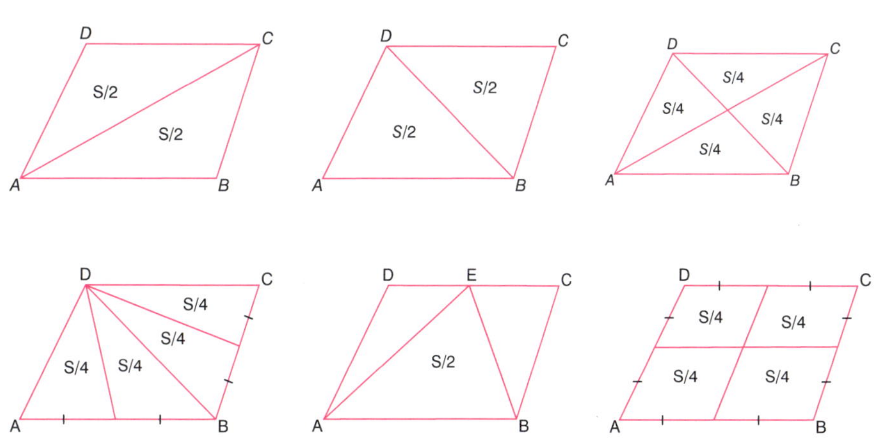

Example 12.5

ABCD is a parallelogram with $DE \perp EC$.

Given that $EC = 7$ and $DE = 6$.

What is total area of the shaded regions?

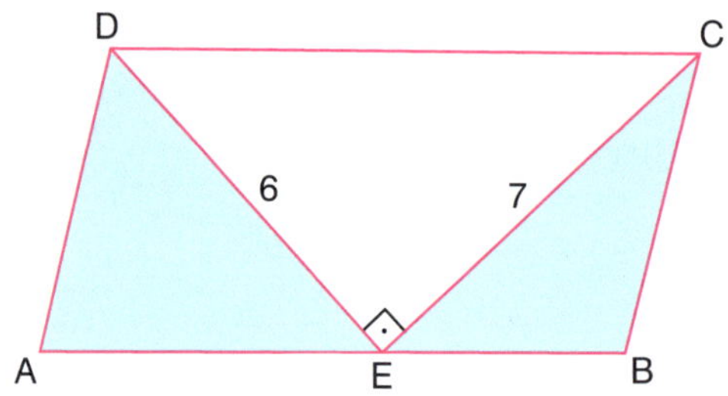

Solution 12.5

Area $\Delta DEC = \frac{6 \cdot 7}{2} = 21$.

Using remark 12.3, the area of the parallelogram is $2 \cdot 21 = 42$.

Therefore, the shaded part is 21.

Example 12.6

In the figure, *ABCD* is a parallelogram with

$AE = EF = FB$ and

$DG = GH = HL = LC$.

Given that Area$(ABCD) = 120$.

What is the area of quadrilateral *EFLG*?

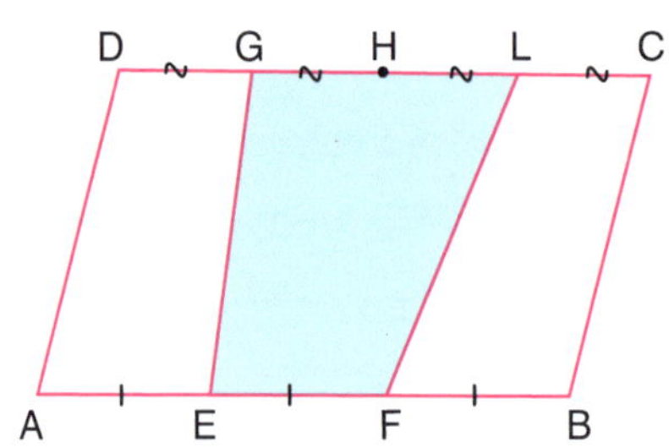

Solution 12.6

By using the area and base ratios, we get the total area of parallelogram 24*S*.

So, $24S = 120 \Rightarrow S = 5$

The total shaded part is

$10S = 10 \cdot 5 = 50$.

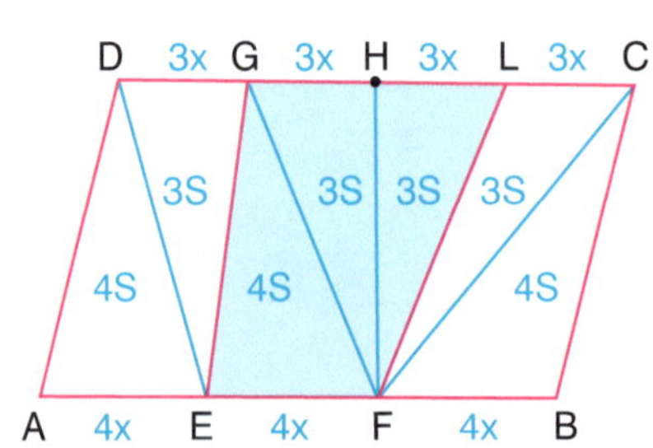

Prove This

ABCD is a parallelogram. Given that points *E* and *F* are the mid-points *AB* and *CB*.

Area $KLFBE = \frac{1}{3}$Area $ABCD$.

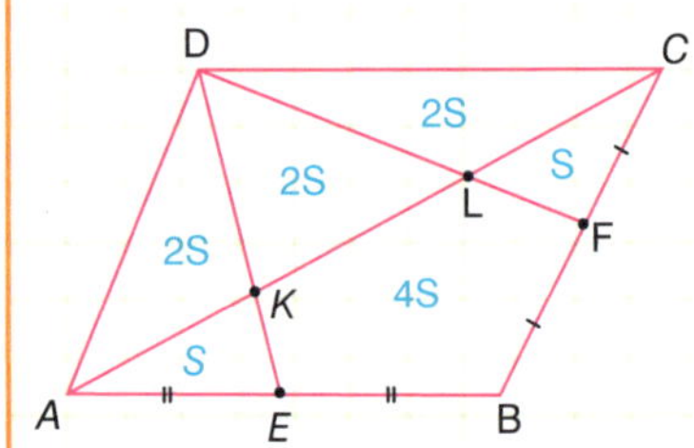

My Notes

My Notes

Example 12.7

In the figure, *ABCD* is a parallelogram with $AE \perp BF$

Given that $AE = 8$ cm and $BF = 7$ cm.

What is the area of parallelogram *ABCD*?

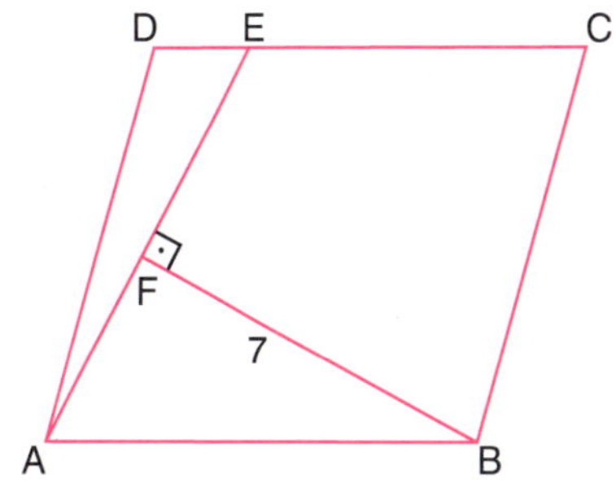

Solution 12.7

Area of $\Delta AEB = \dfrac{8 \cdot 7}{2} = 28 \text{ cm}^2$

Then the area of the whole figure is
$2 \cdot 28 = 56 \text{ cm}^2$.

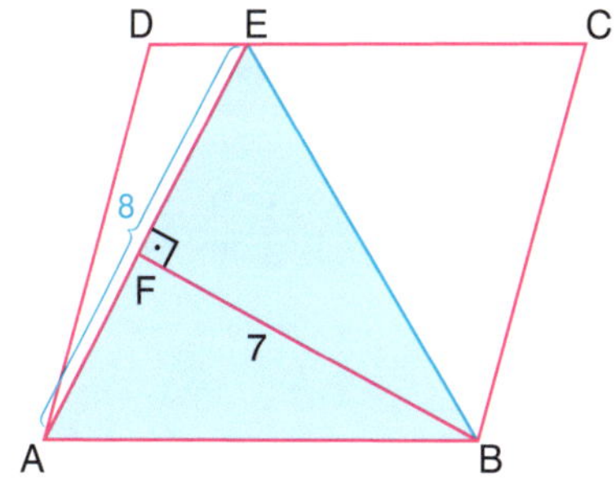

Example 12.8

In the figure, *ABCD* is a parallelogram. *E* is the midpoint of *AB* and *BD* is a diagonal.

Given that area ΔFEB is 6 cm^2, find the area of parallelogram *ABCD.*

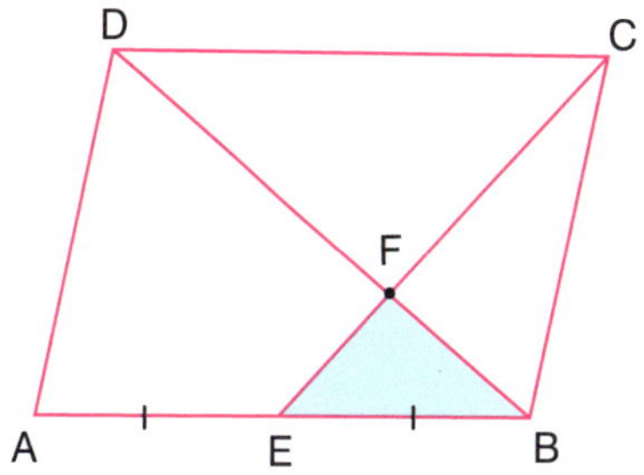

Solution 12.8

Let $FB = k$ and the area of $\Delta FEB = S$.

By the BOW-TIE similarity, $DF = 2k$ and $\Delta DEF = 2S$. Since $AE = EB$, then the area of $\Delta DAE = 3S$.

If, $S = 6 \text{ cm}^2$ then the area of *ABCD* is
$12S = 12 \cdot 6 = 72 \text{ cm}^2$.

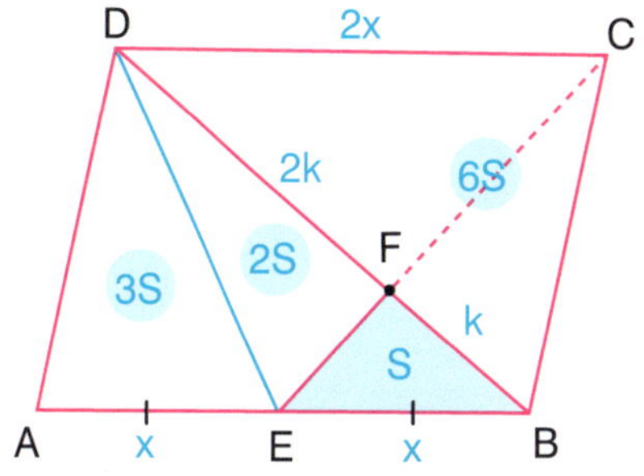

Area of Rhombus

Theorem 12.5

The area of rhombus *ABCD* is

$a^2 \cdot \sin A$ or $a^2 \cdot \sin B$. [$\sin A = \sin B$]

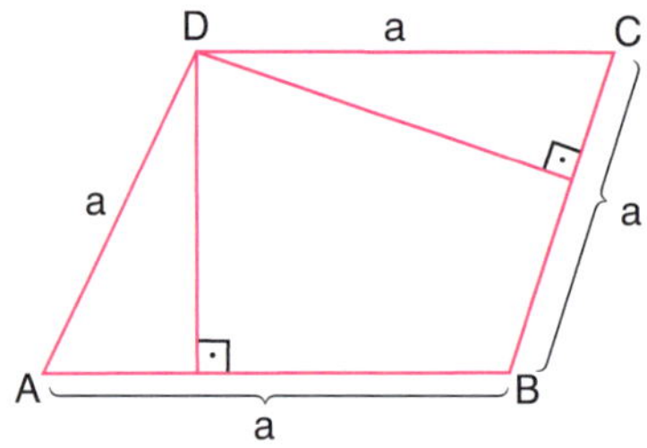

Proof 12.3

Area 1 + Area 2

$$= \frac{1}{2} a \cdot a \sin\alpha + \frac{1}{2} a \cdot a \sin\alpha$$

$$= \frac{1}{2} a^2 \sin\alpha + \frac{1}{2} a^2 \sin\alpha$$

$$= a^2 \sin\alpha$$

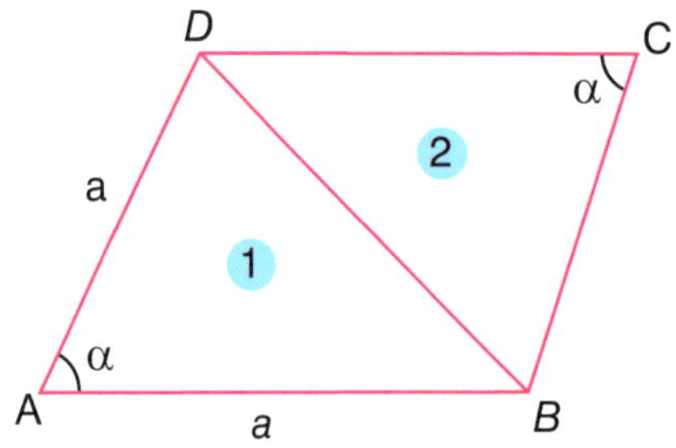

Theorem 12.6

If *ABCD* is a rhombus, then

$$\text{Area } ABCD = \frac{DB \cdot AC}{2}$$

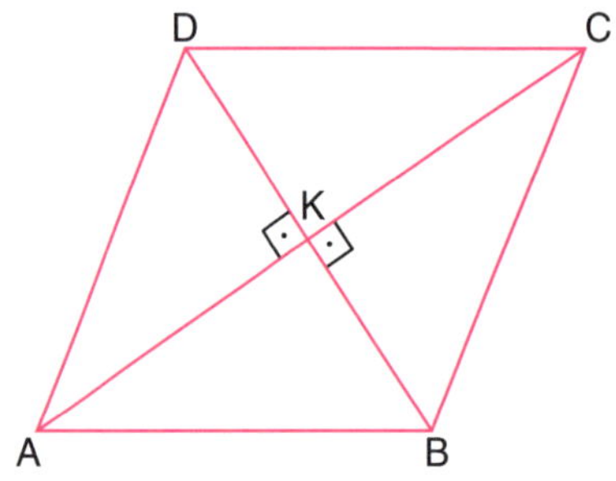

Proof 12.4

Area 1 + Area 2 + Area 3 + Area 4

$$= \frac{1}{2} \frac{e}{2} \frac{f}{2} + \frac{1}{2} \frac{e}{2} \frac{f}{2} + \frac{1}{2} \frac{e}{2} \frac{f}{2} + \frac{1}{2} \frac{e}{2} \frac{f}{2}$$

$$= 4 \frac{1}{2} \frac{e}{2} \frac{f}{2} = \frac{e \cdot f}{2}$$

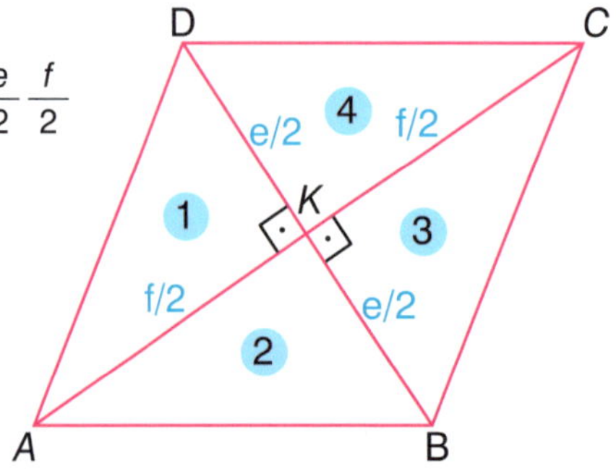

Area of Rhombus

My Notes

My Notes

Example 12.9

The diagonals of a rhombus measure 12 cm and 16 cm.

What is the length of the altitude of the rhombus?

Solution 12.9

Area of the rhombus can be expressed as $\frac{12 \cdot 16}{2}$ and $10 \cdot h$.

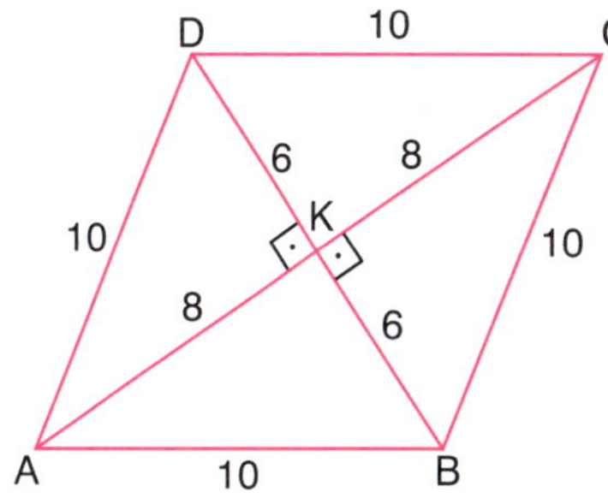

So, $\frac{12 \cdot 16}{2} = 10 \cdot h$

$$h = \frac{12 \cdot 16}{20} = 9.6 \text{ cm}$$

Example 12.10

Given that $ABCD$ is a rhombus and point E is the midpoint of AB, $DE \perp AB$, and $DE = 6$ cm.

What is the area of $ABCD$?

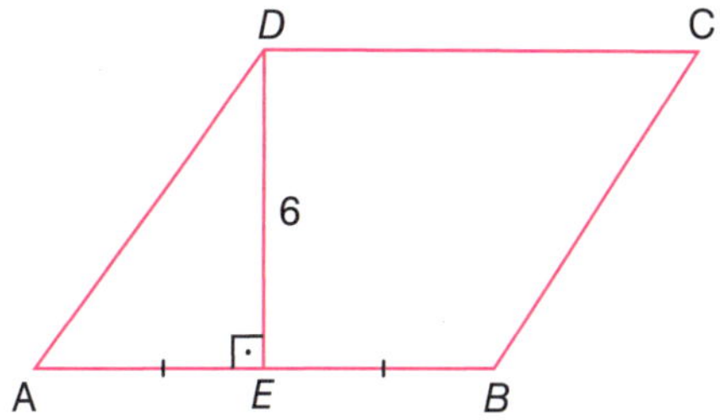

Solution 12.10

By using $30°-60°-90°$ triangle we get

$$a = \frac{6}{\sqrt{3}} = 2\sqrt{3}.$$

So, the side length of the rhombus is $4\sqrt{3}$.
Therefore, the area of $ABCD$ is $6 \cdot 4\sqrt{3} = 24\sqrt{3}$.

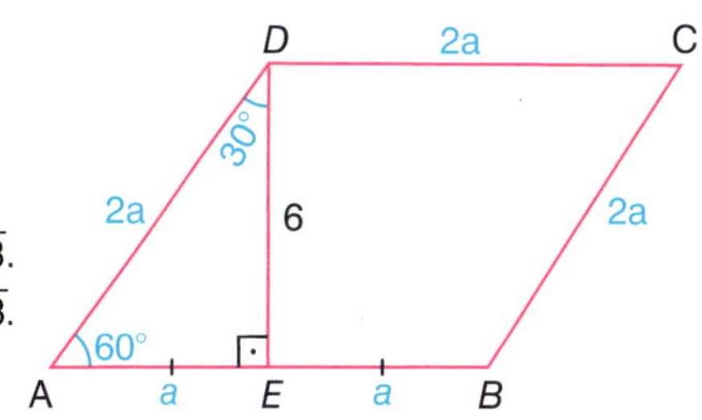

My Notes

Example 12.11

In the figure, $ABCD$ is a rhombus with $DE \perp AB$ and AC is a diagonal.

Given that $AC = 4\sqrt{5}$ and $DE = 4$.

What is the area of $ABCD$?

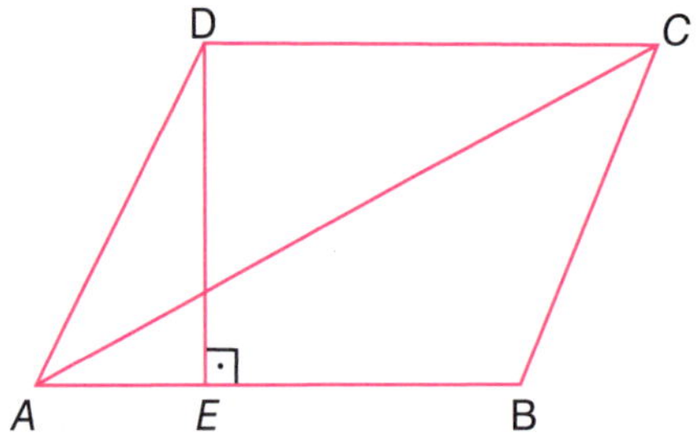

Solution 12.11

Let the length of AB be a. By creating right triangle CBH and expressing BH as $8-a$.

We get $a^2 = 4^2 + (8-a)^2$

$$a^2 = 16 + 64 - 16a + a^2$$

$$16a = 80 \Rightarrow a = 5.$$

Therefore, the area of $ABCD$ is $5 \cdot 4 = 20$ unit^2.

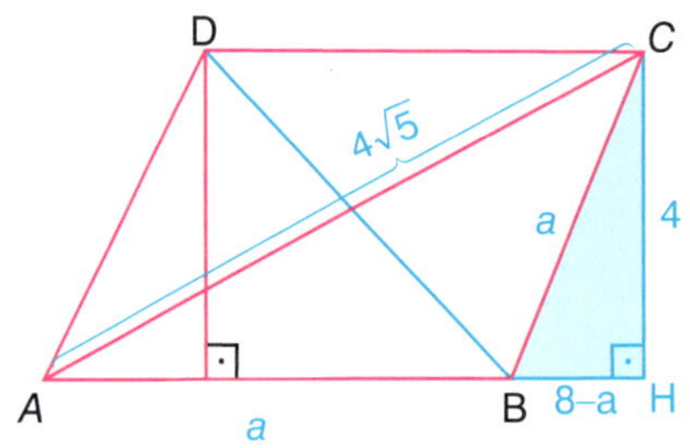

Example 12.12

In the figure, $ABCD$ is a rhombus.

Given that BK and CK are angle bisectors and the area of ΔCKB is 6 cm^2, find the area of rhombus.

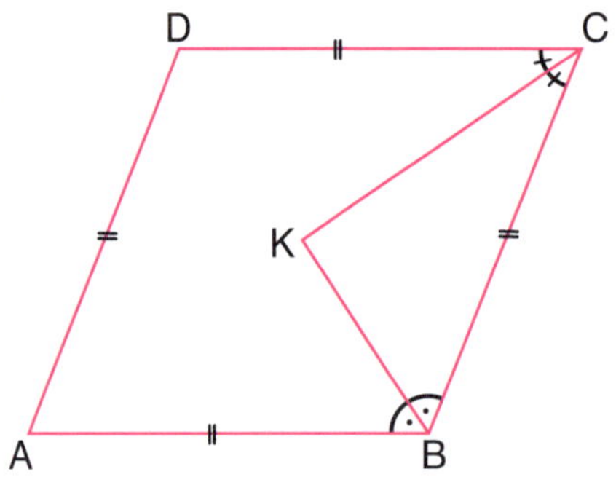

Solution 12.12

We draw three perpendicular feet from point K to E, G and H. From anthe angle bisector theorem $KE = KH = KG = a$

Area $\Delta CKB = \dfrac{a \cdot x}{2} = 6 \Rightarrow a \cdot x = 12.$

So, the area of rhombus is also $x \cdot 2a = 2ax$ which is equals to $2 \cdot 12 = 24$.

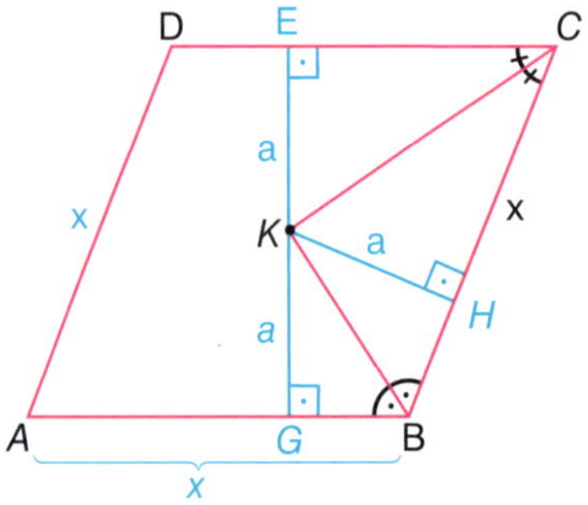

D Area of Squares/Rectangles

Try This

The figure shows three squares where the areas of the two smaller ones are 1 and 16, respectively. What is the area of the large square?

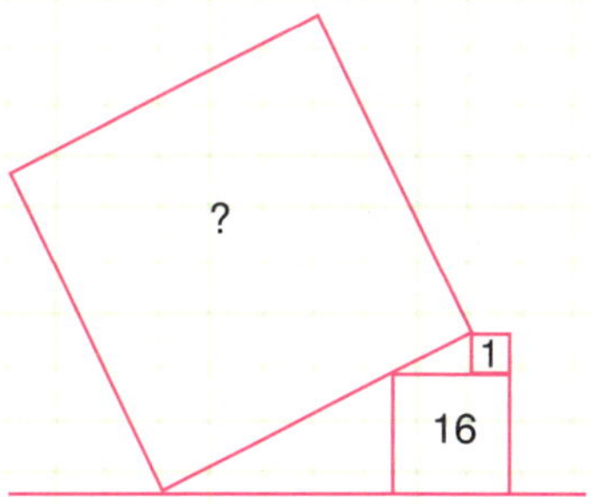

My Notes

Area of Squares/Rectangles

1. Sides and the area connections

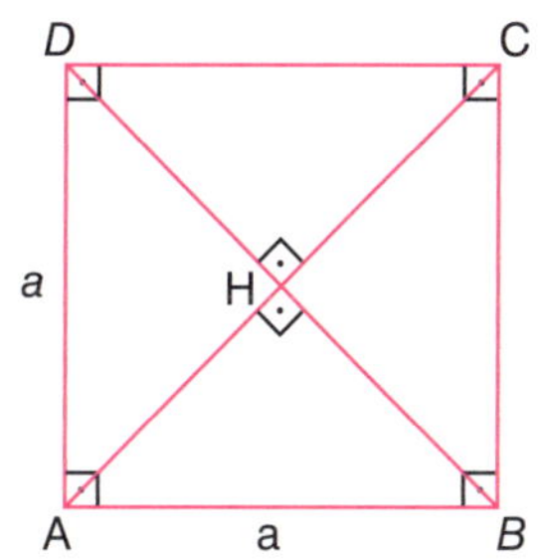

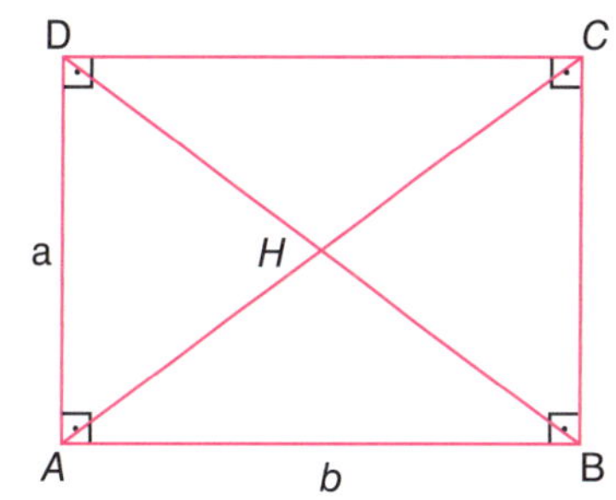

$AB = BC$, Area $ABCD = a \cdot a = a^2$ Area $ABCD = a \cdot b$

2. Area $ABCD = \frac{e^2}{2}$ (e is the length of diagonal of the square)

Example 12.13

A square has an area of 50 cm^2.

What is the length of the diagonal?

Solution 12.13

From the diagonal and area connections

$$\frac{e^2}{2} = 50 \Rightarrow e^2 = 100 \Rightarrow e = 10.$$

Example 12.14

ABCD is a square, given that

$DE = 4$ cm,

$m\angle AEC = 120°$.

What is the area of the square?

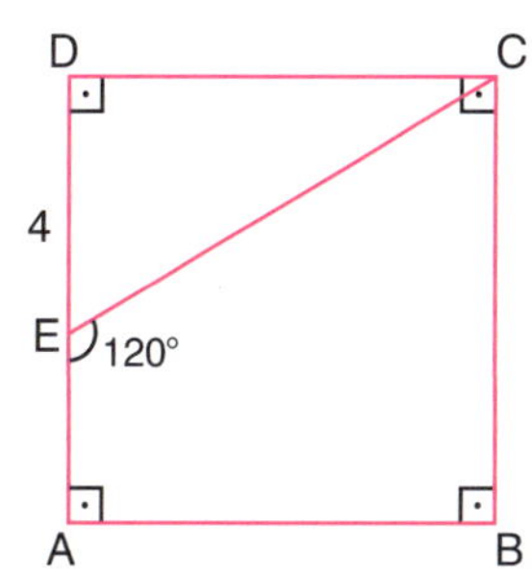

Solution 12.14

From the 30°–60°–90° triangle one side length is $4\sqrt{3}$ then the area is$(4\sqrt{3})\cdot(4\sqrt{3}) = 48$ unit^2.

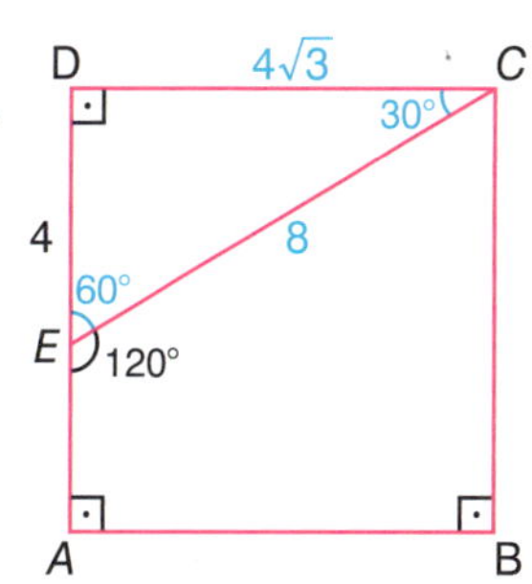

Example 12.15

ABCD is a square with $AF \perp DE$.

Given that

$CE = EB$ and

$AF = 4$ cm,

what is the area of *ABCD*?

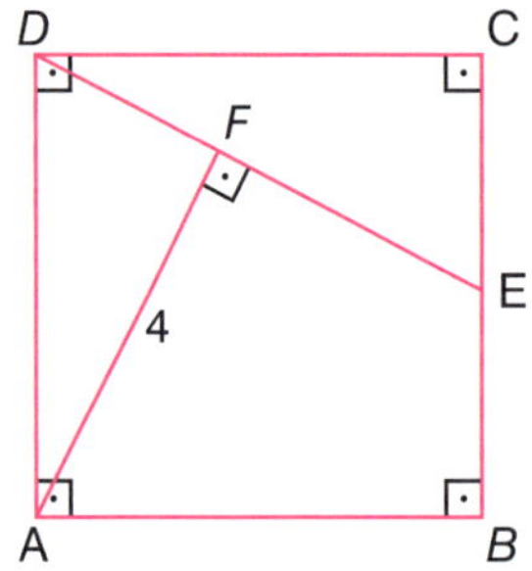

Solution 12.15

Step 1: Make an extension of the segment *DE*.

Step 2: From the IPT.

$$\frac{1}{4^2} = \frac{1}{(2a)^2} + \frac{1}{(4a)^2}$$

$$\frac{1}{16} = \frac{1}{4a^2} + \frac{1}{16a^2}$$

$$\frac{1}{16} = \frac{5}{16a^2} \Rightarrow a^2 = 5$$

Area of $ABCD = 4a^2 = 20$ cm^2.

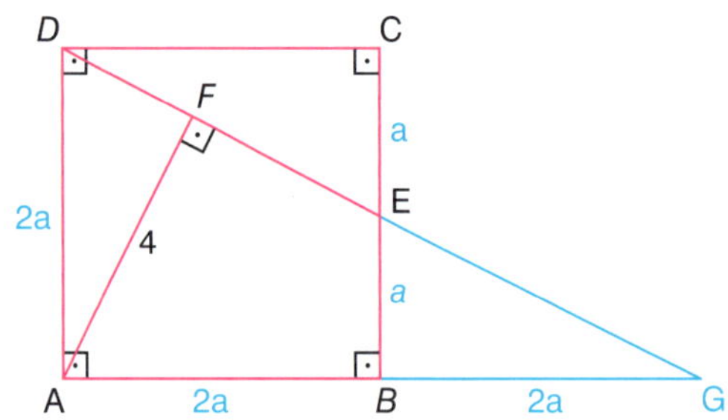

Example 12.16

In the figure, *ABCD* is a square. Given that

$AD = EC$,

$EB = 8$ cm, and

$m\angle AEB = 90°$.

What is the area of ΔBCE?

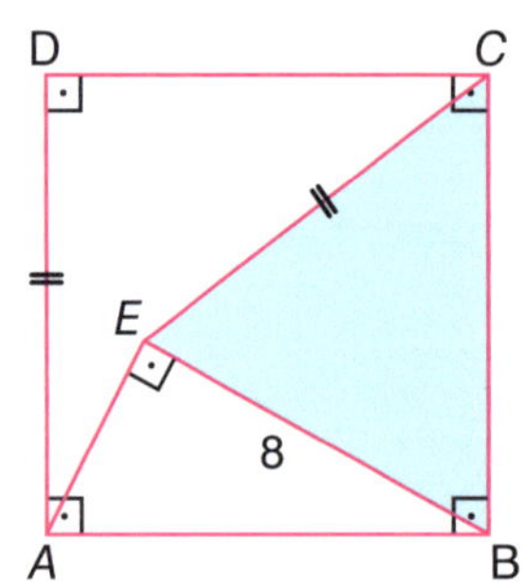

Solution 12.16

We make a right triangle by drawing a perpendicular foot from point *C* to side *EB*. Notice that ΔCHB and ΔBEA are congruent and $EH = HB$.

So, $EB = CH = 8$ cm

Area of $\Delta BCE = \dfrac{8 \cdot 8}{2} = 32$ unit^2

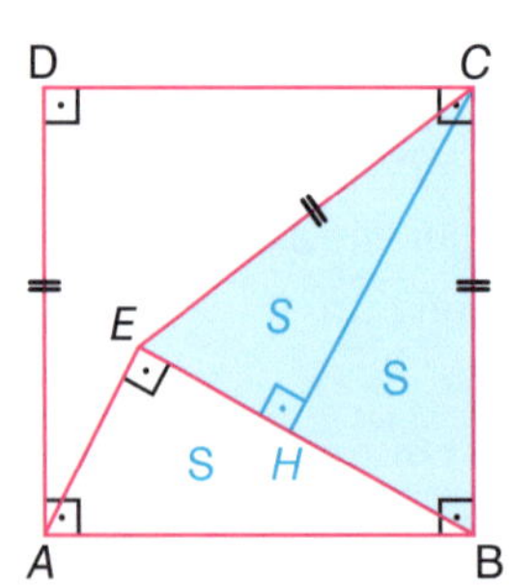

Note

Inverse Pythagorean Theorem (IPT)

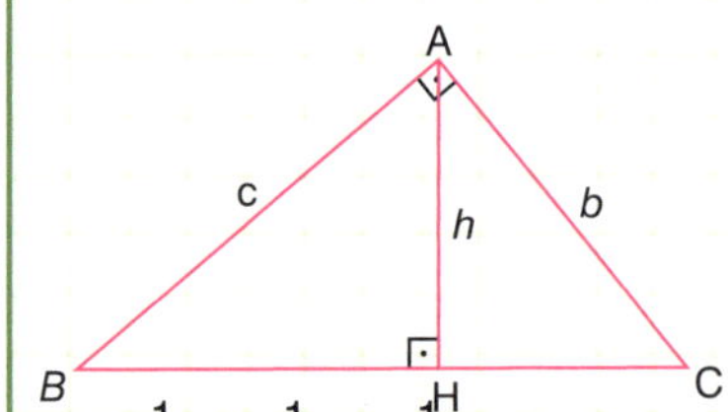

$$\frac{1}{h^2} = \frac{1}{c^2} + \frac{1}{b^2}$$

My Notes

Try This

ABCD is a square with

$DE = AF = 15$ cm and

$FB = EA = 5$ cm.

Find the area of *KCF*.

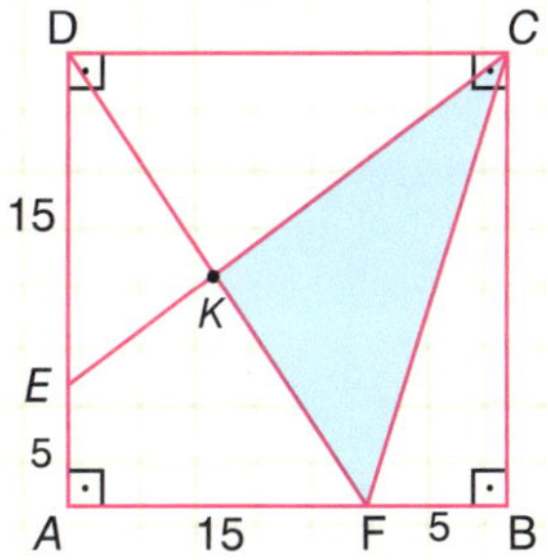

My Notes

Example 12.17

In the figure, *ABCD* is a square with

$DE \perp AE$, $AB = CE$, and

$DE = 6$ cm.

What is the area of *ABCD*?

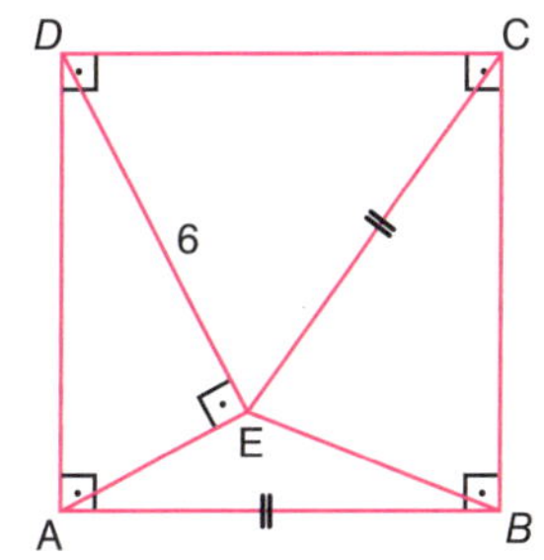

Solution 12.17

From two congruent triangles ($\Delta AED \cong \Delta DHC$)

$DH = AE = 3$ cm.

$a^2 = 3^2+6^2 \Rightarrow a^2 = 9+36$

$a^2 = 45.$

So, the area of *ABCD* is 45 cm^2.

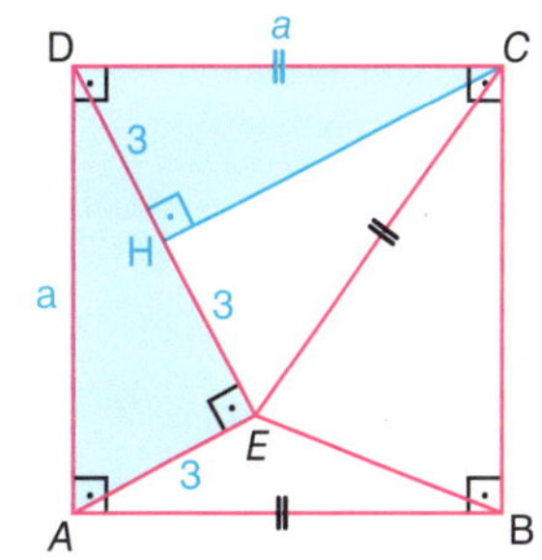

Example 12.18

The diagonal of the square is 8 cm long.

If each sides of the square is extended by 1 cm, by how much is the area of the squared increased?

Solution 12.18

Before

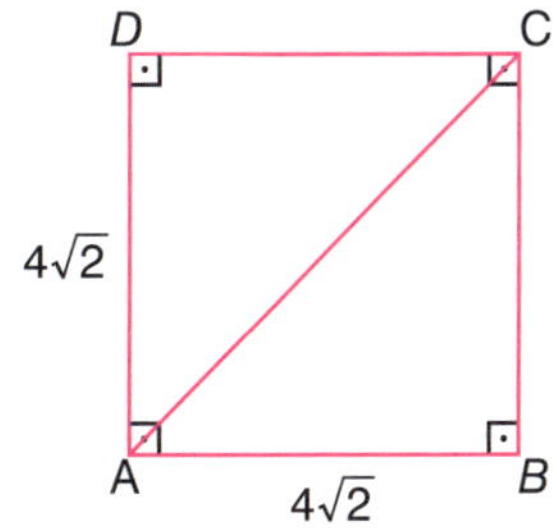

After

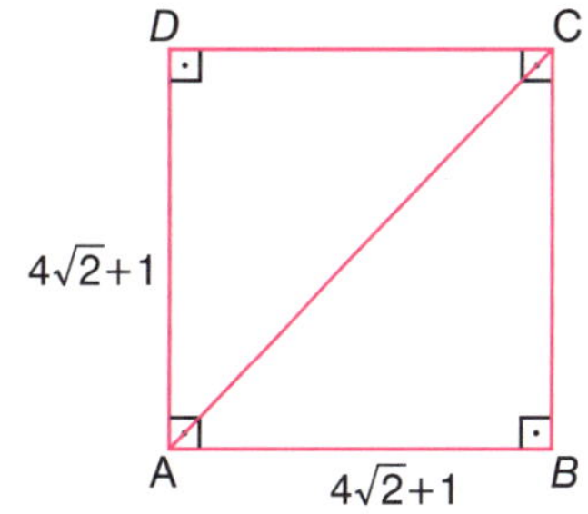

$(4\sqrt{2})\cdot(4\sqrt{2}) = 32$ unit^2 $\quad (4\sqrt{2}+1)\cdot(4\sqrt{2}+1) = 32 + 8\sqrt{2}+1$ unit^2

So, the area is increased by $8\sqrt{2}+1$.

Area of Trapezoid

Theorem 12.8

The area of a trapezoid is the product of the height and half the sum of the base in the figure.

$$\text{Area}(ABCD) = \left(\frac{a+b}{2}\right)\cdot h$$

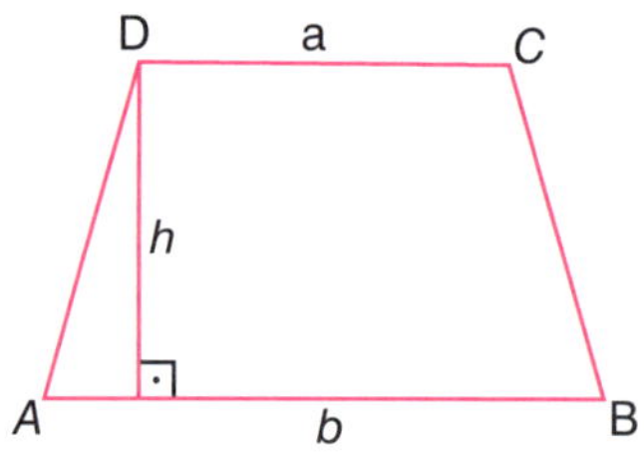

Proof 12.8

$$\text{Area } ① = \frac{a\cdot h}{2}$$

$$\text{Area } ② = \frac{b\cdot h}{2}$$

$$\text{Area}(ABCD) = \frac{a\cdot h}{2} + \frac{b\cdot h}{2} = \left(\frac{a+b}{2}\right)\cdot h.$$

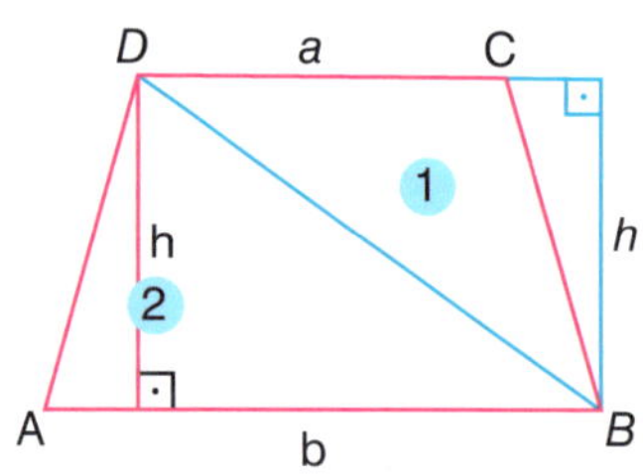

Example 12.19

In the figure, $ABCD$ is an isosceles trapezoid with

$AB = 24$ cm,

$BC = AD = 10$ cm, and

$DC = 12$ cm.

What is the area of this trapezoid?

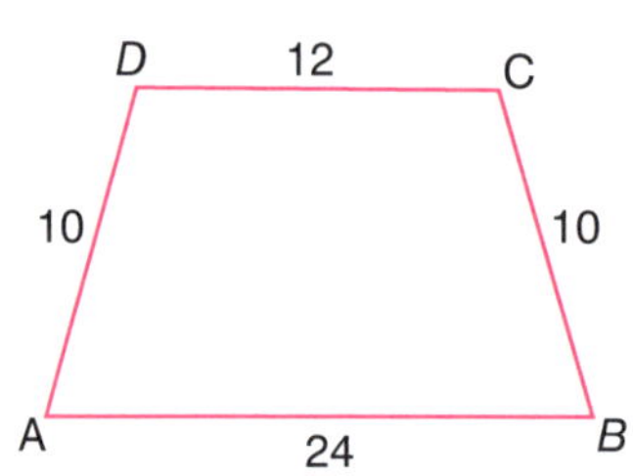

Solution 12.19

By creating two right triangles, the altitude of the trapezoid $h = 8$ (6–8–10 triangle).

$$\text{Area}(ABCD) = \left(\frac{24+12}{2}\right)\cdot 8$$

$$= \frac{36}{2}\cdot 8 = 18\cdot 8 = 144.$$

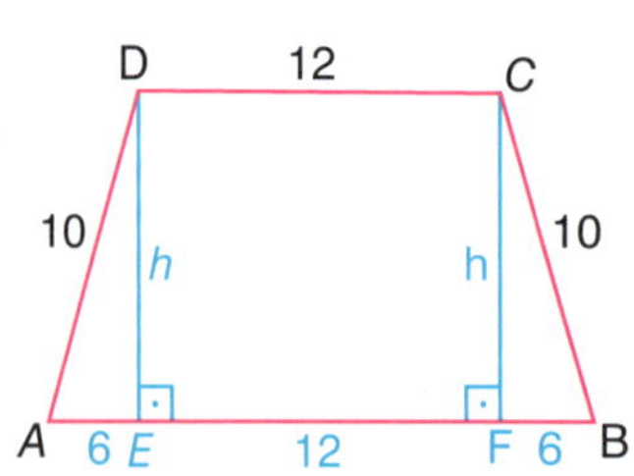

E Area of Trapezoid

Try This

$ABCD$ is a trapezoid.

Area($ADEF$) = Area($ECBF$),

$FB = 13$ cm,

$DC = 9$ cm, and

$AF = 8$ cm.

What is EC?

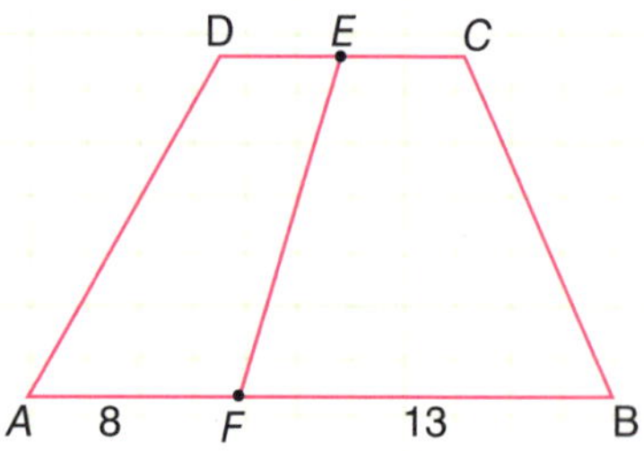

My Notes

Try This

ABCD is a right trapezoid.

Given that
Area ΔBKC = Area $\Delta AKD+6$ and $AB = 4$,
what is DC?

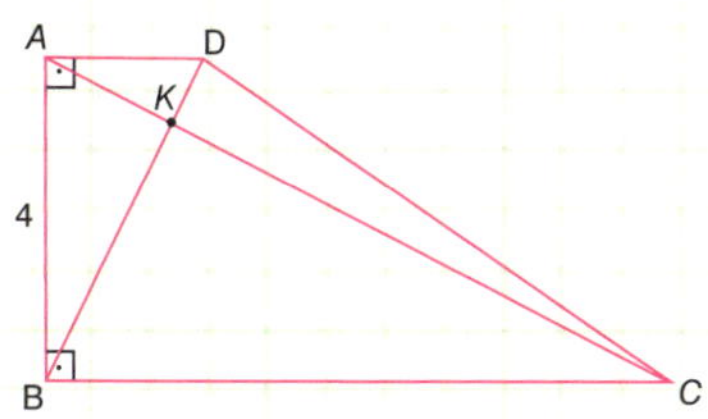

My Notes

Theorem 12.9

Let *ABCD* be a trapezoid

I) Area(ΔAKD) = Area(ΔBKC)

II) $A^2 = X{\cdot}Y$

A^2 = Area(ΔAKB) · Area(ΔDKC)

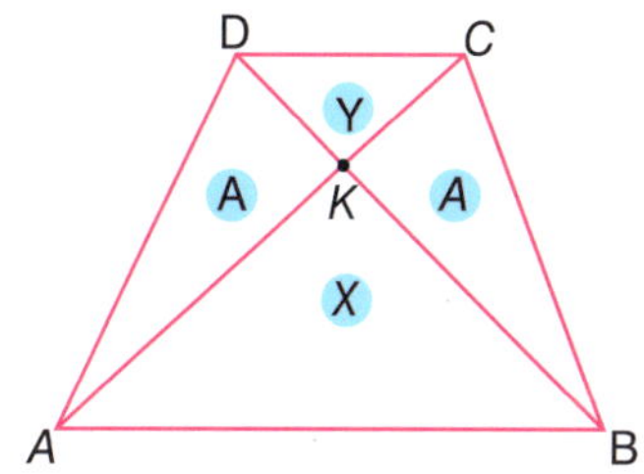

Proof 12.9

I) Since Area ΔADC = *Area* ΔDCB

(They share the same base and the same height.)

$Y+\Delta AKD = Y+BKC$

$\Delta AKD = \Delta BKC$

II) From the area base connections

$\frac{a}{b} = \frac{Y}{A} = \frac{A}{X} \Rightarrow A^2 = X{\cdot}Y.$

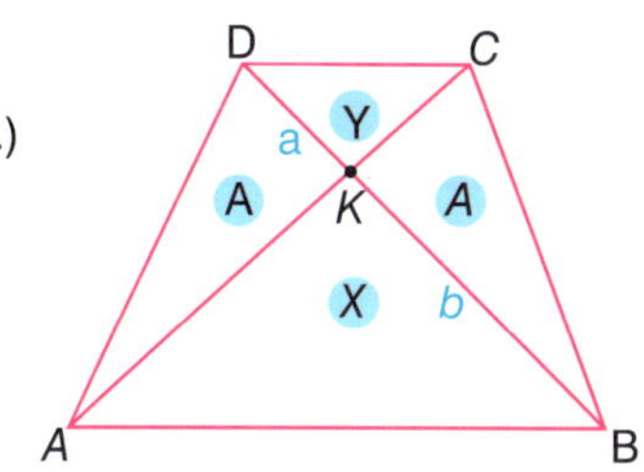

Theorem 12.10

ABCD is a trapezoid, let *E* be the midpoint of side *AD*.

Then Area(ΔBEC) $= \frac{\text{Area}(ABCD)}{2}$.

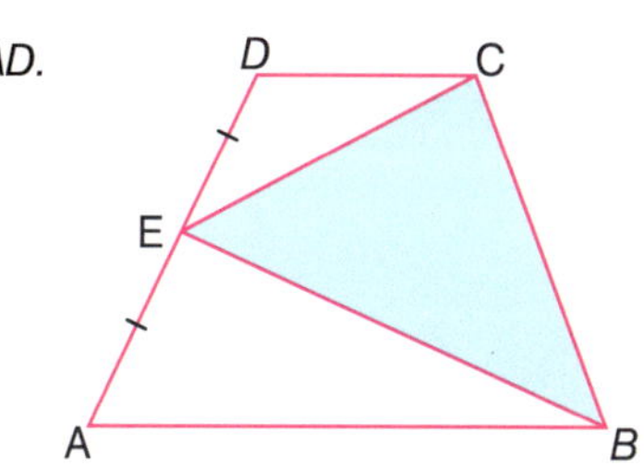

Proof 12.10

We draw *EF* parallel line segment that parallel to two bases.

The shaded area is
$\frac{EF{\cdot}h_1}{2} + \frac{EF{\cdot}h_2}{2} = \frac{EF}{2}(h_1+h_2) = \frac{EF{\cdot}h}{2}$
(*h* is the height of the trapezoid)

Since the area of the trapezoid *ABCD* is $EF{\cdot}h$.

Therefore, Area(ΔBEC) $= \frac{\text{Area}(ABCD)}{2}$.

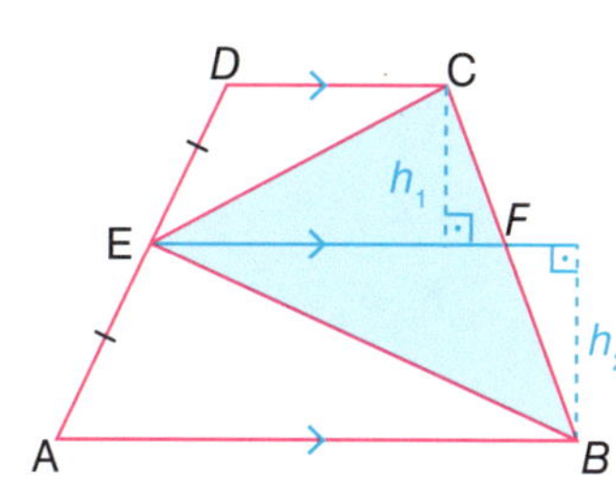

Example 12.20

In the figure, $ABCD$ is a trapezoid.

$AB = 10$, $DC = 5$,

$AD = 3$, and $CB = 4$.

Find the area of $ABCD$.

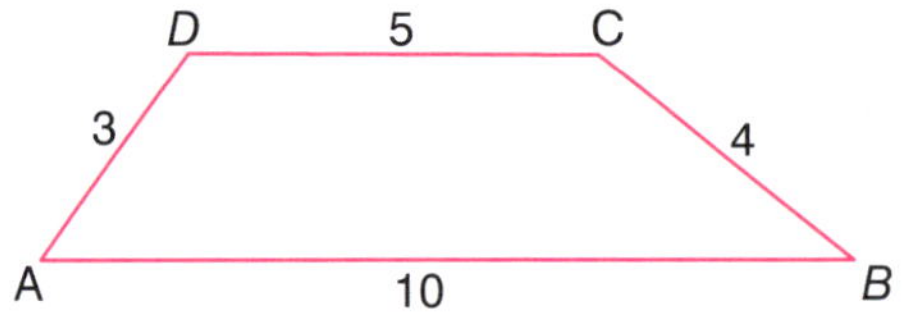

Solution 12.20

We create parallelogram by making a line segment $C'C$. Hence, $\Delta CC'B$ is a right triangle. We can find h by calculating the area of $\Delta CC'B$ in two ways,

$$\frac{3 \cdot 4}{2} = \frac{5 \cdot h}{2} \Rightarrow h = \frac{12}{5}$$

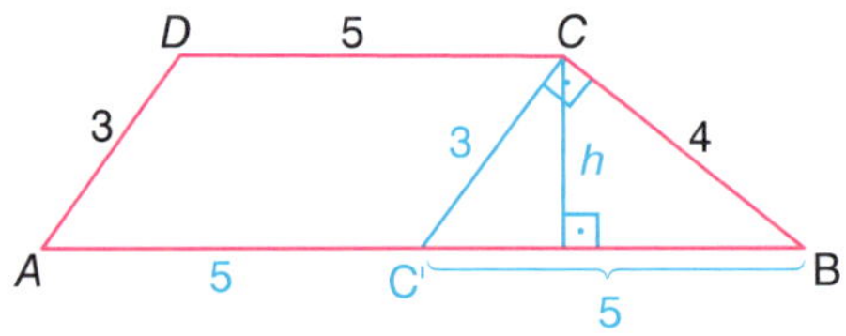

Therefore, the area of $ABCD$ is $\left(\frac{10+5}{2}\right) \cdot \frac{12}{5} = 18 \text{ cm}^2$.

Example 12.21

In the figure, $ABCD$ is a trapezoid with $DC // AB$

Given that $DC = 8$, $AD = 4\sqrt{3}$,

$m\angle ADC = 120°$, and $m\angle ABC = 45°$.

What is the area of $ABCD$?

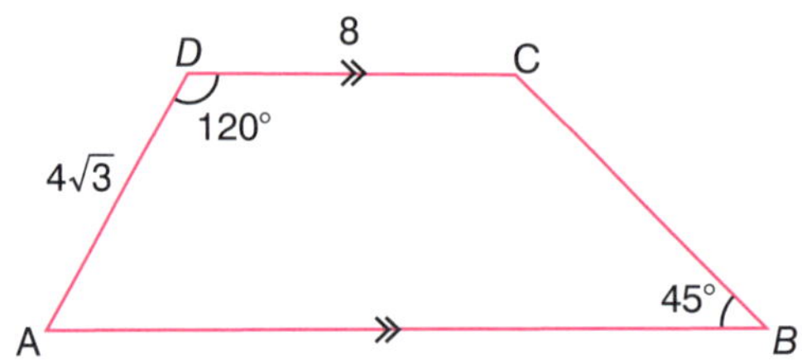

Solution 12.21

We create a parallelogram and use ΔCEB angles to find the longer base and height.

So, the area of

$$ABCD = \left(\frac{8+6+2\sqrt{3}+8}{2}\right) \cdot 6$$
$$= \left(\frac{22+2\sqrt{3}}{2}\right) \cdot 6$$
$$= (11+\sqrt{3}) \cdot 6$$
$$= 66+6\sqrt{3}.$$

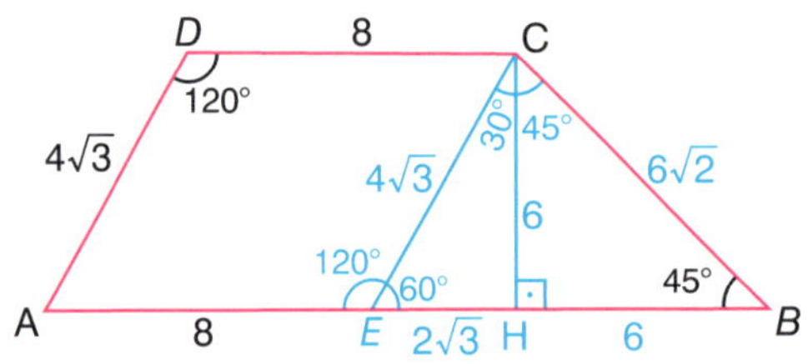

Catriona Agg @Cshearer41

Geometry puzzle

The areas of the three squares are given.

What is the area of the red triangle?

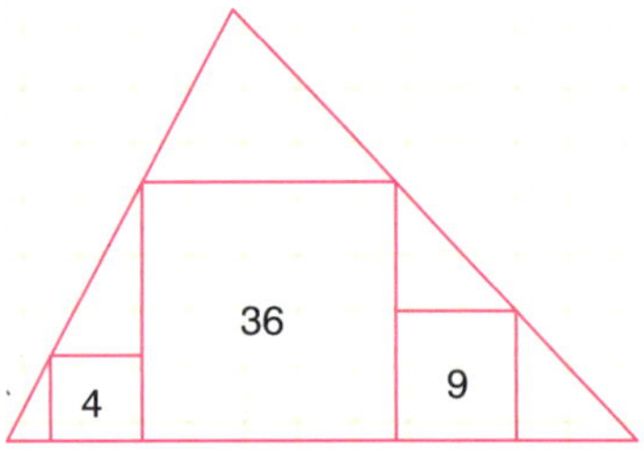

My Notes

F Area of Kite

Catriona Agg @Cshearer41
Geometry puzzle

The total blue area is equal to the purple area.

What fraction of the also called a trapezoid is shaded?

My Notes

Area of Kite

Theorem 12.11

Let $ABCD$ be kite.

$$\text{Area}(ABCD) = \frac{AC \cdot BD}{2}.$$

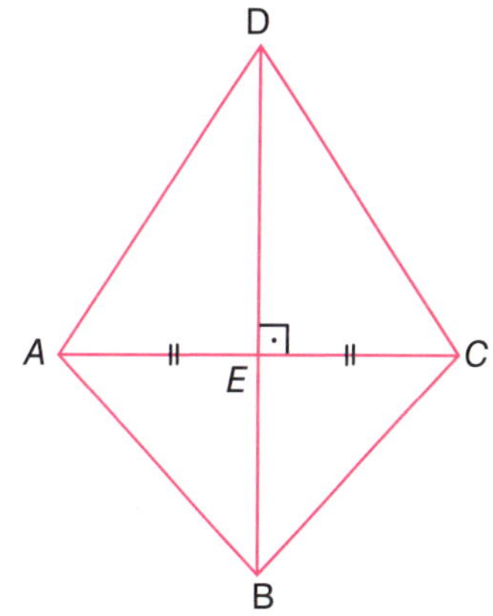

Example 12.22

In the figure, $ABCD$ is a kite.

Given that $BC = 8$ cm,

$DC = 17$ cm, and

$m\angle BAD = m\angle BCD = 90°$.

What is the area of the kite?

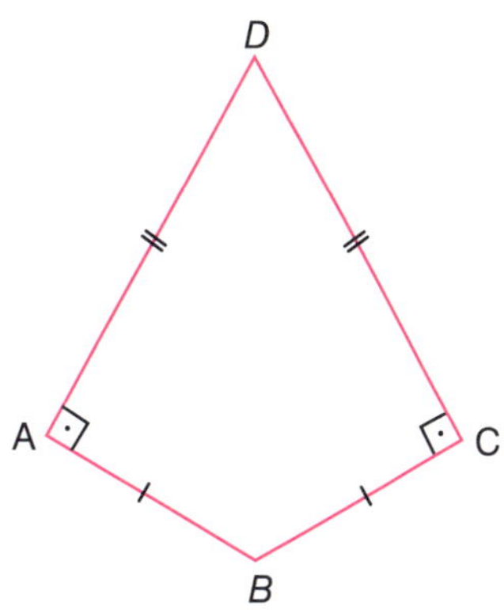

Solution 12.22

Since a kite symmetric about its main diagonal.
Then, Area(ABD) + Area(BED) = Area($ABCD$),

$$\text{Area}(ABCD) = 2 \cdot \frac{8 \cdot 17}{2} = 136 \text{ cm}^2.$$

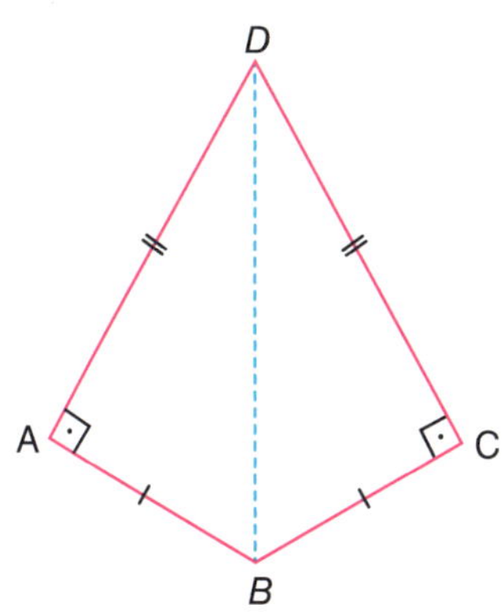

Example 12.23

In the figure, $ABCD$ is a kite with

$AD = DC = 6$, $AB = BC = 5\sqrt{2}$, and

$m\angle ADC = 90°$.

Find the area of $ABCD$.

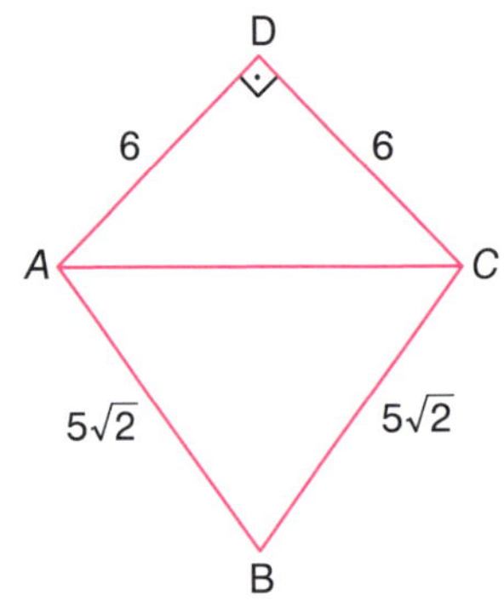

Solution 12.23

We drew the second diagonal DB and it interests the first one at E. $AE = EC = DE = 3\sqrt{2}$ ($45° - 45° - 90°$). By using Pythagorean Theorem $\triangle AEB$ we obtain $EB = 4\sqrt{2}$.

Therefore, the area of

$$ABCD = \frac{AC \cdot BD}{2} = \frac{6\sqrt{2} \cdot (3\sqrt{2} + 4\sqrt{2})}{2}$$

$$= \frac{6\sqrt{2} \cdot 7\sqrt{2}}{2} = 42.$$

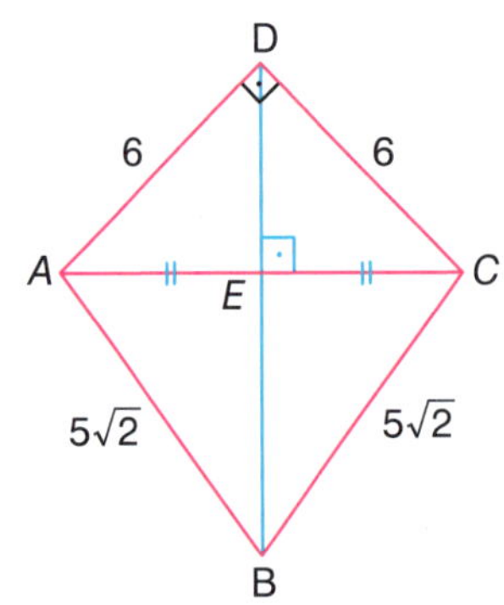

Example 12.24

In the figure, $ABCD$ is a kite with

$AB = BC = 17$ cm, $AD = DC = 25$ cm, and

$BD = 28$ cm.

What is the area of the kite?

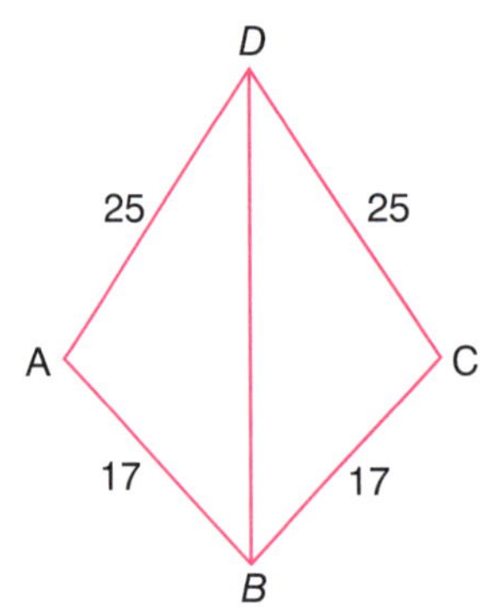

Solution 12.24

$f^2 = 25^2 - (28-e)^2$ (1)

$f^2 = 17^2 - e^2$ (2)

Solving (1) and (2) $17^2 - e^2 = 25^2 - (28-e)^2$

$(28-e)^2 - e^2 = 25^2 - 17^2$

$(28-e-e)(28-e+e) = (25-17)(25+17)$

$(28-2e) \cdot 28 = 8 \cdot 42 \Rightarrow 28-2e = 12$

Hence, $2e = 16 \Rightarrow e = 8$, $f = 15$.

Therefore, Area$(ABCD) = \dfrac{28 \cdot 30}{2} = 420 \text{ cm}^2$.

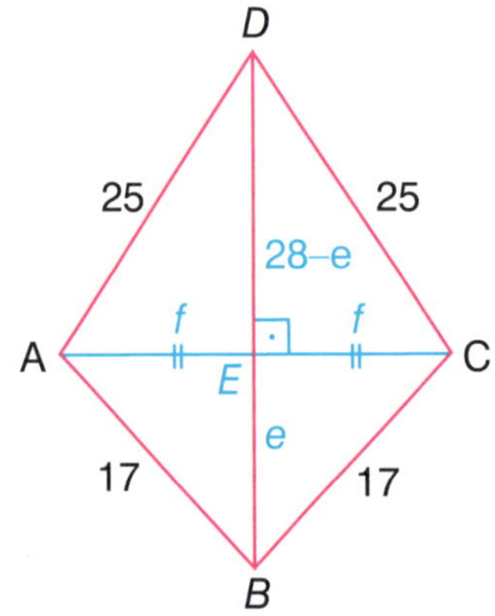

Catriona Agg @Cshearer41
Geometry puzzle

The pink and blue rectangles have the same area.

What is it?

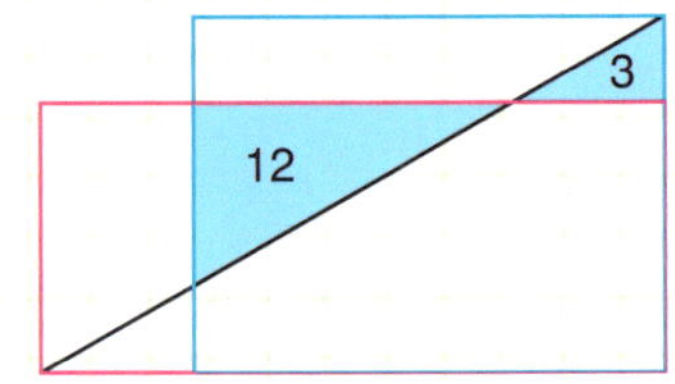

My Notes

1. In the figure, $ABCD$ is a kite.

Given that:

$AB = AD = 4$ cm,

$BC = DC = 8$ cm, and

$m\angle ADC = 45°$,

what is the area of $ABCD$?

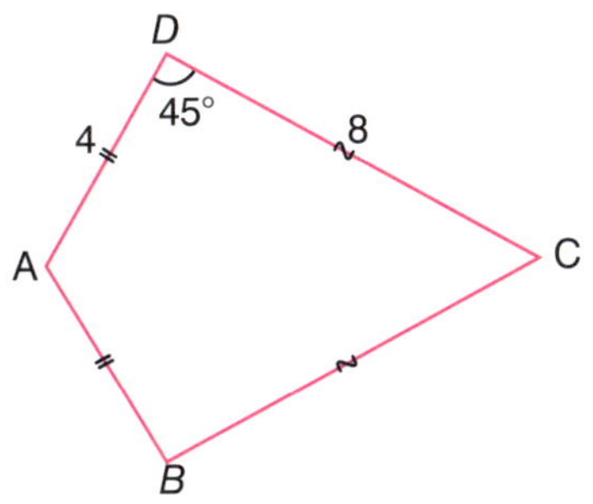

2. In the figure, $ABCD$ is a convex quadrilateral with E is the intersection point of its diagonals.

Given that: $DC = BC$,

$AC = 8$, $BD = 12$,

$m\angle DCA = 23°$, and

$m\angle DBC = 22°$,

what is the area of $ABCD$?

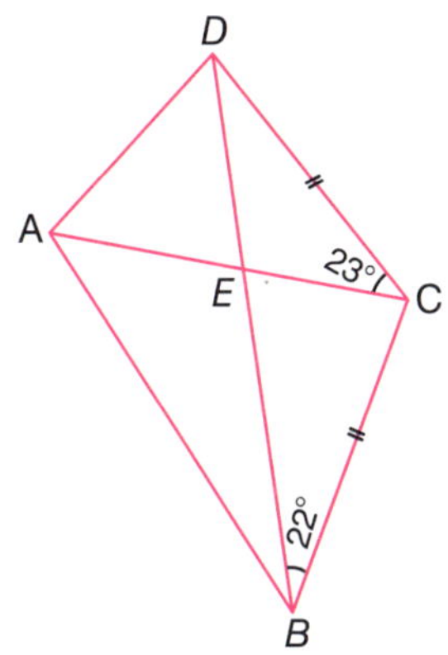

3. $ABCD$ is a convex quadrilateral and E is the intersection point of the diagonals.

If $DE = 4$ cm and, $BE = 8$ cm, find the ratio of area ΔADC to area $ABCD$.

4. $ABCD$ is a convex quadrilateral and E is the intersection point of its diagonals.

Given that:

Area ΔABE = Area $\Delta CDE = 27$ and

Area $\Delta BCE = 9$ Area ΔADE,

what is the area of $ABCD$?

5. $ABCD$ is a parallelogram and H is a point on DC such that $BH \perp DC$.

Given that:

$BH = 8$ cm,

$BC = 12$ cm, and

$DH = 10$ cm,

what is the area of $ABCD$?

6. In the figure, $ABCD$ is a parallelogram. Side AB is divided into five equal parts and CD is divided into four equal parts.

If area of $ABCD$ is 120 cm^2, then find area $FHZY$.

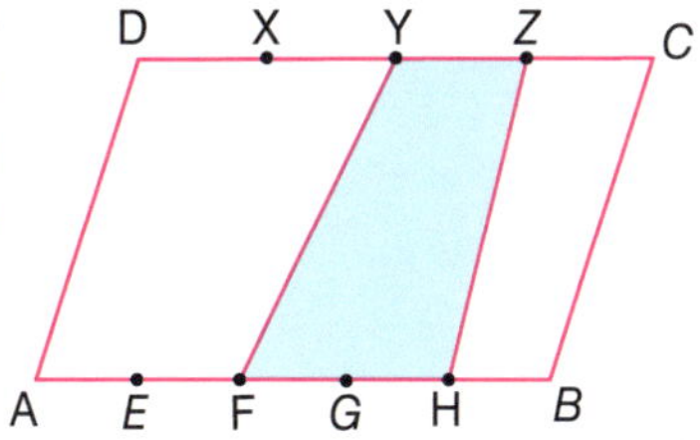

7. In the figure, $ABCD$ is a rhombus with point E is on BD such that $DE = 41$ and $EB = 9$.

What is the area of rhombus?

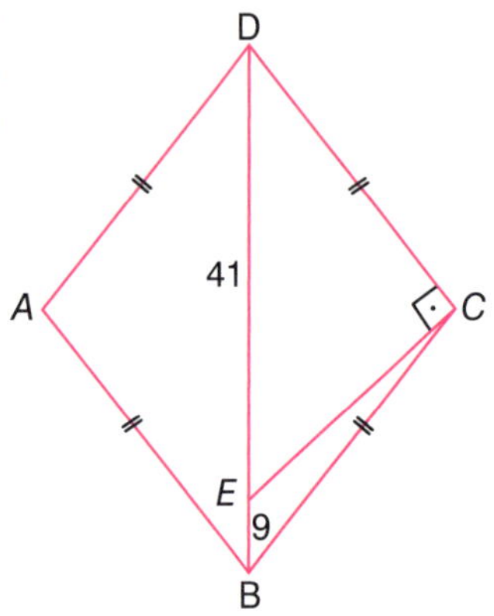

8. In the figure, $ABCD$ is a square and A, B and E are collinear.

If the area of triangle DCE is 20 then find the area of the square.

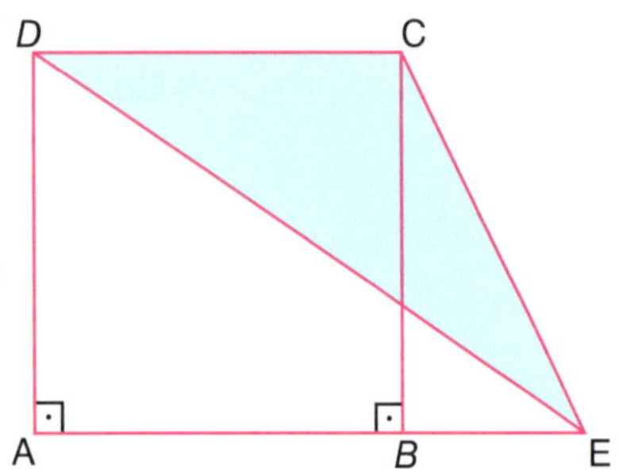

9. In the figure, $ABCD$ is a square and DEC is a right triangle.

If $DE+EC = 12$ and $DC = 10$, then what is the shaded area?

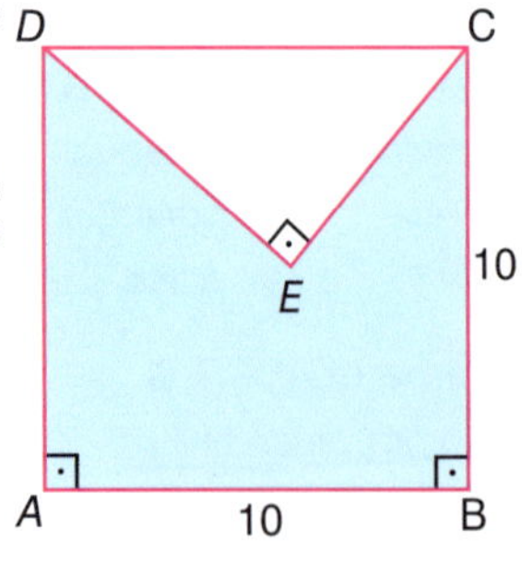

10. In the figure, $ABCD$ is a rhombus with midpoints of E and F.

Given that

$EF = 10$ and

$AF = 13$.

Find the area of $ABCD$.

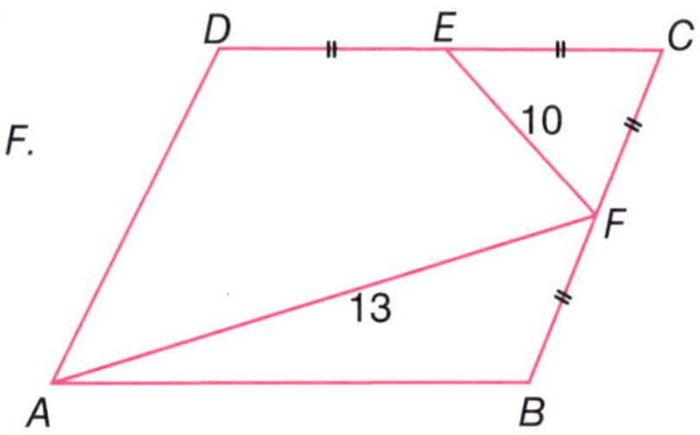

11.

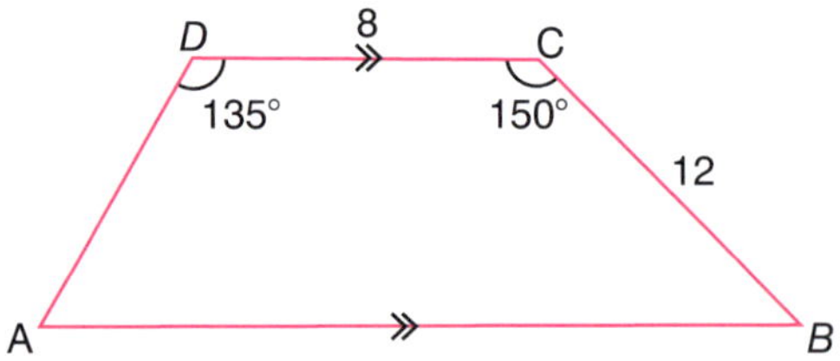

In the figure, $ABCD$ is a trapezoid. Given that $BC = 12$, $DC = 8$, $m\angle ADC = 135°$, and $m\angle DCB = 150°$.

What is the area of $ABCD$?

12. In the figure, $ABCD$ is a trapezoid and E is the midpoint of AD.

If $EH \perp BC$, $EH = 8$, and $BC = 18$, what is the area of $ABCD$?

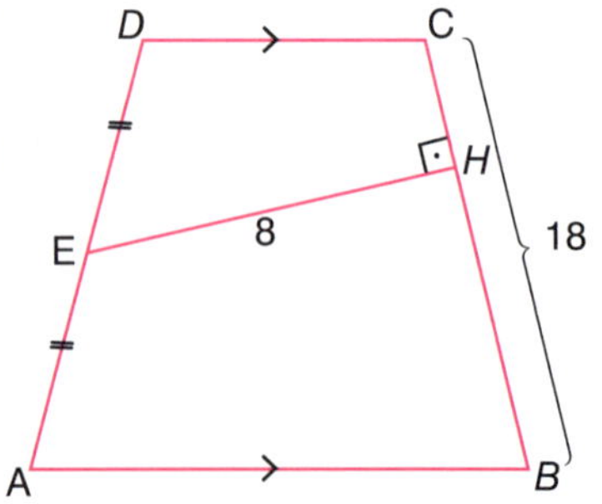

1. The perimeter of a rhombus is 20 and the sum of its diagonals is 12.

Find the area rhombus.

2. $ABCD$ is a trapezoid with bases AB and DC.

Given that $AB = 26$, BC = 16, CD = 6 and AD = 12.

What is the area of trapezoid?

3.

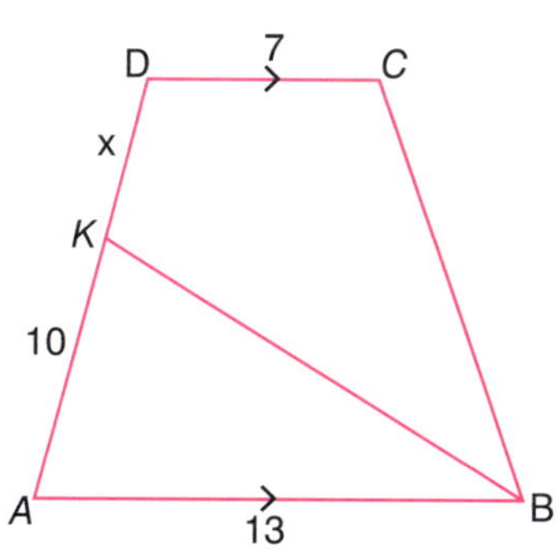

In the figure, $ABCD$ is a trapezoid and K is a point on AD.

Given that Area ΔABK = Area $BCDK$, $AB = 13$, $DC = 7$, and $AK = 10$.

What is the length of $KD = x$?

4.

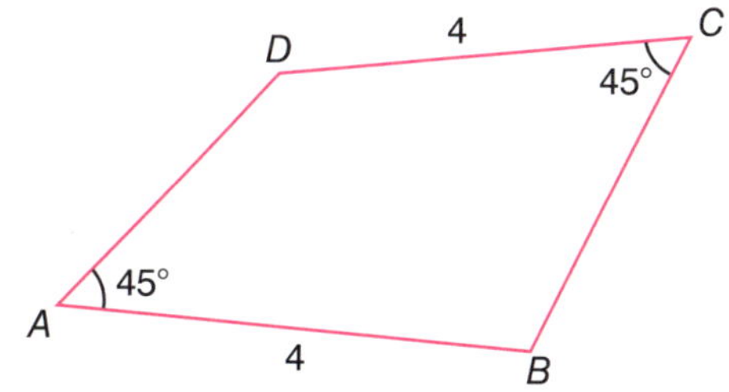

$ABCD$ is a quadrilateral with $m\angle A = m\angle C = 45°$, $DC = AB = 4$, and $AD + BC = 8$.

What is the area of $ABCD$?

5.

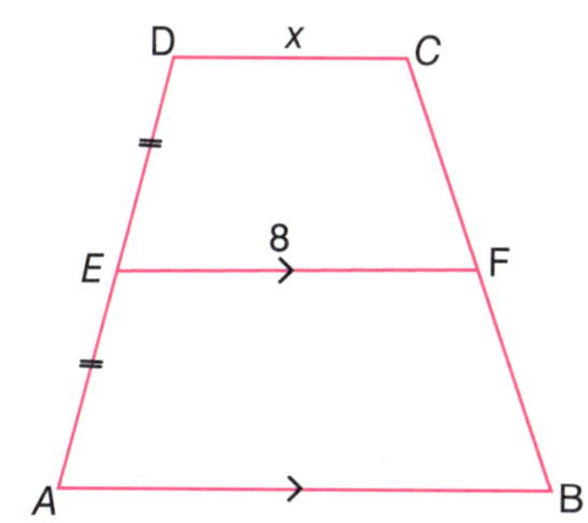

In the figure, $ABCD$ is a trapezoid.

Given that $EF // AB$, $DE = EA$, and $EF = 8$ cm.

If Area $ABFE$ = 2 Area $DEFC$, then find $DC = x$.

6. $ABCD$ is a trapezoid and E is the midpoint of AB.

If $AB = 4DC$, then find $\dfrac{\text{Area } ABCD}{\text{Area } \Delta DEC}$.

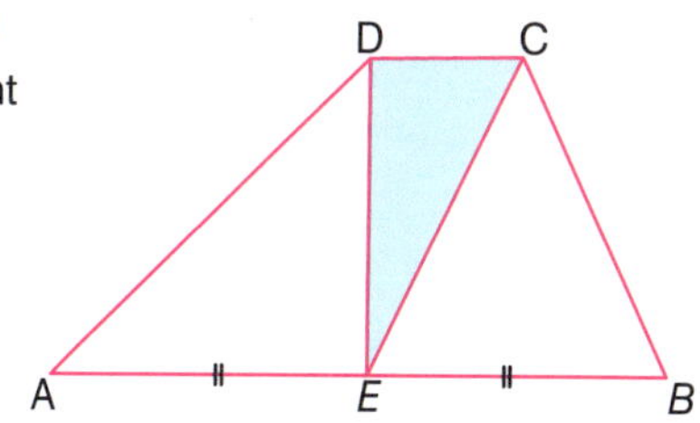

7. *ABCD* is a trapezoid with perpendicular diagonals. If one length of these diagonals is 5 unit and the height of this trapezoid is 4 unit, then find the area of trapezoid.

8. In the figure, *ABCD* is a rectangle.

$EK = 4$,

$DE = EC$ and

$KB \perp AC$.

What is the area of *ABCD*?

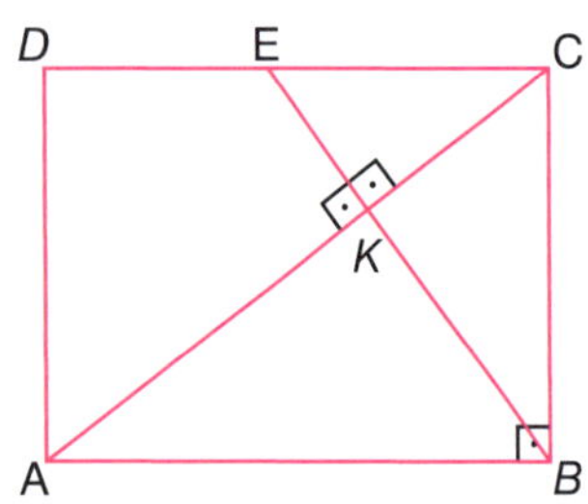

9. *ABCD* is a rectangle.

Given that

$2AD = DC = 2$ and $\angle DAE = 15°$.

Find the area of ΔCEB.

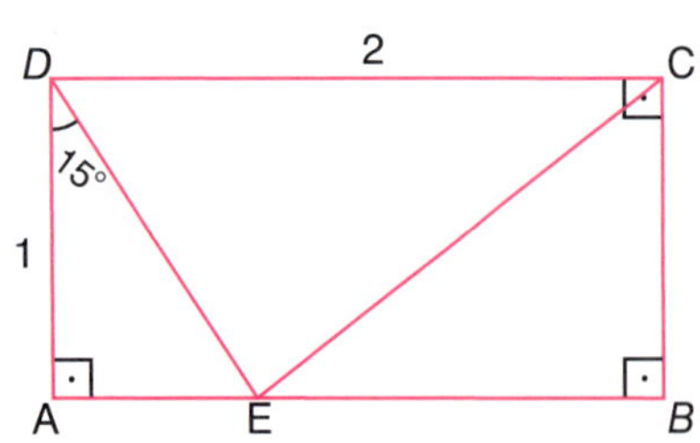

10. *ABCD* is a square. Given that one side of the square is 12.

$DN = NC$, $AE \perp NE$

$FB = 4$, and

$CF = 8$ cm,

What is the area of ΔANE?

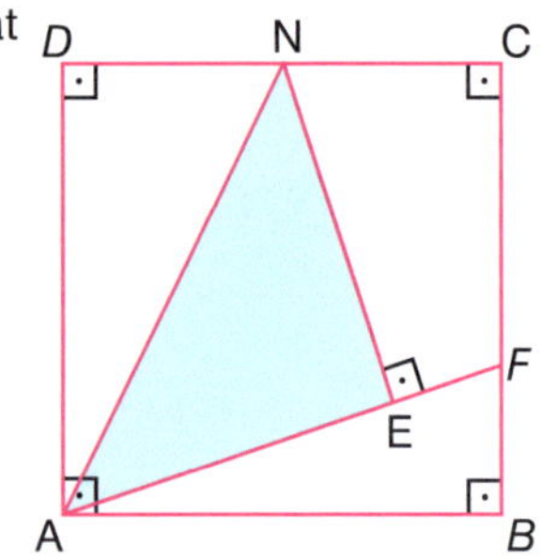

11. A_1, A_2, and A_3 area of their indicated regions and they all are equal to each other.

If $EF = 4$, then find *AD*.

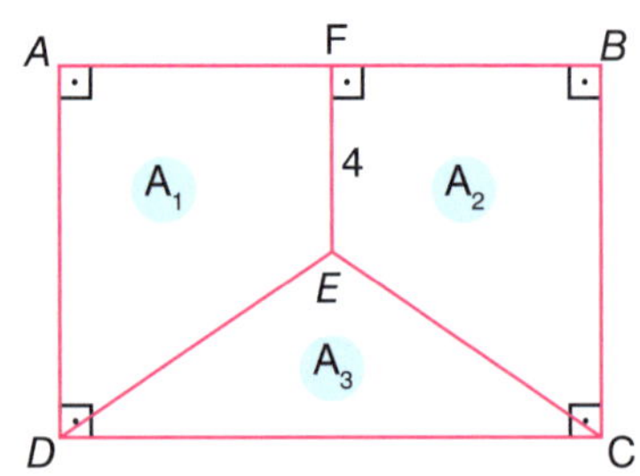

12. *ABCD* is a trapezoid such that $AC = 6$, $BD = 8$. Let S and P be the midpoints of *AD* and *BC*.

If $PS = 5$, find the area of *ABCD*.

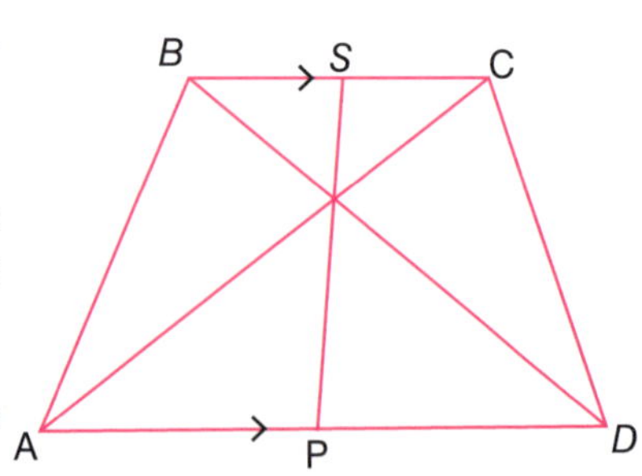

1. **(2018 AMC8 #22)**

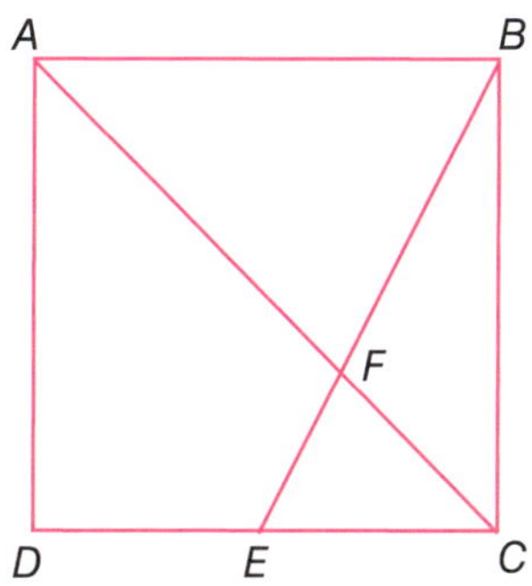

Point E is the midpoint of side CD in square $ABCD$ and, BE meets diagonal AC at F. The area of quadrilateral $AFED$ is 45.

What is the area of $ABCD$?

A) 100 B) 108 C) 120 D) 135 E) 144

2. **(2016 AlphaStar PiMC- # 18)**

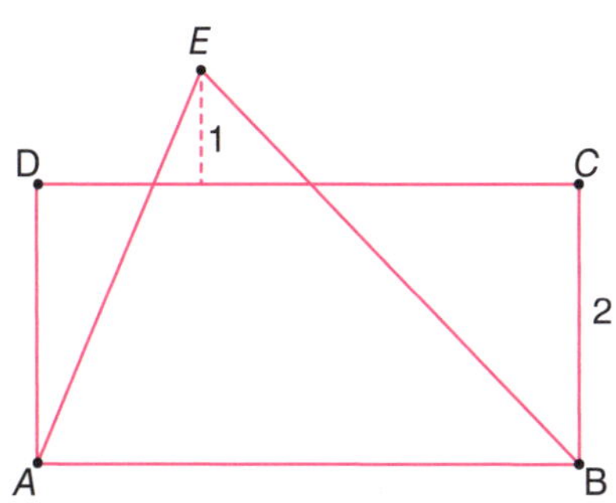

In the image above, $ABCD$ is rectangle with $BC = 2$ units. Point E is above segment CD such that the distance from E to CD such that the distance from E to CD 1 unit.

If the area of rectangle $ABCD$ is 8 square units how many square units is the area of triangle ABE?

3. **(CWML 2014-15-Meet 3 #11)** Stan's rectangular building has a flat roof and is 10 yards wide, 13 yards long and 5 yards high. It is to be painted inside and outside, and on the ceiling, but not on the roof or floor.

Find the total number of square yards to be painted.

4. **(2002 AMC 10B #17)** A regular octagon $ABCDEFGH$ has sides of length two.

Find the area of ΔADG.

A) $4+2\sqrt{2}$ B) $6+\sqrt{2}$ C) $4+3\sqrt{2}$ D) $3+4\sqrt{2}$ E) $8+\sqrt{2}$

5. **(2012 AMC8 #25)**

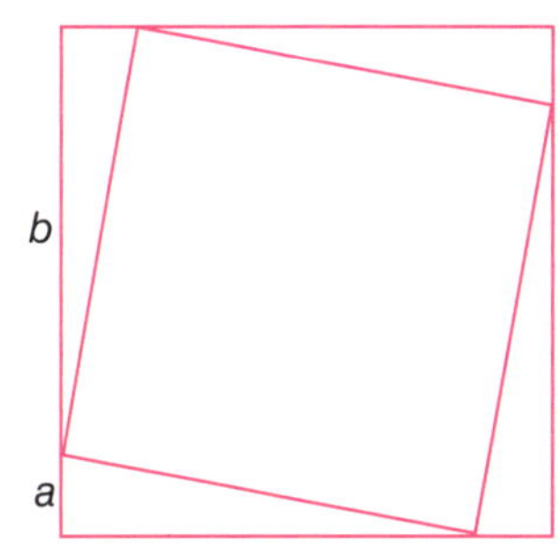

A square with area 4 is inscribed in a square with area 5, with one vertex of the smaller square on each side of the larger square. A vertex of the smaller square divides a side of the larger square into two segments, one of length a, and the other of length b.

What is the value of ab?

A) $\frac{1}{5}$ B) $\frac{2}{5}$ C) $\frac{1}{2}$ D) 1 E) 4

6. **(2009 AMC 12A #20)** Convex quadrilateral $ABCD$ has $AB = 9$ and $CD = 12$. Diagonals AC and BD intersect at E, $AC = 14$, and ΔAED and ΔBEC have equal areas.

What is AE?

A) $\frac{9}{2}$ B) $\frac{50}{11}$ C) $\frac{21}{4}$ D) $\frac{17}{3}$ E) 6

7. **(2002 AMC 10A #25)** In trapezoid $ABCD$ with bases AB and CD, we have $AB = 52$, $BC = 12$, $CD = 39$, and $DA = 5$.

The area of $ABCD$ is

A) 182 B) 195 C) 210 D) 234 E) 260

8. **(MathCON 2019-Final #28)**

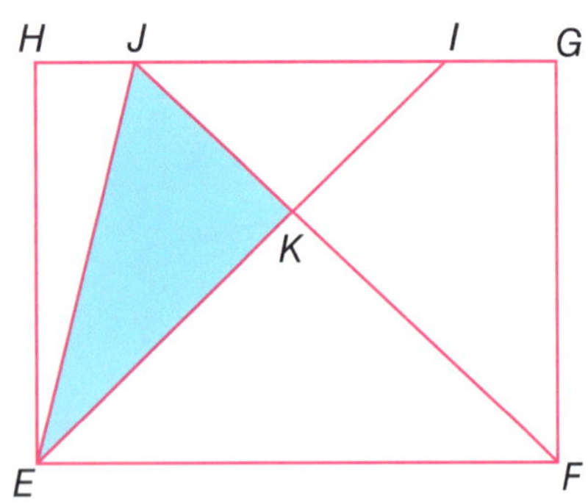

$EFGH$ is a rectangle with $FG = 6$ and $EF = 8$. I and J are the points on $\overline{GH}$ such that $\overline{FJ}$ and $\overline{EI}$ intersect at point K, and the area of triangle EJK is 9 square units.
What is IJ?

A) 2 B) $\frac{3}{5}$ C) 3 D) $\frac{7}{2}$ E) $\frac{24}{5}$

9. **(2004 AMC 10A #20)**

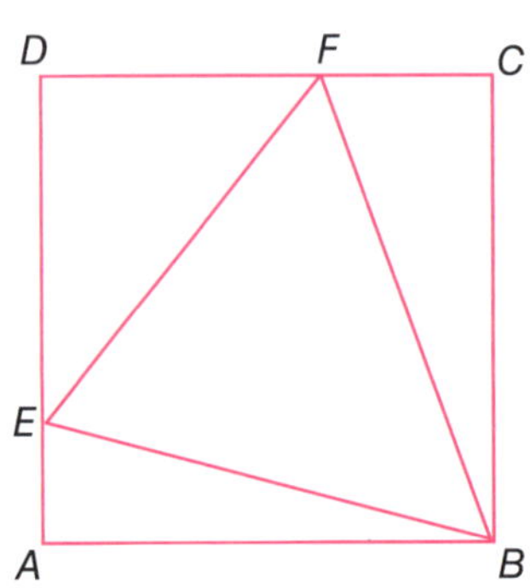

Points E and F are located on square $ABCD$ so that ΔBEF is equilateral.

What is the ratio of the area of ΔDEF to that of ΔABE?

A) $\frac{4}{3}$ B) $\frac{3}{2}$ C) $\sqrt{3}$ D) 2 E) $1+\sqrt{3}$

10. **(MathCON 2017-Final #37)** Let $ABCD$ be a trapezoid such that $AC = 6$, $BD = 8$, and AD parallel BC. Let P and S be the mid-points of AD and BC, respectively.

If $PS = 5$ find the area of the $ABCD$.

11. **(2015 AMC 8 #25)**

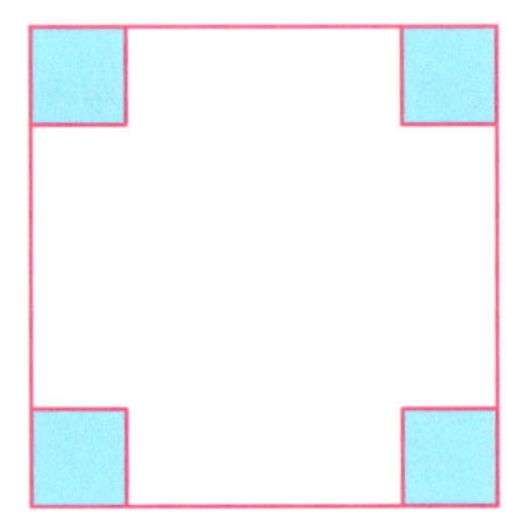

One-inch squares are cut from the corners of this 5 inch square.

What is the area in square inches of the largest square that can be fitted into the remaining space?

A) 9 B) $12\frac{1}{2}$ C) 15 D) $15\frac{1}{2}$ E) 17

12. **(2005 AMC 10A #25)** In ABC we have $AB = 25$, $BC = 39$, and $AC = 42$. Points D and E are on AB and AC respectively, with $AD = 19$ and $AE = 14$.

What is the ratio of the area of triangle ADE to the area of the quadrilateral $BCED$?

A) $\frac{266}{1521}$ B) $\frac{19}{75}$ C) $\frac{1}{3}$ D) $\frac{19}{56}$ E) 1

KEYS

CHAPTER-1 / ANGLES

TEST-1

1.	65°	7.	150°
2.	75°	8.	20°
3.	40°	9.	40°
4.	40°	10.	90°
5.	36°	11.	40°
6.	12°	12.	21°

TEST-2

1.	180°	7.	40°
2.	20°	8.	72°
3.	7°	9.	60°
4.	15°	10.	40°
5.	139°	11.	105°
6.	225°	12.	27°

TEST-3

1.	75°	7.	C) 135
2.	E) None	8.	E) 75°
3.	$\frac{540}{7}$	9.	D) 165
4.	55°	10.	C) 4
5.	4°	11.	D) 50
6.	115°	12.	C) 90°

CHAPTER-2 / SPECIAL TRIANGLES- SIDES

TEST-1

1.	20	7.	15
2.	25	8.	$\frac{7}{10}$
3.	63	9.	2
4.	84	10.	12
5.	12	11.	17
6.	$2\sqrt{21}$	12.	8

TEST-2

1.	$3\sqrt{6}$	7.	$\sqrt{2}$
2.	36	8.	30°
3.	$\frac{5}{2}$	9.	56
4.	15	10.	$AB^2+3BC^2 = 4BD^2$
5.	6	11.	$BD^2+CD^2 = 2AD^2$
6.	$\frac{12}{5}$	12.	28

TEST-3

1.	B) 30	7.	D) $\frac{75}{8}$
2.	D) 34	8.	B) 25
3.	18°	9.	4
4.	D) 13	10.	B) $\frac{59}{4}$
5.	C) 3	11.	43°
6.	B) 31and 32	12.	B) 26

CHAPTER-3/ SPECIAL TRIANGLES- ANGLES

TEST-1

1.	$9+3\sqrt{3}$	7.	$4\sqrt{3}$
2.	$3\sqrt{3}$	8.	$6\sqrt{3}$
3.	3	9.	16
4.	$\sqrt{61}$	10.	13
5.	6	11.	$7-\sqrt{5}$
6.	$4\sqrt{3}-4$	12.	16

TEST-2

1.	$3(\sqrt{3}-1)$	7.	$4\sqrt{3}$
2.	$5+2\sqrt{3}$	8.	$4(\sqrt{3}+2)$
3.	1	9.	45°
4.	$3\sqrt{3}$	10.	$4h$
5.	$\sqrt{2}$	11.	$9\sqrt{3}$
6.	$2\sqrt{2}$	12.	30°

TEST-3

1.	D) $\frac{15}{2}$	7.	D) 13
2.	C) 6 units	8.	120°
3.	C) 3	9.	B) 31 and 32
4.	5	10.	75°
5.	A) 180	11.	C) 8
6.	$16-8\sqrt{3}$	12.	4

CHAPTER-4/ TRIANGLES INEQUALITIES

TEST-1

1.	9	7.	49
2.	9	8.	4
3.	77	9.	C) (10,6)
4.	45	10.	9
5.	10	11.	10
6.	8	12.	84

TEST-2

1.	31	7.	1
2.	7	8.	3
3.	36	9.	2
4.	5	10.	9
5.	11	11.	9
6.	11,12,13	12.	3

TEST-3

1.	D) $5<x<7$	7.	7
2.	71	8.	B) 3
3.	19.6	9.	E) 6
4.	A) 43	10.	D) 18
5.	C) 13	11.	C) 568
6.	A) 49	12.	893

CHAPTER-5 / TRIANGLES SIMILARITIES

TEST-1

1. 13
2. 14
3. 3
4. $\frac{25}{4}$
5. 8
6. 16
7. 6
8. 5
9. 3
10. 9
11. $\frac{24}{5}$
12. 3

TEST-2

1. $\frac{11}{2}$
2. $\frac{1}{4}$
3. $18\sqrt{5}$
4. 5
5. 10
6. $\frac{15}{2}$
7. $2\sqrt{5}$
8. $\frac{20}{3}$
9. 9
10. 8
11. $\frac{37}{5}$
12. 16

TEST-3

1. D) $\frac{4\sqrt{3}}{3}$
2. $\frac{6}{5}$
3. B) $7\sqrt{3}$
4. 24
5. B) 4
6. C) 56
7. A) $\frac{2}{3}$
8. B) 21
9. I, II, and III
10. D) $\frac{15}{8}$
11. B) $\frac{60}{37}$
12. 308

CHAPTER-6 / TRIANGLES-BISECTOR

TEST-1

1. 6
2. $3\sqrt{2}$
3. 3
4. 6
5. 24
6. $3\sqrt{5}$
7. $3\sqrt{2}$
8. 6
9. 6
10. 2
11. 18
12. $5+\sqrt{94}$

TEST-2

1. 22
2. 6
3. $\frac{24\sqrt{2}}{7}$
4. 8
5. 10
6. $2\sqrt{5}$
7. $\frac{5}{2}$
8. $\sqrt{10}$
9. 8
10. 17
11. $8\sqrt{5}$
12. 28

TEST-3

1. 40°
2. 140°
3. 90°
4. 17
5. 8
6. B) $\frac{\sqrt{5}-1}{2}$
7. D) $14\sqrt{5}$
8. D) 75°
9. D) $\frac{8}{15}\sqrt{5}$
10. B) $\frac{5}{3}$
11. D) 75
12. $LP = 72$

KEYS

CHAPTER-7 / TRIANGLE–LENGTH THEOREMS

TEST-1

1. $2\sqrt{11}$
2. $4+2\sqrt{17}$
3. 6
4. 6
5. $\sqrt{10}$
6. 6
7. $5\sqrt{2}$
8. a) 4, b) 6, c) 7
9. 60
10. 16
11. 4
12. 1

TEST-2

1. 6
2. 20
3. $\frac{22}{37}$
4. $\frac{1}{pq}$
5. 6
6. $2\sqrt{39}$
7. $4\sqrt{2}$
8. $6\sqrt{3}$
9. 21
10. 4
11. $\frac{1}{12}$
12. $\frac{24}{13}$

TEST-3

1. 4
2. C) $\frac{12}{25}$
3. B) 30
4. C) 2:1
5. A) $\frac{2}{3}$
6. E) 20
7. C) $\frac{135}{2}$
8. C) 5
9. D) $\frac{4}{11}$
10. D) 9
11. 20
12. 177

CHAPTER-8 / TRIANGLE-AREAS

TEST-1

1. $28\sqrt{3}$
2. $4\sqrt{5}$
3. $27\sqrt{3}$
4. 16
5. 20
6. 8
7. 60
8. $12\sqrt{3}$
9. 7
10. $20\sqrt{3}$
11. $18\sqrt{2}$
12. 64

TEST-2

1. $\frac{43}{180}$
2. $54-18\sqrt{3}$
3. 114
4. 30°
5. 6
6. $48\sqrt{3}$
7. 28
8. $18\sqrt{3}+54$
9. 12
10. 64
11. 3
12. 72

TEST-3

1. 3
2. C) $\frac{\sqrt{3}}{2}$
3. 18
4. D) $\frac{75}{8}$
5. C) 200
6. E) 24
7. 12
8. 32
9. E) 37:1
10. B) 13.5
11. D) 18
12. D) $\frac{19}{56}$

CHAPTER-9 / POLYGONS-QUADRILATERALS-ANGLES

TEST-1

1.	18°	7.	12
2.	140°	8.	18°
3.	80°	9.	10
4.	30°	10.	$\sqrt{31}$
5.	13	11.	1
6.	135°	12.	9

TEST-2

1.	8	7.	$\sqrt{5}$
2.	54°	8.	$3\sqrt{31}$
3.	54°	9.	$4\sqrt{7}$
4.	$\sqrt{17}$	10.	54°
5.	150°	11.	12
6.	81°	12.	10°

TEST-3

1.	B) 23	7.	B) $\frac{\sqrt{2}}{2}$
2.	B) 30°	8.	C) 85
3.	155°	9.	A) $\frac{\sqrt{2}-1}{2}$
4.	100°	10.	E) 6
5.	B) $\frac{(n-4)180}{n}$	11.	A) $\frac{7}{2}$
6.	$\frac{540}{7}$	12.	D) $\frac{1}{2}$

CHAPTER-10 / QUADRILATERALS-LENGTHS-1

TEST-1

1.	9	7.	26
2.	$3\sqrt{3}$	8.	18°
3.	10	9.	$\sqrt{34}$
4.	14	10.	11
5.	4	11.	$6\sqrt{2}$
6.	75°	12.	$\frac{9}{2}$

TEST-2

1.	10	7.	$16\sqrt{17}$
2.	14	8.	$\frac{40}{9}$
3.	$2\sqrt{13}$	9.	$2\sqrt{13}$
4.	17	10.	16
5.	24	11.	30°
6.	6	12.	6

TEST-3

1.	B) $2+\frac{4\sqrt{3}}{3}$	7.	48
2.	96	8.	C) $\frac{\sqrt{5}-1}{2}$
3.	C) 8	9.	A) $\frac{1}{2}$
4.	D) $\frac{10}{91}$	10.	C) 17
5.	B) $2000(\sqrt{2}-1)$	11.	B) 20
6.	C) $\frac{125}{12}$	12.	16

CHAPTER-11 / QUADRILATERALS-LENGTHS-2

TEST-1

1. 60°
2. 7
3. 12
4. 6
5. 10
6. $\frac{48}{5}$
7. 6
8. $\frac{24}{5}$
9. $\frac{5}{3}$
10. 16
11. 22
12. 15

TEST-2

1. $\frac{60}{13}$
2. $\sqrt{61}$
3. 10
4. 18
5. 10
6. 30°
7. 9
8. $\frac{x}{y-x}$
9. $\frac{|x-y|}{2}$
10. $\sqrt{6}$
11. $\frac{120}{17}$
12. 96

TEST-3

1. A) 180
2. D) 34
3. D) 750
4. B) $\frac{4}{5}$
5. B) 25
6. B) 28
7. C) 5
8. C) 210
9. C) $\frac{21}{13}$
10. B) $30\sqrt{3}$
11. D) 63
12. 11

CHAPTER-12 / QUADRILATERALS-AREAS

TEST-1

1. $16\sqrt{2}$
2. $24\sqrt{2}$
3. $\frac{1}{3}$
4. 144
5. $80+32\sqrt{5}$
6. 39
7. 1000
8. 40
9. 89
10. 160
11. $66+18\sqrt{3}$
12. 144

TEST-2

1. 11
2. $\frac{768}{5}$
3. 3
4. $8\sqrt{2}$
5. $\frac{8}{3}$
6. 5
7. $\frac{100}{6}$
8. $96\sqrt{2}$
9. $\frac{3+2\sqrt{3}}{4+2\sqrt{3}}$
10. 45
11. 12
12. 24

TEST-3

1. B) 108
2. 6
3. 590
4. C) $4+3\sqrt{2}$
5. C) $\frac{1}{2}$
6. E) 6
7. C) 210
8. E) $\frac{24}{5}$
9. D) 2
10. 24
11. C) 15
12. D) $\frac{19}{56}$

Made in the USA
Monee, IL
12 September 2026

63115050R00155